Photosynthetic Nitrogen Assimilation and Associated Carbon and Respiratory Metabolism

Advances in Photosynthesis and Respiration

VOLUME 12

The scope of our series, beginning with volume 11, reflects the concept that photosynthesis and respiration are intertwined with respect to both the protein complexes involved and to the entire bioenergetic machinery of all life. *Advances in Photosynthesis and Respiration* is a book series that provides a comprehensive and state-of-the-art account of research in photosynthesis and respiration. Photosynthesis is the process by which higher plants, algae, and certain species of bacteria transform and store solar energy in the form of energy-rich organic molecules. These compounds are in turn used as the energy source for all growth and reproduction in these and almost all other organisms. As such, virtually all life on the planet ultimately depends on photosynthetic energy conversion. Respiration, which occurs in mitochondrial and bacterial membranes, utilizes energy present in organic molecules to fuel a wide range of metabolic reactions critical for cell growth and development. In addition, many photosynthetic organisms engage in energetically wasteful photorespiration that begins in the chloroplast with an oxygenation reaction catalyzed by the same enzyme responsible for capturing carbon dioxide in photosynthesis. This series of books spans topics from physics to agronomy and medicine, from femtosecond processes to season long production, from the photophysics of reaction centers, through the electrochemistry of intermediate electron transfer, to the physiology of whole orgamisms, and from X-ray christallography of proteins to the morphology or organelles and intact organisms. The goal of the series is to offer beginning researchers, advanced undergraduate students, graduate students, and even research specialists, a comprehensive, up-to-date picture of the remarkable advances across the full scope of research on photosynthesis, respiration and related processes.

Photosynthetic Nitrogen Assimilation and Associated Carbon and Respiratory Metabolism

Edited by

Christine H. Foyer
Crop Performance and Improvement Division,
IACR-Rothamsted, Harpenden, U.K.

and

Graham Noctor
Université Denis Diderot Paris VII,
Institut de la Biotechnologie des Plantes, Orsay, France

SPRINGER-SCIENCE+BUSINESS MEDIA, B.V.

A C.I.P. Catalogue record for this book is available from the Library of Congress.

ISBN 978-90-481-5469-2 ISBN 978-0-306-48138-3 (eBook)
DOI 10.1007/978-0-306-48138-3

Printed on acid-free paper

Editorial

Advances in Photosynthesis and Respiration

It gives me great pleasure to announce the publication of Volume 12, *Photosynthetic Nitrogen Assimilation and Associated Carbon and Respiratory Metabolism*, edited by Christine H. Foyer and Graham Noctor in our Series. This volume is the second one to appear under the new title of *Advances in Photosynthesis and Respiration*. Further, a new beginning has already been made with the appointment of new members of the Board of Consulting Editors. They are: Christine Foyer, UK; Elisabeth Gantt, USA; John H. Golbeck, USA; Susan Golden, USA; Wolfgang Junge, Germany; Hartmut Michel, Germany; and Kimiyuki Satoh, Japan. James Siedow, USA, has joined our Board to provide leadership and strength in the area of 'respiration' in this Series. Several volumes on respiration (both plant and bacterial) are already in production or being contracted.

Published Volumes

The present volume is a sequel to the following eleven volumes in the "*Advances in Photosynthesis and Respiration*" (AIPH) series.

(1) *Molecular Biology of Cyanobacteria* (D.A. Bryant, editor, 1994);
(2) *Anoxygenic Photosynthetic Bacteria* (R.E. Blankenship, M.T. Madigan and C.E. Bauer, editors, 1995);
(3) *Biophysical Techniques in Photosynthesis* (J. Amesz and A.J. Hoff, editors, 1996);
(4) *Oxygenic Photosynthesis: The Light Reactions* (D.R. Ort and C.F. Yocum, editors, 1996);
(5) *Photosynthesis and the Environment* (N.R. Baker, editor, 1996);
(6) *Lipids in Photosynthesis: Structure, Function and Genetics* (P.-A. Siegenthaler and N. Murata, editors, 1998);
(7) *The Molecular Biology of Chloroplasts and Mitochondria in Chlamydomonas* (J.-D. Rochaix, M. Goldschmidt-Clermont and S. Merchant, editors, 1998);
(8) *The Photochemistry of Carotenoids* (H.A. Frank, A.J. Young, G. Britton and R.J. Cogdell, editors, 1999);
(9) *Photosynthesis: Physiology and Metabolism* (R.C. Leegood, T.D. Sharkey and S. von Caemmerer, editors, 2000);
(10) *Photosynthesis: Photobiochemistry and Photobiophysics* (B. Ke, author, 2001);
(11) *Regulation of Photosynthesis* (E-M. Aro and B. Andersson, editors, 2001).

See http://www.wkap.nl/prod/s/AIPH for further information and to order these books. Please note that the members of the International Society of Photosynthesis Research (ISPR) (http://www.photosynthesisresearch.org/) receive special discounts.

Photosynthetic Nitrogen Assimilation and Associated Carbon and Respiratory Metabolism, edited by Christine H. Foyer and Graham Noctor, Volume 12 in our series, is a great book that bridges the basics of photosynthesis and respiration with ecology and agriculture. Plant growth and biomass production require the assimilation of nitrogen into organic compounds using energy and carbon skeletons produced by photosynthesis and respiration. Placing nitrogen assimilation firmly at the heart of photosynthesis, this volume provides an original and innovative appraisal of the metabolic co-operation that is required. Unique perspectives are presented in sixteen key areas of current research, each discussing the latest data and critically examining the most important developing concepts. Key themes are the underlying cooperation between organelles (chloroplasts and mitochondria) and pathways (photosynthesis and respiration), as well as the extensive metabolic crosstalk that dictates appropriate gene expression. This book is essential reading for those seeking to understand the details of carbon-nitrogen interactions and the importance of these relationships in determining photosynthetic biomass production.

Future Books

The readers of the current series are encouraged to watch for the publication of the forthcoming books:

(1) *Light-harvesting Antennas in Photosynthesis* (Editors: B.R. Green and W.W. Parson);
(2) *Photosynthesis in Algae* (Editors: A.W.D. Larkum, S. Douglas, and J.A. Raven);
(3) *Respiration in Archaea and Bacteria*, 2 volumes (Editor: D. Zannoni);
(4) *Biochemistry and Biophysics of Chlorophylls* (Editors: B. Grimm, R. Porra, W. Rüdiger, and H. Scheer).
(5) *Chlorophyll Fluorescence* (Editors: G. Papageorgiou and Govindjee);
(6) *Photosystem II: The Water/Plastoquinone Oxido-reductase in Photosynthesis* (Editors: T. Wydrzynski and K. Satoh);

In addition to these contracted books, invitations are out for several books. Topics planned are: Plant Respiration; Protein Complexes of Photosynthesis and Respiration; Photoinhibition and Photoprotection; Photosystem I; Protonation and ATP Synthesis; Global Aspects of Photosynthesis; Functional Genomics; History of Photosynthesis; The Cytochromes; The Chloroplast; Laboratory Methods for Studying Leaves and Whole Plants. In view of the interdisciplinary character of research in photosynthesis and respiration, it is my earnest hope that this series of books will be used in educating students and researchers not only in Plant Sciences, Molecular and Cell Biology, Integrated Biology, Biotechnology, Agricultural Sciences, Microbiology, Biochemistry, and Biophysics, but also in Bioengineering, Chemistry, and Physics.

I take this opportunity to thank Christine Foyer and Graham Noctor; all the authors of volume 12; Larry Orr; Jacco Flipsen, Lanette Setkoski; and my wife Rajni Govindjee for their valuable help and support that made the publication of *Photosynthetic Nitrogen Assimilation and Associated Carbon and Respiratory Metabolism* possible.

Readers are requested to send their suggestions for future volumes, authors or editors to me by E-mail (gov@uiuc.edu) or fax (1-217-244-7246).

Govindjee
Series Editor
Advances in Photosynthesis and Respiration
University of Illinois at Urbana-Champaign
Departments of Biochemistry and Plant Biology
And Center of Biophysics and Computational Biology
265 Morrill Hall, 505 South Goodwin Avenue
Urbana, IL 61801-3707, USA
URL: http://www.life.uiuc.edu/govindjee

Govindjee

The Series Editor of *Advances in Photosynthesis and Respiration*, Govindjee, uses one name only. He has been Professor Emeritus of Biophysics, Biochemistry and Plant Biology, at the University of Illinois at Urbana-Champaign (UIUC), since 1999. He was born in the city of Allahabad (Uttar Pradesh, India) in 1932. Govindjee graduated from the University of Allahabad, India in 1952 with a B.Sc. degree in Chemistry, Botany and Zoology, and obtained his M.Sc. (also from the University of Allahabad) in Botany (specializing in Plant Physiology) under Professor Shri Ranjan, in 1954. He subsequently served as a lecturer in Botany, at the same university, from 1954-1956. Govindjee came to the United States of America in 1956, to pursue his doctoral studies at UIUC. He worked, first with Robert Emerson, then with Jan B. Thomas and Eugene Rabinowitch, and obtained his Ph.D. in 1960, in Biophysics. After postdoctoral research on a US Public Health Service Award, he was appointed as Assistant Professor of Botany at UIUC in 1961; in 1965, he became an Associate Professor, and then in 1969, a Professor of Biophysics and Plant Biology, at the same institution.

Govindjee is co-author of *Photosynthesis* (John Wiley and Sons, New York, 1969), and co-editor of eight volumes on photosynthesis including (1) *Concepts in Photobiology: Photosynthesis and Photomorphogenesis* (Narosa Publishers, New Delhi/ Kluwer Academic Publishers, Dordrecht, 1999), (2) *Molecular Biology of Photosynthesis* (Kluwer Academic Publishers, Dordrecht, 1988), and (3) *Light Emission by Plants and Bacteria* (Academic Press, NY, 1986). Govindjee has edited (1) *Photosynthesis Vol. 1: Energy Conversion by Plants and Bacteria*; and *Photosynthesis Vol. 2: Development, Carbon Metabolism, and Plant Productivity* (Academic Press, NY, 1982. Russian Version, 1987), and (2) Bioenergetics of Photosynthesis (Academic Press, NY. 1975).

In collaboration with others, Govindjee's early research established the participation of a short-wavelength form of chlorophyll *a* in Photosystem II (PS II), that the two-light effect of Robert Emerson was in photosynthesis, not in respiration, and that it could be studied through chlorophyll fluorescence and delayed fluorescence. Over the years, his research, again with many collaborators, has focused on the mechanisms of PS II, including the first studies on its primary charge separation; the specific role of bicarbonate on the acceptor side of PS II, the demonstration that excess light indeed quenches the lifetime of PS II chlorophyll fluorescence (and thus diminishes the quantum yield of fluorescence); and on the theory for the mechanism of thermoluminescence in plants. Currently, however, he focuses on the history of photosynthesis research, and is equally concerned with photosynthesis education (see http://www.life.uiuc.edu/govindjee).

Contents

Editorial	**v**
Contents	**viii**
Preface	**xiii**
Color Plates	**CP-1**
1 Photosynthetic Nitrogen Assimilation: Inter-Pathway Control and Signaling	**1–22**
Christine H. Foyer and Graham Noctor	
Summary	1
I. Introduction	2
II. Control of Leaf Amino Acid Contents	4
III. Integration and Control of Nitrogen and Carbon Metabolism	11
IV. The Carbon-Nitrogen Signal Transduction Network: Interactions Between Nitrate, Sugars and Abscisic Acid	16
V. Conclusions and Perspectives	18
References	19
2 Photosynthesis and Nitrogen-Use Efficiency	**23–34**
P. Ananda Kumar, Martin A. J. Parry, Rowan A. C. Mitchell, Altaf Ahmad and Yash P. Abrol	
Summary	23
I. Introduction	24
II. Nitrogen in the Photosynthetic Apparatus	24
III. Optimization of Amounts of Photosynthetic Components for Different Environments	26
IV. Role of Regulation of Rubisco Activity	29
V. Approaches to Improving Nitrogen-Use Efficiency in Crops	30
Acknowledgments	31
References	31
3 Molecular Control of Nitrate Reductase and Other Enzymes Involved in Nitrate Assimilation	**35–48**
Wilbur H. Campbell	
Summary	35
I. Introduction	36
II. Transcriptional Control of Nitrate Reductase and Other Nitrogen Metabolism Genes	39
III. Post-Translational Control of Nitrogen Metabolism Enzymes	41

IV. Protein Kinases and Control of Carbon and Nitrogen Metabolism 44
V. Future Prospects for the Control of Nitrogen Metabolism 45
Acknowledgment 46
References 46

4 Soluble and Plasma Membrane-bound Enzymes Involved in Nitrate and Nitrite Metabolism 49–62
Christian Meyer and Christine Stöhr

Summary 49
I. Introduction 50
II. Nitrate Reduction at the Plasma Membrane 50
III. Nitrite Transport and Reduction 54
IV. Conclusions 59
Acknowledgments 60
References 60

5 What Limits Nitrate Reduction in Leaves? 63–70
Werner M. Kaiser, Maria Stoimenova and Hui-Min Man

Summary 63
I. Introduction 64
II. Nitrate Reduction and Nitrate Reductase Activity in Photosynthesizing Leaves 64
III. Nitrate Reduction after Artificial Activation of Nitrate Reductase 65
IV. Is Cytosolic Nitrate Concentration Rate-Limiting? 66
V. Is Nitrate Reduction Limited by NAD(P)H? 68
VI. Conclusions 68
Acknowledgments 70
References 70

6 The Biochemistry, Molecular Biology, and Genetic Manipulation of Primary Ammonia Assimilation 71–92
Bertrand Hirel and Peter J. Lea

Summary 71
I. Introduction: Glutamine Synthetase and Glutamate Synthase, Two Enzymes at the Crossroads Between Carbon and Nitrogen Metabolism 72
II. Glutamine Synthetase 72
III. Glutamate Synthase 79
IV. Glutamate Dehydrogenase 85
References 86

7 Regulation of Ammonium Assimilation in Cyanobacteria 93–113
Francisco J. Florencio and José C. Reyes

Summary 93
I. Introduction 94
II. Ammonium Uptake 94
III. The Glutamine Synthetase/Glutamate Synthase Pathway 96

IV. Regulation of Ammonium Assimilation 103
V. Future Perspectives 109
Acknowledgments 109
References 109

8 Photorespiratory Carbon and Nitrogen Cycling: Evidence from Studies of Mutant and Transgenic Plants 115–134
Alfred J. Keys and Richard C. Leegood

Summary 115
I. Introduction 116
II. Entry of Carbon into the Photorespiratory Pathway 119
III. Recycling of Carbon to the Reductive Pentose Phosphate Pathway 120
IV. Recycling of Nitrogen Associated with Photorespiration 124
V. Feedback from Photorespiration on Other Processes 127
VI. Role of Photorespiration During Stress 129
Conclusions 130
References 130

9 The Regulation of Plant Phosphoenolpyruvate Carboxylase by Reversible Phosphorylation 135–150
Jean Vidal, Nadia Bakrim and Michael Hodges

Summary 135
I. Introduction 136
II. Properties of Phosphoenolpyruvate Carboxylase 136
III. The Enzyme's Physiological Context 137
IV. Reversible Modulation in vivo by a Regulatory Phosphorylation Cycle 139
V. Significance of Regulatory Phosphorylation of the Photosynthetic Isoform 144
VI. Regulatory Phosphorylation of the C_3 Form: Importance in Anaplerosis 145
VII. Conclusions and Perspectives 148
References 148

10 Mitochondrial Functions in the Light and Significance to Carbon-Nitrogen Interactions 151–172
Per Gardeström, Abir U. Igamberdiev and A. S. Raghavendra

Summary 152
I. Introduction 152
II. Export of Photosynthate from the Chloroplast 153
III. Mitochondrial Products of Photorespiration 154
IV. Products of Glycolysis in the Light 155
V. Operation of the Tricarboxylic Acid Cycle 157
VI. Electron Transport and Redox Levels in Plant Mitochondria 160
VII. Participation of Mitochondria in the Regulation of Metabolism during Transitions between Light and Darkness 163
VIII. Mitochondrial Respiration and Photoinhibition 164
IX. The Role of Mitochondria in C_4 Photosynthesis 165

X. Glycolate Metabolism in Algal Mitochondria 165
XI. Concluding Remarks 166
Acknowledgments 166
References 167

11 Alternative Oxidase: Integrating Carbon Metabolism and Electron Transport in Plant Respiration 173–191
Greg C. Vanlerberghe and Sandi H. Ordog

Summary 173
I. Integration in Plant Respiration 174
II. The Alternative Oxidase in Plant Mitochondrial Electron Transport 174
III. Regulation of Alternative Oxidase 176
IV. Physiological Function of Alternative Oxidase 181
Acknowledgments 188
References 188

12 Nitric Oxide Synthesis by Plants and its Potential Impact on Nitrogen and Respiratory Metabolism 193–204
A. Harvey Millar, David A. Day and Christel Mathieu

Summary 193
I. Nitric Oxide as a Biological Messenger Molecule 194
II. Evidence of Nitric Oxide Synthesis and Accumulation in Plants 194
III. Evidence of Nitric Oxide Modulation of Plant Signaling, Metabolism and Development 196
IV. So What is the Role of Nitric Oxide in Plants? 201
Acknowledgments 202
References 202

13 Nitrogen and Signaling 205–225
Anne Krapp, Sylvie Ferrario-Méry and Bruno Touraine

Summary 206
I. Introduction 206
II. Processes Regulated by Nitrate and Reduced Nitrogen-Compounds 206
III. Molecular Mechanisms of Nitrogen Signal Perception and Transduction 216
IV. Concluding Remarks 220
Acknowledgments 220
References 220

14 Regulation of Carbon and Nitrogen Assimilation Through Gene Expression 227–238
Tatsuo Sugiyama and Hitoshi Sakakibara

Summary 227
I. Introduction 228
II. Physiological and Biochemical Nature of Plant Response to Nitrogen 228

III. Regulation of Nitrogen-Responsive Genes for Carbon Assimilation 230
IV. Regulation of Nitrogen-Responsive Genes for Assimilation and Subsequent Metabolism of Nitrogen 231
V. Regulation of Partitioning of Nitrogen into Proteins: A Model for Sensing and Signaling 234
Acknowledgments 235
References 235

15 Intracellular And Intercellular Transport Of Nitrogen And Carbon 239–263
Gertrud Lohaus and Karsten Fischer

Summary 239
I. Introduction 240
II. Transport Processes of Plastids 240
III. Transport Processes Involved in Phloem Loading 246
IV. Concluding Remarks 257
Acknowledgments 257
References 258

16 Optimizing Carbon-Nitrogen Budgets: Perspectives for Crop Improvement 265–274
John A. Raven, Linda L. Handley and Mitchell Andrews

Summary 265
I. Introduction 266
II. The Nature of Crops 268
III. What Are We Seeking to Optimize in Carbon-Nitrogen Budgets? 269
IV. How Can We Change Carbon-Nitrogen Budgets? 269
V. What are the Outcomes of Changing Carbon-Nitrogen Budgets? 271
VI. Prospects and Conclusions 272
Acknowledgments 272
References 272

Index 275

Preface

According to many textbooks, carbohydrates are the unique final products of plant photosynthesis. However, the photoautotrophic production of organic nitrogenous compounds may be just as old, in evolutionary terms, as carbohydrate synthesis. In the algae and plants of today, the light-driven assimilation of nitrogen remains a key function, operating alongside and intermeshing with photosynthesis and respiration. Photosynthetic production of reduced carbon and its reoxidation in respiration are necessary to produce both the energy and the carbon skeletons required for the incorporation of inorganic nitrogen into amino acids. Conversely, nitrogen assimilation is required to sustain the output of organic carbon and nitrogen. Together, the sugars and amino acids produced by the pigments and enzymes of the photosynthetic apparatus form the building blocks for plant development, growth, and biomass production. Complex interactions between photosynthetic carbon and nitrogen metabolism must, therefore, have evolved long ago and can be regarded as the expression of a truly ancient principle.

Perhaps more than any other major physiological process, nitrogen assimilation weds together photosynthesis and respiration into a unified network of interdependent processes. In C_3 plants especially, this network is further complicated by the concomitant operation of photorespiratory metabolism. The numerous interactions between carbon and nitrogen metabolism have been intensively studied at multiple levels of complexity and plant anatomy. Within the cell, extensive co-operation is required between different compartments, including chloroplasts, peroxisomes, cytosol, and mitochondria, while changes in carbon and nitrogen status influence organ physiology and root/shoot relationships. Ultimately, carbon/nitrogen relationships are whole plant phenomena but many of the primary interactions of key importance occur in the photosynthetic heart of green cells, the chloroplast, in co-operation with the mitochondrion. A multitude of interconnections are required between chloroplasts and mitochondria that function to achieve optimal energy balance and partitioning of assimilate, and hence avoid undue perturbation of cellular redox balance. Rates of photosynthesis and mitochondrial respiration fluctuate in a circadian manner in almost every organism studied. In addition, external triggers and environmental influences necessitate precise and appropriate re-adjustment of relative flux rates, to prevent excessive swings in energy/resource provision and use. This requires integrated control of the expression and activity of numerous key enzymes in photosynthetic and respiratory pathways, in order to co-ordinate carbon partioning and nitrogen assimilation.

This volume has two principal aims. The first is to provide a comprehensive account of the very latest developments in our understanding of how green cells reductively incorporate nitrate and ammonium into the organic compounds required for growth. From the partitioning of organic nitrogen within the photosynthetic apparatus, through the primary processes of nitrate reduction and ammonia assimilation and cycling in photorespiration, to the intracellular and intercellular transport of carbon and nitrogen, the processes involved in photosynthetic nitrogen assimilation are described and exciting new developments such as nitric oxide production evaluated. The second aim is to provide a comprehensive account of the mechanisms of crosstalk between carbon and nitrogen metabolism. A key theme of this volume is the close co-ordination of photosynthetic and respiratory processes in nitrogen assimilation. Emerging concepts of the interdependence of chloroplasts and mitochondria are described, and essential communication, transport and signaling processes are highlighted. We are becoming more aware that photosynthesis uses light and changes in redox state, as well as carbon and nitrogen metabolites, not only to drive assimilatory metabolism but also to signal 'current status' at the level of control of gene expression. Recent data on carbon/nitrogen interactions suggest that, from the capture of light to the synthesis of amino acids and export of carbon and nitrogen, numerous substrates, intermediates and products are monitored by the cell and the information transduced into regulation at the levels of gene expression and enzyme activity. Effective regulation ultimately determines the fate of the photosynthetic system, as loss of metabolic balance can trigger precocious senescence. These considerations must lead us to consider the term 'homeostasis,' rarely considered in relation to photosynthesis. This term does not imply a static situation but rather describes a dynamic equilibrium between the provision and use of energy within the regulated limits of carbon and nitrogen assimilation capacity. To ensure homeostasis, several key molecules play major roles in signaling appropriate changes in gene expression.

This is the first comprehensive treatise that places nitrogen assimilation firmly within the context of photosynthesis. Volume 7 of this series covered the molecular biology of chloroplasts and mitochondria in algae while Volume 9 provided a comprehensive overview of photosynthetic carbon metabolism in plants. The content of the present book reflects our view that we are at the beginning of an era in which new genomic and related profiling techniques will allow metabolism to be examined more holistically than previously. The present volume therefore reviews the new developments that are uncovering the significance of nitrogen metabolism in photosynthesis and the importance of carbon metabolism for nitrogen assimilation. For the first time in this series, equal emphasis is placed on photosynthetic and respiratory metabolism. A major theme of the book is the intricate relationship between metabolic processes that requires researchers to take a broader view than ever before in examining the enormous complexity of plant metabolism. Written by a multinational team of experts, this work will be an invaluable tool for students at final-year undergraduate and graduate level, as well as essential and engaging reading for all those whose enthusiasm is fired by the intricate metabolic networks that support the growth of photosynthetic organisms on earth.

As editors of this volume, we wish to acknowledge the considerable efforts of all involved in the production of this work. In particular, we wish to thank the authors, who have made the most important contribution of all in providing their unique insights and personal perspectives. We are also deeply indebted to Govindjee and Larry Orr for their invaluable advice, patience and good humor, without which this volume could not have been assembled.

Christine H. Foyer
Rothamsted Research, UK
christine.foyer@bbsrc.ac.uk

Graham Noctor
Institut de Biotechnologie des Plantes, France
graham.noctor@ibp.u-psud.fr

Christine H. Foyer

Christine Foyer is a visiting Professor at the University of Newcastle, U.K. and the head of the Stress Biology Programme in the Crop Productivity and Improvement Division at Rothamsted Research, Harpenden, Hertfordshire, U.K. She was born in the town of Gainsborough (UK) in 1952. She graduated from the University of Portsmouth, UK in 1974 with a B.Sc. degree in Biology (Hons), and obtained her Ph.D. in 1977 from Kings College, University of London, U.K., working with Barry Halliwell. After post-doctoral research at King's College (with David Hall), she moved to the Research Institute for Photosynthesis, University of Sheffield, U.K. In 1988, she became a Directeur de Recherche at the Laboratoire du Métabolisme et de la Nutrition des Plantes, at INRA (Institut National de la Recherche Agronomique), Versailles, France. In 1994, she became Head of the Environmental Biology Department at the Institute of Grassland and Environmental Research, Aberystwyth, Wales. In 1998 she moved to her present position at the Institute of Arable Crops Research, Rothamsted, U.K., where she was Head of the Biochemistry and Physiology Department until 2001. She is author of *Photosynthesis*, Bittar E. E. series ed., *Cell Biology: A Series of Monographs*, John Wiley and Sons, New York, 219 pp, 1984, and co-editor of *Causes of Photooxidative Stress in Plants and Amelioration of Defense Systems*, 1994, CRC Press, 416 pp; and *A molecular Approach to Primary Metabolism in Plants*, Taylor and Francis , London, UK, 347 pp, 1998. Christine's current research interests concern the regulation of primary and intermediary metabolism in optimal and stress conditions. She is particularly interested in the metabolic crosstalk that controls assimilate partitioning between sucrose and amino acid biosynthesis in leaves. Moreover, she is a specialist in the field of oxidative stress in plants having published extensively on plant antioxidant metabolism and its role in stress signaling

Graham Noctor

Graham Noctor is a Professor at the Institut de Biotechnologie des Plantes, Paris, France. He was born in Manchester, UK in 1963. He obtained his first degree (B.Sc.) from the University of Essex, UK and his Ph.D from the University of Keele, UK in 1988, for work with John Mills on the control of photosynthetic metabolism by thiol-regulation. Then followed post-doctoral research in Peter Horton's laboratory at the University of Sheffield, UK, on relationships between photosynthetic light-harvesting efficiency and metabolism, focusing particularly on the mechanisms that underlie non-photochemical quenching of chlorophyll fluorescence. It was while at Sheffield that he became involved in work on carbon-nitrogen interactions, during two visits in 1990 and 1991 to Christine Foyer's laboratory at the Institut National de la Recherche Agronomique (INRA), Versailles, France. He subsequently returned to INRA Versailles, where he worked for four years on the control of the synthesis of the tripeptide, glutathione. From 1998 to 2001, he was a research scientist at the Institut of Arable Crops Research (Rothamsted, UK), where he was involved in several projects, notably investigating the role of mitochondria in photosynthesis and in carbon/nitrogen interactions, and in the relationship between oxidant production and antioxidant metabolism in leaves. He continues to explore these themes in his present post, which he took up in 2001.

Chapter 1

Photosynthetic Nitrogen Assimilation: Inter-Pathway Control and Signaling

Christine H Foyer*
Crop Performance and Improvement Division, Institute of Arable Crops Research, Rothamsted, Harpenden, Herts AL5 2JQ, UK

Graham Noctor
Institut de Biotechnologie des Plantes, Bâtiment 630, Université de Paris XI, 91405 Orsay cedex, France

Summary 1
I. Introduction 2
II. Control of Leaf Amino Acid Contents 4
A. Effect of Nitrogen Nutrition on Amino Acid Contents 4
B. Light-dependent Changes in Leaf Amino Acids 4
C. A Closer Look at the Impact of Photorespiration 5
D. Diurnal Changes in Leaf Amino Acid Contents and Cross-Family Co-ordination of Minor Amino Acids 9
III. Integration and Control of Nitrogen and Carbon Metabolism 11
A. Nitrate Reduction: A Key Control-Point 12
B. Supply of Carbon Skeletons for Amino Acid Synthesis 12
C. Governing the Carbon-Nitrogen Balance: Roles for Amino Acids and Organic Acids? 14
IV. The Carbon-Nitrogen Signal Transduction Network: Interactions Between Nitrate, Sugars and Abscisic Acid 16
V. Conclusions and Perspectives 18
References 19

Summary

The leaf is the predominant site of nitrogen assimilation in many crop species, and the stimulation of nitrogen assimilation by light reveals a close dependence on photosynthesis. Light controls the activity of nitrate reductase and, directly or indirectly, provides the reducing power necessary for the reductive incorporation of nitrate into amino groups. Photosynthetic and respiratory carbon metabolism is also required to generate the carbon skeletons necessary for amino acid synthesis. Amino acids represent the hub around which revolve the processes of nitrogen assimilation, associated carbon metabolism, photorespiration, export of organic nitrogen from the leaf, and the synthesis of nitrogenous end-products. Specific major amino acids are modulated differentially by photorespiration and nitrogen assimilation, even though these processes are tightly intermeshed. Minor amino acids show marked diurnal rhythms and their contents fluctuate in a co-ordinated manner. We discuss how regulation of the expression and activity of key enzymes allows co-ordination of carbon and nitrogen assimilation, and we assess the relative roles of key 'sensors' of Carbon-Nitrogen status. Analysis

*Author for correspondence, email: christine.foyer@bbsrc.ac.uk

Christine H. Foyer and Graham Noctor (eds): Photosynthetic Nitrogen Assimilation and Associated Carbon and Respiratory Metabolism, pp. 1–22.

reveals a complex network of controls brokered by an interplay of signals emanating from nitrate, carbohydrates, key metabolites such as glutamine, and plant hormones. In particular, abscisic acid is clearly implicated in the sensing of sugars and nitrate and associated signaling in higher plants. These controls act not only to orchestrate the activities of carbon and nitrogen assimilation at the intracellular level, but also influence plant development. The integrated perception of signals from hormones, nitrate, sugars, organic acids, and amino acids permits the plant to tailor its capacity for nitrogen assimilation to nutrient availability and requirements.

I. Introduction

Nitrogen is required by plants in greater quantities than any other mineral element. Much of this high demand reflects the large amount of Nitrogen (N) invested in the photosynthetic apparatus (Chapter 2, Kumar et al.). The availability of N is thus a significant determinant of both photosynthetic capacity and crop yield. Plants can absorb and assimilate various forms of N, though high amounts of added nitrate and the presence of nitrifying bacteria mean that nitrate is the principal form available to the roots of crop plants in agricultural conditions. Even in species capable of harboring nitrogen-fixing bacteria, nodules do not form in the presence of abundant nitrate.

In higher plants, nitrate can be reductively assimilated in both roots and shoots. The leaf is the major organ of N assimilation in many species, especially when N is plentiful (Foyer et al., 2001), and this is reflected in the high leaf capacity of enzymes such as nitrate reductase (NR) (e.g., Scheible et al., 1997a). In photosynthetic cells, eighty per cent of the reductant required for N assimilation comes directly from ferredoxin (Fig. 1). Reductant for NR activity is supplied either by the photosynthetic electron transport chain, through the operation of redox shuttles, or via the respiratory oxidation of fixed C (Fig. 1). In total, nitrate assimilation into amino acids requires 10 mole electrons per mole, 2.5 times more reductant than the reduction of CO_2 to carbohydrate. Although leaf nitrate reduction is usually only a fraction of the rate of C fixation, nitrate assimilation may be a variable and potentially significant sink for photosynthetic energy (Lewis et al., 2000). Under many conditions, therefore, N assimilation is a true photosynthetic process, in which light energy is used to power the reductive incorporation of a simple inorganic molecule into organic compounds (Fig. 1). The requirement for photosynthetic energy is reflected in the marked stimulation of nitrate reduction by light in many species (Aslam et al., 1979; Reed et al., 1983).

Nitrogen assimilation is also integrated with respiratory activity. Organic acids are required, first, in order to regulate cellular pH balance during the reduction of nitrate and, second, as amino group acceptors in amino acid synthesis. Numerous articles established that N re-supply to N-starved higher plant cells or unicellular algae is followed by marked stimulation of respiratory C flow (Larsen et al., 1978; Paul et al., 1981; Huppe and Turpin, 1994). Subsequent work has firmly established the concept that not only must assimilated C be partitioned, in a controlled manner, between starch and sucrose: C flow must also be regulated to ensure sufficient supply of organic acids for amino acid synthesis (Fig. 2). This regulation occurs both transcriptionally and post-translationally (Champigny and Foyer, 1992; Scheible et al., 1997b).

The photorespiratory cycle is a further key interaction between photosynthetic C and N metabolism, involving flux of ammonia through leaf pools of Gly, Ser, Gln and Glu (Keys et al., 1978). Under most conditions, in C_3 species at least, throughput of ammonia in photorespiration must be much more rapid than ammonia production originating from the reduction of nitrate.

The interactions between N assimilation, photosynthesis and respiration turn about a central axis constituted by leaf amino acid pools. This chapter reviews recent developments in understanding these interactions. Emphasis will initially be placed on the relationship between photosynthetic processes and leaf amino acid contents. Secondly, the principal

Abbreviations: 2-OG – 2-oxoglutarate; ABA – abscisic acid; Ala – alanine; Asp – aspartate; Fd – ferredoxin; GCN – general control non-reversible; GDH – glutamate dehydrogenase; Gln – glutamine; Glu – glutamate; GOGAT – glutamine:2-oxoglutarate aminotransferase (glutamate synthase); GS – glutamine synthetase; GS1 – cytosolic glutamine synthetase; GS2 – chloroplastic glutamine synthetase; ICDH – isocirate dehydrogenase; LR – lateral root; NR – nitrate reductase; OAA – oxaloacetate; PDH – pyruvate dehydrogenase; PEPc – phosphoenolpyruvate carboxylase; PK – pyruvate kinase; Ser – serine; SNF1 – yeast sucrose non-fermenting control gene encoding a protein kinase; SnRK – SNF-related protein kinase; SPS – sucrose phosphate synthase; TCA – tricarboxylic acid

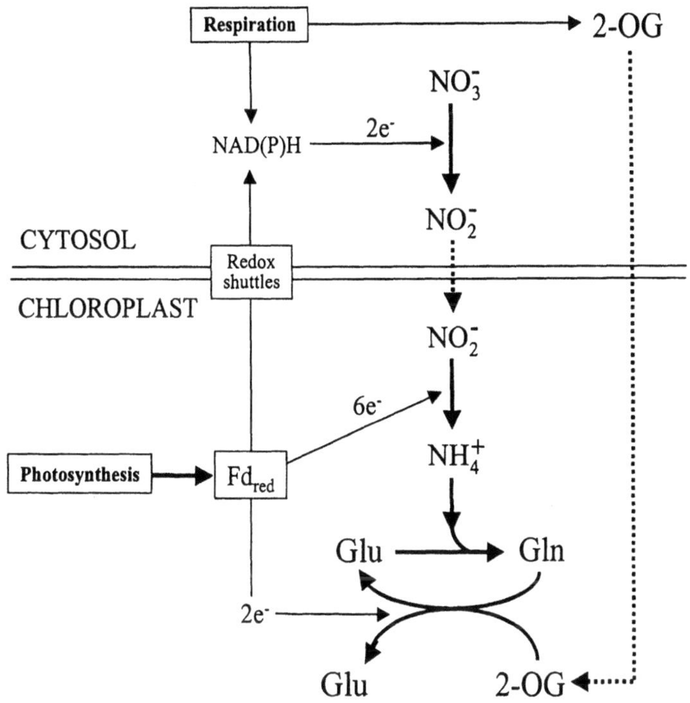

Fig. 1. Simplified scheme of interactions between photosynthesis, respiration and N assimilation. Formation of 2-OG may occur in the cytosol or mitochondria. In either case, the necessary sequences involve both compartments. Fd_{red}, reduced ferredoxin, 2-OG, 2-oxoglutarate.

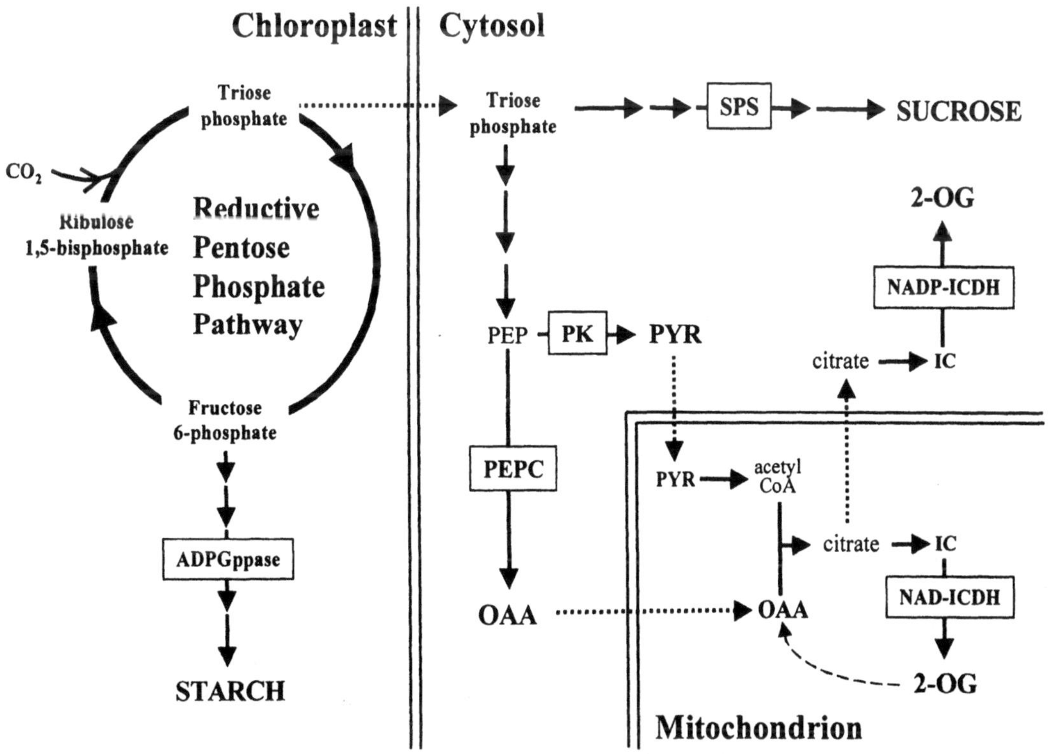

Fig. 2. Carbon partitioning between carbohydrate synthesis and respiratory generation of 2-oxoglutarate (2-OG) and other organic acids (OAA, PYR). The scheme shows the shortest path for net 2-OG generation for glutamate synthesis. Likely key regulatory enzymes are boxed. ADPGppase, ADP-glucose pyrophosphorylase; IC(DH), isocitrate (dehydrogenase); OAA, oxaloacetate; PEP(C), phosphoenolpyruvate (carboxylase); PK, pyruvate kinase; PYR, pyruvate; SPS, sucrose phosphate synthase.

factors that ensure the integration of N and C metabolism will be analyzed. Lastly, we will discuss the significance of nitrate, metabolites and hormones in the co-ordination of C and N assimilation, and in sensing and transmitting information on whole plant C/N status.

II. Control of Leaf Amino Acid Contents

It is established that the major pathway for ammonia incorporation occurs through glutamine synthetase (GS) and GOGAT (Miflin and Lea, 1982; Chapters 6 (Hirel and Lea) and 7 (Florencio and Reyes)). Amino groups are then transferred out of this cycle, predominantly via Glu, to other amino acids, such as Asp and Ala. The GS-GOGAT pathway is also responsible for the reincorporation of NH_3 released by Gly decarboxylation during photorespiration (Keys et al., 1978; Chapter 8, Keys and Leegood).

Leaf amino acid contents are determined by a complex interplay of factors, including: 1. Flux of C into amino acid pools, either from glycolate C generated in photorespiration or in C flux through glycolysis and the TCA cycle; 2. Flux of N into these pools, from NH_3 generated principally by Gly decarboxylation, by translocation from the roots or by leaf nitrate reduction; 3. Exchange between amino acid pools in transamination reactions; 4. Removal of both C and N from the pool in utilization of amino acids for synthesis of end-products such as proteins, pigments and nucleotides; 5. Other processes, such as amino acid catabolism, protein degradation, or non-photorespiratory ammonia cycling (e.g., from the reaction catalysed by phenylalanine ammonia lyase). 6. Import and export of amino acids. Because all of these processes may be influenced by nutrition, developmental stage, and by environmental conditions (temperature, light, stress, etc), it is extremely difficult to predict the influence any given process will have on a given leaf amino acid pool. This conclusion renders two research goals problematic: first, the use of amino acid measurements as physiological indicators, an undertaking facilitated by the development of increasingly powerful analytical tools; second, identification of which amino acid concentrations, if any, are likely to have been recruited during evolution to act as 'signals' that transmit information on the plant's metabolic status.

A. Effect of Nitrogen Nutrition on Amino Acid Contents

It is clear that leaf amino acid contents increase with enhanced supply of N during growth (Khamis et al., 1990; Scheible et al., 1997a). Short-term changes have also been demonstrated. Leaf amino acid contents were markedly increased by supplying nitrate or ammonia to excised maize leaves (Foyer et al., 1994a). Short-term effects probably mainly reflect increased substrate supply and/or enzyme activation while longer-term changes are also due to modified expression of enzymes such as NR (Scheible et al., 1997a). Interestingly, unlike phosphoenolpyruvate carboxylase (PEPc) and sucrose phosphate synthase (SPS), the NR activation state (Chapters 3–5) does not respond to nitrate (Huber et al., 1992; Ferrario et al., 1996). In the short-term at least, increases in amino acid contents are not general. In maize, enhancement of leaf amino acids on feeding N was almost entirely due to accumulation of Gln (Foyer et al., 1994a). The key response of the Gln pool has also been demonstrated in tobacco mutants and transformants with a wide range of NR activities. When these plants were grown on different nitrate concentrations, leaf amino acid contents varied more than four-fold and correlated with NR activity (Scheible et al., 1997a). The clearest correlation with NR was Gln, which showed an almost linear, proportional increase from less than 0.1 to more than 4 $\mu mol.g^{-1}$fresh weight as NR activity increased (Scheible et al., 1997a). Such observations suggest that Gln contents may reflect the balance between the capacity for C and N assimilation, being low when C:N is high and high when C:N is low. This interpretation would fit with the known effects of Gln on expression of enzymes such as NR (Vincentz et al., 1993), and is discussed further below.

B. Light-dependent Changes in Leaf Amino Acids

Besides long-term developmental effects on gene expression, light is known to exert a marked and rapid effect on NR activation state (Chapters 3-5). In addition, light might be expected to favor N assimilation due to increased availability of reductant (discussed further in Foyer et al., 2001). The effect of light on organic acid synthesis is unclear, since respiration may be partially inhibited in the light (Chapters 10-12). Whether there is preferential

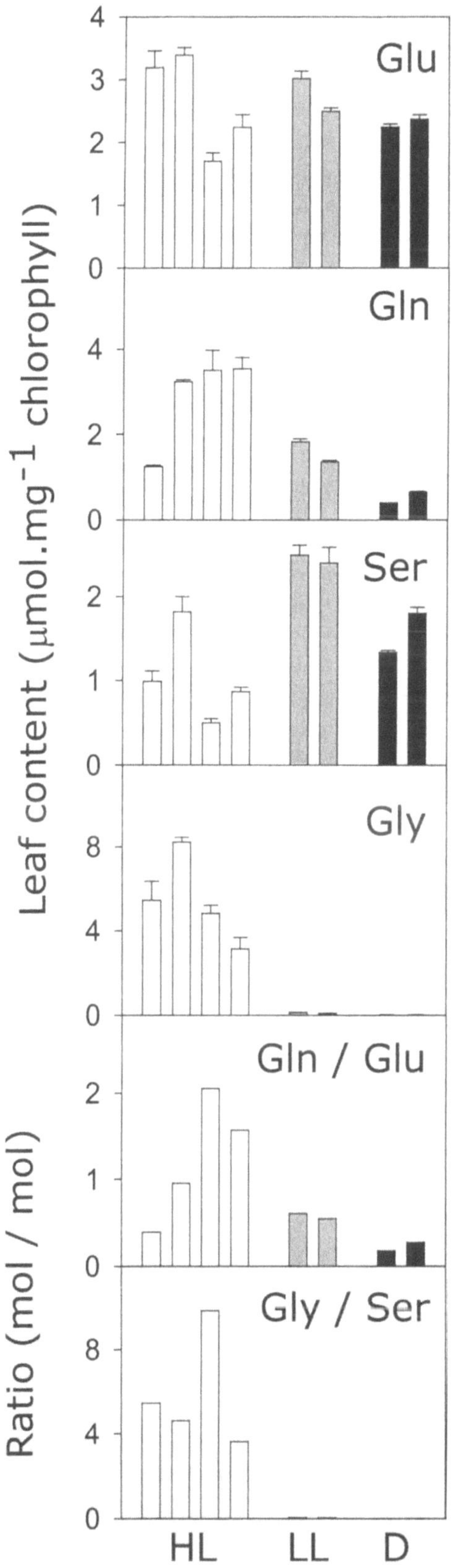

Fig. 3. Short-term effects of light on amino acids in young wheat leaves. Analyses were performed on the fourth leaf of plants five weeks after seed sowing. Three samples were taken by a rapid-quench method from the same leaf, and the data are expressed as means ± SE of extracts from these three samples. Each bar is a different plant, either placed in the dark (D) for 120 min (20 °C: black bars, right), kept at the low irradiance (LL) of the growth chamber (20°C, 250 µmol. quanta $m^{-2}s^{-1}$: gray bars, middle), or incubated for 200 min in a high light (HL) chamber (25 °C, 1000 µmol.$m^{-2}s^{-1}$: white bars, left). All samples were taken within 1 h of each other in the middle of the photoperiod. Data are authors' previously unpublished results.

inhibition of dissimilatory (primarily reductant-generating) respiration over assimilatory (primarily precursor-generating) pathways remains unclear, though the light activation of the C_3 isoform of PEPc (Champigny and Foyer, 1992) strongly suggests that illumination affects the balance between these two processes. At values up to light saturation of photosynthesis, higher irradiance will drive higher rates of the photorespiratory pathway, especially in C_3 plants, and this might impact on several leaf amino acid pools.

Figure 3 shows short-term effects of irradiance on the major wheat leaf amino acids involved in N assimilation and photorespiration. Whereas leaf Glu was relatively stable, Gln generally increased with irradiance though Gln pools were variable, particularly at high light. Like Glu, Ser was also relatively stable. Gly, however, was negligible in the dark, not much higher at low light, but present at very high values at high light, where it typically represented 40–50% of total amino acids. While Gln/Glu showed a tendency to increase with irradiance, this ratio was variable and much less clearly affected by light than Gly/Ser, which increased from less than 0.01 in the dark to about 0.1 at low light, and reached values between 4 and 8 at high light (Fig. 3). Effects such as those shown could be due to photorespiration or primary N assimilation, since both are stimulated by light within the irradiance range used.

C. A Closer Look at the Impact of Photorespiration

In photosynthetically active leaves, particularly young expanding leaves where active protein synthesis is ongoing with CO_2 fixation, GS2 and Fd-GOGAT have to assimilate N at the same time as recycling photorespiratory ammonia at potentially much higher rates. Since there is as yet no indication of metabolic partitioning to maintain distinct pools of metabolites that participate in photorespiration and N assimilation, these enzymes can be assumed to play overlapping roles. This notion is supported by results of studies of *Arabidopsis* mutants deficient in Fd-GOGAT (Coschigano et al., 1998) and by a thorough investigation of age-related changes in N metabolism in tobacco (Masclaux et al., 2000). The effects of N assimilation and photorespiration on amino acids in leaves with C_3 photosynthesis have been considered inextricable (Stitt and Krapp, 1999; Stitt et al., 2002), and must certainly be tightly linked. Evidence in favor of this notion comes from experiments in tobacco leaves where not only Gln, but also Gly and Ser accumulate throughout the day in plants growing at constant irradiance (Scheible et al., 2000). At first sight, the higher fluxes of photorespiration might be considered likely to exert the greater influence on leaf contents of these amino acids. Assuming fairly constant steady-state stomatal conductance, however, photorespiratory flux should follow overall rates of CO_2 fixation and should therefore reach a steady-state rate at the same time as photosynthesis. For example, the Gly/Ser ratio in wheat and barley leaves increases markedly during the induction period of photosynthesis but thereafter remains stable (C. H. Foyer, G. Noctor, unpublished). We have recently examined the influence of photorespiration on leaf amino acids by incubation of attached leaves of wheat and potato in gas-exchange chambers under controlled conditions followed by rapid-quench sampling on attainment of the steady-state rate of photosynthesis (Novitskaya et al., 2002). By illuminating leaves at different irradiance and partial pressures of CO_2 and O_2, and simple modeling of the rate of photorespiration, three parameters were found to show a good correlation with photorespiratory flux (Novitskaya et al., 2002). These were Gly/Ser (positive correlation) and the fraction of amino acids accounted for by Asp and Ala, both correlating negatively with photorespiration. No clear relationship was observed between Glu or Gln and photorespiration. In contrast, Gln (and Gln/Glu) showed a variable but evident correlation with the rate of net CO_2 uptake in wheat leaves (Novitskaya et al., 2002). This suggests that non-photorespiratory ammonia assimilation impacts more strongly on leaf Gln than does photorespiration.

Why should high rates of ammonia recycling in photorespiration impact less on Gln pools than lower rates of non-photorespiratory ammonia production? One possibility is the key role played by the availability of 2-oxoglutarate (2-OG; Fig. 4). If, during steady-state rates of photorespiration, 2-OG is formed at the same rate as ammonia is released from Gly (and GS and GOGAT are not limiting), no accumulation of Gln occurs. Thus, the photorespiratory system is constructed to allow a smooth cycling of amino groups. The pathway is controlled by the rate of glycolate production, and so glyoxylate is generally available to accept amino groups and regenerate 2-OG. This is evidenced by the following effects of faster photorespiration: (1) build-up of 2-OG; (2) depletion of leaf pools of Ala, probably due to direct use in glyoxylate transamination; (3) marked decreases in Asp, probably via Asp aminotransferase and Glu:glyoxylate aminotransferase (Novitskaya et al., 2002). By contrast, comparatively low rates of net N assimilation may cause accumulation of leaf Gln because of relatively loose coupling to anaplerotic 2-OG formation. If this interpretation is correct, N assimilation impacts strongly on Gln concentrations against a much higher rate of photorespiratory ammonia recycling. Importantly, Gln contents would then reflect not the absolute ammonia supply but the *balance between ammonia and 2-OG availability* (Novitskaya et al., 2002). Figure 5 illustrates the potential importance of 2-OG deficit in determining Gln accumulation and, consequently, the Gln/Glu ratio (panels A and B).

A strong influence of N assimilation on leaf Gln is consistent with the data discussed in Section II,B and also with other literature studies. In barley, Gln/Glu decreased in mildly droughted leaves even though photorespiration was probably increased under these conditions (Wingler et al., 1999). Although no changes on growth were observed, overexpression of NR in tobacco resulted in significantly increased Gln (Foyer et al., 1994b). There is also evidence that high Gln levels may reflect insufficient supply of 2-OG. When 2-OG was supplied to tobacco leaves, Gln fell even though the extractable activity of NR was increased (Müller et al., 2001). Control of ammonia assimilation by modulation of GS and GOGAT activities could also be important, and some evidence has been presented that is consistent with in vivo modulation of GS activity when fluxes are changed at low light (Morcuende et al., 1998). However, it remains unclear how modulation of the capacities of GS2 and Fd-GOGAT exerts appreciable control over N assimilation when, in C_3 leaves at least, the activities of these enzymes must be sufficient to cope with much higher rates of ammonia release during

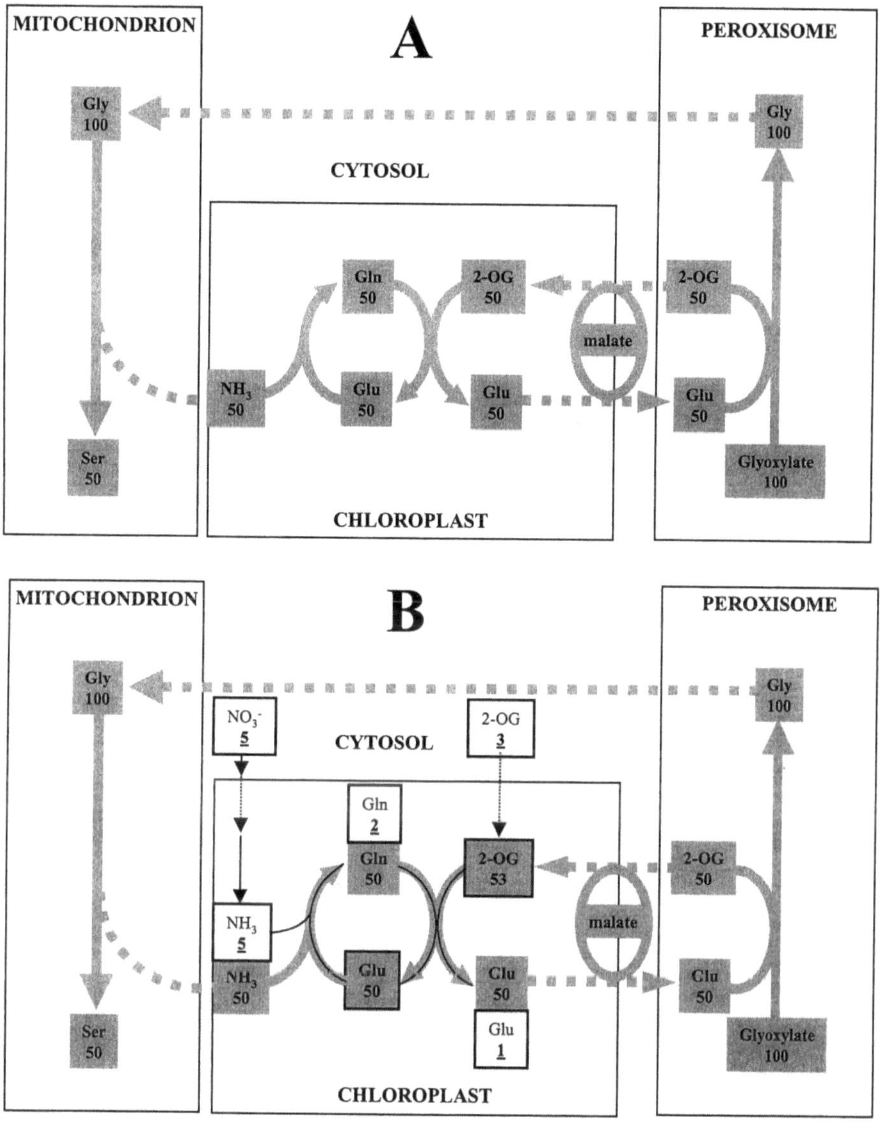

Fig. 4. Simplified scheme showing possible effects of photorespiratory and non-photorespiratory ammonia generation on the chloroplastic Gln pool during typical rates of photosynthesis and photorespiration at intermediate to high irradiance in air. Numbers in the boxes give typical rates of formation or supply of each compound (μmol.mg^{-1}chl.h^{-1}) during fast photosynthesis. A. All ammonia is produced by photorespiration. Net CO_2 fixation and RuBP carboxylation:RuBP oxygenation are taken to be 200 μmol.mg^{-1}chl.h^{-1} and 2.5 (Sharkey 1988; Keys 1999; Noctor et al. 2002b). The rate of glyoxylate formation is therefore 100 μmol.mg^{-1}chl.h^{-1} and photorespiratory ammonia release is 50 μmol.mg^{-1}chl.h^{-1}. Assuming sufficient energy supply and capacities of enzymes and transporters capacity, Gln only accumulates if the supply of NH_3 from Gly oxidation is faster than the rate of 2-oxoglutarate (2-OG) regeneration by Glu:glyoxylate aminotransferase (50% of glyoxylate is transaminated by other amino acids, chiefly Ser). B. Influence of a rate of non-photorespiratory NH_3 formation 10-fold lower than photorespiratory ammonia production, if 2-OG supply is not sufficient. For simplicity, metabolism of Glu and Gln in downstream transamination reactions is not considered. Reduction of nitrate to ammonia occurs at 5 μmol.mg^{-1}chl.h^{-1}, equivalent to 2.5% the rate of net CO_2 fixation (shown by black lines and boxes). The hypothetical case shown assumes a rate of anaplerotic 2-OG formation equal to 60% of the rate of nitrate assimilation, i.e., 3 μmol.mg^{-1}chl.h^{-1} (15 μmol.C mg^{-1}chl.h^{-1} or 6 triose phosphate equivalents, i.e., more than 9% of the rate of net CO_2 fixation (net CO_2 fixation will be lower than in (A) if anaplerotic 2-OG generation increases 'day' respiration). Within 1 h, non-photorespiratory ammonia assimilation leads to *net formation* of 1 μmol.Glu and 2 μmol.Gln per mg chl (boxes). See also Color Plate 1.

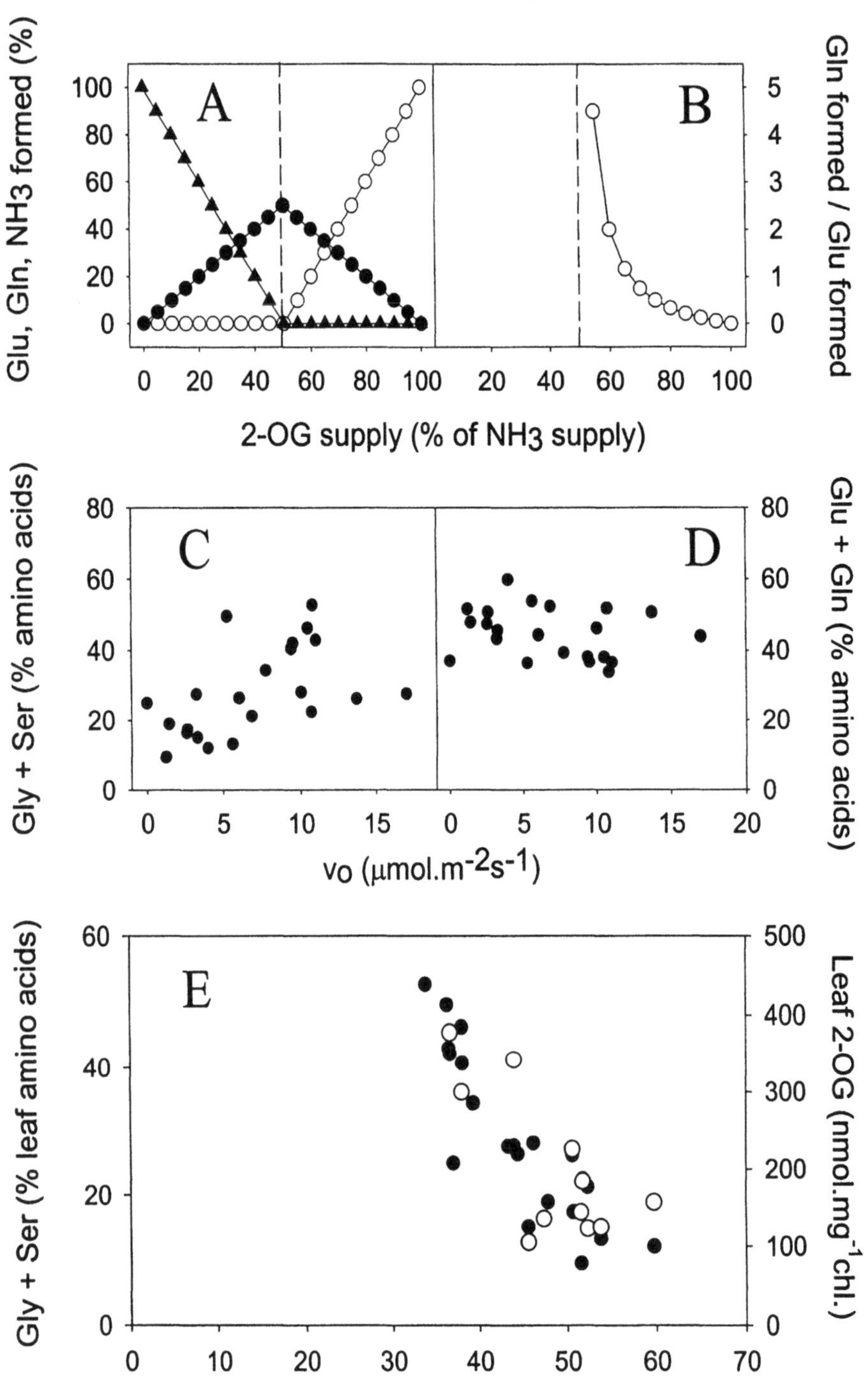

Fig. 5. Theoretical and observed relationship between leaf Gln, Glu, ammonia, Gly plus Ser and 2-oxoglutarate (2-OG). A,B, Predicted effects of the relative supply of ammonia and 2-OG on synthesis of Gln and Glu. Enzyme capacity and substrate concentrations are assumed to be kinetically non-limiting, and so relationships are purely theoretical predictions due to the stoichiometry of the GS-GOGAT pathway. The rate of ammonia supply is arbitrarily set at 100. Net formation of Gln and Glu can be predicted as follows. Gln formed = NH_3 supplied - 2-OG supplied. Glu formed = 2-OG supplied - Gln formed. When 2-OG supply equals the supply of ammonia, all net output is Glu (A, empty circles). As 2-OG deficit increases, the net production of Gln rises linearly (A, filled circles) and in inverse fashion to the net production of Glu. Because of its two-fold higher C/N ratio, the rate of Glu generation decreases twice as fast as Gln generation increases. Thus, Gln formed/Glu formed rises increasingly steeply with increasing 2-OG deficit (B), even if downstream reactions do not preferentially remove Glu. Theoretically, the GS-GOGAT (glutamine synthetase-glutamate synthase) cycle can continue to assimilate all

photorespiration.

In wheat and potato leaves analyzed at different irradiance and gas composition, Gly increased markedly with photorespiration while Ser generally decreased, though less strongly. However, the absolute amounts of both these amino acids were variable, even when expressed as a proportion of total amino acids (C. H. Foyer, G. Noctor, unpublished). Thus, similar rates of photorespiratory flux can proceed at very different Gly and Ser concentrations, even though Gly/Ser remains constant, at least in the short term. It seems that the Gly/Ser ratio is controlled mainly by the rate of photorespiration (i.e., the flux of C) whereas the amounts of Gly and Ser are influenced not only by the supply of C but also by N assimilation. This may explain the accumulation of Gly and Ser throughout the day in tobacco, as well as the much higher Gly and Ser contents in plants growing on abundant nitrate compared to limiting nitrate (Scheible et al., 2000). It is interesting that Gly plus Ser, which shows a loose positive correlation with photorespiratory flux (Fig. 5,C), mainly due to large increases in Gly, and Glu plus Gln (which shows no obvious correlation with photorespiratory flux (Fig. 5,D)) are in negative correlation with each other (Fig. 5,E, closed circles). Leaf contents of 2-OG, which generally increase with the rate of photorespiration (Novitskaya et al., 2002), also show an inverse correlation with Glu plus Gln (Fig. 5,E, open circles). As noted previously (Scheible et al., 2000), transfer of amino groups from the GS-GOGAT cycle could act to dampen changes in the Gln pool. When the C acceptor is scarce, amino groups become increasingly trapped in Gln, ammonia incorporation ceases (Fig. 5,A), or both. Transfer of reduced N to amino acids with low C/N ratios would provide short-term alleviation of this potential problem. In particular, accumulation of Gly would allow 2-OG regeneration to proceed at higher rates than photorespiratory ammonia release. It was estimated that up to 10% of Gly formed during the induction period of photosynthesis (approx. 30 min illumination of dark-adapted leaves) was accumulated rather than metabolized (Novitskaya et al., 2002). More long-term accumulation of Gly, subsequent to the attainment of steady-state photorespiratory cycling, would probably represent only a small fraction of total Gly generated, which is of the order of 20–50 μmol.g^{-1}fresh weight.h^{-1} at intermediate rates of photosynthesis. Accumulation of Gly would therefore represent only a small imbalance between 2-OG recycling via glyoxylate transamination and ammonia release by Gly decarboxylation. However, even a slight imbalance might free up a significant amount of 2-OG to support N assimilation. For example, if photorespiratory 2-OG regeneration exceeds Gly deamination by only 1%, this is enough to provide 20% of the 2-OG demanded by a rate of N assimilation equal to about 5% of the rate of photorespiratory ammonia production. This view implies that rather than favoring Gln accumulation, photorespiration might serve to *attenuate or defer* rises in Gln. Such an effect could explain the slow increase in Gly/Ser observed in tobacco (Scheible et al., 2000), though there may also be some contribution from possible increases in mitochondrial NADH/NAD$^+$ or, perhaps, gradual increase in photorespiratory flux throughout the day due to decreasing stomatal conductance. Another process that could damp down rises in Gln and allow ammonia assimilation to continue is transfer of amino groups to Asn, though no correlation was observed between Gln and Asn in short-term experiments in wheat or potato (Novitskaya et al., 2002).

D. Diurnal Changes in Leaf Amino Acid Contents and Cross-Family Co-ordination of Minor Amino Acids

Light stimulation of N assimilation produces a diurnal rhythm in total leaf amino acids (Fig. 6). The exact nature of these changes is likely to be species- and condition-specific, and it cannot be excluded that growth in constant environment chambers entrains or accentuates such fluctuations. Equally, such diurnal fluctuations could be less marked in younger leaves, where local sinks for amino acids are relatively powerful. The data shown in Fig. 6 were obtained for

(Fig. 5, continued) the available ammonia as long as 2-OG supply equals at least half the rate of ammonia supply (dashed vertical line). At this value, all net output is Gln. Below this value, the balance between C/N supplied and the C/N of the product would have to be adjusted by preferential removal of Gln, either by export or by use via enzymes other than GOGAT. Otherwise, there would be a progressive decrease in the Gln output and an accumulation of ammonia (A, triangles), net consumption of Glu, or both. C,D, Relationship between the calculated rate of photorespiration and Gly plus Ser (C) and Glu plus Gln (D), measured in wheat leaves during steady-state photosynthesis under different conditions. E, Gly plus Ser (filled circles, data replotted from C and D) and 2-oxoglutarate (empty circles) against Glu plus Gln. Further modeling and experimental details in Novitskaya et al. (2000).

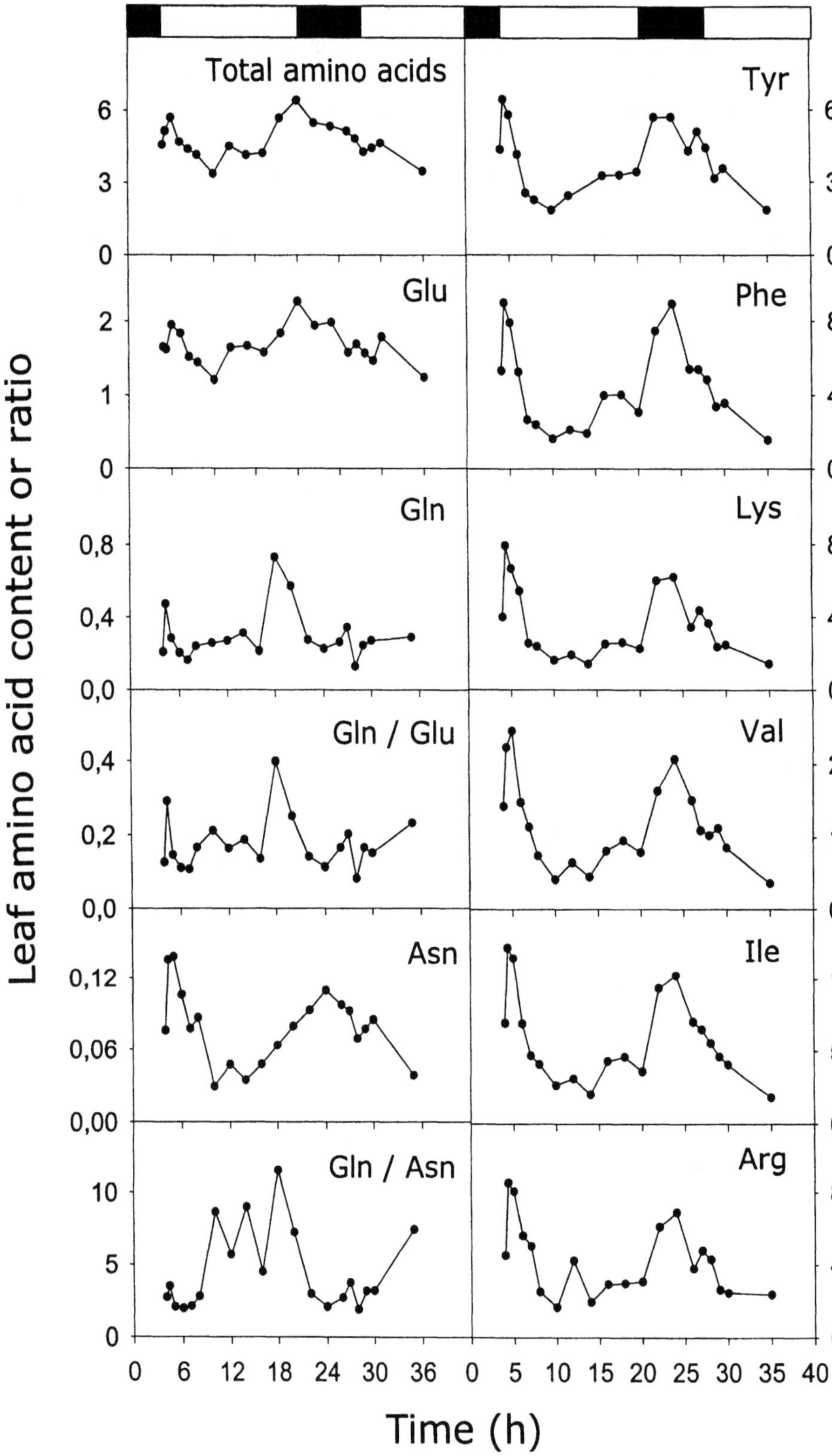

Fig. 6. Day-night rhythms in the amino acid contents of wheat leaves. Plants were grown in soil in field conditions and density until 6.5 weeks after seed sowing. Growth conditions were 16 h light (20 °C, 250 µmol. quanta $m^{-2}s^{-1}$), 8 h dark (16 °C). Samples were taken between 3.30 a.m. (time = 3.5 h) and 11 a.m. on the following day (time = 35 h), light on at 4 a.m. and off at 8 p.m. Data are means of two independent extractions from the fourth or fifth leaves on different plants. Contents of leaf amino acids are expressed as µmol.mg^{-1}chlorophyll (left panels) or nmol.mg^{-1}chlorophyll (right panels). Ratios are mol/mol. (Data are authors' previously unpublished results.)

wheat leaves approaching the end of the sink/source transition, i.e., leaves with high rates of photosynthesis nearing full expansion. In these leaves, the rise in total leaf amino acids was well correlated with Glu (Fig. 6). Gln only increased markedly towards the very end of the light period (Fig. 6). Surprisingly, perhaps, Asn also accumulated in the light and was much lower in the dark (Fig. 6). Ser contents followed a similar pattern to Glu, while Gly was rather low and variable throughout (data not shown), probably because of a generally low irradiance and shading effects, since the plants were grown at field density. A very clear effect in wheat was a concerted rhythm in minor amino acids (Fig. 6). Changes in minor amino acids were more marked than overall modulation of total amino acids, so that minor amino acids increased significantly during the second half of the light period, both with respect to chlorophyll (Fig. 6) and as a fraction of total amino acids. Minor amino acid contents were lowest in the middle of the day and highest during the first part of the dark period (Fig. 6).

What causes the changes in minor amino acids? Studies in tobacco have highlighted the potential significance of leaf carbohydrate metabolism in influencing amino acid contents. Minor amino acid contents were found to correlate with starch and sucrose, when carbohydrate accumulation was manipulated by day length (Matt et al., 1998). In an alternative approach, excised tobacco leaves were fed sucrose, which led to a general increase in several minor amino acids (Morcuende et al., 1998). Compared to tobacco, wheat leaves accumulate more sucrose and less starch. Nevertheless, accumulation of both carbohydrates occurs in wheat leaves during the light period (e.g., Trevanion, 2000), so that the steep rise in carbohydrates during the second half of the light period just precedes the increase in minor amino acids (Fig. 6). The day-night changes in minor amino acids may therefore be influenced by build-up of carbohydrates and/or amides.

Minor amino acids are synthesized through distinct pathways that are under specific control by key enzymes (Morot-Gaudry et al., 2001). Contents are unlikely to be markedly affected by short-term changes in either photosynthetic C supply or photorespiratory rates, and this idea is confirmed by analysis of amino acids in wheat and potato leaves sampled under conditions of widely differing rates of photosynthesis and photorespiration (Novitskaya et al., 2002). No correlation with photosynthetic parameters was observed, even though minor amino acids varied more than 20-fold between leaves under the different conditions (Noctor et al., 2002a). The most striking aspect of the data was the good correlations between minor amino acids synthesized via different pathways (Noctor et al., 2002a). Leaf contents of Tyr, for example, correlated not only with leaf Phe but also with leaf Val and leaf Arg (Noctor et al., 2002a). Only part of this correlation could reflect the concerted diurnal rhythm in minor amino acids, since the experiments were carried out within a window of 4–6 h in the middle of the photoperiod, during which period the mean contents of minor amino acids varied about two-fold (Fig. 6). However, it is clear that the same mechanisms that are responsible for the diurnal rhythm in minor amino acids may also produce the correlations observed in leaves incubated in different short-term conditions. Carbohydrate contents are known to be vary considerably, even between leaves sampled in identical conditions. Though processes other than synthesis may contribute to the changes in minor amino acids observed in tobacco, wheat and potato, it is an intriguing possibility that genes involved in minor amino acid synthesis might be controlled, at least partly, by carbohydrates or associated factors (Noctor et al., 2002a). Control factors influenced by sulfhydryl status cannot be excluded. In poplars with enhanced capacity for glutathione synthesis in the chloroplast, where most of the minor amino acids are produced, the increase in leaf glutathione contents correlated with high contents of several minor amino acids (Noctor et al., 1998). General control of minor amino acid synthesis is discussed further in Section IV.

III. Integration and Control of Nitrogen and Carbon Metabolism

Several studies indicate homeostatic co-ordination of C and N assimilation in higher plants through effects on nitrate uptake by the roots, nitrate translocation in the xylem, and nitrate reduction in the leaves (Foyer et al., 2001). Increases in CO_2 fixation with increasing irradiance are accompanied by enhanced N uptake (Gastal and Saugier, 1989), while N assimilation is decreased at low CO_2 (Pace et al., 1990) or during depletion of carbohydrates in extended darkness (Rufty et al., 1989). Another important limitation over the rate of incorporation of nitrate (and ammonia) into downstream products is the leaf's capacity to supply C skeletons. In unicellular algae, switching from limiting to abundant N causes a marked inhibition of photosynthesis and concomitant stimulation of respiration (Elrifi and Turpin, 1986). In plants, such changes are much less marked (Foyer et al., 1994a). The next sections briefly outline control of NR before discussing the potential importance of anaplerotic C flow in controlling N assimilation in the leaves of higher plants.

A. Nitrate Reduction: A Key Control-Point

Two key observations suggested that regulation of leaf NR activity is important in co-ordinating nitrate reduction and CO_2 fixation. First, the capacity for nitrate reduction (extractable NR activity) increases in the light. Second, the activity of NR is decreased at low CO_2 (Kaiser and Förster, 1989; Pace et al., 1990). Further work has established that changes in extractable activity reflect multilevel control mediated by multiple factors, of which the most important are light, nitrate, Gln, and sugars (Campbell, 1999; Stitt et al., 2002). Expression of *nia* genes, encoding NR, is induced by nitrate and sugars, and suppressed by Gln (Hoff et al., 1994). Transcript abundance is also under light-dark control, but is often out of phase with protein abundance, and in vitro NR activity does not correlate tightly with *nia* transcript levels (Vincentz and Caboche, 1991). At the post-translational level, NR activity is inhibited through protein phosphorylation (Kaiser and Huber, 1994). Phosphorylation per se does not significantly affect enzyme activity, but it allows binding of small proteins belonging to the 14-3-3 class of inhibitor proteins found in all major classes of eukaryotes (Bachman et al., 1996). The light-activation of NR activity presumably involves de-phosphorylation followed by dissociation of enzyme and inhibitor protein, though the intermediates that link light/dark to changes in NR phosphorylation status remain obscure. While nitrate does not affect the phosphorylation status of NR, it does prevent decreases in NR capacity due to protein turnover (Ferrario et al., 1995, 1996). Compensatory modifications at several levels (including transcription, protein turnover, and phosphorylation status) dampened the impact of fewer *nia* genes on nitrate assimilation in tobacco (Scheible et al., 1997c). Several of these controls over NR activity seem to be affected by carbohydrate supply. As well as increasing *nia* transcripts, exogenous sugars influence the post-translational regulation of NR (Kaiser and Huber, 1994). Feeding sugars to tobacco leaves markedly stimulated nitrate reduction (from a relatively low control rate), an effect correlated both with greater stability of the NR protein and with an increase in NR activation state (Morcuende et al., 1998). Both increased stability and increases in activation state may be linked to sugar-induced decreases in NR phosphorylation status. Further work is required to identify the C metabolite(s) most important in controlling NR activity at these levels, and the physiological significance of nitrate, Gln and sugars has recently been critically discussed (Stitt et al., 2002).

Control over NR should be distinguished from control over nitrate reduction. Constitutive expression of NR in tobacco does not affect chlorophyll contents, protein levels, photosynthesis or biomass, although some effect on flux is indicated by increased Gln contents (Foyer et al., 1994b; Ferrario et al., 1995). Whenever the in vivo rate of nitrate reduction has been compared with NR, it has almost always been found to be lower than the extractable activity, even when only active (i.e., dephosphorylated) enzyme is assayed. One explanation is allosteric control over NR that operates in vivo but not during standard enzyme assays. A second explanation is that NR is substrate-limited or, in the presence of sufficient nitrate, reductant-limited (Chapter 5, Kaiser et al.). Nevertheless, current knowledge suggests that control of NR is probably one of the key factors co-ordinating C and N assimilation, even if further work is required to establish how changes in NR protein or activation state exert dynamic control over the rate of nitrate reduction.

B. Supply of Carbon Skeletons for Amino Acid Synthesis

Leaves must allocate a significant proportion of fixed C to amino acid synthesis. The shortest sequence through which 2-OG can be produced (Fig. 2) can be summarized as follows:

Sugar-P (6C) = 2 PEP; PEP + CO_2 = OAA; PEP = acetyl group + CO_2; OAA + acetyl group = citrate; citrate = 2-OG + CO_2.

This sequence involves flux through the lower part of glycolysis, PEPc, and three enzymes of the TCA cycle (or cytosolic isoforms of aconitase and ICDH).

In reality, the metabolic conversions are likely to be more complex (Huppe and Turpin, 1994; Krömer, 1995) and there is cycling of C between carbohydrate synthesis, photorespiratory and respiratory pathways (Pärnik and Keerberg, 1995).

The proportion of fixed C required by N assimilation will change markedly according to developmental stage, N availability and the nature of the products (Lewis et al., 2000). For each molecule of NH_3 assimilated, one molecule of 2-OG is required to form the product of the GS-GOGAT pathway, Glu (C/N = 5). Once ammonia is incorporated into Gln and Glu, other C skeletons are required for amino acid synthesis, chiefly via use of Glu in transamination reactions. The principal physiological fates of assimilated N are protein synthesis (young expanding 'sink' leaves) and export (older 'source' leaves). Even if export proceeds chiefly via Asn and Gln (which may not be the case in many species; Chapter 15, Lohaus and Fischer), at least 2–2.5 C would still be required per N exported. Protein synthesis would require more C per N (mean C/N of the protein amino acids = 4.35). Other nitrogenous products synthesized primarily in young leaves typically have higher C/N ratios (e.g., chlorophyll, C/N = 13.75). The C demand linked to N assimilation will be particularly high, therefore, in young tissues, because both the rates of N assimilation and the C/N of the ultimate product are relatively high. Even in older tissues the demand for C is likely to be significant. For instance, taking N assimilation as 5% of the rate of net CO_2 fixation, 12.5% (export of amino acids as Gln) or around 22% (production of protein amino acids in equal amounts—an approximation) of fixed C needs to be allocated to amino acid synthesis. These values underline the considerable respiratory fluxes that must operate in tandem with N assimilation. Such fluxes involve only partial oxidation and therefore minimal net release of CO_2, i.e. one-sixth of that linked to complete oxidation in the TCA cycle. If net production of all protein amino acids occurs at N assimilation of about 5% of net CO_2 fixation, the minimum required respiratory C release would probably be around 4% of net C fixed.

From a physiological point of view, it can be predicted that a shortfall in C skeletons should be signaled back to nitrate reduction, reining in N assimilation and avoiding excessive production of nitrite and/or ammonia. Insufficient anaplerotic C in the light, where N assimilation is relatively fast, could result from (1) incapacity of the system to divert enough sugar-P to oxidation, e.g., inadequate activity of enzymes such as pyruvate kinase (PK), pyruvate dehydrogenase (PDH) or PEPc; (2) further oxidation of key acceptors such as 2-OG. Negative control of PK and PDH by effectors and phosphorylation may play a significant part in the general partial inhibition of respiration that is observed in the light (Pärnik and Keerberg, 1995). By contrast, PEPc is known to be activated in the light by phosphorylation, and is also activated by Gln (Champigny and Foyer, 1992). The activity of this enzyme correlates well with ammonia assimilation in algae (Vanlerberghe et al., 1990) and with leaf Gln in plants (Murchie et al., 2000). Changes in PEPc activity produce a significant shift from dissimilatory respiration in the dark to anaplerotic respiratory flow in the light, although the latter is likely only one part of overall respiratory rates and still requires PK, PDH, citrate synthase, aconitase, and ICDH activities. An important process preventing oxidation of 2-OG in the TCA cycle may be low activities of the mitochondrial ICDH so that 2-OG is formed principally in the cytosol. This thinking is consistent with the observed export of citrate and isocitrate by isolated leaf mitochondria supplied with substrates at ratios likely to be found in the cytosol in the light (Hanning and Heldt, 1993). Formation of 2-OG in the cytosol would ensure its availability to ammonia assimilation rather than to further respiration (Chen and Gadal, 1990). An important role for the cytosolic ICDH is indicated by the induction of this enzyme via a nitrate-linked signal transduction pathway (Scheible et al., 1997b), though it should be noted that the relative roles of the mitochondrial and cytosolic ICDHs remain to be clearly established (Lancien et al., 2000).

Leaf starch accumulates when N is in short supply (Rufty et al., 1988). High nitrate promotes organic acid synthesis via enhanced expression of C_3-type PEPc (Scheible et al., 1997b) and this enzyme is activated by light and enhanced N supply (Champigny and Foyer, 1992; Foyer et al. 1994a). It is less clear whether, in the short-term, high rates of nitrate reduction are strictly co-ordinated with anaplerotic production of C skeletons. This question has been investigated in the youngest fully expanded leaves of tobacco plants during the day-night cycle (Scheible et al., 2000). In mutants and transformants with low NR activity, which accumulate high nitrate, changes in NR transcripts correlated with PEPc transcripts though less well with those for PK, citrate synthase, and cytosolic ICDH (Scheible et al., 2000). Changes were generally less evident in wild-type tobacco. In wild-type tobacco on high nitrate, however, a clear antiparallel correlation was observed in the light period between the extractable activities of NR and PK and those of ICDH (Scheible et al., 2000). In contrast, light-dependent changes in PEPc activity were small and were in phase with NR activity. On the basis of these data and metabolite analysis, the authors suggested that early in the light period PEPc activity may serve primarily to generate malate as a counterion to balance pH during high rates of nitrate assimilation, whereas the production of C skeletons for incorporation of assimilated ammonia occurs principally towards the end of the light period (Scheible et al., 2000). This is consistent with the accumulation, throughout the light period, of malate and of assimilated ammonia in Gln, Gly and Ser, as discussed above in Section II.C. Although the exact timing of these changes cannot be generalized to other species in different conditions, they do suggest that N assimilation and anaplerosis can be temporally offset. Such a property may be one way the plant manages to allocate enough C to amino acid synthesis, particularly at high rates of nitrate assimilation. In sink leaves, the system could be constructed to allow significantly higher allocation of C to amino acid synthesis during high rates of N assimilation.

Redox coupling could influence the integration of N assimilation and C metabolism. The net formation of one 2-OG from sugar phosphate would involve net production of four NAD(P)H (two at glyceraldehyde-3-phosphate dehydrogenase, one at PDH, one at ICDH). Fifty to 75% of this reductant could be formed in the cytosol, depending on the location of isocitrate oxidation. Even if nitrate acts as an electron acceptor for one quarter of the reductant formed, there is still an excess that must be oxidized by other means, presumably via the mitochondrial electron transport chain. Mitochondrial electron transport and NR can compete for reductant (Foyer et al., 2001). Insufficient reductant sinks could hamper the production of 2-OG and lead to accumulation of Gln and, possibly, NH_3 in the chloroplast. When the potential constraints of mitochondrial and cytosolic ATP sinks on respiratory electron flow are considered, numerous potential interactions between photosynthesis and N assimilation can be described that could operate in the cytosol and mitochondria, and these are discussed more fully in Chapters 10–12 of this volume. Even in the light, anaplerotic 2-OG formation may be a fairly small proportion of total rates of respiratory CO_2 release, O_2 consumption and oxidative phosphorylation. Anaplerosis will contribute a larger fraction of the total amount of cytosolic reductant generated, however, particularly if isocitrate is oxidized via the cytosolic ICDH. Limitation of NR by cytosolic reductant could be one factor that ensures that nitrate reduction does not proceed at rates that are much faster than the supply of 2-OG.

C. Governing the Carbon-Nitrogen Balance: Roles for Amino Acids and Organic Acids?

Nitrate induction of NR and enzymes involved in organic acid synthesis is a feed-forward activation of downstream pathways signaled by increased substrate availability. What are the other factors that feed-back on nitrate reduction and feed-forward on organic acid synthesis to adjust imbalances in C and N assimilation? Short-term effects of these factors could be less influential than previously thought, if temporal decoupling of N assimilation and anaplerotic organic acid synthesis is a general phenomenon (Scheible et al., 2000). There is, however, good in vitro evidence that Gln can be important in controlling nitrate reduction and organic acid synthesis, e.g., repression of *nia* transcripts, activation of PEPc. As discussed above, Gln should be ideally placed to act as an indicator of the balance between the availability of ammonia and 2-OG. However, although supplying Gln by the transpiration stream caused repression of *nia* transcripts in *Arabidopsis*, no repression was associated with the accumulation of Gln in Fd-GOGAT mutants (Dzuibany et al., 1998). These observations were reconciled by invoking an indirect effect of exogenous Gln on NR expression, mediated via decreased nitrate concentrations (Dzuibany et al., 1998). This hypothesis cannot, however, account for repression of NR transcripts in sulfur-limited tobacco, where Gln and Asn accumulated but leaf nitrate remained relatively high (Migge et al., 2000). An alternative explanation is that the effectiveness of Gln as a signal is amplified by the plant cell's possible capacity to sense 2-OG, so that the Gln/2-OG ratio is an important regulatory parameter, as in bacteria and fungi. This would explain the data of Dzuibany et al. (1998), where 2-OG was not measured, if supplying Gln brought about a decrease in 2-OG whereas both metabolites increased together in the mutant. The role of 2-OG has recently been investigated by two groups. In tobacco lines where a range of Fd-GOGAT activities was produced by antisense technology, both Gln and 2-OG increased as GOGAT capacity decreased (Ferrario-Méry et al., 2000). In agreement with the data of Dzuibany et al. (1998), the rise in Gln did not repress NR transcripts. In fact, NR transcripts increased as Fd-GOGAT capacity decreased (Ferrario-Méry et al., 2001). Feeding experiments showed that Gln decreased NR expression while sucrose had an opposing effect (Ferrario-Méry et al., 2001). A new observation was that supplying 2-OG, which caused a two-fold increase in leaf 2-OG contents, had a similar effect on NR transcripts to supplying sucrose (Ferrario-Méry et al., 2001). These data suggest that antagonistic effects of Gln and 2-OG are able to explain the lack of NR suppression accompanying Gln accumulation in plants with low GOGAT activity (Dzuibany et al., 1998; Ferrario-Méry et al., 2001). Feeding experiments in the Stitt laboratory also produced some evidence for NR induction by 2-OG, although effects were small and it was shown that interpretation may be complicated by accompanying changes in other metabolites such as Gln and malate (Müller et al., 2001). In fact, an interesting observation was the repression of NR by feeding malate (Müller et al., 2001), which is consistent with the notions that accumulation of this organic acid is closely coupled to nitrate synthesis primarily due to its role as a counterion and that it may be a marker for the rate of nitrate reduction. While 2-OG feeding had no effect on extractable NR activities in Ferrario-Méry et al. (2001), these were slightly but significantly increased in Müller et al. (2001).

It is becoming clear that both the rate of N assimilation and the co-ordination of C and N assimilation may be under multifactorial control by a repertoire of signals, including nitrate, sugars, Gln, and organic acids such as 2-OG and malate. Ammonia may also be important and has been shown to increase GS expression (Brechlin et al., 2000). One advantage of control by Gln may be anticipation of comparable increases in ammonia (Fig. 5,A), thereby allowing the system to adjust to minimize deleterious ammonia accumulation or loss. For example, the data of Scheible et al. (1997) show that the accumulation of ammonia in Gln during the day is 6-8 times higher than the rise in free ammonium. The suitability of Gln as a regulatory metabolite has been questioned, because of its involvement in photorespiration (Stitt and Krapp, 1999; Müller et al., 2001). As suggested above, photorespiration may be less influential than a simple comparison of fluxes would suggest, and Gln is theoretically well placed to signal an imbalance in ammonia and 2-OG supply. Nevertheless, the question must be addressed: How stringent is control by Gln? In the extreme case, if feed-back regulation were immediately and totally effective, no Gln accumulation would occur. Thus, an important role for Glu has been suggested, given the relative stability of overall leaf pools of this amino acid (Stitt et al., 2002). Repression of NR transcripts was observed on supplying either Glu or Gln to detached tobacco leaves (Vincentz et al., 1993). In the absence of characterization of the effects of feeding on leaf metabolite contents, such effects are difficult to interpret. Knock-on effects of Gln accumulation, such as increases in Asn, could also transmit information on the C/N balance (Migge et al., 2000). It is unlikely that any one factor, be it nitrate or a specific metabolite, will exert total control. Rather, each factor that is sensed will work in the context of changes induced by several others. The permitted elasticity in the Gln pool, which is effectively the inverse of the system's sensitivity to changes in Gln concentration, will be the subject of future studies that may have to address the difficult problem of compartmentation of leaf amino acids, as well as the role of antagonistic factors such as 2-OG (Ferrario-Méry et al., 2001). At present, the physiological significance of feedback control by amino acids on nitrate reduction remains unclear.

The tobacco transformants with decreased Fd-GOGAT have been used to analyze other aspects of the C/N interaction (Ferrario-Méry et al., 2002a, 2002b). Following transfer of these plants from high CO_2 to air, accumulation of ammonia, Gln and 2-OG was observed during the second part of the light period (Ferrario-Méry et al., 2002a). The nocturnal decrease in these compounds was accompanied by an increase in Asn, suggesting that this amino acid serves as a temporary storage compound for the elimination of excess photorespiratory ammonia (Ferrario-Méry et al., 2002a). Most interestingly, the direction of the glutamate dehydrogenase (GDH) reaction varied during the day/night cycle such that a higher ratio of aminating to deaminating activity occurred in the first half of the light period (Ferrario-Méry et al., 2002a). This was correlated with the decline in NH_4^+ and 2-OG concentrations, consistent with an increase in aminating GDH activity in vivo. Such observations suggest that the ammonia assimilation pathway may be very flexible, and that pathways alternative to GS-GOGAT can be activated as required. Transfer to photorespiratory conditions also led to activation of anaplerosis, as evidenced by increases in PEPc and ICDH protein amounts and activities (Ferrario-Méry et al., 2002b). By contrast, transcripts for PEPc were unaltered, as were those for both cytosolic and mitochondrial ICDHs (Ferrario-Méry et al., 2002b). PEPc activity correlated well with PEPc protein and with leaf Gln, suggesting that Gln may affect translation or protein stability (Ferrario-Méry et al., 2002b). It is interesting that PEPc protein should be induced under conditions where 2-OG accumulates, emphasizing the influence of increases in Gln. From a physiological point of view, the expression of PEPc and other anaplerotic enzymes might be expected to be regulated by C/N metabolites in an inverse fashion to NR (i.e., induced by Gln and other amino acids, repressed by organic acids). If so, the lack of induction of PEPC and ICDH isoforms at the gene level may be explained by compensatory increases in both Gln and 2-OG.

IV. The Carbon-Nitrogen Signal Transduction Network: Interactions Between Nitrate, Sugars and Abscisic Acid

The above discussion has indicated that effective metabolic cross-talk between the pathways of C and N assimilation involves the concerted action of a repertoire of signals. Surprisingly few signal molecules have been identified to date but nitrate is clearly a key regulator of gene expression and plant development. Shoot nitrate contents regulate C partitioning (Scheible et al., 1997b) and also shoot-root allocation (Scheible et al., 1997a). Sugars elicit transcriptional and post-translational controls that limit the rate of nitrate assimilation, amino acid metabolism and photosynthesis (Kaiser and Huber, 1994; Morcuende et al., 1998). Thus, the pools of major end-products hold vital information on the plant's C and N status. This information provides an estimate of metabolic resource capacity that allows the plant to adapt in response to environmental changes.

The concept that sugars and amino acids participate in extensive metabolic cross talk, by modulating gene expression and thereby regulating rates of photosynthetic C and N assimilation, is central to current thinking on plant assimilate partitioning and utilization. These metabolites also participate in the signaling cascades that activate defense and development responses. The genes activated in response to sugars and amino acids are often overlapping, at least in part, indicating extensive molecular and metabolic cross-talk. Sugars control the expression of key genes in most, if not all, developmental processes including seed germination and development, as well as leaf and root morphogenesis. Most importantly, sugars orchestrate carbohydrate metabolism in source and sink tissues and balance supply and demand in carbohydrate-producing and consuming cells over a wide range of environmental conditions.

Concepts of sugar-mediated regulation of gene transcription in plants are largely based on the pathways of signal transduction found in *Saccharomyces cerevisiae,* where transcriptional regulation of a large group of genes involved in glucose fermentation has been described (Koch, 1996; Jang and Sheen, 1997; Smeekens and Rook, 1997). In plants, sugars generally induce genes involved in C metabolism and storage, while repressing those involved in photosynthesis and mobilization of stored reserves (Koch, 1996; Pego et al., 2000). For example, sugar-mediated regulation of Rubisco large and small sub-unit gene expression has been unequivocally demonstrated (van Oosten and Besford, 1996; Smeekens and Rook, 1997; Gesch et al., 1998). Hexokinases are important in sugar sensing in plants as they are in yeast (Graham et al., 1994; Jang and Sheen, 1997), but hexokinase-independent pathways have also been shown to be involved in the transcription of many genes (Martin et al., 1997). Homologues of SNF1 and other interacting proteins have been described in plants. SNF1-related protein kinases (SnRKs) are considered to be global regulators of plant metabolism (Halford and Hardie, 1998). In particular, SnRKs were shown to be involved in the regulation of the activities of NR and SPS (Sugden et al., 1999). These enzymes are phosphorylated by SnRKs and the phosphorylated enzymes become targets for 14-3-3 proteins which render them inactive (Bachmann et al., 1996; Moorhead et al., 1999). SnRKs are therefore involved in the regulation of the C partitioning between the pathways of carbohydrate synthesis and N assimilation.

Various homologs of components known to be involved in N signaling in bacteria have been suggested to play similar roles in plants. The functions of components such as PII-like proteins and 'two-component regulatory systems' or 'multistep His-

Asp phosphorelay' systems in the sensing of N status are fully discussed in chapters 13 (Krapp et al.) and 14 (Sugiyama) of this volume. Amino acid-mediated regulation of gene transcription in plants may have similarities to the yeast system. Amino acid deficiency in yeast decreases protein synthesis and increases the expression of a number of amino acid biosynthetic genes. This process, involving at least 35 genes in the twelve different amino acid synthesis pathways, is known as 'general amino acid control' (Hinnebusch, 1994). As with sugars, the yeast pathway of amino acid sensing involves protein kinases. In particular, the GCN2 (General Control Non-reversible 2) factor is a kinase of major importance in amino acid signaling (Wek et al., 1989). GCN2-mediated phosphorylation of eIF-2 under conditions of amino acid deprivation increases expression of amino acid biosynthesis genes through the action of a transcriptional activator, GCN4 (Hinnebusch, 1997). In turn, the amount of GCN4 protein appears to be regulated by translational controls (Hinnebusch, 1994).

Homologues of GCN2 have been identified in *Drosophila melanogaster* (Santoyo et al., 1997) and *Neurospora crassa* (Sattleger et al., 1998) but no reports on equivalent clones from plants have appeared to date. As discussed in Section II.D. there is growing evidence from metabolite measurements that amino acids could be under some form of general control in plants, but there is relatively limited evidence of co-ordinated regulation of genes encoding enzymes of amino acid biosynthesis (Noctor et al., 2002a). Blocking histidine biosynthesis in *A. thaliana* for example increases the expression of eight genes involved in the synthesis of the aromatic amino acids, histidine, lysine and purines (Guyer et al., 1995). Similarly, genes encoding tryptophan biosynthesis pathway enzymes *A. thaliana* have also been shown to be induced by amino acid deficits (Zhao et al., 1998).

Many of the advances in our current understanding of sugar-signaling have been made via the characterization of mutants impaired in the sugar sensing process. Genetic screens for such mutants have been generally based on either sugar-regulated gene expression or on the arrest of development imposed by high sugar concentrations. A large number of sugar-hypersensitive or sugar-insensitive mutants have been isolated (Boxall et al., 1996; Dijkwel et al., 1997; Martin et al., 1997; Mita et al., 1997a,b; Pego et al., 1998). Such mutants are generally selected at the germination stage by the ability to grow on concentrations of glucose, sucrose or mannose that inhibit wild-type *A. thaliana* seedling development (Jang et al., 1997). These 'metabolic arrest' screens have yielded mutants that are glucose insensitive (*gin*) (Areanas-Huertero et al., 2000), glucose oversensitive (*glo*), carbohydrate insensitive (*cai*), sucrose insensitive (*sis*) and the mannose insensitive germination (*mig*) type (Smeekens and Rook, 1997). Other mutants that have proved useful in elucidating the sugar signaling process are: a) reduced sucrose response (*rsr*) (Martin et al., 1997), b) sucrose uncoupled (*sun*) mutants, (Dijkwel et al., 1996, 1997; Van Oosten et al., 1997); c) low and high β-amylase (*lba* and *hba*) (Mita et al., 1997a,b) mutants.

The molecular and metabolic analysis of these mutants has revealed the existence of a signal transduction network that co-ordinates information from carbohydrate and N assimilation via the phytohormone, abscisic acid (ABA). ABA regulates plant development, seed dormancy, germination, cell division, and facilitates cell survival during environmental stresses such as drought, cold, salt, pathogen attack, and UV radiation. It has long been recognized that ABA regulates defense gene expression. Following the characterization of *A. thaliana* mutants, five genes in the ABA signaling pathway have been cloned. Of these *ABI1* and *ABI2* encode protein phosphatases while *ABI3-5* encode putative transcription factors. In particular, the *ABI4* gene encodes a putative AP2 domain transcription factor (Finkelstein et al., 1998). There is now considerable evidence that the ABI4 protein is involved in sugar signaling. For example, the *sun6-2* mutant (carrying an insertion in the 5′ untranslated region of the *ABI4* gene) is insensitive to both mannose-induced inhibition of seed germination and to repression of photosynthetic genes by sucrose (Huijser et al., 2000). The *gin6* mutant, which carries a T-DNA insertion 2.0 kb upstream of the *ABI4* gene and has lost the expression of the *ABI4* mRNA, is found to be less sensitive to high glucose (Arenas-Huertero et al., 2000). Recently the ABI4 protein has also been found in the signaling pathway by which lateral roots (LR) sense nitrate (Signora et al., 2001)

The formation of LR is a major post-embryonic developmental event in plants. This process is under hormonal control, notably by auxin, but also displays enormous plasticity in response to environmental triggers, particularly nitrate. In *Arabidopsis*, stimulation of LR formation by low concentrations of nitrate is localized and is mediated by a nitrate-inducible transcription factor, ANR1 (Zhang et al., 1999). Inhibition of root branching occurs at high concentrations of nitrate, an effect which is delocalized and which is mitigated by increasing sugar availability so that the inhibitory effect of high nitrate is significantly reduced when the sucrose supply in the medium is increased (Zhang et al., 1999). The roots clearly combine information derived from sugar (C) and nitrogen (N) signals (Zhang et al., 1999; Zhang and Forde, 2000). There is evidence that ABA signaling components are key intermediaries that link C/N status to LR development. The inhibitory effect of high soil nitrate on LR development was significantly decreased in two ABA-insensitive mutants (*abi4* and *abi5*: Signora et al., 2001) which are also insensitive to sugar (Huijser et al., 2000; Arenas-Huertero et al., 2000). This would indicate that the transcription factors, ABI4 and ABI5, are involved in the co-ordinate sensing of sucrose, glucose and nitrate (Finkelstein, 1998; Huijser et al., 2000; Signora et al., 2001) and stress responses (Arenas-Huerto et al., 2000; Finkelstein and Lynch, 2000). These similarities may reflect overlap in the signal transduction pathways linking the sugar-dependent inhibition of seedling development to nitrate-dependent inhibition of LR development. Furthermore, mutants with impaired ABA synthesis (*aba1-1*, *aba2-1*, *aba3-1*) also show sugar-insensitive phenotypes (*sun* and *gin*), implicating ABA itself in the mechanisms relaying information on C status.

Down-regulation of *ANR1* expression resulted in a negative linear relationship between nitrate concentration and LR growth (Zhang et al., 1999). This led to the hypothesis that the inhibitory effect of nitrate on LR growth was dose-dependent and occurred over the range of all nitrate concentrations studied (Zhang et al., 1999, 2000). However, the results of Signora et al. (2001) indicate that *aba* mutations have the opposite effect to *ANR1* down-regulation within the lower KNO_3 concentration range, 0.1–1.0 mM. The *aba* mutations increase LR growth with increasing nitrate concentrations, indicating that ABA plays a major role in mediating the inhibitory effect of nitrate. Clearly, the inhibitory effect of nitrate requires ABA synthesis and it is possible that nitrate induces ABA synthesis. It must be noted, however, that nitrate-dependent inhibition of LR formation was not completely absent from the *aba* mutants. While this may be due to residual capacity for ABA synthesis in the mutants, it is also probable that there is also an ABA-independent pathway involved in the nitrate inhibition response (Signora et al., 2001). In conclusion, the integration of information arising from nitrate signaling at the whole plant level involves at least three plant hormones: ABA, auxin and cytokinin, and the significance of the last is discussed in Chapter 14 (Sugiyama and Sakakibara). The evidence presented above would suggest that, of these, ABA has a key role in signaling imbalances in C/N status. Figure 7 depicts how ABA may be involved in the integration of various signals to regulate photosynthetic N assimilation.

V. Conclusions and Perspectives

Nitrogen assimilation is integrated with photosynthetic and respiratory carbon metabolism at intracellular, intercellular and interorgan levels. The response of plants to C and N status, mediated through modulation of hormones and hormone-signaling pathways, highlights the plasticity of plant development. Given the autotrophic and sedentary nature of plants, it is not surprising that development should be very responsive to nutritional and metabolic status. Progress in understanding C/N interactions has been greatly accelerated by analysis of transformed plants and mutants, including mutants perturbed in hormone perception or synthesis. The combination of these approaches with the new genomic techniques is likely to produce an even greater flood of illuminating (or potentially confusing) data. In particular, the integration of C/N metabolism provides an excellent system for study by metabolomic approaches. The application of hypothesis-independent approaches is likely to throw up a number of surprises, and to reveal the complexity of 'C/N interactions'. A full understanding of metabolic control at the molecular level is likely to require development of refined techniques that are able to measure and track metabolites in situ, to generate accurate data on intercompartmental traffic in vivo, and to identify the extent to which metabolite channeling occurs.

A key development over the last decade has been the identification of signals in the C/N interaction. Key questions for the next decade are: What are the mechanisms that interpret and relay these signals? How are the concentrations of metabolites such as Gln and organic acids sensed? Does the transduction

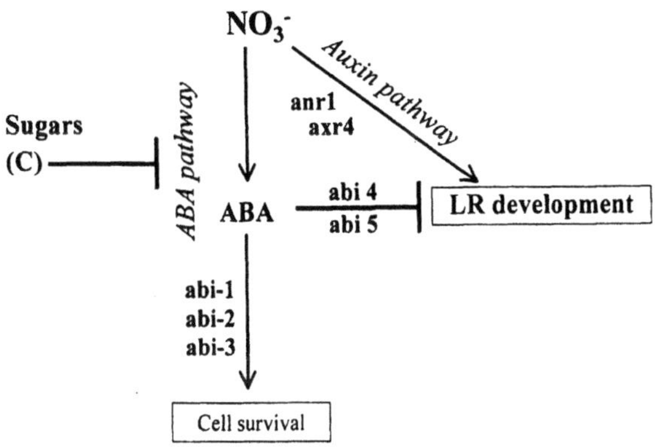

Fig. 7. Interactions between auxin- and abscisic acid (ABA)-mediated pathways in signaling nitrogen and carbon status. Nitrate signaling via the auxin pathway involves anr1 and axr4. The scheme shows that some components involved in ABA-signaling related to stress may be different from those necessary for ABA-mediated relay of information of carbon/nitrogen status.

from sensor to modification of gene expression transit exclusively via kinases, or are there unknown types of components that await discovery? How many factors are common to the integration of different signals? The clear evidence for the modulation of NR by multiple compounds begs the question of signal transduction hierarchies, both with regard to the proximity or remoteness of transduction (intracellular, interorgan) and to the relative influence of different perceived compounds. Present evidence suggests that nitrate and sugars produce a gross control of the C/N interaction, acting both at the intracellular and interorgan levels. It remains to be established (1) whether metabolites such as Gln and organic acids act effectively as 'fine-tuners' of C and N assimilation, and (2) whether such metabolites act only at the intracellular level or also transmit information between organs on whole-plant C and N status.

References

Arenas-Huertero F, Arroyo A, Zhou L, Sheen J and Leon P (2000) Analysis of *Arabidopsis* glucose insensitive mutants, gin 5 and gin 6, reveals a central role of the plant hormone ABA in the regulation of plant vegetative development by sugar. Genes and Development 14: 2085–2096

Aslam M, Huffaker RC, Rains DW and Rao KP (1979) Influence of light and ambient carbon dioxide concentration on nitrate assimilation by intact barley seedlings. Plant Physiol 63: 1205–1209

Bachmann M, Huber J, Liao PC, Gage DA and Huber SC (1996) The inhibitor protein of phosphorylated nitrate reductase from spinach (*Spinacia oleracea*) leaves is a 14-3-3 protein. FEBS Lett 387: 127–131

Boxall SF, Martin TR, Regad F and Graham IA (1997) *Arabidopsis thaliana* mutants that are carbohydrate insensitive. Plant Physiol 114: 1266

Brechlin P, Unterhalt A, Tischner R and Mäck G (2000) Cytosolic and chloroplastic glutamine synthetase of sugarbeet (*Beta vulgaris*) respond differently to organ ontogeny and nitrogen source. Physiol Plant 108: 263–269

Campbell WH (1999) Nitrate reductase structure, function and regulation: Bridging the gap between biochemistry and physiology. Annu Rev Plant Physiol Plant Mol Biol 50: 277–303

Champigny ML and Foyer CH (1992) Nitrate activation of cytosolic protein kinases diverts photosynthetic carbon from sucrose to amino acid biosynthesis. Basis for a new concept. Plant Physiol 100: 7–12

Chen RD and Gadal P (1990) Do the mitochondria provide 2-oxoglutarate needed for glutamate synthesis in higher plant chloroplasts? Plant Physiol Biochem 28: 141–145

Coschigano KT, Melo-Olivieira R, Lim J and Coruzzi GM (1998) *Arabidopsis gls* mutants and distinct ferredoxin-GOGAT genes: Implications for photorespiration and primary nitrogen assimilation. Plant Cell 10: 741–752

Dijkwel PP, Kock PAM, Bezemer R, Weisbeek PJ and Smeekens SCM (1996) Sucrose represses the developmentally controlled transient activation of the plastocyanin gene in *Arabidopsis thaliana* seedlings. Plant Physiol 110: 455–463

Dijkwel PP, Huijser C, Weisbeek PJ, Chua NH and Smeekens SCM (1997) Sucrose control of phytochrome A signaling in *Arabidopsis*. Plant Cell 9: 583–595

Dzuibany C, Haupt S, Fock H, Biehler K, Migge A and Becker T (1998) Regulation of nitrate reductase transcript level by glutamine accumulating in the leaves of a ferredoxin-dependent glutamate synthase-deficient *gluS* mutant of *Arabidopsis thaliana*, and by glutamine provided via the roots. Planta 20: 515–522

Elrifi IR and Turpin DH (1986) Nitrate and ammonium induced photosynthetic suppression in N-limited *Selenastrum minutum*. Plant Physiol 81: 273–279

Ferrario S, Valadier MH, Morot-Gaudry JF and Foyer CH (1995) Effects of constitutive expression of nitrate reductase in transgenic *Nicotiana plumbaginifolia* L. in response to varying nitrogen supply. Planta 196: 288–294

Ferrario S, Valadier MH and Foyer CH (1996) Short-term modulation of nitrate reductase activity by exogenous nitrate in *Nicotiana plumbaginifolia* and *Zea mays* leaves. Planta 199: 366–371

Ferrario S, Valadier MH and Foyer CH (1998) Overexpression of nitrate reductase in tobacco delays drought-induced decreases in nitrate reductase activity and mRNA. Plant Physiol 117: 293–302

Ferrario-Méry S, Suzuki A, Kunz C, Valadier MH, Roux Y, Hirel B and Foyer CH (2000) Modulation of amino acid metabolism in transformed tobacco plants deficient in Fd-GOGAT. Plant and Soil 221: 67–79

Ferrario-Méry S, Masclaux C, Suzuki A, Valadier MH, Hirel B and Foyer CH (2001) Glutamine and α-ketoglutarate are metabolite signals involved in nitrate reductase gene transcription in untransformed and transformed tobacco plants

deficient in ferredoxin-glutamine-α-ketoglutarate aminotransferase. Planta 213: 265–271
Ferrario-Méry S, Valadier MH, Godefroy N, Miallier D, Hirel B, Foyer CH and Suzuki A (2002a) Diurnal changes in ammonia assimilation in transformed tobacco plants expressing ferredoxin-dependent glutamate synthase mRNA in the antisense orientation. Plant Sci 163: 59–67
Ferrario-Méry S, Valadier MH, Hodges M, Hirel B and Foyer CH (2002b) Photorespiration-dependent increases in phosphoenolpyruvate carboxylase, isocitrate dehydrogenase and glutamate dehydrogenase in transformed tobacco plants deficient in ferredoxin-dependent glutamine α-ketoglutarate aminotransferase. Planta 214: 877–886
Finkelstein RR and Lynch TJ (2000) Abscisic acid inhibition of radicle emergence but not seedling growth is suppressed by sugars. Plant Physiol 122: 1179–1186
Finkelstein RR, Wang ML, Lynch TJ, Rao S and Goodman HM (1998) The *Arabidopsis* abscisic acid response locus ABI4 encodes an APETALA2 domain protein. Plant Cell 10: 1043–1054
Foyer CH, Noctor G, Lelandais M, Lescure JC, Valadier MH, Boutin JP and Horton P (1994a) Short-term effects of nitrate, nitrite and ammonium assimilation on photosynthesis, carbon partitioning and protein phosphorylation in maize. Planta 192: 211–220
Foyer CH, Lescure JC, Lefebvre C, Morot-Gaudry JF, Vincentz M and Vaucheret H (1994b) Adaptations of photosynthetic electron transport, carbon assimilation, and carbon partitioning in transgenic *Nicotiana plumbaginifolia* plants to changes in nitrate reductase activity. Plant Physiol 104: 171–178
Foyer CH, Ferrario-Méry S and Noctor G (2001) Interactions between carbon and nitrogen assimilation. In Lea PJ and Morot-Gaudry JF (eds) *Plant Nitrogen*, pp. 237–254. Springer, Berlin
Gastal F and Saugier B (1989) Relationships between nitrogen uptake and carbon assimilation in whole plants of tall fescue. Plant Cell Environ 12: 407–418
Gesch RW, Boote KJ, Vu JCV, Allen LH and Bowes G (1998) Changes in growth CO_2 result in rapid adjustments of ribulose-1,5-bisphosphate carboxylase oxygenase small subunit gene expression in expanding and mature leaves of rice. Plant Physiol 118: 521–529
Graham IA, Denby KJ and Leaver CJ (1994) Carbon catabolite repression regulates glyoxylate cycle gene-expression in cucumber. Plant Cell 6: 761–772
Guyer D, Patton D and Ward E (1995) Evidence for cross-pathway regulation of metabolic gene expression in plants. Proc Natl Acad Sci USA 92: 4997–5000
Halford N and Hardie D (1998) SNF1-related protein kinases: Global regulators of carbon metabolism in plants? Plant Mol Biol 37: 735–748
Hinnebusch AG (1994) Translational control of GCN4—An In-Vivo Barometer of initiation-factor activity. Trends Biochem Sci 19: 409–414
Hinnebusch AG (1997) Translational regulation of yeast GCN4—A window on factors that control initiator-tRNA binding to the ribosome. J Biol Chem 272: 21661–21664
Hoff T, Truong HN and Caboche M (1994) The use of mutants and transgenic plants to study nitrate assimilation. Plant Cell Environ 17: 489–506
Huber JL, Huber SC, Campbell WH and Redinbaugh MG (1992) Reversible light/dark modulation of spinach leaf nitrate reductase activity involves protein phosphorylation. Arch Biochem Biophys 296: 58–65
Huijser C, Kortstee A, Pego J, Weisbeek P, Wisman E and Smeekens S (2000) The *Arabidopsis* sucrose-uncoupled 6 gene is identical to abscisic acid-insensitive 4: Involvement of abscisic acid in sugar responses. Plant J 23: 577–585
Huppe HC and Turpin DH (1994) Integration of carbon and nitrogen metabolism in plant and algal cells. Annu Rev Plant Physiol Plant Mol Biol 45: 577–607
Jang JC and Sheen J (1997) Sugar sensing in higher plants. Trends Plant Sci 2: 208–214
Jang JC, Leon P, Zhou L and Sheen J (1997) Hexokinase as a sugar sensor in higher plants. Plant Cell 9: 5–19
Kaiser WM and Förster J (1989) Low CO_2 prevents nitrate reduction in leaves. Plant Physiol 91: 970–974
Kaiser WM and Huber SC (1994) Post-translational regulation of nitrate reductase in higher plants. Plant Physiol 106: 817–821
Keys AJ (1999) Biochemistry of photorespiration and the consequences for plant performance. In Bryant JA, Burrell MM and Kruger NJ (eds) *Plant Carbohydrate Biochemistry*, pp. 147-161. BIOS Scientific, Oxford
Keys AJ, Bird IF, Cornelius MJ, Lea PJ, Miflin BJ and Wallsgrove RM (1978) Photorespiratory nitrogen cycle. Nature 275: 741–743
Khamis S, Lamaze T, Lemoine Y and Foyer CH (1990) Adaptation of the photosynthetic apparatus in maize leaves as a result of nitrogen limitation. Relationships between electron transport and carbon assimilation. Plant Physiol 94: 1436–1443
Koch KE (1996) Carbohydrate-modulated gene expression in plants. Annu Rev Plant Physiol Plant Mol Biol 47: 509–540
Lancien M., Gadal P and Hodges M (2000) Enzyme redundancy and the importance of 2-oxoglutarate in higher plant ammonium assimilation. Plant Physiol 123: 817–824
Larsen PO, Cornwell KL, Gee SL and Bassham JA (1981) Amino acid synthesis in photosynthesizing spinach cells. Effects of ammonia on pool sizes and rates of labeling from $^{14}CO_2$. Plant Physiol 68: 292–299
Lewis E Noctor G, Causton D and Foyer CH (2000) Regulation of assimilate partitioning in leaves. Aust J Plant Physiol 27: 507—517
Martin T, Hellmann H, Schmidt R, Willmitzer L and Frommer WB (1997) Identification of mutants in metabolically regulated gene expression. Plant J 11: 53–62
Masclaux C, Valadier MH, Brugière N, Morot-Gaudry JF and Hirel B (2000) Characterization of the sink/source transition in tobacco (*Nicotiana tabacum* L.) shoots in relation to nitrogen management and leaf senescence. Planta 211: 510–518
Matt P, Geiger M, Walch-Liu P, Engels C, Krapp A and Stitt M (2001) The immediate cause of the diurnal changes of nitrogen metabolism in leaves of nitrate-replete tobacco: A major imbalance between the rate of nitrate reduction and the rates of nitrate uptake and ammonium metabolism during the first part of the light period. Plant Cell Environ 24: 177–190
Miflin BJ and Lea P (1982) Ammonia assimilation. In: Miflin BJ (ed), The Biochemistry of Plants, pp. 169–202, Vol 5. Academic press, New York
Migge A, Bork C, Hell R and Becker TW (2000) Negative regulation of nitrate reductase gene expression by glutamine or asparagine accumulating in leaves of sulfur-deprived tobacco. Planta 211: 587–595
Mita S, Hirano H and Nakamura K (1997a) Negative regulation

in the expression of a sugar-inducible gene in *Arabidopsis thaliana*—A recessive mutation causing enhanced expression of a gene for beta-amylase. Plant Physiol 114: 575–582

Mita S, Murano N, Akaike M and Nakamura K (1997b) Mutants of *Arabidopsis thaliana* with pleiotropic effects on the expression of the gene for beta-amylase and on the accumulation of anthocyanin that are inducible by sugars. Plant J 11: 841–851

Moorhead G, Douglas P, Cotelle V, Harthill J, Morrice N, Meek S, Deiting U, Stitt M, Scarabel M, Aitken A and MacKintosh C (1999) Phosphorylation-dependent interactions between enzymes of plant metabolism and 14-3-3 proteins. Plant J 18: 1–12

Morcuende R, Krapp A, Hurry V and Stitt M (1998) Sucrose feeding leads to increased rates of nitrate assimilation, increased rates of α-oxoglutarate synthesis, and increased synthesis of a wide spectrum of amino acids in tobacco leaves. Planta 206: 394–409

Morot-Gaudry JF, Job D and Lea PJ (2001) Amino acid metabolism. In: Lea PJ and Morot-Gaudry JF (eds) Plant Nitrogen, pp 167–211. Springer Verlag, Berlin

Müller C, Scheible WR, Stitt M and Krapp A (2001) Influence of malate and 2-oxoglutarate on the NIA transcript and nitrate reductase activity in tobacco leaves. Plant Cell Environ 24: 191–203

Murchie EH, Ferrario-Méry S, Valadier MH and Foyer CH (2000) Short-term nitrogen-induced modulation of phospho-*enol*pyruvate carboxylase in tobacco and maize leaves. J Exp Bot 51: 1349–1356

Noctor G, Arisi ACM, Jouanin L and Foyer CH (1998) Manipulation of glutathione and amino acid biosynthesis in the chloroplast. Plant Physiol 11: 471–482

Noctor G, Novitskaya L, Lea PJ and Foyer CH (2002a) Co-ordination of leaf minor amino acid contents in crop species: Significance and interpretation. J Exp Bot 53: 939–945

Noctor G, Veljovic-Jovanovic S, Driscoll S., Novitskaya L and Foyer CH (2002b) Drought and oxidative load in wheat leaves: A predominant role for photorespiration? Ann Bot 89: 841–850

Novitskaya L, Trevanion S, Driscoll SD, Foyer CH and Noctor G (2002) How does photorespiration modulate leaf amino acid contents? A dual approach through modelling and metabolite analysis. Plant Cell Environ 25: 821–835

Pace GH, Volk RJ and Jackson WA (1990) Nitrate reduction in response to CO_2-limited photosynthesis. Relationship to carbohydrate supply and nitrate reductase activity in maize seedlings. Plant Physiol 92: 286–292

Pärnik T and Keerberg O (1995) Decarboxylation of primary and end products of photosynthesis at different oxygen concentrations. J Exp Bot 46: 1439–1447

Paul JS, Cornwell KL and Bassham JA (1978) Effects of ammonia on carbon metabolism in photosynthesizing isolated mesophyll cells from *Papaver somniferum* L. Planta 142: 49–54

Pego JV, Kortstee AJ, Huijser G and Smeekens SGM (2000) Photosynthesis, sugars and the regulation of gene expression. J Exp Bot 51: 407–416

Reed AJ, Canvin DT, Sherrard JH and Hageman RH (1983) Assimilation of [^{15}N]nitrate and of [^{15}N]nitrite in leaves of five plant species under light and dark conditions. Plant Physiol 71: 291–294

Rufty TW, Huber SC and Volk RJ (1988) Alterations in leaf carbohydrate metabolism in response to nitrogen stress. Plant Physiol 88: 725–730

Rufty TW, MacKown CT and Volk RJ (1989) Effects of altered carbohydrate availability on whole plant assimilation of $^{15}NO_3^-$. Plant Physiol 89: 457–463

Sattlegger E, Hinnebusch AG and Barthelmess IB (1998) *cpc-3*, the *Neurospora crassa* homologue of yeast *GCN2*, encodes a polypeptide with juxtaposed eIFα kinase and histidyl-tRNA synthetase-related domains required for general amino acid control. J Biol Chem 273: 20404–20416

Scheible WR, Lauerer M, Schulze ED, Caboche M and Stitt M (1997a) Accumulation of nitrate in the shoot acts as a signal to regulate shoot-root allocation in tobacco. Plant J 11: 671–691

Scheible WR, Gonzáles-Fontes A, Lauerer M, Müller-Röber B, Caboche M and Stitt M (1997b) Nitrate acts as a signal to induce organic acid metabolism and repress starch metabolism in tobacco. Plant Cell 9: 783–798

Scheible WR, Gonzáles-Fontes A, Morcuende R, Lauerer M, Geiger M, Glaab J, Gojon A, Schulze ED, Caboche M and Stitt M (1997c) Tobacco mutants with a decreased number of functional *nia* genes compensate by modifying the diurnal regulation of transcription, post-translational modification and turnover of nitrate reductase. Planta 203: 304–319

Scheible WR, Krapp A and Stitt M (2000) Reciprocal diurnal changes of phosphoenolpyruvate carboxylase expression and NADP-isocitrate dehydrogenase expression regulate organic acid metabolism during nitrate assimilation in tobacco leaves. Plant Cell Environ 23: 1155–1167

Sharkey TD (1988) Estimating the rate of photorespiration in leaves. Physiol Plant 73: 147–152

Signora L, De Smet, I, Foyer CH and Zhang H (2001) ABA plays a central role in mediating the regulatory effects of nitrate on root branching in *Arabidopsis*. Plant J 28: 655–662

Smeekens S and Rook F (1997) Sugar sensing and sugar mediated signal transduction in plants. Plant Physiol 115: 7–13

Somerville SC and Ogren WL (1983) An *Arabidopsis thaliana* mutant defective in chloroplast dicarboxylate transport. Proc Natl Acad Sci USA 80: 1290–1294

Stitt M and Krapp A (1999) The interaction between elevated carbon dioxide and nitrogen nutrition: The physiological and molecular background. Plant Cell Environ 22: 583–621

Stitt M, Müller C, Matt P, Gibon Y, Carillo P, Morcuende R, Scheible WR and Krapp A (2002) Steps towards an integrated view of nitrogen metabolism. J Exp Bot 53: 959–970

Sudgen C, Donaghy P, Halford N and Hardie D (1999) Two SNF1-related protein kinases from spinach leaf phosphorylate and inactivate3-hydroxy-3-methylglutaryl-Coenzyme A reductase, nitrate reductase, and sucrose phosphate synthase in vitro. Plant Physiol 120: 257–274

Trevanion SJ (2000) Photosynthetic carbon metabolism in wheat (*Triticum aestivum* L.) leaves: Optimization of methods for determination of fructose 2,6-bisphosphate. J Exp Bot 51: 1037–1045

Vanlerberghe GC, Schuller KA, Smith RG, Feil R, Plaxton WC and Turpin DH (1990) Relationship between NH_4^+ assimilation rate and in vivo phospho*enol*pyruvate carboxylase activity. Regulation of anaplerotic carbon flow in the green alga *Selenastrum minutum*. Plant Physiol 94: 284–290

Van Oosten JJ and Besford RT (1996) Acclimation of photosynthesis to elevated CO_2 through feedback regulation of gene expression: Climate of opinion. Photosynth Res 48: 353–365

Van Oosten JJ, Gerbaud A, Huijser C, Dijkwel PP, Chua N-H

and Smeekens SCM (1997) An *Arabidopsis* mutant showing reduced feedback inhibition of photosynthesis. Plant J 12: 1011-1020

Vincentz M and Caboche M (1991) Constitutive expression of nitrate reductase allows normal growth and development of *Nicotiana plumbaginifolia* plants. EMBO J 10: 1027–1035

Vincentz M, Moureaux T, Leydecker MT Vaucheret H and Caboche M (1993) Regulation of nitrate and nitrite reductase expression in *Nicotiana plumbaginifolia* leaves by nitrogen and carbon metabolites. Plant J 3: 315–324

Wek RC, Jackson BM and Hinnebusch, AG (1989) Juxtaposition of domains homologous to protein kinases and histidyl-transfer RNA-synthetases in GCN2 protein suggests a mechanism for coupling GCN4 expression to amino-acid availability. Proc Natl Acad Sci USA 86: 4579–4583.

Zhang H, Jennings A, Barlow PW and Forde BG (1999) Dual pathways for regulation of root branching by nitrate. Proc Natl Acad Sci USA 96: 6529–6534

Zhang H and Forde BG (2000) Regulation of *Arabidopsis* root development by nitrate availability. J Exp Bot 51: 51–59

Zhao J, Williams CC and Last RL (1998) Induction of *Arabidopsis* tryptophan pathway enzymes and camalexin by amino acid starvation, oxidative stress, and an abiotic elicitor. Plant Cell 10: 359–370

Chapter 2

Photosynthesis and Nitrogen-Use Efficiency

P. Ananda Kumar[1], Martin A. J. Parry[2], Rowan A. C. Mitchell[2], Altaf Ahmad[3] and Yash P. Abrol[3]*
[1]*National Research Centre for Plant Biotechnology and* [3]*Division of Plant Physiology, Indian Agricultural Research Institute, New Delhi – 110012, India;* [2]*Biochemistry and Physiology Department, IACR-Rothamsted, Harpenden, Hertfordshire AL5 2JQ U.K.*

Summary 23
I. Introduction 24
II. Nitrogen in the Photosynthetic Apparatus 24
III. Optimization of Amounts of Photosynthetic Components for Different Environments 26
A. Nitrogen Supply 26
B. Growth Irradiance 28
C. CO_2 Enriched Environment 29
IV. Role of Regulation of Rubisco Activity 29
V. Approaches to Improving Nitrogen-Use Efficiency in Crops 30
Acknowledgments 31
References 31

Summary

In C_3 crop plants about 60–80% of leaf nitrogen (N) is invested in the photosynthetic apparatus, and N nutrition plays a crucial role in determining photosynthetic capacity. The proportion of leaf N invested in photosynthetic components is fairly constant. By contrast, both N per unit leaf area and the allocation of N between the component photosynthetic processes depend on environmental factors such as N availability, irradiance and CO_2 concentration. Light-harvesting and electron transport components often show a co-ordinated and equivalent response to N nutrition. In contrast, most studies have shown disproportionately large changes in ribulose-1,5-bisphosphate carboxylase/oxygenase (Rubisco) in response to N supply, demonstrating the importance of this protein in leaf N economy. At low light, for a given N availability, more protein is allocated towards light harvesting components in order to maximize light capture and, expressed per unit Chl, electron transport and carboxylation capacities are relatively small. High irradiance tends to alter the partitioning of N away from thylakoid protein to soluble proteins, particularly Rubisco. Growth at elevated CO_2 often leads to decreases in the amounts of Rubisco and other photosynthetic components on a leaf area basis. This is explicable in terms of greater N sinks elsewhere in the plant as a result of increased carbohydrate availability and acclimatory changes. Models predict that in order to arrive at optimal N use efficiency (NUE) at likely future ambient CO_2 concentrations, leaves will need to achieve a redistribution of N so that the ratio between the capacities for regeneration of ribulose-1,5-bisphosphate and carboxylation increases by 30–40%. Human intervention to improve the NUE of crops would have economic and environmental benefits, reducing pollution of water supply by nitrates. The NUE of photosynthesis could be increased either through manipulation of Rubisco amounts or properties, or by decreasing photorespiration. While decreasing Rubisco content could enhance NUE by only about 5%, eliminating photorespiration could produce a change of more than 50%.

*Author for correspondence, email: ypabrol@vsnl.com

Christine H. Foyer and Graham Noctor (eds): Photosynthetic Nitrogen Assimilation and Associated Carbon and Respiratory Metabolism, pp. 23–34.

I. Introduction

Nitrogen nutrition plays a crucial role in determining plant photosynthetic capacity in both natural and agricultural environments (Abrol, 1993; Abrol et al., 1999). Leaf growth is a key determinant of plant N demand because photosynthetic function accounts for a high proportion of the plant's reduced N requirements (Novoa and Loomis, 1981). In C_3 plants, approximately 60–80% of leaf N is invested in the chloroplasts. (Evans and Seemann, 1989; Makino and Osmond, 1991). Two important expressions can be used to describe how effectively plants use N to sustain growth and photosynthesis. The first, N use efficiency (NUE), is defined as the increase in plant biomass over a given time period per unit plant N content. The second parameter is photosynthetic N use efficiency (PNUE), which is the rate of C assimilation per unit leaf N. The relationship between these two parameters is complex, not least because it depends on the extent to which PNUE measurements, performed at a given point in time, can be related to growth of the whole plant integrated over a much longer period of time.

The rate of photosynthesis depends on (i) light harvesting capacity, (ii) the rate at which NADPH and ATP can be regenerated; (iii) the capacity for the carboxylation of ribulose-1,5-bisphosphate (RuBP) by ribulose-1,5-bisphosphate carboxylase/oxygenase (Rubisco); (iv) the rate of utilization of sugar phosphate, the product of the RPP pathway. The degree to which each of these processes determines photosynthetic rate varies depending on external conditions as well as on internal regulation and resource distribution. Because the photosynthetic apparatus often accounts for a large part of plant N, the availability of N is a key external factor regulating photosynthetic capacity and plant growth. This chapter will discuss in turn: 1) The contribution of the different photosynthetic components to leaf N economy; 2) the effect of environmental factors on allocation of N between these components; 3) the role of the regulation of Rubisco amounts and activity in leaf N economy; and 4) the novel approaches that are being used to improve plant NUE.

II. Nitrogen in the Photosynthetic Apparatus

The N-containing components responsible for photosynthesis are (i) the light harvesting Chl-protein complexes (LHC); (ii) the electron transport and photophosphorylation membrane complexes; (iii) the enzymes of the RPP pathway and carbohydrate synthesis. The last category notably includes Rubisco. Nitrogen associated with proteins of the photosynthetic apparatus can be divided into two major pools, representing components associated with the 'light' and 'dark' reactions. The first encompasses thylakoid membrane-bound proteins associated with light harvesting, electron transport and photophosphorylation. The second pool consists of soluble proteins, and includes those involved in CO_2 assimilation, photorespiration, RuBP regeneration, and starch and sucrose synthesis.

The thylakoid components typically account for about 25% of leaf N (Fig 1; Evans and Seemann, 1989; Sivasankar et al., 1993) and consist of five major complexes: the light-harvesting Chl *a/b* protein complexes (LHC), Photosystem I (PS I), Photosystem II (PS II), the cytochrome *b/f* complex and the coupling factor (CF_0-CF_1). The total cost of light harvesting, i.e. PS I, PS II and LHC, has been estimated as 17% of leaf N (Evans and Seemann, 1989). The other components required for photophosphorylation account for a further 6–8% (Table 1).

The RPP pathway consists of 11 enzymes but, in terms of protein amounts, it is dominated by Rubisco. Typically, Rubisco accounts for 25–30% of leaf N in sun leaves (Fig. 1) and about 12% of total N in C_3 plants during vegetative growth (Osaki et al., 1993). In C_4 plants, the contribution of Rubisco to leaf protein is lower: these species contain less Rubisco protein per unit leaf area than C_3 plants (Sage et al., 1987) and require less N to sustain a given rate of photosynthesis (Sinclair and Horie, 1989). These observations underline the importance of Rubisco in leaf N economy and its influence on NUE in C_3 plants.

In addition to catalysis of CO_2 assimilation, Rubisco catalyses a competing oxygenase reaction which leads to photorespiration. Large amounts of Rubisco are required because of its slow catalytic

Abbreviations: CA1P – 2′-carboxy-D-arabinitol 1-phosphate; CF_0-CF_1 – coupling factor; Chl – chlorophyll; FNR – ferredoxin-$NADP^+$ reductase; LHC – light harvesting chlorophyll-protein complexes; NUE – nitrogen use efficiency; PEP – phosphoenolpyruvate; PNUE – photosynthetic nitrogen use efficiency; PRK – phosphoribulokinase; PS I – Photosystem I (reaction center and antennae); PS II – Photosystem II (reaction center and antennae); RPP – reductive pentose phosphate (RPP pathway = Calvin cycle); Rubisco – ribulose-1,5-bisphosphate carboxylase/oxygenase; SBPase – sedoheptulose-1,7-bisphosphatase

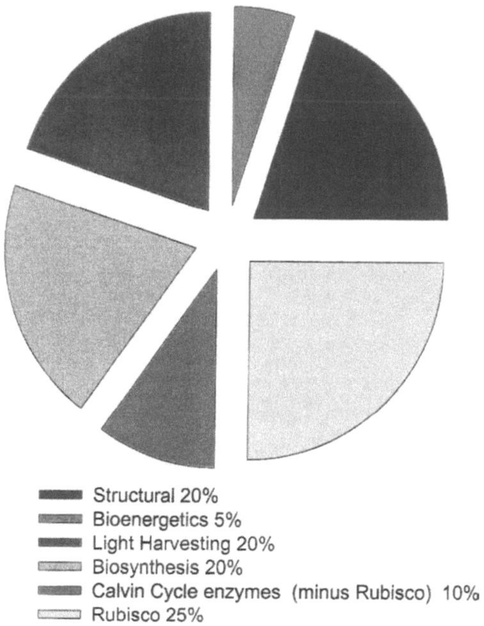

Fig. 1. Partitioning of protein N in a leaf of C_3 plant (after Evans and Seemann, 1989; reproduced with permission).

rate (e.g. 1.5 mol CO_2 mol active site^{-1} s^{-1} at 25 °C and 36 Pa CO_2), its low affinity for CO_2, and the engagement of active sites in catalyzing the non-assimilatory oxygenation reaction. Thus, in C_4 plants, where Rubisco operates under conditions of saturating CO_2, and the oxygenation reaction is suppressed, active site turnover per mol CO_2 is 3.3 s^{-1}. As a result, C_4 plants have greater photosynthetic rates and lower leaf N contents at high light (Sinclair and Horie, 1989). This explains their much greater NUE in warm, high-irradiance environments (Sage, 1999).

In C_3 plants, even with Rubisco active site concentrations of 4 mM (Woodrow and Berry, 1988), this enzyme activity exerts considerable limitation over the rate of CO_2 fixation under many conditions (Hudson et al., 1992). Under other conditions, as discussed in the next section, photosynthesis is limited by electron transport capacity or availability of Pi, and Rubisco is regulated to decrease activation state (Farquhar and Sharkey, 1994).

In comparison to Rubisco, many other RPP pathway enzymes are often in apparent excess. However, these enzymes may become limiting in certain conditions (e.g. phosphoribulokinase (PRK); Banks et al., 1999). The photosynthetic rate was little changed in transgenic plants with moderately (up to 30%) decreased amounts of other highly regulated RPP pathway enzymes: fructose-1,6-bisphosphatase (Kossman et al., 1994); glyceraldehyde-3-phosphate dehydrogenase (Price et al., 1995); PRK (Paul et al., 1995) and sedoheptulose-1,7-bisphosphatase (SBPase) (Harrison et al., 1998). Of these the highest control coefficient was estimated for SBPase, though such coefficients are dependent on leaf age and environment. In terms of contribution to leaf N

Table 1. Nitrogen content of photosynthetic components for a typical C_3 sun leaf with high leaf N per unit leaf area

Component	Relative to N in Rubisco	% leaf N	Source
Rubisco	1	23	Evans and Seemann, 1989
Rubisco Activase	0.04	1	Robinson et al., 1988
3-phosphoglycerate kinase	0.007–0.01	< 1	Morrow and Bradbeer, 1990
Glyceraldehyde-3-phosphate dehydrogenase	0.002–0.021	< 1	Trost et al., 1993; Leegood, 1990
Triose phosphate isomerase	0.002	< 1	Leegood, 1990
Aldolase	0.079	2	Leegood, 1990
Fructose-1,6-bisphosphatase	0.001–0.01	< 1	Leegood, 1990
Transketolase	0.003–0.004*	< 1	Leegood, 1990
Sedoheptulose-1,7-bisphosphatase	0.002–0.006	< 1	Leegood, 1990
Ribose-5-phosphate isomerase	0.001*	< 1	Leegood, 1990
Ribulose-5-phosphate epimerase	0.001*	< 1	Leegood, 1990
Phosphoribulokinase	0.003–0.004	< 1	Leegood, 1990
Carbonic anhydrase	0.01–0.05	< 1–1	Badger and Price, 1994
Electron transport excluding PS I, PS II and LHC	0.25–0.34	6–8	Evans, 1987; Makino et al., 1994; Evans and Seemann, 1989
PS I, PS II and LHC	0.74	17	Evans and Seemann, 1989

The estimates for reductive pentose phosphate enzymes were mostly derived from activities per unit chlorophyll and specific activities of purified enzymes. Values marked * assume a 1:1 stoichiometry with phosphoribulokinase.

economy, RPP pathway enzymes other than Rubisco are probably less than 7% (Table 1; Fig 1). Other proteins involved in C metabolism may be more significant in leaf N balance. Chief among these are proteins involved indirectly in the primary carboxylation process, such as Rubisco activase and carbonic anhydrase. These two proteins can account for more than 2% of leaf N (Makino et al., 1992).

III. Optimization of Amounts of Photosynthetic Components for Different Environments

Much of the variation in PNUE between species differing in specific leaf area can be accounted for by their investment of N in Rubisco and other photosynthetic components. PNUE is greater in species with additional layers of palisade cells and high specific leaf area than in those with a small specific leaf area (Poorter and Evans, 1998). Although the proportion of leaf N invested in photosynthetic components is fairly constant within a species, the amount per unit leaf area is highly variable. The amount of N per unit leaf area depends on both environmental and metabolic factors (e.g. N supply, irradiance, CO_2 concentration and sink regulation) (Evans, 1989; Quick et al., 1992; Krapp et al., 1993; Laurer et al., 1993). Two variables influenced by the environment can be distinguished: 1) the total amount of N invested in photosynthetic components per unit leaf area; 2) the distribution of N between the component processes of photosynthesis.

A. Nitrogen Supply

Numerous studies have indicated that net photosynthesis and the amounts of photosynthetic components are correlated with the N content of the leaf (Figs. 2 and 3; Field and Mooney, 1986; Evans, 1989; Lawlor et al, 1989; Nakano et al., 1997; Ahmad and Abdin, 2000). The relationship between the components of the photosynthetic system may change over the range of N content, reflecting adaptation of the photosynthetic system.

Leaf N affects the size and morphology of chloroplasts. Ample N increases the number of chloroplasts per mesophyll cell and their cross-sectional area and length compared to N-deficient chloroplasts, which have slightly more thylakoid membrane but lower stromal volume (Sivasankar et al., 1998a). Also, the density of protein (predominantly Rubisco) in the stroma is greater with high N supply (Kutik et al., 1995).

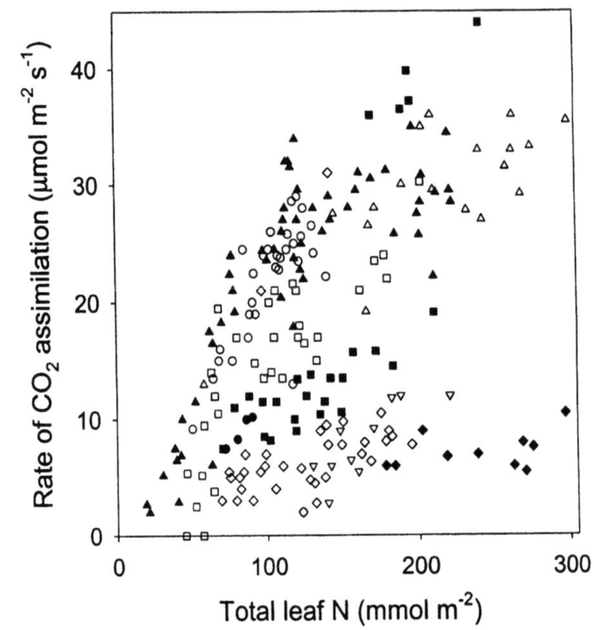

Fig. 2. Relationship between leaf N content and rate of CO_2 assimilation expressed on per unit leaf area. Empty diamonds, Californian evergreen trees and shrubs and rainforest trees; filled diamonds, *Prunus ilicifolia*; inverted empty triangles, South African shrubs; filled triangles, *Triticum aestivum*; empty triangles, Death valley annuals; empty squares, Illinois annuals; filled squares, *Lepechinia calycin;* filled circles, *Alocasia macrorrhiza*; empty circles, *Oryza*. (After Evans, 1989; reproduced with permission).

Nitrogen deficiency induces equivalent decreases in the LHC, reaction centers, the plastoquinone pool, cytochrome *f* and CF_0-CF_1 (Leong and Anderson, 1984a). In spinach, N nutrition affected electron transport capacities *via* modifications in the amount of thylakoids per unit leaf area, and the composition of the membranes was not affected (Evans and Terashima, 1987; Terashima and Evans, 1988). Nitrogen deprivation does not strongly alter CO_2 assimilation rates at low light intensities: thus, the apparent quantum yield of CO_2 assimilation is only slightly altered by N-deficiency, probably due to decreased light absorption (Lawlor et al., 1987). In contrast, photosynthetic rates at high irradiance can be markedly decreased by N limitation, due to a reduction in photosynthetic components, particularly Rubisco. A proportionately greater reduction in Rubisco than in electron transport capacity was found in spinach (Medina, 1971), cotton (Wong, 1979), potato (Ferrar and Osmond, 1986) and wheat (Jain et

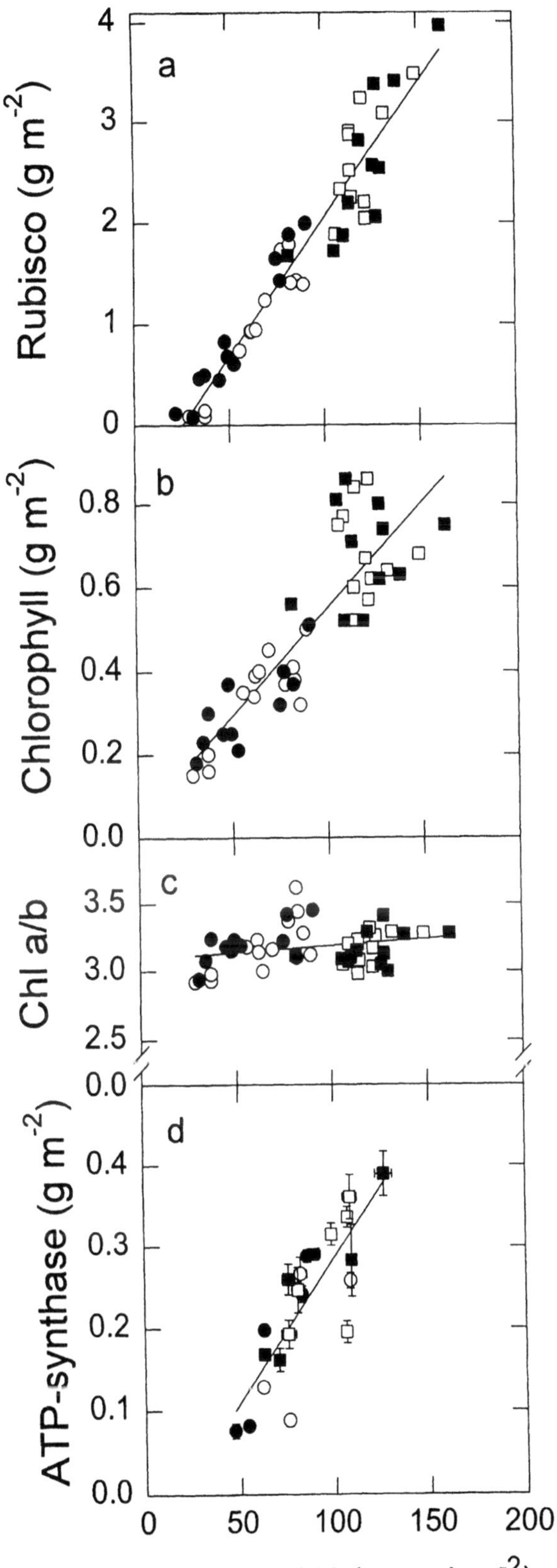

Fig. 3. Relationship between leaf N and the photosynthetic apparatus in wheat (After Theobald et al., 1998).

al., 1999). However, in *Phaseolus*, the two processes were decreased to the same extent at low N (Caemmerer and Farquhar, 1981). Where Rubisco activity has been shown preferentially to decrease, N deficiency increases the intercellular CO_2 concentration at which CO_2 assimilation changes from Rubisco limitation to an electron transport limitation (Evans and Terashima, 1987; Nakano et al., 1997; Mitchell et al., 2000).

The relationship between N supply and Rubisco content is complex, and depends on species and habitat. In trees, the proportion of N allocated to Rubisco can be independent of N supply (Brown et al., 1996). In C_4 leaves, N deficiency decreases photosynthesis with a selective reduction not only in Rubisco, but also in the activities of phosphoenolpyruvate (PEP) carboxylase and pyruvate orthophosphate dikinase (Khamis et al., 1990). Intensive study of C_3 crops has shown that Rubisco content increases linearly (but not proportionately) with N uptake and leaf N content (Evans, 1983; Sage et al., 1987; Makino et al., 1992, 1994, 1997a; Nakano et al., 1997; Theobald et al., 1998; Sivasankar et al., 1998a). In these plants, photosynthetic rate increases curvilinearly with respect to the amount of Rubisco (Evans, 1983; Evans and Seemann, 1984; Makino et al., 1985, 1992, 1994, 1997a; Sage et al., 1987; Lawlor et al., 1989, Nakano et al., 1997). This response is at least partly due to the increased resistance to diffusion of CO_2 from the intercellular spaces to the site of carboxylation in the chloroplast stroma (Makino et al., 1988). At the high absolute photosynthetic rates in leaves with high Rubisco contents, this resistance results in lower CO_2 concentrations at the Rubisco active sites, thereby partly offsetting the predicted higher carboxylation rates due to increased active site concentration.

The increased amount of leaf N invested in Rubisco at high leaf N in crops (Makino et al., 1988; Theobald et al., 1998; Sivasankar et al., 1998a) may act to mitigate the lower CO_2 concentration due to the increased internal resistance. This notion is consistent with changes in chloroplast morphology (Kutik et al., 1995; Sivasankar et al., 1998b). Provided that N is not limiting, an over-investment in Rubisco not only acts as a reserve of N (Millard, 1988), but also increases the leaf's ability to exploit short periods of intense illumination. A high carboxylation capacity also contributes to increased water-use efficiency. These considerations may explain why, at high N supply, Rubisco often appears to be in excess of requirements for the growth environment (Lawlor et al., 1989; Theobald et al., 1998).

Nitrogen deficiency decreases the point at which light saturates photosynthesis. This increases the likelihood of photoinhibitory damage. 'Shade' plants such as *Solanum dulcamara* are particularly susceptible to photoinhibition under excess light and this susceptibility is increased when N is scarce (Ferrar and Osmond, 1986). However, N-limited plants do not appear to suffer photoinhibitory damage following short-term exposure to light above saturating for photosynthesis: this is probably related to stimulation of zeaxanthin contents at low N (Khamis et al., 1990; Verhoeven et al., 1997). Production of zeaxanthin in the xanthophyll cycle provides a mechanism for protection of PS II function (Demmig-Adams and Adams, 1992). It is not known whether, under long-term high-light stress, N limitation affects the ability of the chloroplast to sustain replenishment of polypeptides such as the PS II reaction center protein D1, which turns over rapidly and which must be continuously replaced if photoinhibition is to be avoided.

B. Growth Irradiance

Leaves in natural environments can experience a range of irradiance from darkness to full sunlight (1500–2000 μmol.quanta.$m^{-2}s^{-1}$). The absolute amount of N in photosynthetic apparatus per unit leaf area is influenced by the irradiance at which the leaves are grown. Gradients in leaf N content with respect to the irradiance available to the leaves have been observed in *Prunus persica* (DeJong and Doyle, 1985), *Cymopsis tetragonoloba* (Charles-Edwards et al., 1987), *Lysmachia vulgaris* (Hirose et al., 1988) and *Nothofagus solandri* (Hollinger, 1989). When photosynthesis is measured at low light, PNUE is increased by the decrease in the absolute leaf N content because N does not limit photosynthetic capacity under these conditions. Leaves that are not acclimated to low light can therefore be considered to have an excess of N under these conditions. In a canopy, where the average light intensity decreases with depth, a theoretical optimum distribution of leaf N can be calculated. The lower N contents of leaves observed deep in the canopy do not exactly match this distribution, although they approach optimal values (Field, 1983; Hirose and Werger, 1987; Evans, 1993). In a variable light environment, leaves with low N and low Rubisco contents will tend to have higher PNUE. This will only benefit whole plant NUE if extra N available can be usefully employed elsewhere, e.g., extra light interception by greater leaf area or increased root production for ongoing N capture.

In addition to changes in overall amounts of the photosynthetic apparatus, acclimation to irradiance also involves specific changes in composition and structure of the chloroplasts (Björkman, 1981). This is associated with changes in the composition of the thylakoid membranes that are easily measurable, notably modifications of the Chl *a*:*b* ratio. At low irradiances, plants tend to maximize light absorption and more protein is allocated towards the LHC components, which contain Chl *b*. The Chl *a*:*b* ratio decreases due to an increase in the proportion of Chl associated with LHC at the expense, primarily, of PS II (Leong and Anderson, 1984a,b). This may be significant in N economy since LHC units hold more pigment molecules per unit N than the core PS I and II complexes (Leong and Anderson, 1984b; Chow and Hope, 1987).

The division of photosynthetic components into the traditional categories of 'light' and 'dark' reactions is not very useful when considering acclimation to growth irradiance. Experimental data clearly indicate that it is better to distinguish between components involved in light capture (LHC) and those that use the captured light (electron transport components, CF_0-CF_1, Rubisco). Thus, acclimation to low irradiance involves a decrease in amounts of Rubisco and other RPP Pathway enzymes, when expressed relative to Chl (Boardman, 1977; Björkman, 1981), as well as diminished electron transport. Decreased electron transport capacity at low light was shown to be associated with a proportional decrease in the cytochrome *f* content per unit Chl, such that there was no effect of growth irradiance when the capacity was expressed per unit cytochrome *f* (Evans, 1988). Conversely, acclimation to high light involves increased electron transport capacity, chiefly due to a relative increase in the amounts of the cytochrome *b*/*f* complex and CF_0-CF_1 (Berzborn et al., 1981; Davies et al., 1987). Consequently, at high light, a greater amount of thylakoid N per unit Chl is associated with higher rates of oxygen evolution (Terashima and Evans, 1988). When grown at high light, therefore, plants generally favor high capacities of electron transport and carboxylation; at lower light, available N is preferentially allocated to light capture and there is a drop in electron transport and carboxylation capacities per unit Chl. The dependence on irradiance is particularly significant in leaf canopies, where leaves lower in the canopy are older and experience lower irradiance. High growth irradiance tends to alter the partitioning of N away from thylakoid protein to soluble protein (Evans, 1988). This extra investment in light-harvesting components can markedly affect the PNUE, which was reported to decrease by up to two-fold in the leaves of rice acclimated to low irradiance (Makino et al. 1997a)

C. CO_2 Enriched Environment

The PNUE increases in C_3 crop plants grown at elevated CO_2. This increase results from an increase in photosynthetic rate as higher CO_2 concentrations, first, compensate for the poor affinity of Rubisco for CO_2 and, second, suppress oxygenase activity. Light-saturated photosynthesis at elevated CO_2 levels is limited by electron transport or Pi-regeneration capacity (Farquhar et al., 1980; Sharkey, 1985). Furthermore, photosynthesis is more likely to be light-limited at elevated CO_2 for a given leaf N content. In response to growth at elevated CO_2 the amounts of Rubisco and other photosynthetic components sometimes decrease on a leaf area basis (Lawlor and Mitchell, 1991; Bowes, 1993). This is often explicable entirely in terms of greater N sinks elsewhere due to greater growth at elevated CO_2 (Farage et al., 1998).

In theory, in addition to this general decrease in photosynthetic components, optimal N use predicts that the balance of investment should shift from Rubisco to favor components determining the light-saturated capacity for RuBP regeneration (e.g. CF_0-CF_1 and cytochrome *b/f* complex) (Farquhar and Sharkey, 1994; Sage, 1994; Medlyn, 1996). For example, a 30–40% increase in the ratio of light-saturated RuBP-regeneration capacity to carboxylation capacity is needed for optimal N-use efficiency at twice the current ambient CO_2 concentration (Medlyn, 1996; Mitchell et al., 2000). Since N is likely to be more limiting to growth at elevated CO_2, such an increased ratio would most likely result from a specific decrease in the amount of Rubisco (Sage et al., 1989; Rogers et al., 1996; Theobald et al., 1998).

There is, however, conflicting evidence concerning the response to elevated CO_2 of the ratio between Rubisco and the components that determine RuBP regeneration capacity. The extent to which redistribution of components is observed appears to be dependent on N supply. In wheat and rice there is no evidence for redistribution in young leaves of plants grown at elevated CO_2 (Nie et al., 1995; Nakano et al., 1997; Theobald et al., 1998). However, as discussed in Section III.A, leaves with lower N contents often have decreased values of Rubisco:electron transport components, regardless of CO_2 environment. In older leaves, elevated CO_2 often induces a decrease in leaf N content and, therefore, a partial redistribution. In wheat and rice this appears to be the main process behind reallocation at elevated CO_2: redistribution is still much less than the predicted optimum (Nakano et al., 1997; Theobald et al., 1998). However, in soybean and sunflower at elevated CO_2, data were close to the predicted optimal redistribution of leaf components (Woodrow, 1994; Simms et al., 1998). In summary, while there is evidence that some reallocation occurs under certain conditions (particularly nutrient deficiency), it is usually less than the predicted optimum and often there is none (Sage, 1994).

Elevated CO_2 can also affect the partitioning of C between plant organs. However, this appears to depend on the nutritional status of the plants (Stulen and Den Hertog, 1993). Changes in the N concentration of different plant tissues under elevated CO_2 can lead to a different response of N partitioning compared with C partitioning between plant organs (Lutze and Gifford, 1998).

IV. Role of Regulation of Rubisco Activity

Regulation of Rubisco is generally thought to be such that activity which is not needed to maintain photosynthetic rate for the current environment is deactivated (Sage et al., 1990). Often the increased investment of N in Rubisco is associated with a decrease in activation state (Machler et al., 1998). Conversely, plants compensate for a decreased amount of Rubisco by using the residual Rubisco more effectively by increasing activation state (Quick et al., 1991). Moreover, the control coefficient of Rubisco for photosynthesis is greater in plants growing on limited rather than surplus N (Quick et al., 1992). However, the changes in Rubisco activation state are dependent on both growth and measurement conditions (Evans and Terashima, 1988; Hudson et al., 1992; Stitt and Schulze, 1994).

The reasons for deactivation are not yet clear. One

possibility is that it decreases the turnover of the Rubisco protein. This idea is supported by data showing that deactivation and association with CA1P or RuBP can protect Rubisco against protease activity (Khan et al., 1999). Since Rubisco is so abundant in C_3 plants, the turnover of Rubisco protein represents a significant energetic cost which would be expected to be manifested as increased maintenance respiration. Thus regulation of Rubisco activity may serve to increase NUE by decreasing maintenance respiration (Mitchell, Andralojc and Parry, unpublished).

V. Approaches to Improving Nitrogen-Use Efficiency in Crops

Rubisco is a major sink for N supplied in the form of fertilizers. Manipulation of crops to improve NUE would have economic and environmental benefits, reducing pollution of water supply by nitrates. It has been suggested that improvement of cereal NUE is not a useful goal because the N is needed in the grain (Sinclair and Sheehey, 1999). However, if NUE is improved, N uptake could benefit from greater root growth and grain quality can be maintained by adding N fertilizer during grain development, when it is less polluting.

Recombinant DNA technology offers the possibility to increase the NUE of photosynthesis by genetic manipulation. Various strategies can be tested to manipulate Rubisco amounts or properties to improve NUE. Questions of optimization of amounts of Rubisco are of particular importance for C_3 crop species, since if they do not optimize investment in photosynthetic components in response to growth conditions they could be improved by genetic manipulation. Makino et al. (1997b) demonstrated that PNUE in rice, measured on short-term exposure to elevated CO_2 concentrations, was greater in lines that had been transformed to specifically reduce the amount of Rubisco, compared to wild-type.

Another approach to decrease the amount of Rubisco and nitrogen required would be to decrease photorespiration. The benefits of this for NUE are potentially much greater than for decreasing Rubisco content. The latter approach could benefit only NUE by about 5% (Mitchell, unpublished) while eliminating photorespiration could increase NUE by more than 50%.

While the use of three-dimensional models of Rubisco and site-directed mutagenesis has greatly extended our understanding of the catalytic process, knowledge-based alterations in Rubisco structure have not yet succeeded in altering the enzyme properties to increase photosynthetic performance. However, there is considerable natural variation in the kinetic properties of Rubisco from diverse sources (Bainbridge et al., 1995). Rubisco from the red alga *Galderia partita* has a specificity factor (i.e. the ratio of Vc.Ko/Vo.Kc where Kc and Ko are the Michaelis constants for CO_2 and O_2 and Vc and Vo are the maximal velocities for carboxylation and oxygenation, respectively) almost three-fold greater than for Rubisco from most crop plants (Uemura et al., 1996). Transformation of both nuclear and plastid genomes is already possible and initial attempts to produce novel Rubiscos *in planta* have been encouraging (Getzoff et al., 1998, Kanevski et al., 1999). This suggests that decreased oxygenase activity and thereby photorespiration is in the long term an achievable goal.

Another approach to decrease photorespiration in C_3 plants is to introduce C_4 characteristics by genetic manipulation to ensure Rubisco operates under conditions of super-ambient CO_2. A number of groups have tried to improve CO_2 fixation in C_3 plants in this way. The three-fold overexpression of PEP carboxylase activity in potato decreased the compensation point (Hausler et al., 1999). Similarly an 80-fold overexpression of maize PEP carboxylase in rice resulted in decreased O_2 inhibition of photosynthesis (Ku et al., 1999). In addition, overexpression of the NADP-dependent malic enzyme in the potato line already overexpressing PEP carboxylase led to a significantly decreased electron requirement for apparent CO_2 assimilation at higher temperature. A likely explanation of this observation is that overexpression of both these enzymes led to significant suppression of Rubisco oxygenase activity and consequent photorespiratory metabolism (Lipka et al., 1999). While there remain many hurdles still to be overcome (e.g. overexpression of PEP carboxylase also increased dark respiration; Hausler et al. (1999)), such results are encouraging support for the eventual successful introduction of the C_4 pathway of photosynthesis into C_3 plants (Mann, 1999). A strategy to reduce photorespiration by manipulating catalase activity in tobacco has also been reported (Brisson et al., 1998).

Acknowledgments

Thanks are due to Council of Scientific and Industrial Research for financial support under the Emeritus Scientist scheme to YPA and AA. Institute of Arable Crops Research receives grant-aided support from the Biotechnology and Biological Sciences Research Council of the United Kingdom.

References

Abrol YP (ed) (1993) Nitrogen-Soils, Physiology, Biochemistry, Microbiology, Genetics. Indian National Science Academy, New Delhi

Abrol YP, Chatterjee SR, Kumar PA and Jain V (1999) Improvement in nitrogen use efficiency: Physiological and molecular approaches. Curr Sci 76: 1357–1364

Ahmad A and Abdin MZ (2000) Photosynthesis and its related physiological variables in the leaves of *Brassica* genotypes as influenced by sulphur fertilization. Physiol Plant 110: 144–149

Badger MR and Price GD (1994) The role of carbonic anhydrase in photosynthesis. Annu Rev Plant Physiol Plant Mol Biol 45: 369–392

Bainbridge G, Madgwick P, Parmar S, Mitchell R, Paul M, Pitts J, Keys AJ and Parry MAJ (1995) Engineering Rubisco to change its catalytic properties. J Exp Bot 46: 1269–1296

Banks FM, Driscoll SP, Parry MAJ, Lawlor DW, Knight JS, Gray JC and Paul MJ (1999) Decrease in phosphoribulokinase activity by antisense RNA in transgenic tobacco. Relationship between photosynthesis, growth, and allocation at different nitrogen levels. Plant Physiol 119: 1125–1136

Berzborn RJ, Muller D, Roos P and Anderson B (1981) Significance of different quantitative determinations of photosynthetic ATP synthetase, CF_1, for heterogenous CF_2 distribution and grana formation. In: Akoyunoglou G (ed) Proceedings of Vth International Photosynthetic Congress, Vol 3, pp 107–120. Balaban International Science Services, Philadelphia.

Björkman O (1981) Responses to different quantum flux densities. In: Lange OL, Nobel PS, Osmond CB and Ziegler H (eds) Physiological Plant Ecology. 1. Responses to the Physical Environment, pp 57–107. Springer-Verlag, Berlin

Boardman NK (1977) Comparative photosynthesis of sun and shade plants. Annu Rev Plant Physiol 13: 221–237

Bowes G (1993) Facing the inevitable: Plants and increasing atmospheric carbon dioxide. Annu Rev Plant Physiol Plant Mol Biol 44: 309–332

Brisson LF, Zelitch I and Havir EA (1998) Manipulation of catalase levels produces altered photosynthesis in transgenic tobacco plants. Plant Physiol 116: 259–269

Brown KR, Thompson WA, Camm EL, Hawkins BJ and Guy RD (1996) Effects of N addition rates on the productivity of *Picea sitchensis*, *Thuja plicarta* and *Tsuga heterophylla* seedlings. II Photosynthesis, ^{13}C discrimination and N partitioning in foliage. Trees 10: 198–205

Caemmerer S von and Farquhar GD (1981) Some relationships between the biochemistry of photosynthesis and the gas exchange of leaves. Planta 153: 376–387

Charles-Edwards DA, Sturzel H, Ferraris R and Beech DF (1987) An analysis of spatial variation in the nitrogen content of leaves from different horizon with a canopy. Ann Bot 60: 421–426

Chow WS and Hope AB (1987) The stoichiometries of supermolecular complexes in thylakoid membranes from spinach chloroplasts. Aust J Plant Physiol 14: 21–28

Davies EC, Jordan BR, Partis MD and Chow WS (1987) Immunochemical investigation of thylakoid coupling factor protein during photosynthetic acclimation to irradiance. J Exp Bot 38: 1517–1527

DeJong TM and Doyle JF (1985) Second relationship between leaf nitrogen content (photosynthesis capacity) and leaf canopy light exposure in peach (*Prunus persia*). Plant Cell Environ 8: 701–706.

Demmig-Adams B and Adams WW (1992) Photoprotection and other responses of plants to high light stress. Annu Rev Plant Physiol Plant Mol Biol 43: 599–626

Drake BG, Gonzalez-Meler MA and Long SP (1997) More efficient plants: A consequence of rising atmospheric CO_2. Annu Rev Plant Physiol Plant Mol Biol 48: 609–639

Evans JR (1983) Nitrogen and photosynthesis in the flag leaf of wheat (*Triticum aestivum* L.). Plant Physiol 72: 297–302

Evans JR (1988) Acclimation by the thylakoid membranes to growth irradiance and the partitioning of nitrogen between soluble and thylakoid proteins. In: Evans JR, von Caemmerer S and Adams WW III (eds) Ecology of photosynthesis in sun and shade, pp 93–106. CSIRO, Melbourne

Evans JR (1989) Photosynthesis and nitrogen relationship in leaves of C_3 plants. Oecologia 78: 9–19

Evans JR (1993) Photosynthetic acclimation and nitrogen partitioning within a lucerne canopy. 2. Stability through time and comparison with a theoretical optimum. Aust J Plant Physiol 20: 69–82

Evans JR and Seemann JR (1984) Differences between wheat genotypes in specific activity of ribulose-1,5-bisphosphate carboxylase and the relationship to photosynthesis. Plant Physiol 74: 759–765

Evans JR and Seemann JR (1989) The allocation of protein nitrogen in the photosynthetic apparatus: Costs, consequences, and control. In: Briggs WR (ed) Photosynthesis, pp 183–205. Alan R Liss Inc., New York

Evans JR and Terashima I (1987) Effect of nitrogen nutrition on electron transport components and photosynthesis in spinach. Aust J Plant Physiol 14: 59–68

Evans JR and Terashima I (1988) Photosynthetic characteristics of spinach leaves grown with different nitrogen treatments. Plant Cell Physiol 29: 157–165

Farage PK, McKee IF and Long SP (1998) Does a low nitrogen supply necessarily lead to acclimation of photosynthesis to elevated CO_2? Plant Physiol 118: 573–580

Farquhar GD and Sharkey TD (1994) Photosynthesis and carbon assimilation. In: Boote KJ, Bennett JM, Sinclair TR and Paulson GM (eds) Physiology and Determination of Crop Yield, pp 187–210. American Society of Agronomy, Madison

Farquhar GD, von Caemmerer S and Berry JA (1980) A biochemical model of photosynthetic CO_2 assimilation in leaves of C_3 species. Planta 149: 78–90

Ferrar PJ and Osmond CB (1986) Nitrogen supply as a factor

influencing photoinhibition and photosynthetic acclimation after transfer of shade grown *Solanum dulcamara* to bright light. Planta 168: 563–570

Field C (1983) Allocating leaf nitrogen for the maximization of carbon gain: Leaf age as a control on the allocation program. Oecologia 56: 341–347

Field C and Mooney HA (1986) The photosynthesis-nitrogen relationship in wild plants. In: Givnish TJ (ed) On the Economy of Plant Form and Function, pp 25–55. Cambridge University Press, Cambridge

Getzoff TP, Zhu GH, Bohnert HJ and Jensen RG (1998) Chimeric *Arabidopsis thaliana* ribulose-1,5-bisphosphate carboxylase/oxygenase containing a pea small subunit protein is compromised in carbamylation. Plant Physiol 116: 695–702

Harrison EP, Willingham NM, Lloyd JC and Raines CA (1998) Reduced sedoheptulose-1,7-bisphosphatase levels in transgenic tobacco lead to decreased photosynthetic capacity and altered carbohydrate accumulation. Planta 204: 27–36

Häusler RE, Kleines M, Uhrig H, Hirsch HJ and Smets H (1999) Overexpression of phosphoenolpyruvate carboxylase from *Corynebacterium glutamicum* lowers the CO_2 compensation point (Γ^*) and enhances dark and light respiration in transgenic potato. J Exp Bot 50: 1231–1242

Hirose T and Werger MJA (1987) Maximizing daily canopy photosynthesis with respect to the leaf nitrogen allocation pattern in the canopy. Oecologia 72: 520–526

Hirose T, Werger MJA, Pons TL and Van Rheenem JWA (1988) Canopy studies and leaf nitrogen distribution in a stand of *Lysimachia vulgaris* L. as influenced by stand density. Oecologia 77: 145–150.

Hollinger DY (1989) Canopy organisation and foliage photosynthetic capacity in a broad-leaved evergreen montane forest. Forest Ecol 3: 53–62

Hudson GS, Evans JR, von Caemmerer S, Arvidsson YB and Andreus TJ (1992) Reduction of ribulose-1,5-phosphate carboxylase/oxygenase content by antisense RNA reduces photosynthesis in tobacco plants. Plant Physiol 983: 294–302

Jain V, Pal M, Lakkineni KC and Abrol YP (1999) Photosynthetic characteristics in two wheat genotypes as affected by nitrogen nutrition. Biol Plant 42: 217–222

Kanevski I, Maliga P, Rhoades D and Gutteridge S (1999) Plastome engineering of ribulose-1,5-bisphosphate carboxylase/oxygenase in tobacco to form a sunflower large subunit and tobacco small subunit hybrid. Plant Physiol 119: 133–141

Khamis S, Lamaze T, Lemoine Y and Foyer C (1990) Adaptation of the photosynthetic apparatus in maize leaves as a result of nitrogen limitation. Plant Physiol 94: 1436–1443

Khan S, Andralojc P, Lea P and Parry MAJ (1999) 2′-carboxy-D-arabinitol 1-phosphate (CA1P) protects ribulose-1,5-bisphosphate carboxylase/oxygenase against proteolytic breakdown: A possible role in vivo. Eur J Biochem 266: 840–847

Kossmann J, Sonnewald U and Willmitzer l (1994) Reduction of the chloroplastic fructose-1,6-bisphosphatase in transgenic potato plants impairs photosynthesis and plant growth. Plant J 6: 637–650

Krapp A, Hofmann B, Schafir C and Stitt M (1993) Regulation of the expression of *rbcS* and other photosynthetic genes by carbohydrates: A mechanism for the sink-regulation of photosynthesis. Plant J 3: 817–828

Ku MSB, Agarie S, Nomura M, Fukayama H, Tsuchida H, Ono K, Hirose S, Toki S, Miyao M and Matsuoka M (1999) High-level expression of maize phosphoenolpyruvate carboxylase in transgenic rice plants. Nature Biotech 17: 76–80

Kutik J, Natr L, Demmers-Derks HH and Lawlor DW (1995) Chloroplast structure of sugar beet (*Beta vulgaris* L.) cultivated in normal and elevated CO_2 concentrations with two contrasted nitrogen supplies. J Exp Bot 46: 1797–1802

Laurer M, Saptic D, Quick WP, Labate C, Fichtner K, Schulze ED, Rodermal SP, Bogorad L and Stitt M (1993) Decreased ribulose-1,5-bisphosphate carboxylase/oxygenase in transgenic tobacco transformed with 'antisense' *rbc*S. VI. Effect on photosynthesis in plants grown at different irradiance. Planta 190: 332–345

Lawlor DW and Mitchell RAC (1991) The effects of increasing CO_2 on crop production and productivity: A review of field studies. Plant Cell Environ 14: 807–818

Lawlor DW, Boyle FA, Young AT, Keys AJ and Kendall AC (1987) Nitrate nutrition and temperature effects on wheat: photosynthesis and photorespiration of leaves. J Exp Bot 38: 393–408

Lawlor DW, Kontturi M and Young AT (1989) Photosynthesis by flag leaves of wheat in relation to protein, ribulose-1,5-bisphosphate carboxylase activity and nitrogen supply. J Exp Bot 40: 43–52

Leegood RC (1990) Enzymes of the Calvin cycle. In: Lea PJ (ed) Enzymes of primary metabolism, pp 15–37. Academic Press, London

Leong TY and Anderson JM (1984a) Adaptation of the thylakoid membranes of pea chloroplast to light intensities. I. Study on the distribution of chlorophyll-protein complexes. Photosynth Res 5: 105–115

Leong TY and Anderson JM (1984b) Adaptation of the thylakoid membranes of pea chloroplast to light intensities. II. Regulation of electron transport capacities, electron carriers, coupling factor (CF1) activity and rates of photosynthesis. Photosynth Res 5: 117–128

Lipka V, Häusler RE, Rademacher T, Li J, Hirsch HJ and Kreuzaler F (1999) *Solanum tuberosum* double transgenic expressing phosphoenolpyruvate carboxylase and NADP-malic enzyme display reduced electron requirement for CO_2 fixation. Plant Sci 144: 93–105

Lutze JL and Gifford RM (1998) Acquisition and allocation of carbon and nitrogen by *Danthonia richardsonii* in response to restricted nitrogen supply and CO_2 enrichment. Plant Cell Environ 21: 1133–1141

Machler F, Oberson A, Grub A and Noseberger J (1988) Regulation of photosynthesis in nitrogen deficient wheat seedlings. Plant Physiol. 87: 46–49

Makino A and Osmond B (1991) Solubilization of ribulose-1,5-bisphosphate carboxylase from the membrane fraction of pea leaves. Photosynth Res 29: 79–85

Makino A, Mae T and Ohira K (1985) Photosynthesis and ribulose-1,5-bisphosphate carboxylase oxygenase in rice leaves from emergence through senescence—quantitative-analysis by carboxylation oxygenation and regeneration of ribulose 1,5-bisphosphate. Planta 166: 414–420

Makino A, Mae T and Ohira K (1988) Differences between wheat and rice in the enzymic properties of ribulose-1,5-bisphosphate carboxylase oxygenase and the relationship to photosynthetic gas-exchange. Planta 174: 30–38

Makino A, Sakashita H, Hidema J, Mae T, Ojima K and Osmond B

(1992) Distinctive responses of riblulose-1,5-bisphosphate decarboxylase and carbonic anhydrase in wheat leaves to nitrogen nutrition and their possible relationships to CO_2 transfer resistance. Plant Physiol 100: 1737–1743

Makino A, Nakama H and Mae T (1994) Responses of ribulose-15-bisphosphate decarboxylase, cytochrome *f* and sucrose synthesis in rice leaves to leaf nitrogen and their relationship to photosynthesis. Plant Physiol 105: 173–179

Makino A, Sato T, Nakano H and Mae T (1997a) Leaf photosynthesis, plant growth and nitrogen allocation in rice under different irradiances. Planta 203: 390–398

Makino A, Shimada T, Takumi S, Kaneko K, Matsuoka M, Shimamoto K, Nakano H, Miyao-Tokutomi M, Mae T and Yamamoto N (1997b) Does decrease in ribulose-1,5-bisphosphate carboxylase by 'antisense' RbcS lead to a higher nitrogen-use efficiency of photosynthesis under conditions of saturating CO_2 and light in rice plants? Plant Physiol 114: 483–491

Mann CC (1999) Genetic engineers aim to soup up crop photosynthesis. Science 283: 314–316

McMorrow EM and Bradbeer JW (1990) Separation, purification, and comparative properties of chloroplast and cytoplasmic phosphoglycerate kinase from barley leaves. Plant Physiology 93: 374–383

Medina E (1971) Effects of nitrogen supply and light intensity during growth on the photosynthetic capacity and carboxylase activity of leaves of *Atriplex patula* sp. Hastate. Carnegie Inst Wash Year Book 70: 551–559

Medlyn BE (1996) The optimal allocation of nitrogen within C_3 photosynthetic system at elevated CO_2. Aust J Plant Physiol 23: 593–603

Millard P and Catt JW (1988) The influence of nitrogen supply on the use of nitrate and ribulose-1,5-bisphosphate carboxylase oxygenase as leaf nitrogen stores for growth of potato-tubers (*Solanum tuberosum* L). J Exp Bot 39: 1–11

Mitchell RAC, Theobald JC, Parry MAJ and Lawlor DW (2000) Is there scope for improving balance between RuBP-regeneration and carboxylation capacities in wheat at elevated CO_2? J Exp Bot 5: 391–397

Nakano H, Makino A and Mae T (1997) The effects of elevated partial pressures of CO_2 on the relationship between photosynthetic capacity and N content in rice leaves. Plant Physiol 115: 191–198

Nie GY, Long SP, Garcia RL, Kimball BA, Lamorte RL, Pinter PJ, Wall GW and Webber AN (1995) Effects of free air CO_2 enrichment on the development of the photosynthetic apparatus in wheat, as indicated by changes in leaf proteins. Plant Cell Environ 18: 855–864

Novoa R and Loomis RS (1981) Nitrogen and plant production. Plant Soil 58: 177–204

Osaki M, Shinano T and Tadono T (1993) Effect of nitrogen application on the accumulation of ribulose-1,5-bisphosphate carboxylase oxygenase and chlorophyll in several field crops. Soil Sci Nut 39: 427–436

Paul MJ, Knight JS, Habash D, Parry MAJ, Lawlor DW, Barnes SA, Loynes A and Gray JC (1995) Reduction in phosphoribulokinase activity by antisense RNA in transgenic tobacco—effect on CO_2 assimilation and growth in low irradiance. Plant J 7: 535–542

Poorter H and Evans JR (1998) Photosynthetic nitrogen use efficiency of species that differ inherently in specific leaf area. Oecologia 116: 26–37

Price GD, Evans JR, Caemmerer S von, Yu JW and Badger MR (1995) Specific reduction of chloroplast glyceraldehyde-3-phosphate dehydrogenase activity by antisense RNA reduces CO_2 assimilation via a reduction in ribulose bisphosphate regeneration in transgenic tobacco plants. Planta 195: 369–378

Quick WP, Schurr U, Scheibe R, Schulze ED, Rodermel SR, Bogorad L and Stitt M (1991) The impact of decreased Rubisco on photosynthesis, growth, allocation and storage in tobacco plants which have been transformed with antisense *rbc*S. Plant J 1:51–58

Quick WP, Fichtner K, Schulze ED, Wendler R, Leegood RC, Mooney H, Rodermel SR, Bogorad L and Stitt M (1992) Decreased ribulose-1,5-bisphosphate carboxylase-oxygenase in transgenic tobacco transformed with 'antisense' *rbc*S. IV. Impact on photosynthesis in ambient growth conditions of altered nitrogen supply. Planta 188: 522–531

Robinson SP and Walker DA (1981) Photosynthetic carbon reduction cycle (Calvin cycle) (including fluxes and regulation). In: Stumpf PK and Conn EE (eds) The Biochemistry of Plants. A Comprehensive Treatise, Vol 8, pp 193–226. Academic Press, London

Robinson SP, Streusand VJ, Chatfield JM and Portis AR Jr. (1988) Purification and assay of Rubisco activase in leaves. Plant Physiol 88: 1008–1014

Rogers GS, Milham PJ, Gillings M and Conroy JP (1996) Sink strength may be the key to growth and nitrogen responses in N-deficient wheat at elevated CO_2. Aust J Plant Physiol 23: 253–264

Roumet C, Bel MP, Sonie L, Jardon F and Roy J (1996) Growth response of grasses to elevated CO_2: A physiological plurispecific analysis. New Phytol 133: 595–603

Sage RF (1994) Acclimation to photosynthesis to increasing atmospheric CO_2: The gas exchange perspective. Photosynth Res 39: 351–368

Sage RF (1999) Why C_4 photosynthesis? In: Sage RF and Monson RK (eds) C_4 Plant Biology, pp 3–16. Academic Press, London

Sage RF, Pearcy RW and Seemann JR (1987) The nitrogen use efficiency of C_3 and C_4 plants. I. Leaf nitrogen effects on the gas exchange characteristics of *Chenopodium album* (L.) and *Amaranthus retroflexus* (L.). Plant Physiol 84: 959–963

Sage RF, Sharkey TD and Seemann JR (1989) Acclimation of photosynthesis to elevated CO_2 in five C_3 species. Plant Physiol 89: 590–596

Sage RF, Sharkey TD and Seemann JR (1990) Regulation of ribulose-1,5-bisphosphate carboxylase activity in response to light-intensity and CO_2 in the C-3 annuals *Chenopodium album* L and *Phaseolus vulgaris* L. Plant Physiol 94: 1735–1742

Sharkey TD (1985) Photosynthesis in intact leaves of C3 plants: Physics, physiology and rate limitations. Bot Rev 51: 53–105

Sims DA, Luo Y and Seemann JR (1998) Comparison of photosynthetic acclimation to elevated CO_2 and limited nitrogen supply in soybean. Plant Cell Environ 21: 945–952

Sinclair TR and Horie T (1989) Leaf nitrogen, photosynthesis and crop radiation use efficiency: A review. Crop Sci 42: 90–98

Sinclair TR and Sheehy JE (1999) Erect leaves and photosynthesis in rice. Science 283: 1456–1457

Sivasankar A, Bansal KC and Abrol YP (1993) Nitrogen in relation to leaf area development. In: YP Abrol (ed) Nitrogen, pp 75–84. Proc Ind Nat Sci Acad, New Delhi

Sivasankar A, Lakkineni KC, Jain V and Abrol YP (1998a) Differential response of two wheat genotypes to nitrogen supply. I. Ontogenic changes in laminae growth and photosynthesis. J Agron Crop Sci 181: 21–27

Sivasankar A, Lakkineni KC, Jain V, Kumar P and Abrol YP (1998b) Differential response of two wheat genotypes to nitrogen supply. II. Mesophyll cell characteristics and photosynthesis of lamina at full expansion. J Agron Crop Sci 181: 65–70

Stitt M and Schulze ED (1994) Does Rubisco control the rate of photosynthesis and plant growth? An exercise in molecular ecophysiology. Plant Cell Environ 17: 518–552

Stulen I and Den Hertog J (1993) Root growth and functioning under atmospheric CO_2 enrichment. Vegetation 104/105: 99–115

Terashima I and Evans JR (1988) Effects of light and nitrogen nutrition on the organization of the photosynthetic apparatus in spinach. Plant Cell Physiol 29: 143–155

Theobald JC, Mitchell RAC, Parry MAJ and Lawlor DW (1998) Estimation of excess investment in ribulose-1,5-bisphosphate carboxylase/oxygenase in leaves of spring wheat grown under elevated CO_2. Plant Physiol 118: 945–955

Trost P, Scagliarini S, Valentini V and Pupillo P (1993) Activation of spinach chloroplast glyceraldehyde-3-phosphate dehydrogenase—effect of glycerate 1,3-bisphosphate. Planta 190: 320–326

Uemura K, Suzuki Y, Shikanai T, Wadano A, Jensen RG, Chmara W and Yokota A (1996) A rapid and sensitive method for determination of relative specificity of Rubisco from various species by anion exchange chromatography. Plant Cell Physiol 37: 325–331

Verhoeven AS, Demmig-Adams B and Adams WW (1997) Enhanced employment of the xanthophyll cycle and thermal energy dissipation in spinach exposed to high light and N stress. Plant Physiol 113: 817–824

Wong SC (1979) Elevated atmospheric pressure of CO_2 and plant growth. I. Interactions of nitrogen nutrition and photosynthetic capacity in C_3 and C_4. Oecologia 44: 68–74

Woodrow IE (1994) Control of steady state photosynthesis in sunflowers growing in enhanced CO_2. Plant Cell Environ 17: 277–286

Woodrow IE and Berry JA (1988) Enzymatic regulation of photosynthetic CO_2 fixation in C_3 plants. Annu Rev Plant Physiol Plant Mol Biol 39: 533–594

Chapter 3

Molecular Control of Nitrate Reductase and Other Enzymes Involved in Nitrate Assimilation

Wilbur H. Campbell*
Department of Biological Sciences, Michigan Technological University, Houghton, MI 49931-1295 U.S.A.

Summary 35
I. Introduction 36
A. Overview of Nitrogen Metabolism and Its Regulation 36
B. Nitrate Reductase Structure and Function 36
II. Transcriptional Control of Nitrate Reductase and Other Nitrogen Metabolism Genes 39
A. Genes Encoding Nitrate Reductase and Other Proteins 39
B. Control of Nitrate Reductase Gene Expression 40
C. The Nitrate Box 41
III. Post-Translational Control of Nitrogen Metabolism Enzymes 41
A. Nitrate Reductase Biosynthesis and Turnover 41
B. Nitrate Reductase Phosphorylation and Inhibition by 14-3-3 Binding Protein 42
C. Mechanism of Inhibition of Nitrate Reductase by 14-3-3 43
IV. Protein Kinases and Control of Carbon and Nitrogen Metabolism 44
V. Future Prospects for the Control of Nitrogen Metabolism 45
Acknowledgment 46
References 46

Summary

Nitrate acts as both a nutrient and a signal in plants. Nitrate induces gene expression of enzymes for its metabolism into amino acids but also has other effects on plant metabolism and development. Familiar nitrate-induced enzymes are nitrate and nitrite reductases, nitrate transporters, glutamine synthetase, glutamate synthase, ferredoxin and ferredoxin $NADP^+$ reductase. Microarray analysis of nitrate-stimulated gene expression has identified 40 transcripts including hemoglobin, transaldolase, regulatory and stress proteins, several protein kinases and several methyltransferases. Coordinated expression of these nitrate-stimulated genes is probably due to a single 'nitrate-transacting factor' and a 'nitrate' box has been elucidated for nitrate and nitrite reductase genes with constitutively expressed nuclear proteins which bind to the box. A MADS transcription factor is nitrate-induced in roots and involved in development of lateral roots. However, accumulation of nitrate overcomes this signal and halts lateral root development. Post-translational inhibition of nitrate reductase activity illustrates a complex control mechanism involving protein phosphorylation and binding of the ubiquitous binding protein called 14-3-3. Protein kinases catalyzing phosphorylation have been identified and 14-3-3 binding elucidated, which includes activation of 14-3-3 by polycations such as polyamines. Using molecular modeling, it was shown that one 14-3-3 binding site can bind to the nitrate reductase dimer. Nitrate reductase-14-3-3 complexes could bind via the second binding site on 14-3-3 to another enzyme/protein with a 14-3-3 binding site or nitrate reductase aggregation could result in rapid degradation. Two types of protein kinases are involved in nitrate reductase phosphorylation: calcium-dependent protein kinases and SnRKs

Email: wcampbel@mtu.edu

Christine H. Foyer and Graham Noctor (eds): Photosynthetic Nitrogen Assimilation and Associated Carbon and Respiratory Metabolism, pp. 35–48.

(enzymes related to yeast sucrose non-fermenting (SNF1) protein kinases). Calcium-dependent protein kinases are activated by environmental and development signals via changes in intracellular calcium level, which impacts many plant metabolic pathways. SnRKs are less well understood and may be responding to more general metabolic signals.

I. Introduction

A. Overview of Nitrogen Metabolism and Its Regulation

Nitrogen (N) is the most limiting element for crop plant growth, after factors like light and water. Nitrogen fertilizer had a major impact on agricultural crop productivity in the second half of the Twentieth Century (Smil, 1997) and is likely to be a big factor in the Twenty-First. Another impact of N fertilizers has been the contribution to nutrient run-off and pollution of natural ecosystems with excess N, which has upset the natural balance especially in coastal estuaries (Vitousek, 1997; National Research Council, 2000). For all these reasons, understanding the regulation of N metabolism in plants, especially for major crop plants like corn, wheat, barley, and rice where N fertilizer has had the greatest impact on productivity, is crucial for successfully feeding the growing world population of humans and the animals they consume, with decreased impact on the world's ecosystems.

While much of the N fertilizer is applied to crops as ammonium, this reduced form of N is converted to nitrate by bacteria in the soil and most plants utilize nitrate as essentially the sole N source. Although some plants such as legumes utilize symbiotic N_2 fixation to meet N needs, it appears unlikely that this capacity will be extended to important crop plants in the near future. A few plants growing in manure-rich soils take up ammonium directly from the soil, but this is less common and not the case for major crop plants. In addition, ammonium metabolism in leaves of most plants is dominated by recycling of N released in photorespiratory metabolism (Chapter 8, Keys and Leegood) with recently acquired N representing only 10 to 15% of the total flux. Hence, the focus of this review is on nitrate utilization and its regulation by plants.

The greatest attention will be on the three steps in plants where external or environmental nitrate is converted to ammonium: 1) nitrate uptake mediated by energy-dependent nitrate transport systems; 2) nitrate reduction to nitrite catalyzed by nitrate reductase (NR; EC 1.6.6.1-3); and 3) nitrite reduction to ammonium catalyzed by nitrite reductase (NiR; EC 1.7.7.1). Nitrate transporters, especially those in the epidermal cells of roots, have now been cloned and studied at the molecular level (Forde, 2000; Vidmar et al., 2000). A number of studies have shown that both low and high affinity nitrate transporters of roots are expressed in response to nitrate treatment (Stitt, 1999). NR and NiR are the classic enzymes and genes which are induced by nitrate (Campbell, 1988; Redinbaugh and Campbell, 1991).

Recent developments in understanding regulation of the catalyst for nitrate reduction, namely NR, have opened up new avenues for gaining understanding of how N metabolism is integrated with C metabolism in plants (Huber et al., 1996; Moorhead et al., 1996; Campbell, 1996). At the same time, most evidence suggests that NiR which catalyzes nitrite conversion to ammonium in the chloroplast, is coordinately regulated with NR at the transcription level and not so tightly regulated post-transcriptionally, with the main control being the N source (Chapter 4, Meyer and Stöhr). As a consequence, this chapter will largely focus on regulation of NR and the signal pathways mediating this regulation.

B. Nitrate Reductase Structure and Function

The structure and function of NR will be only briefly discussed since it was recently reviewed in considerable detail (Campbell, 1999). NR catalyzes NAD(P)H-dependent reduction of nitrate to nitrite which is essentially irreversible since the reaction is accompanied by the release of a large amount of free

Abbreviations: BPB – bromophenol blue; CbR – Cyt *b* reductase fragment of NR; CDPK – calcium-dependent protein kinase; Cyt – cytochrome; GOGAT – glutamate synthase; GS – glutamine synthetase; MADS – MCM1, AGAMOUS, DEFICIENS, and serum response factor genes; Mo-MPT – molybdenum-molybdopterin cofactor; MoR – molybdenum reductase fragment of NR; MV – methyl viologen; NiR – nitrite reductase; NR – nitrate reductase; SNF1 – sucrose non-fermenting control gene encoding a protein kinase; SnRK1 – plant protein kinases related to yeast SNF1 protein kinase energy. The enzyme has two active sites in its monomeric subunit—one for NAD(P)H to donate electrons and one for reduction of nitrate to nitrite (Fig. 1a). The steady-state kinetic mechanism is two-site ping-pong reflecting the two independent active sites and the redox nature of NR, which can exist in both oxidized and reduced enzyme forms. NR has a polypeptide chain with about 900 amino acid residues and contains the cofactors FAD, heme-Fe, and Mo-MPT (Fig. 1b), which are bound into structurally

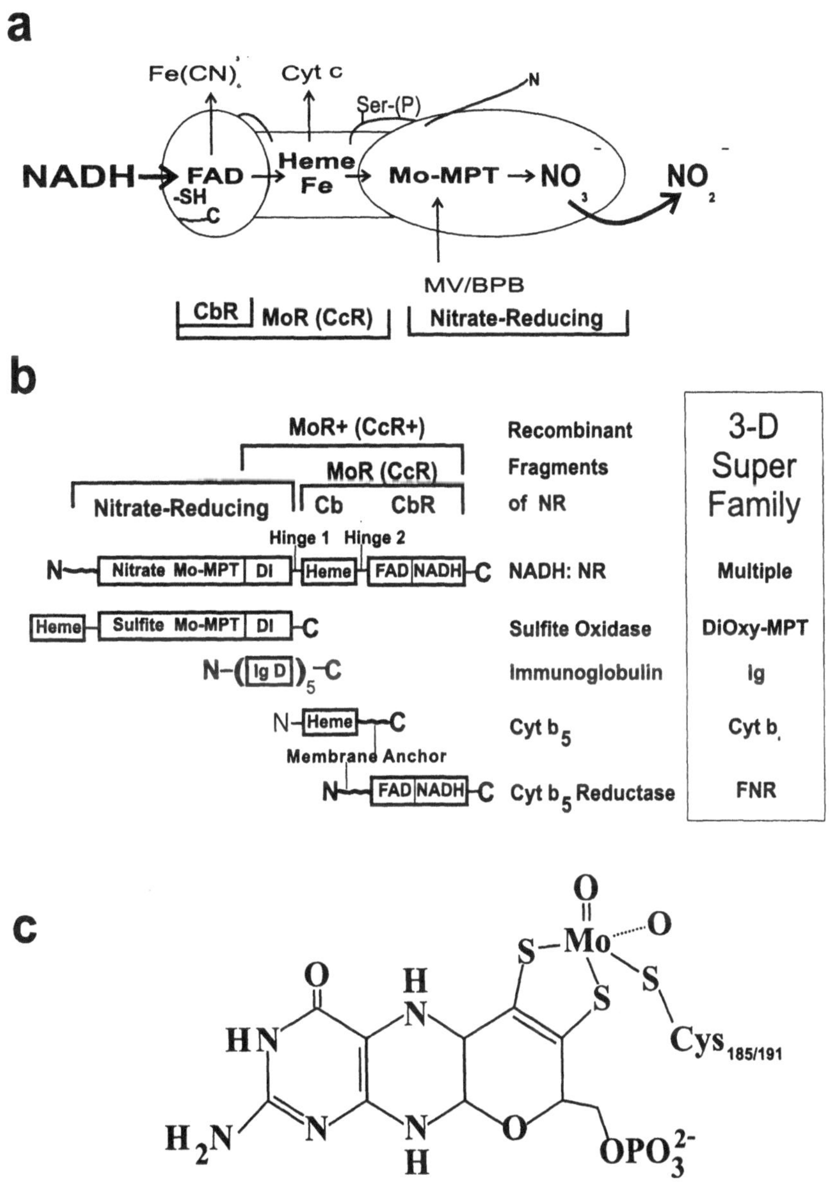

Fig. 1. Structure and function models for NR. A. NR catalyzes NADH-dependent nitrate reduction by internal electron transfer from the NADH reaction site via FAD, heme-Fe and Mo-MPT where nitrate is reduced. Partial activities of NR are NADH:ferricyanide reductase involving only the FAD; NADH:Cyt *c* reductase activity involving only the FAD and heme-Fe; and reduced MV: and BPB:NR activity involving only the Mo-MPT. The phospho-Ser in Hinge 1 is involved in regulation of NR activity by 14-3-3. Cbr, Ccr, MoR = NR protein fragments able to reduce cytochrome *b*, cytochrome *c*, and the molybdenum co-factor, respectively; BPB = bromophenol blue; MPT = molybdopterin; MV = methyl viologen. B. NR has five function domains which have been identified by recombinant expression and comparisons of sequences and 3-D structures with other enzymes and proteins. The 3-D super families are taken from the Protein Data Base. DI is the dimer interface domain. FNR stands for ferredoxin $NADP^+$ reductase. C. The structure of Mo-MPT in NR and sulfite oxidase is based on the structure of the latter (Kisker et al., 1997). The fifth ligand to Mo is a thiol from Cys residue 185 in sulfite oxidase and Cys residue 191 in NR (Su et al., 1997).

independent domains (Campbell and Kinghorn, 1990). The ~100-kDa subunit dimerizes to form the active enzyme but tetrameric forms also exists as dimers of the homodimer. The two active sites of NR are formed between domains at each end of the monomer: 1) the nitrate-reducing active site between the Mo-MPT and dimer interface domains; and 2) the pyridine nucleotide electron-donor active site between the FAD and NADH domains (Fig. 1a). While there is not yet a 3-D structure for NR, a working model of its 3-D conformation was derived from the 3-D structure of mammalian sulfite oxidase (Kisker et al., 1997; Campbell, 1999) by docking on it the structure of the Cyt *b* reductase or CbR fragment of NR (Lu et al., 1994, 1995).

The working 3-D model of NR identified five structural domains with functional significance: Mo-MPT binding, the dimer interface, Cyt *b*, FAD and NADH (Fig. 1b). NR has three other sequence regions with no known structural similarity to any other protein: 1) N-terminal extension preceding the Mo-MPT domain; 2) Hinge 1 between the dimer interface and the Cyt *b* domains; and 3) Hinge 2 between the Cyt *b* and FAD domains. The two 'hinge' regions are expected to be flexible structurally and are known to contain proteolytically sensitive sites (Campbell, 1996, 1999). In addition, Hinge 1 has been shown to contain the Ser residue which is phosphorylated in a reversible process involved in NR activity regulation in vivo (Bachmann et al., 1996; Su et al., 1996). The N-terminal extension is longest in plant NR forms and shortest in algal and fungal NR forms. Recent evidence suggests that the N-terminal region may play a role in NR regulation and this will be discussed below in more detail.

The Mo-MPT cofactor is the most unique component of NR, and appears to be identical in sulfite oxidase (Fig. 1c). Recent x-ray absorption spectroscopic analysis of the coordination ligands of Mo in NR demonstrates that NR and sulfite oxidase have identical sulfur and oxygen ligands in turnover forms (George et al., 1999). Two of the sulfur ligands are from the MPT, while the other comes from a thiol of the protein, which is Cys 191 in *Arabidopsis* NR (Su et al., 1997). However, resting NR has a longer bond length for one sulfur ligand relative to the turnover form of the enzyme, which suggests NR undergoes a conformation change from resting to active enzyme.

Functionally, NR is a highly efficient catalyst with NADH-dependent nitrate reduction to nitrite having a k_{cat} = ~200 s^{-1} and low K_M values of 1 to 7 μM NADH and 20 to 40 μM nitrate (Campbell, 1999; Chapter 5, Kaiser et al.). Holo-NR catalyzes various partial reactions which reflect the modular structure of the enzyme to some extent (Fig. 1a). The CbR fragment catalyzes NADH: ferricyanide reductase activity while the MoR fragment with the Cyt *b* domain added to CbR via Hinge 2 is an efficient catalyst of NADH: Cyt *c* reductase activity. The CbR and MoR fragments, which are the C-terminal portion of the NR monomer (Fig. 1b), have been recombinantly expressed and studied in considerable detail (Dwivedi et al., 1994; Ratnam et al., 1995, 1997; Mertens et al., 2000).

Holo-NR also catalyzes various reduced dye-dependent nitrate reductase activities, which require only the presence of the Mo-MPT and dimer interface domains in most cases. For example, reduced BPB and MV appear to directly donate electrons to the Mo-MPT to drive nitrate reduction, while $FADH_2$ and $FMNH_2$ probably require the presence of the Cyt *b* domain's heme-Fe for activity. Since the MV: and BPB:NR activities are greater than NADH:NR activity, it has been suggested that internal electron transfer is rate-limiting the catalytic activity (Campbell, 1999). Indeed, NADH:ferricyanide and Cyt *c* reductase activities are much greater than the NR activities (Mertens et al., 2000). Recent preliminary pre-steady-state analysis of electron transfer from NADH to NR indicates that both the FAD and heme-Fe cofactors are rapidly reduced with rates sufficient to support the nitrate reduction activity of the enzyme (Mertens et al., 1999). While more detailed studies of internal electron transfer rates are in progress, NR catalysis appears to be limited by electron transfer from reduced heme-Fe to Mo.

Until recently, the presence of a functional heme-Fe in the Cyt *b* domain was thought to be required for reduced MV: NR activity. This was based on a mutant NR with Asn substituted for one of the His ligands of the heme-Fe in the Cyt *b*. This NR retained BPB:NR activity but lacked activity with reduced MV as electron donor (Meyer et al., 1991). Recently we were able to cleave a recombinant form of NADH:NR into a 60-kDa fragment with MV:NR activity and a 40-kDa fragment with ferricyanide and Cyt *c* reductase activity using mild trypsin digestion (R. Dubois-Dauphin and W. H. Campbell, unpublished). The key was the separation of the trypsin-digestion fragments of NR by immunoaffinity chromatography on monoclonal antibody Zm2,69 Sepharose, which binds the Cyt *b* of holo-NR and its CbR and MoR fragments (Mertens et al., 2000). These results strongly refute the earlier conclusions that the heme-Fe is involved in MV:NR activity and have important implications for the mechanism of reversible NR inhibition by the in vivo regulatory system which will be discussed below.

II. Transcriptional Control of Nitrate Reductase and Other Nitrogen Metabolism Genes

A. Genes Encoding Nitrate Reductase and Other Proteins

Nitrate is the key regulator of NR and NiR transcription in most plants. This makes metabolic sense because there is no need to produce enzymes for metabolism of nitrate unless the plant detects this substrate in its environment. Plants maintain a highly sensitive nitrate detection system, which is constitutively expressed as shown by various experiments with protein synthesis inhibitors (Gowri et al., 1992). This system has been called the primary response (Redinbaugh and Campbell, 1991). It has been shown that the NR and NiR genes in roots and leaves are specifically induced by nitrate concentrations as low as 1 μM within 15 min exposure to the ion. Also induced in roots are plastidic components associated with the provision of reducing power to assimilate nitrate and nitrite, i.e., ferredoxin, ferredoxin $NADP^+$ reductase, glucose-6-phosphate dehydrogenase, and 6-phosphogluconate dehydrogenase (Redinbaugh and Campbell, 1993; Ritchie et al., 1994; Matsumara et al., 1997; Redinbaugh and Campbell, 1998).

In addition, genes associated with assimilation of ammonium into amino acids in plant roots are also induced to higher levels of gene expression by nitrate. GS, GOGAT, phosphoenolpyruvate carboxylase, pyruvate kinase, citrate synthase, and isocitrate dehydrogenase are the key genes activated in response to nitrate (Redinbaugh and Campbell, 1993; Sakakibara et al., 1997; Stitt, 1999). This provides for an enhanced capacity to synthesize glutamate and glutamine probably both by an increase in the levels of GS and GOGAT activity and an increase in capacity to produce organic acids, especially 2-OG. Stitt and coworkers (Scheible et al., 1997; Stitt, 1999) have also suggested that starch synthesis may be slowed in leaves by nitrate suppressing the gene expression level for ADP-glucose pyrophosphorylase, a key enzyme of starch biosynthesis. In fact, considerable evidence exists for a linkage between nitrate stimulation of gene expression and the effects of sugar levels on transcripts (McMichael et al., 1995; MacKintosh, 1998; Stitt, 1999; Cotelle et al., 2000). Although the regulation of sugar and nitrate responses have some similarity, the molecular mechanisms governing interaction of two systems of gene control are not completely clear at this time.

A recent microarray analysis of nitrate-induced gene expression in *Arabidopsis* plants using more than 5000 genes and clones has revealed that there are 40 transcripts strongly induced (Wang et al., 2000). Among the 40 transcripts induced by short-term low nitrate treatment were those expected, including a nitrate transporter, NR, NiR, GS, GOGAT, ferredoxin, ferredoxin $NADP^+$ reductase, glucose-6-phosphate dehydrogenase, 6-phosphogluconate dehydrogenase, phosphoenolpyruvate carboxylase, and uroporphyrin methyltransferase involved in siroheme biosynthesis (a cofactor of NiR). Also induced, however, were transcripts encoding proteins less obviously involved in N assimilation. These included a senescence-associated protein, putative sugar transporter and auxin-induced protein, histidine decarboxylase, transaldolase, calcium antiporters, chloroplast malate dehydrogenase, hemoglobin, a transcription factor, several protein kinases, and several transferases and methyltransferases. Induction of oxidative pentose phosphate cycle enzymes may suggest that the biosynthesis of nucleotides is up regulated in the presence of nitrate, which is probably needed to support developmental changes. The hemoglobin induced by nitrate may be related to the potential involvement of nitric oxide (NO) with regulation of plant functions since NO can be synthesized from nitrite produced from nitrate by NR (Yamasaki and Sakihama, 2000; Chapter 4, Meyer and Stöhr). Upon longer term treatment at higher nitrate, a number of other unexpected transcripts were detected at elevated levels (Wang et al., 2000). Different induction patterns were observed for the various genes, which suggests complexity of the plant's response to nitrate. Microarray analysis provides a snap shot of a metabolic state with respect to transcript levels and complements analysis of enzyme activity. Obviously, the power of microarray analysis in evaluating many genes simultaneously is adding to the list of nitrate-responsive genes: this development will considerably extend the complexity of reactions considered to participate in 'N metabolism.'

It is logical that nitrate acts via a transacting protein factor in stimulating expression of these genes (Campbell, 1988). Coordinated transcriptional expression of such a large set of genes indicates that the 'nitrate-response' transacting protein is activating each of these promoters (Redinbaugh and Campbell, 1991). However, this has not yet been shown nor has a nucleotide sequence in these genes been shown to have a common sequence or motif, except for NR and NiR, as will be described below. Although a number of suggestions have been made concerning the signal transduction pathway between nitrate and nitrate-response gene expression, little is known of this system. There are perhaps three possible 'mechanisms' by which nitrate could act. One is that nitrate binds to a receptor protein at the cell surface, which transmits the signal to the cell possibly via a G-protein and protein phosphorylation (Chandok and Sopory, 1996), resulting in activation of the nitrate transacting factor protein. A second is that nitrate enters the cell and binds to an internal 'receptor' which activates the transacting factor protein. A third possibility, involving NO production at the plasma membrane, is discussed in Chapter 4 (Meyer and Stöhr) of this volume. These aspects of nitrate signaling require additional investigation.

Furthermore, there are a number of changes in plants in addition to induction of metabolic genes when nitrate is detected, especially in roots. These include an increase in respiration and stimulation of root branching as well as root hair development (Redinbaugh and Campbell, 1991; Stitt, 1999). These secondary responses suggest that nitrate also induces the expression of regulatory proteins which stimulate general changes in the plant adapting it to a 'nitrate mode' or metabolic state. One such regulatory factor may be the MADS-type of transcription factor which is induced by nitrate in *Arabidopsis* roots (Zhang and Forde, 1998). However, it should be noted that, in the recent microarray analysis of nitrate-induced gene expression in *Arabidopsis*, this gene was suppressed by high nitrate treatment along with an ammonium transporter (Wang et al., 2000).

MADS-type transcription factors are mainly involved with flower development, which suggests the nitrate-induced root protein is a candidate for regulation of lateral root growth. Transgenic anti-sense plants lacking the transcription factor failed to response to nitrate by making lateral roots (Zhang et al., 1999). Lateral root growth is also governed by a signal factor from plant shoots which is inhibitory and prevents the emergence of the lateral roots when high nitrate is present in the plants (Stitt, 1999). Thus, it appears that the local effect of nitrate as a signal is to stimulate lateral root development, but an overriding effect is found if high tissue nitrate is present.

Apparently not under the control of nitrate are genes for the enzymes involved in production of Mo-MPT by plants, which appear to be constitutively expressed (Mendel, 1997). There are four enzymes which require Mo-MPT or a modified version of this cofactor: NR, xanthine dehydrogenase, aldehyde oxidase, and sulfite oxidase, which has recently been cloned from plants. The aldehyde oxidase catalyzes a step in abscisic acid biosynthesis and the important role of this phytohormone in regulation of wilting may account for the constitutive expression of Mo-MPT genes. There are seven genes involved in the Mo-MPT biosynthetic system and considerable progress in the characterization of the enzymes of the pathway has been made recently (Schwarz et al., 1997).

B. Control of Nitrate Reductase Gene Expression

Beyond the primary control of NR expression by nitrate, there are many factors modulating the transcript level. These factors include phytohormones, light, nutritional status, and drought (Crawford, 1995). Some of these factors assert strong enough control to overcome the stimulation of gene expression by nitrate. For example, NR gene expression in etiolated plants is not induced by nitrate; however, this has been attributed more to carbohydrate availability than to the requirement for a light signal, as shown by stimulation of gene expression by supplying sugar along with nitrate (Cheng et al., 1992). On the other hand, a role for phytochrome in NR gene expression has been shown. In addition, NR gene expression may be regulated by the circadian rhythm of the plant, which suggests that 'clock' genes influence the response to nitrate. All of these factors stimulating or suppressing the level of transcription require that the NR transcript be labile and rapidly degraded, so that changes in the rate of transcription can effectively control the steady-state level of NR mRNA. However, detailed studies of NR transcript stability have not been done. Thus, plants growing in a natural environment appear to have a rhythmic response to nitrate which may be controlled to some extent by a rhythm in nitrate uptake but also by the system controlling the general rhythm of the plant's metabolism. Metabolic control of NR gene expression appears to be governed by the ratio of Gln to Glu, which is linked to the photosynthetic capacity of the plant and the availability of C skeletons for biosynthesis of amino acids (Stitt, 1999).

As mentioned above, NR is a highly efficient catalyst and plants appear to produce more of the enzyme than is required to meet their needs for reduced N. Thus, it appears that regulation of the expression of the NR gene is tuned to the need of the plant for an 'adequate' level of NR enzyme by a combination of factors. However, post-translation regulation of NR activity is also complex, as described below, and this system is superimposed on the control at the transcription level, which illustrates the sophistication of the plant in controlling N metabolism in relation to photosynthesis and other environmental conditions for optimum growth. To summarize, transcriptional control of NR gene expression and the steady-state level of the NR transcript appear to be set up to provide an excess of NR enzyme which permits fine control of the enzyme's activity to adapt the level of nitrate reduction to the needs of the plant in various metabolic states encountered over a given day. This concept is illustrated by the result that constitutive expression of NR is compatible with normal growth and development of a plant system (Vincentz and Caboche, 1991; Kaye et al., 1997).

C. The Nitrate Box

Defining the nucleotide sequence where a nitrate-stimulated transacting factor binds in the promoter of NR and NiR genes has been difficult. The first identification of a nitrate box was achieved in studies of the promoter of spinach NiR (Rastogi et al., 1993, 1997). Parallel studies with NR genes from *Arabidopsis* using transgenic tobacco plants revealed regions with nitrate-response sequences in the 5′ region of both genes (Lin et al., 1994). More detailed analysis of these nucleotide sequences in the *Arabidopsis* NR genes by linker scanning has resulted in a definition of the nitrate box, which is referred to as the NP motif (Hwang et al., 1997). In the *Arabidopsis* NR1 gene promoter, nucleotides –57 to –46, TTTATTTACTCA, and nucleotides –110 to –99, ATTAAAAGTCA, and in the NR2 promoter, nucleotides –162 to –151, TTAATTAAGTCA, were shown to specifically bind proteins from nuclear extracts of nitrate-induced tobacco leaves. In addition, the nuclear protein(s) binding to the first NR1 nitrate box sequence were constitutively expressed in tobacco leaves (Hwang et al., 1997). This nitrate box motif was identified in NR and NiR promoters from a variety of plants. The general nitrate box sequence is a series of A or T nucleotides followed by ACTCA or AGTCA.

III. Post-Translational Control of Nitrogen Metabolism Enzymes

A. Nitrate Reductase Biosynthesis and Turnover

The regulation of the activity of an enzyme can be achieved by activation of existing protein or biosynthesis of new protein. Substrate levels, of course, also influence the activity of an enzyme and for NR the cytoplasmic level of nitrate and NADH are important in determining the amount of nitrate reduced to nitrite (Chapter 5, Kaiser, et al.). Prior to the isolation of the NR gene, it was shown that NR was synthesized de novo in plants in response to application of nitrate (Remmler and Campbell, 1986). Subsequently, it was shown that the NR transcript was first made in response to nitrate and then the active enzyme was synthesized. For effective regulation, NR protein must be rapidly degraded by proteolysis. Since NR is well known to be labile in vitro, many studies have been done of NR degradation in vivo. Thus, the basic level of NR activity is controlled by de novo biosynthesis of NR protein on a daily basis and subsequent degradation, at least in part, of this protein. This might appear to be highly wasteful of plant resources and energy; however, it must be remembered that NR is a highly effective catalyst and so plants contain only very small amounts of NR protein. While this irreversible biosynthesis/protein turnover mechanism can account, at least in part, for post-translational regulation of NR activity, it did not rule out the possibility of additional controls such as reversible inhibition and activation. In fact, evidence was presented in early studies for a reversible light-mediated mechanism controlling NR activity (Remmler and Campbell, 1986), and current knowledge of this control is discussed in the next section

B. Nitrate Reductase Phosphorylation and Inhibition by 14-3-3 Binding Protein

Early work by Kaiser and Spill (1991) suggested that NR was phosphorylated in vivo in response to rapid changes in physiological conditions. Two groups then showed that indeed NR was phosphorylated in the dark to a greater extent than in light (Huber et al., 1992; MacKintosh, 1992). This was followed by identification of a Ser residue in Hinge 1 as the site of regulatory phosphorylation (Douglas et al., 1995; Bachmann et al., 1996a; Su et al., 1996). This Ser residue is at position 534 in the amino acid sequence of *Arabidopsis* NR2 and 543 in spinach NR. Specific protein kinases involved in NR phosphorylation have been identified and isolated (McMichael et al., 1995; Douglas et al., 1996). Protein phosphatases potentially involved in removal of the regulatory phosphate have been identified as type 2A, which are inhibited by microcystin and okadaic acid (Huber et al., 1992; MacKintosh, 1992).

Eventually, it was shown that phosphorylation of NR was not sufficient to inhibit the activity and proteins in plant extracts were identified as inhibitors of phosphorylated NR. Characterization of these proteins showed they are members of the binding protein family known as 14-3-3, for which 14 different isoforms have been found for *Arabidopsis* (Bachmann et al., 1996b; Huber et al., 1996; Moorhead et al., 1996; Bachmann et al., 1998; MacKintosh, 1998). Inhibition of NR activity by 14-3-3 requires a divalent cation such as Mg^{2+} and this was originally thought to be mediating the formation of the complex between the binding protein and NR. However, most recently the metal ion was shown to bind to 14-3-3 and activate the binding protein to form a complex with NR (Athwal et al., 1998, 2000). In fact, polyamines were shown to be good replacements for the metal ion with spermidine being more effective than Mg^{2+} at activating 14-3-3 for binding and inhibition of NR activity (Provan et al., 2000).

The in vivo regulation of NR by phosphorylation and binding of 14-3-3 is proposed to be a light regulated process (Fig. 2). In the dark, NR is phosphorylated by a specific protein kinase. The phosphorylated enzyme then binds 14-3-3 in the presence of polycations, resulting in inhibition of NR activity. Some recent evidence suggests that phospho-NR with 14-3-3 is degraded by a proteinase (Weiner and Kaiser, 1999). However, some inhibited NR remains in leaves and is reactivated in the light

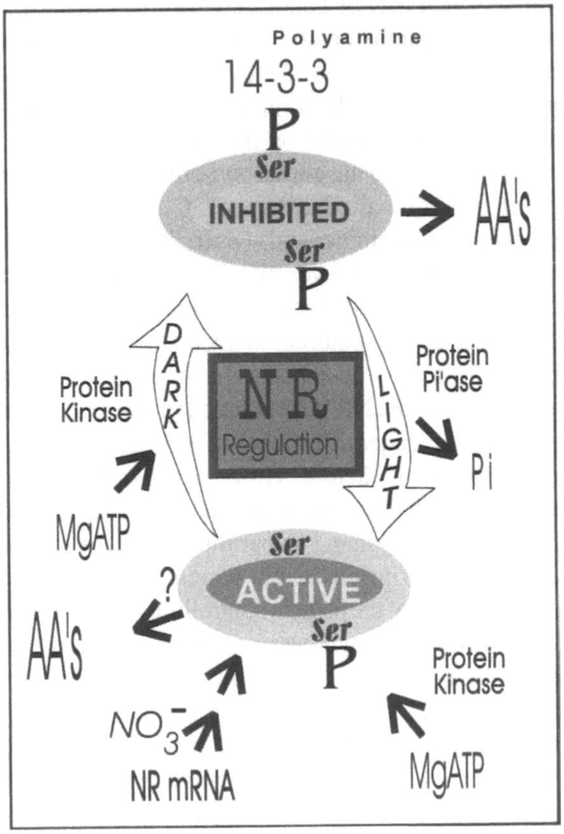

Fig. 2. Model for NR regulation by phosphorylation and binding of activated 14-3-3. In the light, NR is synthesized de novo after nitrate induction and may be phosphorylated on a non-regulatory Ser and this form of the enzyme may be degraded. In the dark, protein kinase(s) catalyze the phosphorylation of the regulatory Ser residue in Hinge 1. After activation by polyamines or other polycations, 14-3-3 binds to NR at the regulatory phospho-Ser, which results in inhibition of NR activity. NR may either be reactivated by release of 14-3-3 and hydrolysis of the regulatory phospho-ester by protein phosphatase 2A or be degraded. AA's, amino acids; Pi'nase, phosphatase; Ser, serine

by the action of a protein phosphatase. In addition, it is clear that NR mRNA levels are increased by light and this results in de novo synthesis of new NR protein which may be phosphorylated on a Ser not associated with regulation (Fig. 2). NR protein not involved in the reversible regulation cycle may also be degraded by a proteinase in leaves.

One of the features of 14-3-3 proteins is that they are homo-dimers with two binding sites which can be filled at the same time, at least by relatively small molecules (Petosa et al., 1998). Thus, one of the questions surrounding the interaction of 14-3-3 with NR is: can both 14-3-3 binding sites in one dimer bind to NR at once? To answer this question the 3-D

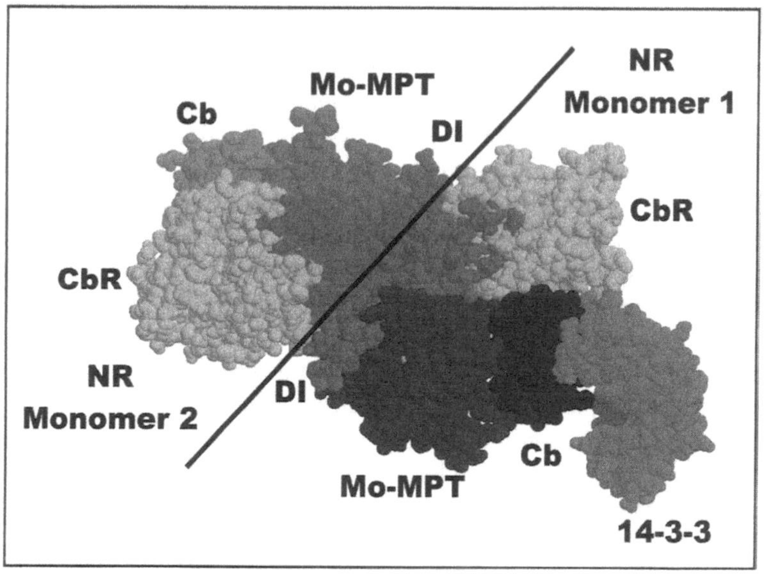

Fig. 3. Model of 14-3-3 bound to phospho-NR. The homo-dimer model of NR was modified to contain a raf peptide sequence and 14-3-3 was docked (Petosa et al., 1998; Campbell, 1999). Only one of the two binding sites in the 14-3-3 dimer is bound to NR. The major NR domains and fragments are: the C-terminal Cyt *b* reductase fragment (CbR, light gray in both monomers), the Cyt *b* domain (Cb, colored more darkly in monomer 1), the dimer interface domain (DI), and the molybdenum-molybdopterin cofactor (Mo-MPT) domain. The N-terminal region of NR is not shown, while the dimer interface is emphasized by a black line. See Color Plate 2 for a version with color-coded domains.

model of an animal 14-3-3 (Petosa et al., 1998) was docked on the working model of NR, which was recently described (Campbell, 1999). The model of the docked 14-3-3-NR complex suggests that only one of the 14-3-3 binding sites can bind to NR at once, but that two independent 14-3-3 molecules could bind to the dimeric NR (Fig. 3). However, this model does not rule out the possibility that the docked 14-3-3 binds to another NR dimer or another protein with a 14-3-3 recognition sequence. Thus, binding of 14-3-3 could result in aggregation of NR, which might result in rapid degradation of the complex (Weiner and Kaiser, 1999). It is also interesting to note that replacement of the regulatory Ser with an Asp residue and additional modifications of the NR result in 14-3-3 binding in the absence of phosphorylation (Kanamaru et al., 1999). The finding that some isoforms of 14-3-3 are more effective as inhibitors of NR than others (Bachmann et al., 1998) suggests that residues outside the binding site for the phospho-Ser sequence, which are on the surface of the 14-3-3 dimer, may also be involved in mediating the binding strength of the complex between phospho-NR and 14-3-3. Thus, secondary interactions between NR and 14-3-3 may be important features to study.

The overall implications for the regulation of NR activity by 14-3-3 are quite substantial. It has been found that 14-3-3 is involved in regulation of many cellular processes including the cell cycle, metabolism, cell signaling and cell survival (Moorhead et al., 1999). In *Arabidopsis* cells, 14-3-3 was found to bind to NR, glyceraldehyde-3-phosphate dehydrogenase, a calcium-dependent protein kinase, sucrose-phosphate synthase and glutamyl-tRNA synthetase. When the cells were starved for sugars, the binding of 14-3-3 was lost and the proteins were proteolytically degraded. These findings and the results of others suggest that 14-3-3 is involved in global regulation of not only N metabolism but also C metabolism in plants (Cotelle et al., 2000). When this concept is combined with the recent finding that polyamines may be involved with activating 14-3-3 for binding to NR (Provan et al., 2000), one begins to recognize 14-3-3 as a link between cell growth and development and the basic metabolic pathways.

C. Mechanism of Inhibition of Nitrate Reductase by 14-3-3

It is interesting to make a more detailed analysis of

the mechanism of inhibition of NR activity by 14-3-3. Assays of the partial activities impacted by 14-3-3 showed that only the MV-NR activity was inhibited along with the NADH:NR activity (Bachmann et al., 1996a). At the time this study was done, the MV-NR activity of NR was thought to depend on electron transfer from the enzyme's Cyt *b* to the Mo-MPT-containing nitrate-reducing active site. Thus, it was suggested that electron transfer from the heme-Fe to Mo-MPT was inhibited by binding of 14-3-3 to phospho-NR. I refined this suggestion by adding that the redox potential of the heme-Fe in NR is conformationally dependent and binding of 14-3-3 might prevent the Cyt *b* from taking on a conformation with the highly negative redox potential observed for this group in holo-NR (Campbell, 1999). Now that it has been definitively shown that MV:NR activity does not depend on the Cyt *b*, the previous explanations of 14-3-3 inhibition are probably inadequate. In the end, the difference between reduced dye electron donors, namely MV is a positively charged electron donor and BPB is negatively charged, probably accounts for situations where BPB works and MV does not. However, it seems likely that binding of 14-3-3 to phospho-NR causes a local conformation change which inhibits electron transfer from the heme-Fe of the enzyme's Cyt *b* to Mo-MPT and this may be the mechanism of 14-3-3 inhibition of NADH:NR activity. The conformation change may also prevent the cationic reduced MV from gaining access to the Mo-MPT, while it has no impact on the access of anionic reduced BPB to the enzyme's nitrate-reducing active site. Thus, the loss of MV:NR activity is only indirectly coupled to the loss of NADH:NR activity in both the inhibition by 14-3-3 binding and the mutant NR previously discussed (Meyer et al., 1991; Bachmann et al., 1996a). Clearly, there is a need for direct experimental investigation of the mechanism of inhibition of NR by 14-3-3 using, for example, fast reaction kinetic analysis. If phospho-NR aggregates in the presence of 14-3-3, as I have suggested above, this could be determined by any of several methods and may offer an excellent explanation of 14-3-3 inhibition of NADH:NR activity and the reason that complete inhibition is not observed even when 14-3-3 is in excess (Bachmann et al., 1996a).

Work has also been done on the impact of the N-terminal region of NR (as defined in Fig. 1) on the inhibition of NR by 14-3-3. Meyer and coworkers (Pigaglio et al., 1999) constructed a tobacco NR form with most of the N-terminal region deleted and found that 14-3-3 was a much less effective inhibitor of NR activity. This implied that the N-terminal region was involved in binding of 14-3-3 and that the N-terminal deleted NR did not bind 14-3-3. Subsequent analysis of the N-terminal deleted NR in purified form has shown that 14-3-3 does bind to the modified NR in the presence of Mg^{2+} (Provan et al., 2000). When the N-terminal deleted NR was purified, it was found to have substantially lower NR activity relative to its Cyt *c* reductase activity when compared to wild type NR. Thus, it appears that N-terminal deletion has an impact on the stability of the nitrate-reducing activity of the enzyme after purification, which is interesting biochemically but implies nothing about the mechanism of 14-3-3 inhibition of NR.

IV. Protein Kinases and Control of Carbon and Nitrogen Metabolism

Originally, phosphorylation of NR was found to be catalyzed by Ca^{2+}-dependent protein kinases (McMichael et al., 1995; Douglas et al., 1996). CDPK are a large group of enzymes with different structures and responses to Ca^{2+} (Harmon et al., 2000). The CDPK are probably the best characterized of all plant protein kinases and are known to be involved in regulation of a number of different plant processes including growth and development. CDPK forms which are more or less specific for NR, sucrose phosphate synthase, 3-hydroxy-3-methylglutaryl-CoA reductase, and sucrose synthase, have been identified, purified and cloned in many cases (Harmon et al., 2000). Thus, a clear regulatory linkage between environmental and developmental signals via cellular Ca^{2+} concentration and CDPK and the regulation of these enzymes has been established (Fig. 4).

However, in the original work on NR phosphorylation, a protein kinase not dependent on Ca^{2+} was also found (McMichael et al., 1995; Douglas et al., 1996). More recent work has shown that this enzyme is of the type known as SNF1-related protein kinases, which are called SnRK1 (Sugden et al., 1999). SnRK1 protein kinases are related to animal and yeast protein kinases which are activated by AMP and respond to cellular depletion of ATP. Their function has been described as acting as a 'fuel gauge' for the cell that springs into action when the cell is stressed by low levels of nutrients to stimulate C metabolism (Halford and Hardie, 1998; Hardie et al., 1998). SnRK1 from

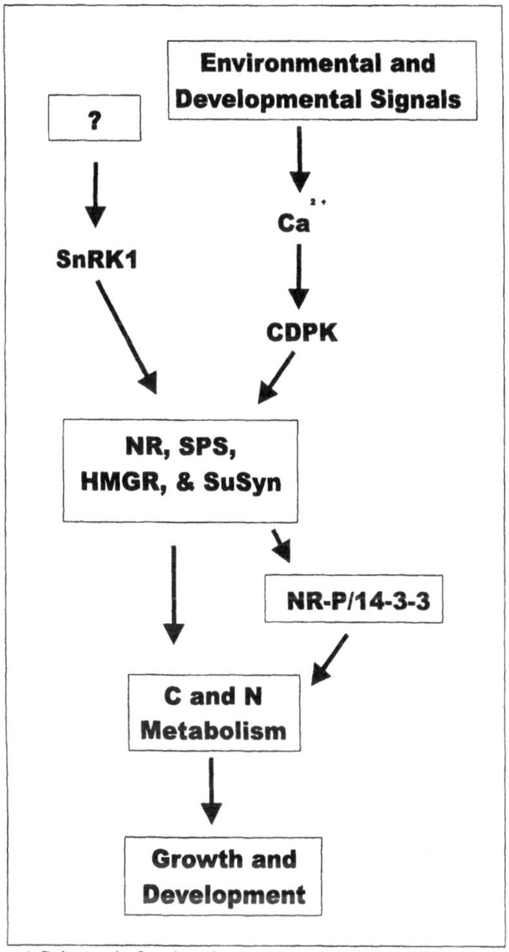

Fig. 4. Schematic for signal transduction pathways for regulation of nitrogen and carbon metabolism in plants. SnRK1 and CDPK are SNF1-related and calcium-dependent protein kinases, respectively. SPS, HMGR and SuSyn are sucrose phosphate synthase, 3-hydroxy-3-methylglutaryl-CoA reductase, and sucrose synthase, respectively. CDPK, calcium-dependent protein kinase; SnRK1, plant protein kinase related to yeast sucrose non-fermenting (SNF1) kinases.

spinach and wheat have now been shown to catalyze phosphorylation of NR, sucrose phosphate synthase, 3-hydroxy-3-methylglutaryl-CoA reductase, and sucrose synthase (Sugden et al., 1999). Thus, the SnRK1 protein kinases appear to be a parallel system for regulation of the same enzymes which are regulated by CDPK (Fig. 4). However, in the case of plants it is not known if the SnRK1 are activated by AMP and it is not clear what cellular signals are involved in turning on these protein kinases. It has been shown that the SnRK1 involved in regulation of sucrose phosphate synthase is inhibited by glucose-6-phosphate (Toroser et al., 2000). It has been suggested that SnRK1 are 'global regulators of carbon metabolism' (Halford and Hardie, 1998; Hardie et al., 1998). Consequently, it appears that two protein kinases systems operate in plant cells for controlling C and N metabolism. Presumably, these two regulatory cascades operate independently and respond to different cellular conditions and signals.

V. Future Prospects for the Control of Nitrogen Metabolism

This chapter has identified several areas in the control of N metabolism where there is a lack of knowledge. One is nitrate signaling where the components of the signal transduction mechanism have not been characterized. Does nitrate bind at the cell surface to a receptor to start the process or must nitrate enter the cell? How is that signal transmitted to the nucleus to turn-on the constitutive transacting factor(s) which bind to the nitrate boxes in the genes activated by nitrate?

Next the mechanism of inhibition of NR activity by 14-3-3 when it binds to the phosphorylated enzyme needs to be studied more to gain understanding of this process. Considerable effort has been focused on the protein kinases and 14-3-3 molecules involved in NR regulation, but this area has many open questions also. In particular, it needs to be established how the two classes of protein kinases which phosphorylate NR interact. This might be best addressed by constructing antisense plants to suppress one type of protein kinase and observe how the transgenic plants regulate NR and related enzymes also controlled by this type of protein kinase. Another interesting area for further investigation lies in the interaction between phospho-NR and 14-3-3: what factor(s) cause the dissociation of the complex which results in dephosphorylation of NR and its reactivation in the light? Or is NR mostly degraded once it has been complexed with 14-3-3?

Finally, the recent demonstration that NR catalyzes in vitro production of nitric oxide (NO) has resulted in the suggestion that NR is perhaps itself involved in cellular regulation (Yamasaki and Sakihama, 2000). NR seems to catalyze NO production when nitrate is largely depleted and nitrite is still available. Since NR is also capable of catalyzing production of superoxide under these same conditions, this could result in the production of the highly toxic peroxynitrite, which has also been detected (Yamasaki and Sakihama, 2000). Furthermore, since there is evidence now accumulating that NO can act as a hormone in plants like it does in animals (Durner and Klessig, 1999), NR may be a source of a regulatory signal. However, tight control of NR activity under the cellular conditions leading to catalysis of NO production would appear to be necessary to avoid the formation of the toxic peroxynitrite byproduct. Thus, the regulation of NR activity may serve the plant in several different ways at the same time. Clearly, we need to gain greater understanding of NO metabolism in plants and the enzymes which catalyze production of this potential hormone, and this subject is discussed further in Chapter 4 (Meyer and Stöhr) and in Chapter 13 (Millar et al.).

Acknowledgment

Research in the author's laboratory is currently supported by National Science Foundation grant MCB-9727982.

References

Athwal GS, Huber JL and Huber SC (1998) Phosphorylated nitrate reductase and 14-3-3 proteins. Site of interaction, effects of ions, and evidence for an AMP-binding site on 14-3-3 proteins. Plant Physiol 118: 1041–1048

Athwal GS, Lombardo CR, Huber JL, Masters SC, Fu H and Huber SC (2000) Modulation of 14-3-3 protein interactions with target peptides by physical and metabolic effectors. Plant Cell Physiol 41: 523–533

Bachmann M, Shiraishi N, Campbell WH, Yoo BC, Harmon AC and Huber SC (1996a) Identification of Ser-543 as the major regulatory phosphorylation site in spinach leaf nitrate reductase. Plant Cell 8: 505–517

Bachmann M, Huber JL, Liao PC, Gage DA and Huber SC (1996b) The inhibitor protein of phosphorylated nitrate reductase from spinach (*Spinacia oleracea*) leaves is a 14-3-3 protein. FEBS Lett 387: 127–131

Bachmann M, Huber JL, Athwal GS, Wu K, Ferl RJ and Huber SC (1998) 14-3-3 proteins associate with the regulatory phosphorylation site of spinach leaf nitrate reductase in an isoform-specific manner and reduce dephosphorylation of Ser-543 by endogenous protein phosphatases. FEBS Lett 398: 26–30

Campbell WH (1988) Nitrate reductase and its role in nitrate assimilation in plants. Physiol Plant 74: 214–219

Campbell WH (1996) Nitrate reductase biochemistry comes of age. Plant Physiol 111: 355–361

Campbell WH (1999) Nitrate reductase structure, function and regulation: Bridging the gap between biochemistry and physiology. Annu Rev Plant Physiol Plant Mol Biol 50: 277–303

Campbell WH and Kinghorn JR (1990) Functional domains of assimilatory nitrate reductases and nitrite reductases. Trends Biochem Sci 15: 315–319

Chandok MR and Sopory SK (1996) Phosphorylation/dephosphorylation steps are key events in the phytochrome-mediated enhancement of nitrate reductase mRNA levels and enzyme activity in maize. Mol Gen Genet 251: 599–608

Cheng CL, Acedo GN, Cristinsin M and Conkling MA (1992) Sucrose mimics the light induction of *Arabidopsis* nitrate reductase gene transcription. Proc Natl Acad Sci USA 89: 1861–1864

Cotelle V, Meek SE, Provan F, Milne FC, Morrice N and MacKintosh C (2000) 14-3-3s regulate global cleavage of their diverse binding partners in sugar-starved *Arabidopsis* cells. EMBO J 19: 2869–2876

Crawford N (1995) Nitrate: Nutrient and signal for plant growth. Plant Cell 7: 859–868

Douglas P, Morrice N and MacKintosh C (1995) Identification of a regulatory phosphorylation site in the hinge 1 region of nitrate reductase from spinach (*Spinacea oleracea*) leaves. FEBS Lett 377: 113–117

Douglas P, Moorhead G, Hong Y, Morrice N and MacKintosh C (1998) Purification of a nitrate reductase kinase from *Spinacea oleracea* leaves, and its identification as a calmodulin-domain protein kinase. Planta 206: 435–442

Durner J and Klessig DF (1999) Nitric oxide as a signal in plants. Curr Opin Plant Biol 2: 369–374

Dwivedi UP, Shiraishi N and Campbell WH (1994) Identification of an 'essential' cysteine of nitrate reductase via mutagenesis of its recombinant cytochrome *b* reductase domain. J Biol Chem 269: 13785–13791

Forde BG (2000) Nitrate transporters in plants: Structure, function and regulation. Biochim Biophys Acta 1465: 219–235

George GN, Mertens JA and Campbell WH (1999) Structural changes induced by catalytic turnover at the molybdenum site of *Arabidopsis* nitrate reductase. J Am Chem Soc 121: 9730–9731

Gowri G, Ingemarsson B, Redinbaugh M and Campbell WH (1992) Nitrate reductase transcript is expressed in the primary response of maize to environmental nitrate. Plant Mol Biol 18: 55–64

Halford NG and Hardie DG (1998) SNF1-related protein kinases: Global regulators of carbon metabolism in plants? Plant Mol Biol 37: 735–748

Hardie DG, Carling D and Carlson M (1998) The AMP-activated/SNF1 protein kinase subfamily: Metabolic sensors of the eukaryotic cell? Annu Rev Biochem 67: 821–855

Harmon AC, Gribskov M and Harper JF (2000) CDPKs—a kinase for every Ca^{2+} signal? Trends Plant Sci 5: 154–159

Huber JL, Huber SC, Campbell WH and Redinbaugh MG (1992) Reversible light/dark modulation of spinach leaf nitrate reductase activity involves protein phosphorylation. Arch Biochem Biophys 296: 58–65

Huber SC, Bachmann M and Huber JL (1996) Post-translational regulation of nitrate reductase activity: A role for Ca^{2+} and 14-3-3 proteins. Trends Plant Sci 1: 432–438

Hwang CF, Lin Y, D'Souza T and Cheng CL (1997) Sequences necessary for nitrate-dependent transcription of *Arabidopsis* nitrate reductase genes. Plant Physiol 113: 853–862

Kaiser WM and Spill D (1991) Rapid modulation of spinach leaf nitrate reductase by photosynthesis. II. In vitro modulation by ATP and AMP. Plant Physiol 96: 368–375

Kanamaru K, Wang R, Su W and Crawford NM (1999) Ser-534 in the hinge 1 region of *Arabidopsis* nitrate reductase is conditionally required for binding of 14-3-3 proteins and in vitro inhibition. J Biol Chem 274: 4160–4165

Kaye C, Crawford NM and Malmberg RL (1997) Constitutive non-inducible expression of the *Arabidopsis thaliana* Nia 2 gene in two nitrate reductase mutants of *Nicotiana plumbaginifolia*. Plant Mol Biol 33: 953–964

Kisker C, Schindelin H, Pacheco A, Wehbi WA, Garrett RM, Rajagopalan KV, Enemark JH and Rees DC (1997) Molecular basis of sulfite oxidase deficiency from the structure of sulfite oxidase. Cell 91: 973–983

Lin Y, Hwang CF, Brown JB and Cheng CL (1994) 5′ proximal regions of *Arabidopsis* nitrate reductase genes direct nitrate-induced transcription in transgenic tobacco. Plant Physiol 106: 477–484

Lu G, Campbell WH, Schneider G and Lindqvist Y (1994) Crystal structure of the FAD-containing fragment of corn nitrate reductase at 2.5 Å resolution: Relationship to other flavoprotein reductases. Structure 2: 809–821

Lu G., Lindqvist Y, Schneider G, Dwivedi UN and Campbell WH (1995) Structural studies on corn nitrate reductase. Refined structure of the cytochrome *b* reductase fragment at 2.5 Å, its ADP complex and an active site mutant and modeling of the cytochrome *b* domain. J Mol Biol 248: 931–948

MacKintosh C (1992) Regulation of spinach-leaf nitrate reductase by reversible phosphorylation. Biochim Biophys Acta 1137: 121–126

MacKintosh C (1998) Regulation of cytosolic enzymes in primary metabolism by reversible protein phosphorylation. Curr Opin Plant Biol 1: 224–229

Matsumara T, Sakakibara H, Nakano R, Kimata Y, Sugiyama T and Hase T (1997) A nitrate-inducible ferredoxin in maize roots: Genomic organization and differential expression of two non-photosynthetic ferredoxin apoproteins. Plant Physiol 114: 653–660

McMichael RW, Bachmann M and Huber SC (1995) Spinach leaf sucrose-phosphate synthase and nitrate reductase are phosphorylated/inactivated by multiple protein kinases in vitro. Plant Physiol 108: 1077–1082

Mendel RR (1997) Molybdenum cofactor of higher plants: Biosynthesis and molecular biology. Planta 203: 399–405

Mertens JA, Campbell WH, Skipper L and Lowe DJ (1999) Electron transfer from FAD to heme-Fe in plant NADH:nitrate reductase. In: Ghisla S, Kroneck P, Macheroux P and Sund H (eds) Flavins and Flavoproteins 1999, pp 131–134. Agency for Scientific Publications, Berlin

Mertens JA, Shiraishi N and Campbell WH (2000) Recombinant expression of molybdenum reductase fragments of plant nitrate reductase at high levels in *Pichia pastoris*. Plant Physiol 123: 743–756

Meyer C, Levin JM, Roussel J-M and Rouze P (1991) Mutational and structural analysis of the nitrate reductase heme domain of *Nicotiana plumbaginifolia*. J Biol Chem 266: 20561–20566

Moorhead G, Douglas P, Morrice N, Scarabel M, Aitken A and MacKintosh C (1996) Phosphorylated nitrate reductase from spinach leaves is inhibited by 14-3-3 proteins and activated by fusicoccin. Curr Biol 6: 1104–1113

Moorhead G, Douglas P, Cotelle V, Harthill J, Morrice N, Meek S, Deiting U, Stitt M, Scarabel M, Aitken A and MacKintosh C (1999) Phosphorylation-dependent interactions between enzymes of plant metabolism and 14-3-3 proteins. Plant J 18: 1–12

National Research Council (2000) Clean Coastal Waters: Understanding and Reducing the Effects of Nutrient Pollution. Committee on the Causes and Management of Eutrophication, National Research Council, National Academy Press, Washington, DC

Petosa C, Masters SC, Bankston LA, Pohl J, Wang B, Fu H and Liddington RC (1998) 14-3-3zeta binds a phosphorylated raf peptide and an unphosphorylated peptide via its conserved amphipathic groove. J Biol Chem 273: 16305–16310

Pigaglio E, Durand N and Meyer C (1999) A conserved acidic motif in the N-terminal domain of nitrate reductase is necessary for the inactivation of the enzyme in the dark by phosphorylation and 14-3-3 binding. Plant Physiol 119: 219–230

Provan F, Aksland L-M, Meyer C and Lillo C (2000) Deletion of the nitrate reductase N-terminal domain still allows binding of 14-3-3 proteins but affects their inhibitory properties. Plant Physiol 123: 757–764

Rastogi R, Back E, Schneiderbauer A, Bowsher CG, Moffat B and Rothstein SJ (1993) A 330 bp region of the spinach nitrite reductase gene promoter directs nitrate-inducible tissue-specific expression in transgenic tobacco. Plant J 4: 317–326

Rastogi R, Bate NJ, Sivasankar S and Rothstein SJ (1997) Footprinting of the spinach nitrite reductase gene promoter reveals the preservation of nitrate regulatory elements between fungi and higher plants. Plant Mol Biol 34: 465–476

Ratnam K, Shiraishi N, Campbell WH and Hille R (1995) Spectroscopic and kinetic characterization of the recombinant wild-type and C242S mutant of the cytochrome *b* reductase fragment of nitrate reductase. J Biol Chem 270: 24067–24072

Ratnam K, Shiraishi N, Campbell WH and Hille R (1997) Spectroscopic and kinetic characterization of the recombinant cytochrome *c* reductase fragment of nitrate reductase: identification of the rate limiting catalytic step. J Biol Chem 272: 2122–2128

Redinbaugh MG and Campbell WH (1991) Higher plant responses to environmental nitrate. Physiol Plant 82: 640–650

Redinbaugh MG and Campbell WH (1993) Glutamine synthetase and ferredoxin-dependent glutamate synthase expression in the maize (*Zea mays*) root primary response to nitrate. Plant Physiol 101: 1249–1255

Redinbaugh MG and Campbell WH (1998) Nitrate regulation of the oxidative pentose phosphate pathway in maize root plastids: Induction of 6-phosphogluconate activity, protein and transcript levels. Plant Sci 134: 129–140

Remmler JL and Campbell WH (1986) Regulation of corn leaf nitrate reductase: II. Synthesis and turnover of the enzyme's activity and protein. Plant Physiol 80: 442–447

Ritchie SW, Redinbaugh MG, Shiraishi N, Verba JM and Campbell WH (1994) Identification of a maize root transcript expressed in the primary response to nitrate: Characterization of a cDNA with homology to ferredoxin-$NADP^+$ reductase. Plant Mol Biol 26: 679–690

Sakakibara H, Kobayashi K, Deji A and Sugiyama T (1997) Partial characterization of the signaling pathway for the nitrate-dependent expression of genes for nitrogen-assimilatory enzymes using detached leaves of maize. Plant Cell Physiol 38: 837–843

Scheible W-R, Gonzales-Fontes A, Lauerer M, Muller-Rober B, Caboche M and Stitt M (1997) Nitrate acts as a signal to induce organic acid metabolism and repress starch metabolism in tobacco. Plant Cell 9: 783–798

Schwarz G, Boxer DH and Mendel RR (1997) Molybdenum

cofactor biosynthesis. The plant protein Cnx1 binds molybdopterin with high affinity. J Biol Chem 272: 26811–26814

Smil V (1997) Global Population and the Nitrogen Cycle. Sci Amer 277: 76–81

Stitt M (1999) Nitrate regulation of metabolism and growth. Curr Opin Plant Biol 2: 178–186

Su W, Huber SC and Crawford NM (1996) Identification in vitro of a post-translational regulatory site in the hinge 1 region of *Arabidopsis* nitrate reductase. Plant Cell 8: 519–527

Su W, Mertens JA, Kanamaru K, Campbell WH and Crawford NM (1997) Analysis of wild-type and mutant plant nitrate reductase expressed in the methylotrophic yeast *Pichia pastoris*. Plant Physiol 115: 1135–1143

Sugden C, Donaghy PG, Halford NG and Hardie DG (1999) Two SNF1-related protein kinases from spinach leaf phosphorylate and inactivate 3-hydroxy-3-methylglutaryl-coenzyme A reductase, nitrate reductase, and sucrose phosphate synthase in vitro. Plant Physiol 120: 257–274

Toroser D, Plaut Z and Huber SC (2000) Regulation of a plant SNF1-related protein kinase by glucose-6-phosphate. Plant Physiol 123: 403–412

Vidmar JJ, Zhuo D, Siddiqi MY, Schjoerring JK, Touraine B and Glass AMD (2000) Regulation of high-affinity nitrate transporter genes and high-affinity nitrate influx by nitrogen pools in roots of barley. Plant Physiol 123: 307–318

Vincentz M and Caboche M (1991) Constitutive expression of nitrate reductase allows for normal growth and development of *Nicotiana plumbaginifolia* plants. EMBO J 10: 1027–1035

Vitousek PM, Mooney HA, Lubchenco J and Melillo JM (1997) Human Domination of Earth's Ecosystems. Science 277: 494–499

Wang R, Guegler K, LaBrie ST and Crawford NM (2000) Genomic analysis of a nutrient response in *Arabidopsis* reveals diverse expression patterns and novel metabolic and potential regulatory genes induced by nitrate. Plant Cell 12: 1491–1509

Weiner H and Kaiser WM (1999) 14-3-3 proteins control proteolysis of nitrate reductase in spinach leaves. FEBS Lett 455: 75–78

Yamasaki H and Sakihama Y (2000) Simultaneous production of nitric oxide and peroxynitrite by plant nitrate reductase: in vitro evidence for the NR-dependent formation of active nitrogen species. FEBS Lett 468: 89–92

Zhang H and Forde BG (1998) An *Arabidopsis* MADS box gene that controls nutrient-induced changes in root architecture. Science 279: 407–409

Zhang H, Forde BG (2000) Regulation of *Arabidopsis* root development by nitrate availability. J Exp Bot 51: 51–59

Zhang H, Jennings A, Barlow PW and Forde BG (1999) Dual pathways for regulation of root branching by nitrate. Proc Natl Acad Sci USA 96: 6529–6534

Chapter 4

Soluble and Plasma Membrane-bound Enzymes Involved in Nitrate and Nitrite Metabolism

Christian Meyer
Unité de Nutrition Azotée des Plantes, INRA, 78026 Versailles, France

Christine Stöhr*
Institut für Botanik, Technische Universität Darmstadt, Schnittspahnstr. 10, D-64287 Darmstadt, Germany

Summary 49
I. Introduction 50
II. Nitrate Reduction at the Plasma Membrane 50
A. Structure of Plasma Membrane-Bound Nitrate Reductase 50
B. Influence of External Factors on the Activity of Plasma Membrane-Bound Nitrate Reductase 52
C. Formation of Nitric Oxide at the Plasma Membrane 53
III. Nitrite Transport and Reduction 54
A. Nitrite Transport Across Membranes 54
B. Nitrite metabolism 54
1. Are There Other Enzymes Involved in Nitrite Metabolism in Plants? 55
2. Nitrite Reduction Catalyzed by Nitrite Reductase 55
a. The Source of the Reducing Power Needed for Nitrite Reduction 55
b. Structure and Function of Nitrite Reductase 56
c. Nitrite Reductase Genes and Mutants 57
d. Regulation of Nitrite Reductase Gene Expression 58
IV. Conclusions 59
Acknowledgments 60
References 60

Summary

Cytosolic nitrate reductase has been the subject of numerous studies because it has long been considered the principal site of the regulation of nitrate assimilation. Recently, specific plasma membrane-bound enzymes have been identified, which are able to reduce nitrate as well as nitrite and which exhibit particularly interesting structural and biochemical properties. Other recent studies have demonstrated that nitrite reductase shares its ability to reduce nitrite with the plasma membrane-bound nitrite:NO oxidoreductase in roots and also with cytosolic nitrate reductase. Nitrite reduction catalysed by these enzymes leads to the production of NO. We assess the physiological significance of these reactions in the detoxification of nitrate and nitrite or in the production of a signaling molecule. We also discuss other enzyme activities that may play significant roles in nitrite detoxification, either by reduction to gaseous species or by oxidation to nitrate. The second reaction of

*Author for correspondence, email: stoehr@bio.tu-darmstadt.de

Christine H. Foyer and Graham Noctor (eds): Photosynthetic Nitrogen Assimilation and Associated Carbon and Respiratory Metabolism, pp. 49–62. © 2002 Kluwer Academic Publishers.

nitrate assimilation, the conversion of nitrite to ammonium, can consume a significant proportion of photosynthetic reducing energy, either directly in chloroplasts or indirectly, via the oxidative pentose phosphate pathway. Evidence is presented that plastidic nitrite reductase, the enzyme that catalyses this conversion, might be as finely regulated as nitrate reductase by endogenous factors. The expression of both reductases appears to respond in a similar fashion to carbon and nitrogen metabolites and also to nitrate, though some differences are discussed. In concert with the regulation of the expression of these components, nitrate also controls expression of the enzymatic machinery needed for the supply of reducing power to nitrite reduction, underscoring the importance of this reaction as a sink for reducing power.

I. Introduction

The tight connection between nitrate assimilation and photosynthesis in plants is primarily caused by the demand for energy and C skeletons during assimilation of inorganic N into organic compounds. The balance of these two important pathways, necessary to ensure organic C and N supply on one hand and to avoid accumulation of toxic N compounds on the other hand, is maintained by regulation of the key enzymes involved. In the case of nitrate assimilation, the enzyme subject to the tightest control is cytosolic nitrate reductase (cNR), which catalyzes a two-electron transfer to nitrate resulting in the formation of nitrite (Chapters 3 (Campbell) and 5 (Kaiser et al.)). The view that nitrate reduction takes place exclusively in the cytosol has been modified, following the identification of a plasma membrane-bound nitrate reductase (PM-NR) at the extracellular surface of plasma membranes (see review by Stöhr, 1998 and references therein). Whether formed in the apoplast or in the cytosol, most of the nitrite is subsequently reduced to ammonium by plastidic nitrite reductase (NiR). To date, NiR has attracted less attention than NR, probably because it has long been assumed that NR is the main regulatory and limiting step of the nitrate assimilation pathway. However, mounting evidence suggests that the metabolism of nitrite is also very finely controlled. This chapter will first review the distribution and physiological significance of plasma membrane-bound and soluble nitrate-assimilating enzymes in plants. In the second half of the chapter, recent developments in the understanding of NiR regulation and nitrite metabolism will be considered, including alternative routes of nitrite reduction or oxidation.

Abbreviations: cNR – cytosolic nitrate reductase; Fd – ferredoxin; FNR – ferredoxin-NADP$^+$ oxidoreductase; Glc-6-P – glucose-6-phosphate; GPI – glycosyl-phosphatidylinositol; NiR – nitrite reductase; NR – nitrate reductase; OPPP – oxidative pentose phosphate pathway; PM-NR – plasma-membrane-bound nitrate reductase; SiR – sulfite reductase

II. Nitrate Reduction at the Plasma Membrane

As the border between the plant cell and the environment, the plasma membrane plays an important role in the acquisition of nutrients such as nitrate. It is the site of specific carriers for the uptake of nitrate into the cell, for which process the plasma membrane H^+-ATPase provides the necessary energy. Therefore, the physiological function of extracellular nitrate reduction is not easy to understand. Several explanations are conceivable for this location: i) In vivo the PM-NR may be primarily involved in electron transfer reactions at the plasma membrane that are irrelevant to nitrate assimilation. ii) Nitrate reduction at the plasma membrane may have no metabolic significance but could be important for the nitrate-sensing process. iii) Apoplastic nitrate reduction may be a protective mechanism that leads to the production of gaseous N-compounds. iv) Apoplastic nitrate reduction may contribute to the organic N pool of a cell under certain physiological conditions. The following sections discuss these possibilities and attempt to assess which of them may have relevance in planta.

A. Structure of Plasma Membrane-Bound Nitrate Reductase

Redox reactions at the plasma membrane are receiving increasing attention (Asard et al., 1998) although only a certain number of enzymes has as yet been identified. It is possible that PM-NR, an oxido-reductase, may participate in one or more of these redox reactions. PM-NR reduces nitrate to nitrite using NADH or succinate, depending on its location in leaves or in roots (Stöhr and Ullrich, 1997), and has been demonstrated to be located and attached at the apoplastic surface of the plasma membrane by a glycosyl-phosphatidylinositol (GPI) anchor in both algae (Stöhr et al., 1995b) and the leaves of higher plants (Kunze et al., 1997). In leaves of sugar beet it has been shown that the attachment of the GPI anchor involves the secretory pathway operating via the ER and Golgi apparatus (Kunze et al., 2000). At the cytoplasmic surface of plasma membranes a further form of NR can usually be detected. This NR is a loosely associated, soluble protein, but displays hydrophobic properties which distinguish it from cNR located in the cytosol (Stöhr et al., 1993). By contrast, the form of NR located at the outer surface of the plasma membrane is a true membrane protein and can easily be separated from the cNR forms by temperature-induced phase partitioning with Triton X-114 or with hydrophobic interaction chromatography (Stöhr et al., 1993).

The cNR is composed of several redox centers with individual domains containing the co-factors FAD, heme and molybdopterin, which can independently catalyze partial reactions. As components of PM-NR these domains may also participate in plasma membrane redox reactions. In plasma membrane vesicles purified from roots or leaves, each of the known partial reactions of cNR was demonstrated, indicating that cNR and PM-NR share a similar composition (Stöhr et al., 1993; Kunze et al., 1997; Wienkoop et al., 1999). However, the PM-NR in roots additionally catalyzes electron transfer from succinate to nitrate, resulting in formation of fumarate and nitrite (Stöhr and Ullrich, 1997). Since this property has been observed neither for the cNR in roots or leaves nor for the PM-NR in leaves, there must be structural differences between these enzymes and root PM-NR. Comparison with the mitochondrial membrane-bound succinate dehydrogenase suggests that this electron transfer is likely to be mediated by an FAD-containing protein or by the FAD domain of NR. Since the succinate-dependent reaction of root PM-NR was almost completely lost upon detergent treatment, whereas the PM-NR of leaves and the cNRs were not particularly sensitive to detergents (Stöhr, 1998), a non-covalent association of the FAD domain seems likely.

Western blot analysis with an antibody specific for the N-terminus of tobacco cNR (molybdopterin cofactor-containing domain) enabled the detection of a 63 kDa polypeptide in the plasma membrane protein fraction of roots, whereas a 98 kDa polypeptide was found in protein extracts from the soluble fraction and from leaf plasma membranes (Stöhr, 1998). This information is also consistent with non-covalent association of the molybdopterin cofactor and heme domains of the root PM-NR with either the FAD domain or another flavoprotein mediating the oxidation of succinate and NADH. This notion was supported by results of northern blot analysis, which were also in agreement with the idea of a shorter NR protein lacking the covalently bound FAD domain in root plasma membranes (Wienkoop et al., 2000). In roots and leaves of tobacco three functional transcripts (3.6 kb, 3.1 kb and 1.8 kb) were found to represent NR mRNA. Using specific probes for the transcripts encoding either the FAD or the molybdopterin co-factor domain of NR, it was demonstrated that the smallest transcript was curtailed in the region coding for the FAD domain and might be the transcript encoding root PM-NR.

To gain further information about the putative non-covalently linked FAD protein, biochemical characterization of the reaction of root PM-NR with NADH and succinate was performed. NADH and succinate did not act additively when both electron donors were present during the assay, suggesting that the reactions were mediated by the same enzyme (Stöhr and Ullrich, 1997). However, NADH and succinate must bind at different sites of the protein since malonate, a succinate analogue, was an effective inhibitor of succinate-dependent, but not of NADH-dependent nitrate reduction. Temperature dependence of the complete and of the partial reactions of root PM-NR also indicate differences in the domain composition of the enzyme. Among the partial reactions, that mediated by the heme domain seems to be the temperature limited step of succinate-dependent PM-NR activity, since it showed a very similar temperature course to the overall reaction with an optimum at 50 °C. In contrast, the temperature course of the NADH-dependent overall reaction followed that of the diaphorase activity (FAD domain) with an optimum at 30 °C. This suggests that the succinate binding site might be related to a flavoprotein different from the NADH binding domain (as shown in Fig. 1). This idea is further supported by the fact that ferricyanide reduction, which is known to be mediated by the NR flavin domain, could not be found with succinate. However, a flavoprotein is presumably involved since a two-electron transfer by the NR heme domain is not likely. Likewise, the reduction of cytochrome *c* by

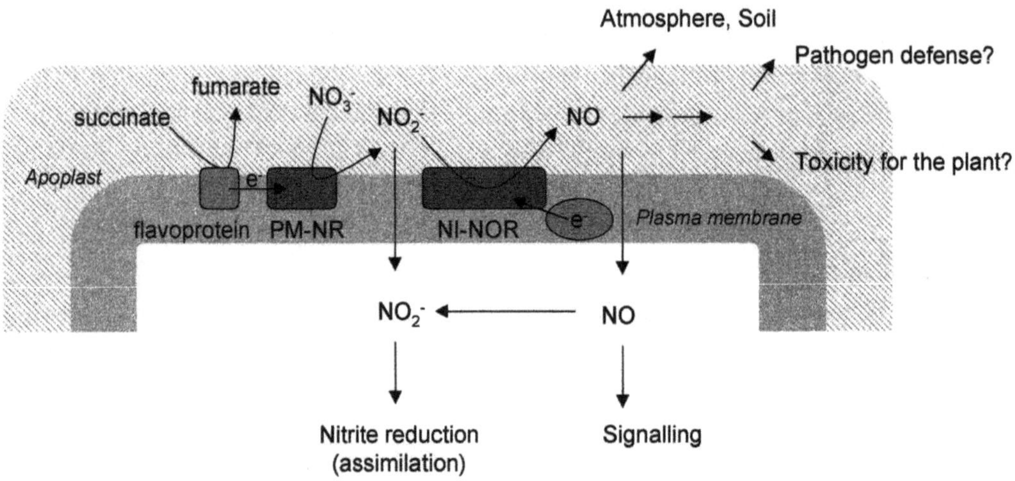

Fig. 1. Model of the proteins and reactions involved in nitrate reduction at the root plasma membrane. PM-NR, plasma-membrane-bound nitrate reductase; NI-NOR, nitrite: nitric oxide oxidoreductase

succinate, which by-passes the NADH-binding subdomain, points to the involvement of a flavoprotein (Mendel and Schwarz, 1999).

The temperature induced changes in activity may be directly caused by conformational changes of the enzyme and, perhaps, also by interaction with other membrane proteins or lipids. The succinate-dependent PM-NR activity showed only one pH optimum of pH 7.0 at 50 °C, but two at the physiological temperature of 30 °C (pH 8.0 and 5.6), contrary to the pH optimum of 7.5 reported for NADH-PM-NR in sugar-beet leaves (Kunze et al., 1997). This suggests a temperature dependent interaction of PM-NR with components of the plasma membrane, resulting in a variation in pH optima and substrate affinities. Thus, the Km of root PM-NR for nitrate varies between 35 and 153 μM at 30 °C depending on pH, with the highest affinity for both substrates (nitrate and succinate) at pH 5.6.

Together, the above data suggest that the FAD-domain of NR does not exist in PM-NR from roots. A likely explanation is that the flavin function is ensured by an unknown non-covalently linked FAD-containing PM-protein (Fig. 1). Thus, the heme or molybdopterin containing domain of PM-NR may interact with other plasma membrane-bound proteins or components involved in electron transfer. Succinate is likely to be a non-limiting electron donor in the root apoplast as roots release organic acids (Marschner and Römheld, 1996; Bar-Yosef, 1996) and succinate was found to be the major organic compound in root exudates (Mench et al., 1988). This indicates that extracellular nitrate reduction is a physiological process in this tissue mainly during the night since our data suggest two independent nitrate assimilating phases in roots, one during daytime in the cytosol by cNR and a second during the dark period in the apoplast by PM-NR (Stöhr and Mäck, 2001). The contribution of cNR and PM-NR to root nitrate assimilation is therefore dependent both on time of day and nitrate supply, and can vary significantly in favor of each NR form.

B. Influence of External Factors on the Activity of Plasma Membrane-Bound Nitrate Reductase

The possible participation of PM-NR in assimilatory nitrate reduction was estimated under specific conditions like nitrate supply at different concentrations and at different times of the day. Changing the external nitrate concentration markedly affected the ratio between cNR and PM-NR in tobacco plants (Stöhr, 1999). Root cNR activity was induced by low nitrate with a maximum specific activity at 5 mM external nitrate concentration (supplied once a day in sand culture), which correlated with the lowest growth rate of the roots but the highest of the shoots. In this condition cNR was the dominating NR in roots. At higher nitrate concentrations the cNR activity in roots was suppressed and the PM-NR activity strongly increased. In roots both NADH-dependent and succinate-dependent PM-NR activity responded to the higher external nitrate concentrations, with a maximum activity at 25 mM nitrate (in the special conditions of sand culture). This high activity level coincided with a lower level of nitrate accumulation in roots and shoots, but also with a decrease in growth parameters. Whereas root PM-NR activity was inversely correlated with the tissue nitrate content, i.e. highest when tissue nitrate was at a minimum, the activity of leaf PM-NR followed the course of nitrate accumulation. The high activity of root PM-NR at high nitrate supply seems to represent a reaction to avoid detrimental nitrate accumulation in the cell rather than to contribute significantly to plant N content.

While cNR is known to reduce nitrate to nitrite at high rates during the light period, the contribution of PM-NR to the organic N pool has been estimated to be of minor importance during daytime. Recently, however, it has been shown that the PM-NR activity of optimally supplied tobacco plants (10 mM nitrate in sand culture) also varies diurnally: the root enzyme shows maximal activity during the night (Stöhr and Mäck, 2001). Since NiR and glutamine synthetase were also highly active in this tissue during the night, PM-NR may contribute significantly to root assimilation of N under these optimal growth conditions.

C. Formation of Nitric Oxide at the Plasma Membrane

Another role of PM-NR could be in nitrate sensing. Indeed, PM-NR has already been shown to be involved in the blue light regulation of nitrate uptake in *Chlorella* (Stöhr et al., 1995a). The succinate-dependent PM-NR in roots of higher plants is a particularly attractive candidate as a nitrate-sensing component since it reacts with both N- and C-compounds, and might thus act as a tuning system between C and N metabolism. Moreover, this enzyme is located in the plasma membrane of root tissue, the first contact zone between the plant and nutrients. A key question is how the signal 'nitrate' might be transduced to the nitrate uptake system or to any other possible target within the cell. Since nitrite is not accumulated under normal conditions, any that is produced by the PM-NR may be metabolized in the cell. Alternatively, the nitrite formed may be consumed by secondary reactions in the apoplast, e.g. reduction to ammonia or gaseous nitrous compounds such as nitric oxide.

During an assay with PM vesicles prepared from tobacco roots the disappearance of nitrite was observed (Stöhr et al., 2001). The possibility that some NiR activity may be associated with the plasma membrane was investigated. Simultaneous reduction of nitrite and formation of ammonia was not found in PM vesicles and the pH dependence of nitrite disappearance was different from that of NiR. The maximum activity of the plasma membrane-bound activity was found at pH 6.1, whereas the soluble nitrite reducing activity was highest at pH 8.0. With regard to the electron donors, both enzymes accepted electrons from the artificial electron donor, reduced methyl viologen. However, with reduced cytochrome *c* as electron donor, only the PM-bound nitrite reduction proceeded, though with a lower rate than with reduced methyl viologen. It was found that the product of this reaction is nitric oxide (NO). Almost all the NO formation activity found in the crude extract was recovered in the microsomal fraction. Further purification of the membrane fraction resulted in a high enrichment of NO formation activity (40-fold) in the plasma membrane fraction (Stöhr et al., 2001), which has never been detected in plasma membrane vesicles from tobacco leaves.

In plants and animals NO is enzymatically produced by NO synthase with NADH, arginine and O_2 as substrates (Durner and Klessig, 1999; Chapter 13, Millar et al.). In addition, it has been shown by several groups that the cNR of plants reduces nitrite to NO in a side-reaction, using NADH as electron donor (Dean and Harper, 1988; Wildt et al., 1997; Yamasaki et al., 1999). However, NADH was inactive in the plasma membrane-associated NO formation. Moreover, using antibodies and size exclusion chromatography it was shown that the NO formation activity of roots was not caused by PM-NR but by a hitherto unknown enzyme, the nitrite:NO oxidoreductase (NI-NOR).

The specific activity of NO formation at the PM (about 300 nmol h^{-1} mg protein^{-1}) would be sufficient to reduce all the nitrite produced by PM-NR at pH 6.0, the apoplastic pH value. Considering losses in activity during plasma membrane preparation, the in vivo activity is probably 10- to 20-fold higher. Due to its apolarity NO can easily enter the cell via diffusion through the plasma membrane and may induce secondary reactions in the cytosol. Thus we hypothesize that, under limited nitrate availability, NO might be one of the primary signals that report the presence of nitrate. Higher concentrations of nitrate in the apoplast would lead to high NO production rates. This would involve a loss of N to the soil and atmosphere as gaseous NO, but it could also result in higher NO concentrations in the cell which might be assimilated to organic N, particularly during the night.

III. Nitrite Transport and Reduction

In plant cells, nitrite is the product of nitrate reduction by NR. Plants can also acquire nitrite from the outside, by taking up either nitrite from the exogenous medium or NO_x gases from the atmosphere. In both cases it seems that nitrite never accumulates to high concentrations within the cell. Indeed, nitrite is highly toxic and its acid form, nitrous acid, is even more so (Sinclair, 1987). As discussed in Section II, nitrite is produced in the cytosol by cNR or at the plasma membrane by PM-NR. It must therefore be transported across the plastid membranes to be further reduced to ammonium by plastidic nitrite reductase (NiR, EC 1.7.7.1). At the forefront of any discussion of the control of nitrite reduction must be recognition of the importance of the supply of reducing power, of which ammonium formation from nitrite requires considerable amounts. In this section, we present some recent data on the transport and reduction of nitrite, with a particular emphasis on alternative enzymatic reactions involved in nitrite metabolism, such as nitrite detoxification or N_2O and NO production from nitrite. For excellent reviews of earlier data, see Wray (1993) and Sivasankar and Oaks (1996).

A. Nitrite Transport Across Membranes

Nitrite can be transported either in its protonated form (nitrous acid, HNO_2) or as an ion. The protonated form (pKa = 3.29) is able to diffuse freely across membranes whereas an active transport system is probably needed for the nitrite anion. At present, little is known about nitrite transport in higher plants. It has been proposed that nitrite transport across the chloroplast membranes occurs mainly through a saturable nitrite transporter which is sensitive to protein modifiers (Brunswick and Cresswell, 1988a,b), while other authors have argued that nitrite transport operates through the diffusion of nitrous acid (Shingles et al., 1996). Molecular data on nitrite transport are still lacking for higher plants but very recent results by Rexach et al. (2000) have allowed a better understanding of this process in the unicellular alga *Chlamydomonas reinhardtii*. These authors have characterized a gene (*Nar1*) which is clustered with other genes involved in nitrate assimilation and whose sequence is homologous to the bacterial FOCA formate transporter and to the putative bacterial NIRC nitrite transporter. The NAR1 protein appears to be a chloroplastic membrane protein and is clearly involved in nitrite transport. Interestingly, a protein sequence derived from an *Arabidopsis* EST (Accession number, N37972) shows some homology to the NAR1 protein (Rexach et al., 2000) but the involvement of this protein in nitrite transport remains to be determined. It seems likely that nitrite enters the chloroplast both by a free diffusion process and by active transport (Shingles et al., 1996; Rexach et al., 2000).

Apoplastic reduction of nitrate produces nitrite, which must enter the cell. That nitrite can cross the plasma membrane rapidly is shown by the ability of nitrite to sustain plant growth when supplied as the sole N source (Aslam and Huffaker, 1989; Siddiqi et al., 1992), and it has long been known that anoxic roots excrete nitrite into the medium (Botrel and Kaiser, 1997). In *C. reinhardtii*, four high affinity nitrate/nitrite transporters have been described (Rexach et al., 1999 and references therein). In higher plants, very little is known about the involvement of the nitrate transporters in nitrite transport. Nitrite has been found to inhibit nitrate influx in a competitive manner which suggests that both ions share at least some transport systems (Siddiqi et al., 1992). Again, it is likely that nitrite influx and efflux involve a combination of nitrous acid and nitrite ions. Indeed, plant cells are much more sensitive to high nitrite concentrations when grown at an acidic pH which favors free diffusion of the acid form (Vaucheret et al., 1992).

B. Nitrite Metabolism

It has been assumed that in most plants nitrite is important only as a substrate for NiR. As discussed in section II, however, recent data suggest that there might be alternate pathways of either nitrite utilization or detoxification involving other enzymes. Indeed, in normal growth conditions, nitrite never accumulates to very high levels in plants, even when the NiR activity is absent and/or the NR activity increased.

1. *Are There Other Enzymes Involved in Nitrite Metabolism in Plants?*

It has long been known that soybean can produce NO_x (NO and NO_2) gases from nitrite by the action of the so-called constitutive NRs (Dean and Harper, 1988; Klepper, 1990). Since then, it has been found that many plants have the ability to emit nitrogen oxide(s) (Wildt et al., 1997), a denitrifying capability closely associated with nitrate availability. Interestingly, Goshima et al. (1999) have found emission of N_2O by transgenic tobacco plants expressing an antisense NiR construct (line 271: Vaucheret et al., 1992), which have very low NiR activities and which accumulate nitrite. Emission of N_2O was not observed in the wild type or in transgenic plants grown on ammonium. Furthermore, when NR activity was blocked, no evolution of N_2O was observed (Goshima et al., 1999). Thus, it appears that in tobacco nitrite is partly detoxified by reduction to N_2O, but whether this step is enzymatic or results from chemical reduction of the nitrite ion remains to be established. Very recently it has also been suggested that cytosolic NADH:NR from maize is capable of reducing nitrite to NO and peroxynitrite, which opens up the possibility that cNR is somehow involved in NO production in plants (Yamasaki et al., 1999; Yamasaki and Sakihama, 2000). This reaction would only take place when nitrite concentrations are high and, indeed, the Km of cNR for nitrite was found to be around 300 μM (Yamasaki and Sakihama, 2000). This would explain why NO production was detected when NiR activity was low, e.g. in the dark or when the photosynthetic electron transfer chain was inhibited (Wildt et al., 1997). Apart from nitrite detoxification, this reaction could also be involved in the production of NO in plants. Indeed, NO has already been implicated in the regulation of several plant processes, including cell damage and the hypersensitive response (Van Camp et al., 1998). However, it is clear that other reactions could account for the production of NO when nitrite concentrations are high. For instance, the respiratory chain of mammalian mitochondria also seems to have the ability to reduce nitrite to NO (Kozlov et al., 1999). Similarly, it has been found that xanthine oxidase, a ubiquitous molybdo-enzyme, catalyzes the reduction of nitrite to NO under hypoxia and in the presence of NADH (Zhang et al., 1998; Godber et al., 2000).

Apart from the detoxification of nitrite by reduction to gaseous nitrogen oxide(s), it has been suggested that plants can oxidize nitrite back to nitrate (Aslam et al., 1987). This would be analogous to the oxidative detoxification of sulfite by sulfite oxidase in mammals. Interestingly, an *Arabidopsis* EST presents significant homologies to the molybdenum cofactor domain of sulfite oxidase (T. Nakamura, C. Meyer et al., unpublished). It could thus be possible that this plant enzyme catalyzes nitrite oxidation to nitrate along with sulfite oxidation to sulfate. Indeed, nitrite and sulfite ions have similar properties and can both be reduced by both sulfite reductase (SiR) and NiR (Mikami and Ida, 1989).

2. *Nitrite Reduction Catalyzed by Nitrite Reductase*

a. The Source of the Reducing Power Needed for Nitrite Reduction

In green leaves, NiR is located within the chloroplast, whereas in roots or heterotrophic tissues it is located within plastids. In both cases, NiR appears to be a soluble enzyme found in the stroma (Dalling et al., 1972; Wray, 1993). NiR is synthesized in the cytosol as a precursor protein with an N-terminal transit peptide which directs the enzyme to these organelles. However, the presence of an extrachloroplastic form has been proposed in cotyledons of mustard (Schuster and Mohr, 1990). Six electrons are needed for the reduction of one nitrite molecule to ammonium. In both the chloroplasts and the non-photosynthetic plastids, reduced Fd supplies the necessary electrons (Matsumura et al., 1997; Emes and Neuhaus, 1997). In the chloroplast, Photosystem I directly provides reduced Fd while in other plastids the oxidation of Glc-6P via the oxidative pentose phosphate pathway (OPPP) generates NADPH which is used by Fd:$NADP^+$ oxidoreductase (FNR) for the generation of reduced Fd (Bowsher et al., 1989). Glc-6P dehydrogenase catalyzes the first step of the OPPP and probably represents a major controlling step for reductant supply to NiR in non-photosynthetic plastids (i.e. in roots or darkened leaves) (Wright et al., 1997). Different isoforms of Fd and FNR are found in leaves and roots, with root FNR showing a higher affinity for root Fd than for the leaf protein (Onda et al., 2000). This difference may be crucial in determining the opposing direction of net electron transport between NADPH and Fd in leaves and roots. Nitrite reduction accounts for 75% of the reducing power required to convert nitrate to ammonium, and nitrate assimilation overall can account for a small but significant proportion of photosynthetic energy (Lewis et al., 2000). It is therefore possible that plants have evolved mechanisms to save photosynthetic energy when nitrate, and thus nitrite, are absent. Recently Wang et al. (2000) have performed a systematic analysis of *Arabidopsis* genes induced by nitrate in whole seedlings. Among the most strongly induced transcripts were those for NiR, along with mRNAs of genes involved in the OPPP and of Fd and FNR genes. These results confirm previous observations in maize (Redinbaugh and Campbell, 1988; Matsumura et al., 1997). It seems, therefore, that the genes most inducible by nitrate are those coding for NiR and for proteins involved in the supply of reducing power to NiR. Another gene displaying a strong induction by nitrate was uroporphyrin III methyltransferase, which codes for an enzyme specifically involved in the synthesis of siroheme, a prosthetic group only found in NiR and SiR (Sakakibara et al., 1996; Wang et al., 2000).

b. Structure and Function of Nitrite Reductase

NiR is thought to be a monomeric enzyme (spinach NiR has a molecular mass of 61 kDa) containing two prosthetic groups, namely a $[Fe_4S_4]$ cluster and siroheme (a reduced porphyrin of the isobacteriochlorin class), which transfer electrons in that order from Fd to nitrite (see Knaff, 1996; Meyer and Caboche, 1998 for reviews). Nitrite is bound and reduced by the siroheme. NiR and SiR catalyze the unusual transfer of six electrons to a single redox center, the siroheme (Fig. 2). This electron transfer involves one-electron carriers (Fd, the $[Fe_4S_4]$ cluster and siroheme) and NiR has thus the highly unusual capability of retaining all reaction intermediates between nitrite and ammonium, releasing only the fully reduced ammonium ion. NiR interacts with Fd through electrostatic binding; this interaction involves positively charged residues of NiR (Arg 375 and 556, Lys 436 in the spinach NiR protein) and negatively charged residues on Fd (Frieman et al., 1992; Dose et al., 1997). The mechanism of Fd binding to NiR seems quite similar to that of Fd to

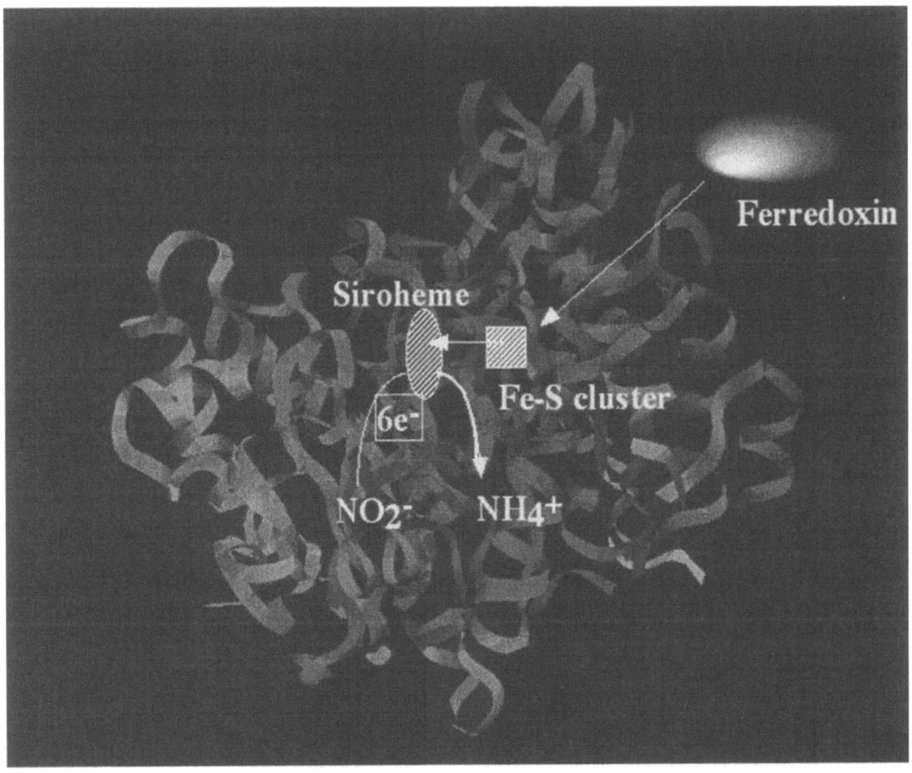

Fig. 2. Structure and organization of higher plant nitrite reductase. The tertiary structure of nitrite reductase is derived from that of bacterial sulfite reductase.

FNR (Knaff, 1996). Indeed the N-terminal part of plant and cyanobacterial NiRs has clear homology to FNRs. Sequence comparison of NiR proteins show a high conservation among plant species (75-80% similarity). However, there is only a small degree of homology between plant NiRs and fungal or bacterial NiRs, except for cyanobacterial NiR, which seems to be more similar to plant NiR than to enterobacterial NiR (Luque et al., 1993). The site of interaction between NiR and Fd is conserved among plant NiRs but is also found on cyanobacterial Fd:NiR as well as on SiRs. The alignment of plant and cyanobacterial NiRs and SiRs also shows that the regions involved in the binding of the siroheme and the [Fe_4S_4] cluster are very well conserved (Crane et al., 1995). In general, NiRs and SiRs sequences are quite conserved, which reflects the fact that these two enzymes catalyze very similar reactions and that both are able to reduce nitrite as well as sulfite, although they show a much better affinity for their physiological substrates. The structure of plant NiR can be deduced from the structure of *Escherichia coli* SiR hemoprotein, which has been solved (Crane et al., 1995), as well as from spectroscopic and biochemical data (reviewed in Knaff, 1996). The two NiR prosthetic groups appear to be very closely arranged in the holoprotein and to be coupled by a conserved cysteine residue (Siegel and Wilkerson, 1989; Crane et al., 1995) of the [Fe_4S_4] cluster. In the *E. coli* SiR structure, the siroheme and the [Fe_4S_4] cluster are found together at the interface of the three SiR protein domains, which are probably conserved in NiRs (Fig. 2), and in each of them a central β-sheet contributes to domain interaction and cofactor binding (Crane et al., 1995). Interestingly, this bacterial SiR seems to be the result of a gene duplication event as two structurally conserved moieties exist, which confers on the protein a pseudo two-fold axis of symmetry. These two subdomains were also found in other NiRs and SiRs. Despite their common features, differences might exist in the enzymatic mechanism between NiR and SiR. For instance, it has been found that nitrite binding induces conformational changes in NiR and that the Km for nitrite is much lower when Fd is used instead of the artificial electron donor methyl viologen (Mikami and Ida, 1989). This suggests that allosteric regulation may occur when the quaternary complex Fd-NiR-nitrite is formed and could explain the higher Km of NiR for sulfite. The spinach (Bellissimo and Privalle, 1995) and tobacco (Crété et al., 1997) NiRs have also been expressed in *E. coli* as active enzymes, allowing the identification of critical residues in the NiR sequence (Bellissimo and Privalle, 1995).

c. Nitrite Reductase Genes and Mutants

It has proved much more difficult to produce mutants deficient in NiR (*nii* mutants) than in NR, probably because accumulation of nitrite caused by NiR deficiency would be more detrimental than accumulation of nitrate. In addition, there are no direct selection methods available for isolating *nii* mutants, like chlorate resistance for NR-deficient mutants. Nevertheless, one *nii* mutant has been isolated in barley (Duncanson et al., 1993) by screening a M_2 population of mutagenized barley seeds for nitrite accumulation. NiR activity was strongly reduced in this mutant line. Since the mutation segregated with RFLP markers associated with the NiR apoenzyme gene, the mutant is probably affected in this gene (Ward et al., 1995). Recently, *nii* mutants were also obtained in the unicellular algae *Chlorella sorokiniana* (Burhenne and Tischner, 2000) and *Chlamydomonas reinhardtii* (Navarro et al., 2000). These mutants were unable to grow with nitrate or nitrite as nitrogen sources and excreted nitrite into the medium in the presence of nitrate. In *Nicotiana tabacum*, a NiR-deficient mutant (line 271) has been constructed (Vaucheret et al., 1992) by introducing an antisense *Nii* coding sequence. This mutant line showed very low NiR activity and NiR mRNA levels and, as a result, accumulated more nitrite than wild type plants. The mutant plants grew normally on ammonium but when grown on nitrate as sole N source, they displayed drastically reduced development, markedly decreased growth rate, chlorotic leaves, and produced seeds only after one or two years of culture. We have tried to complement the NiR deficiency in the 271 line by expressing a fungal NiR cDNA in these mutants (A. Krapp, C. Meyer et al., unpublished). This fungal NADPH:NiR should be expressed and active in the cytosol. Successful transformation would therefore have allowed us to examine the effect of localizing nitrite metabolism and thus ammonium production in this compartment. Unfortunately, the transgenic plants generated displayed very low levels both of NiR activity and phenotype complementation.

Nii cDNAs or *Nii* genes have been cloned from several higher plants, such as spinach, maize, birch, rice, *Arabidopsis* and tobacco (for reviews see Wray, 1993; Hoff et al., 1994; Meyer and Caboche, 1998).

Some higher plants contain only a single *Nii* gene per haploid genome, e.g. *Arabidopsis*, whereas other plant species contain two copies per haploid genome. Tobacco even contains four *Nii* genes, two from each tobacco ancestor (Kronenberger et al., 1993), which are expressed differentially in leaves and roots.

d. Regulation of Nitrite Reductase Gene Expression

The NiR mRNA level is increased in the presence of nitrate and, depending on the plant species, this increase depends on or is augmented by light (Wray, 1993 for a review; Vincentz et al., 1993; Seith et al., 1994; Cabello et al., 1998). Whether nitrate per se or nitrite is the actual inducing factor is difficult to establish unequivocally since, as already discussed, plants can probably oxidize nitrite to nitrate (Aslam et al., 1987) in addition to the ability of NR to catalyze the reduction of nitrate. In NR-deficient mutants, NiR expression was still induced by exogenous nitrate (Faure et al., 1991) which suggests that nitrate is the actual inducing molecule. However, as shown in the first part of this chapter, NR activities not linked to the *Nia* gene may exist in plants and result in the production of nitrite in *nia* mutants. Nevertheless, the *Nii* gene was found as one of the genes that were most induced by nitrate in a survey of nitrate-regulated genes in *Arabidopsis* (Wang et al., 2000). Ammonium and amides (Gln and Asn) inhibit the expression of NiR in detached leaves and roots while sucrose induces NiR expression (Vincentz et al., 1993; Sivasankar et al., 1997). These are also well-known responses of NR expression and, in fact, NiR is often found to be coregulated with NR in response to N- and C-metabolites or light, at least at the transcriptional level (Fig. 3). There are, however, some differences: the NiR mRNA was less induced than that of NR by exogenous sugars in dark-adapted *N. plumbaginifolia* leaves (Vincentz et al., 1993). In maize roots, moreover, sucrose relieved the inhibition of NiR expression by amides, but not that of NR

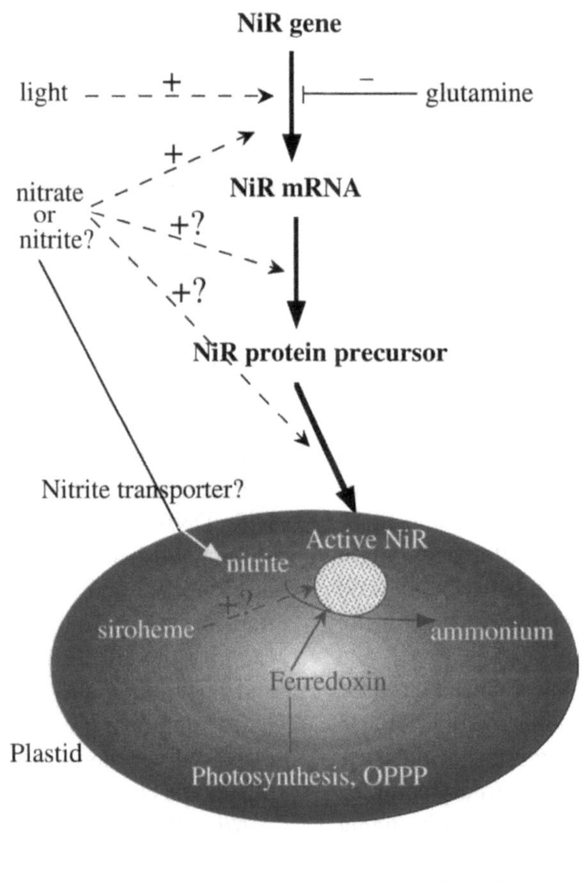

Fig. 3. Regulation of nitrite reductase gene expression. Inducing (+) and repressing (–) factors are shown. OPPP: oxidative pentose phosphate pathway.

(Sivasankar et al., 1997). Ammonium was also found to induce NR and NiR gene expression in the absence of nitrate in *Clematis vitalba* (Bungard et al., 1999). Photooxidative damage to chloroplasts of Norflurazon-treated plants was shown to inhibit NiR gene expression in tobacco (Neininger et al., 1992) and sunflower (Cabello et al., 1998) indicating that some 'plastidic factor' could be required for NiR expression, though light-induced changes in cytosolic components cannot be ruled out. So far the nature of these factors remains undetermined.

The regulation of the NiR mRNA level seems to operate mainly through effects on transcription. Indeed, the *Nii* gene promoter from several plant species was shown to confer nitrate inducibility on a reporter gene fused to it (Back et al., 1991; Neininger et al., 1994; Truong et al., 1994). Moreover, the accumulation of a NiR mRNA derived from the expression of a 35S-NiR construct was not affected by the exogenous N source (Crété et al., 1997). Promoter analysis of the bean *Nii* gene in transgenic tobacco plants showed that the elements involved in nitrate regulation reside in the proximal 0.6 kb region upstream of the translation start (Sander et al., 1995). Experiments in which the *Nii* promoter from spinach was deleted, fused to the *GUS* reporter gene and introduced into tobacco, indicated that the basic elements required for light- and nitrate-dependent expression of the reporter gene were within a 331 bp promoter sequence located 200 bp upstream and 131 bp downstream from the transcription initiation site (Neininger et al., 1994). Furthermore, in vivo footprinting revealed nitrate-inducible binding of proteins to GATA elements in the –230 to –181 bp region of the spinach *Nii* promoter (Rastogi et al., 1997). This suggested that GATA sequences could mediate nitrate regulation of the *Nii* gene, although gain of function experiments with these putative nitrate responsive elements are thus far lacking. In addition, it was shown by analysis of the tobacco *Nii* promoter fused to either a *GUS* or luciferase reporter gene that the sequences required for nitrate induction of the reporter gene expression were retained in the proximal 200 bp fragment of the promoter (Dorbe et al., 1998). Further deletions, however, abolished both promoter activity and nitrate inducibility. So far, it has been very difficult to clearly separate the general transcriptional activity and the nitrate inducibility of the *Nii* gene promoter.

A possible post-transcriptional control of NiR gene expression by nitrate has been evoked by some authors (Gupta et al., 1983; Schuster and Mohr, 1990). In order to study the post-transcriptional regulation of NiR, *N. plumbaginifolia* and *Arabidopsis* plants were transformed with a 35S-NiR construct. The resulting transgenic plants were found to overexpress the NiR activity in the leaves (Crété et al., 1997). When these plants were grown in vitro on media containing either nitrate or ammonium as sole nitrogen source, the level of NiR mRNA derived from transgene expression was unchanged, whereas NiR activity and protein level were strongly reduced on medium containing ammonium. These results suggest that, together with transcriptional control, post-transcriptional regulation by the N source also operates on NiR expression. One explanation for this mechanism could be that a specific enzyme for siroheme synthesis is induced by nitrate (Sakakibara et al., 1996; Wang et al., 2000). This post-transcriptional regulation of NiR expression by the nitrogen source is thus different from the post-translational control of NR by light (Meyer and Caboche, 1998). The reason for this difference is unknown but clearly illustrates the redundancy of the regulation of the nitrate assimilation pathway in plants (Fig. 3). Although NiR from *Candida utilis* has been shown to be regulated by phosphorylation (Sengupta et al., 1997), no clear mechanisms of NiR regulation by post-translational modifications have so far been described in plants.

IV. Conclusions

Exciting developments in the understanding of both nitrate and nitrite reduction have underlined the complexity of N assimilation and its regulation in plants. While the physiological function of PM-NR remains an open question, recent advances suggest that PM-NR can fulfill multiple roles in the root cell and that the importance of these will vary with physiological conditions (Fig. 1). The composition of root PM-NR very much suggests that it is involved in plasma membrane-associated redox reactions, e.g. the postulated interaction with a flavoprotein of the plasma membrane. A role in protection against high nitrate concentrations is suggested by the correlation between high activities, stable nitrate content, and poor growth rates. However, assimilatory nitrate reduction also seems to occur under certain conditions in roots, e.g. during darkness. Without doubt, a significant step forward in understanding redox reactions at the plasma membrane was the elucidation of the involvement of PM-NR in NO production in root plasma membranes. NO thus produced may either act as a signaling molecule for nitrate or it could be the final product of nitrate reduction at high nitrate conditions and be released to the soil and atmosphere. Finally, the apoplastically produced NO could also be involved in the processes observed during pathogen defense (Delledonne et al., 1998). The control of nitrite metabolism may be as finely tuned as nitrate reduction in plants, and seems to operate at both transcriptional and translational levels, although there is scant evidence for direct regulation of NiR activity through, for example, changes in activation states. It is likely that plants have evolved efficient mechanisms to co-ordinate NR and NiR activities, thus avoiding the accumulation of the cytotoxic nitrite ion as well as adjusting consumption of photosynthetic reducing power to the need of the cell for reduced N.

Acknowledgments

This work was partly supported by the European Union contract # BIO4CT97-2231 to C. Meyer and by the Deutsche Forschungsgemeinschaft (SFB 199) to C. Stöhr. We thank T. Nakamura, A. Krapp, T. Moureaux and P. Crété for sharing unpublished results.

References

Aslam M and Huffaker RC (1989) Role of nitrate and nitrite in the induction of nitrite reductase in the leaves of barley seedlings. Plant Physiol 91: 1152–1156

Aslam M, Rosichan JL and Huffaker RC (1987) Comparative induction of nitrate reductase by nitrate and nitrite in barley leaves. Plant Physiol 83: 579–584

Asard H, Bérczi A and Caubergs RJ (eds) (1998) Plasma Membrane Redox Systems and Their Role in Biological Stress and Disease. Kluwer Academic Publishers, Dordrecht

Back E, Dunne W, Schneiderbauer A, de Framond A, Rastogi R and Rosthstein SJ (1991) Isolation of a spinach nitrite reductase gene promoter which confers nitrate inducibility on GUS gene expression in transgenic tobacco. Plant Mol Biol 17: 9–18

Bar-Yosef B (1996) Root excretion and their environmental effects: Influence on availability of phosphorus. In: Waisel Y, Eshel A and Kafkafi U (eds) Plant Roots. The Hidden Half, pp 581–605. Marcel Dekker Inc., New York

Bellissimo DB and Privalle L (1995) Expression of spinach nitrite reductase in *Escherichia coli*: Site-directed mutagenesis of predicted active site amino acids. Arch Biochem Biophys 323:155–163

Botrel A and Kaiser WM (1997) Nitrate reductase activation state in barley roots in relation to the energy and carbohydrate status. Planta 201: 496–501

Bowsher CG, Hucklesby DP and Emes MJ (1989) Nitrite reduction and carbohydrate metabolism in plastids purified from roots of *Pisum sativum* L. Planta 177: 359–366

Brunswick P and Cresswell CF (1988a) Nitrite uptake into intact pea chloroplasts. I. Kinetics and relationship with nitrite assimilation. Plant Physiol 86: 378–383

Brunswick P and Cresswell CF (1988b) Nitrite uptake into intact pea chloroplasts. II. Influence of electron transport regulators, uncouplers, ATPase and anion uptake inhibitors and protein binding reagents. Plant Physiol 86: 384–389

Bungard RA, Wingler A, Morton JD, Andrews M, Press MC and Scholes JD (1999) Ammonium can stimulate nitrate and nitrite reductase in the absence of nitrate in *Clematis vitalba*. Plant Cell Environ 22: 859–866

Burhenne N and Tischner R (2000) Isolation and characterization of nitrite-reductase-deficient mutants of *Chlorella sorokiniana* (strain 211-8k). Planta 211: 440–445

Cabello P, de la Haba P, Gonzales-Fontes A and Maldonado JM (1998) Induction of nitrate reductase, nitrite reductase and glutamine synthetase isoforms in sunflower cotyledons as affected by nitrate, light, and plastid integrity. Protoplasma 201: 1–7

Crane BR, Siegel LM and Getzoff ED (1995) Sulfite reductase structure at 1.6 Å: Evolution and catalysis for reduction of inorganic anions. Science 270: 59–67

Crété P, Caboche M and Meyer C (1997) Nitrite reductase expression is regulated at the post-transcriptional level by the nitrogen source in *Nicotiana plumbaginifolia* and *Arabidopsis thaliana*. Plant J 11: 625–634

Dalling MJ, Tolbert NE and Hageman RH (1972) Intracellular location of nitrate reductase and nitrite reductase. I. Spinach and tobacco leaves. Biochim Biophys Acta 283: 505–512

Dean JV and Harper JE (1988) The conversion of nitrite to nitrogen oxide(s) by the constitutive NAD(P)H-nitrate reductase enzyme from soybean. Plant Physiol 88: 389–395

Delledonne M, Xia Y, Dixon RA and Lamb C (1998) Nitric oxide signal functions in plant disease resistance. Nature 394: 585–588

Dorbe MF, Truong HN, Crété P and Daniel-Vedèle F (1998) Deletion analysis of the tobacco *Nii1* promoter in *Arabidopsis thaliana*. Plant Sci 139: 71–82

Dose MM, Hirasawa M, Kleis-SanFrancisco S, Lew EL and Knaff DB (1997) The ferredoxin-binding site of ferredoxin:nitrite oxidoreductase. Plant Physiol 114: 1047–1053

Duncanson E, Gilkes AF, Kirk DW, Sherman A and Wray JL (1993) *nir1*, a conditional-lethal mutation in barley causing a defect in nitrite reduction. Mol Gen Genet 236: 275–282

Durner J and Klessig DF (1999) Nitric oxide as a signal in plants. Curr Opin Plant Biol 2: 369-374

Emes MJ and Neuhaus HE (1997) Metabolism and transport in non-photosynthetic plastids. J Exp Bot 48: 1995–2005

Faure JD, Vincentz M, Kronenberger J and Caboche M (1991) Co-regulated expression of nitrate and nitrite reductases. Plant J 1: 107–113

Friemann A, Brinkmann K and Hachtel W (1992) Sequence of a cDNA encoding nitrite reductase from the tree *Betula pendula* and identification of conserved protein regions. Mol Gen

Genet 231: 411–416

Godber BLJ, Doel JJ, Sapkota GP, Blake DR, Stevens CR, Eisenthal R and Harrison R (2000) Reduction of nitrite to nitric oxide catalyzed by xanthine oxidoreductase. J Biol Chem 275: 7757–7763

Goshima N, Mukai T, Suemori M, Takahashi M, Caboche M and Morikawa H (1999) Emission of nitrous oxide (N_2O) from transgenic tobacco expressing antisense NiR mRNA. Plant J 19: 75–80

Gupta A, Disa S, Saxena IM, Sarin NB, Guha-Mukerjee S and Sopory SK (1983) Role of nitrate in the induction of nitrite reductase activity during wheat seed germination. J Exp Bot 141: 396–404

Hoff T, Truong HN and Caboche M (1994) The use of mutants and transgenic plants to study nitrate assimilation. Plant Cell Environ 17: 489–506

Klepper L (1990) Comparison between NOx evolution mechanisms of wild-type and nr1 mutant soybean leaves. Plant Physiol 93: 26–32

Knaff DB (1996) Ferredoxin and ferredoxin-dependent enzymes. In: Ort DR and Yocum CF (eds) Oxygenic Photosynthesis: The Light Reactions, pp 333–361. Kluwer Academic Publishers, Dordrecht

Kozlov AV, Staniek K and Nohl H (1999) Nitrite reductase activity is a novel function of mammalian mitochondria. FEBS Letters 454: 127–130

Kronenberger J, Lepingle A, Caboche M and Vaucheret H (1993) Cloning and expression of distinct nitrite reductases in tobacco leaves and roots. Mol Gen Genet 236: 203–208

Kunze M, Riedel J, Lange U, Hurwitz R and Tischner R (1997) Evidence for the presence of GPI-anchored PM-NR in leaves of *Beta vulgaris* and for PM-NR in barley leaves. Plant Physiol Biochem 35: 507–512

Kunze M, Riedel J, Lange U, and Tischner R (2000) Plasma membrane bound nitrate reductase in plants—The current status of research. In: Martins-Loução MA and Lips HS (eds) Nitrogen in a Sustainable Ecosystem—From the Cell to the Plant, pp 19–26. Backhuys, Leiden

Lewis CE, Noctor G, Causton DH, Foyer CH (2000) Regulation of assimilate partitioning in leaves. Aust J Plant Physiol (Special issue) 57: 507–519

Luque I, Flores E and Herrero A (1993) Nitrite reductase gene from *Synechococcus Sp* PCC-7942—Homology between cyanobacterial and higher-plant nitrite reductases. Plant Mol Biol 21: 1201–1205

Marschner H and Römheld V (1996) Root-induced changes in the availability of micronutrients in the rhizosphere. In: Waisel Y, Eshel A and Kafkafi U (eds) Plant Roots. The Hidden Half, pp 557–579. Marcel Dekker Inc., New York

Matsumura T, Sakakibara H, Nakano R, Kimata Y, Sugiyama T and Hase T (1997) A nitrate-inducible ferredoxin in maize roots—Genomic organization and differential expression of two nonphotosynthetic ferredoxin isoproteins. Plant Physiol 114: 653–660

Mench M, Morel JL, Guckert A and Guillet B (1988) Metal binding with root exudates of low molecular weight. J Soil Sci 39: 521–527

Mendel RR and Schwarz G (1999) Molybdoenzymes and molybdenum cofactor in plants. Crit Rev Plant Sci 18: 33–69

Meyer C and Caboche M (1998) Manipulation of nitrogen metabolism. In: Lindsey K (ed) Transgenic Plant Research, pp 125–133. Harwood Academic Publishers, London

Mikami B and Ida S (1989) Spinach ferredoxin-nitrite reductase: characterization of catalytic activity and interaction of the enzyme with substrates. J Biochem 105: 47–50

Navarro MT, Prieto R, Fernández E and Galván A (2000) Nitrite reductase mutants as an approach to understanding nitrate assimilation in *Chlamydomonas reinhardtii*. Plant Physiol 122: 283–289

Neininger A, Kronenberger J and Mohr H (1992) Coaction of light, nitrate and a plastidic factor in controlling nitrite-reductase gene expression in tobacco. Planta 187: 381–387

Neininger A, Back E, Bichler J, Schneiderbauer A and Mohr H (1994) Deletion analysis of a nitrite reductase promoter from spinach in transgenic tobacco. Planta 194: 186–192

Onda Y, Matsumura T, Kimata-Ariga Y, Sakakibara H, Sugiyama T and Hase T (2000) Differential interaction of maize root ferredoxin:$NADP^+$ oxidoreductase with photosynthetic and non-photosynthetic ferredoxin isoproteins. Plant Physiol 123: 1037–1045

Rastogi R, Bate NJ, Sivasankar S and Rothstein S (1997) Footprinting of the spinach nitrite reductase gene promoter reveals the preservation of nitrate regulatory elements between fungi and higher plants. Plant Mol Biol 34: 465–476

Redinbaugh MG and Campbell WH (1988) Nitrate regulation of the oxidative pentose phosphate pathway in maize (*Zea mays* L.) root plastids: induction of 6-phosphogluconate dehydrogenase activity, protein and transcript levels. Plant Sci 134: 129–140

Rexach J, Montero B, Fernández E and Galván A (1999) Differential regulation of the high affinity nitrite transport systems III and IV in *Chlamydomonas reinhardtii*. J Biol Chem 274: 27801–27806

Rexach J, Fernández E and Galván A (2000) The *Chlamydomonas reinhardtii Nar1* gene encodes a chloroplast membrane protein involved in nitrite transport. Plant Cell 12: 1441–1453

Sakakibara H, Takei K and Sugiyama T (1996) Isolation and characterization of a cDNA that encodes maize uroporphyrinogen III methyltransferase, an enzyme involved in the synthesis of siroheme, which is a prosthetic group of nitrite reductase. Plant J 10: 883–892

Sander L, Jensen PE, Back LF, Stummann BJ and Henningsen KW (1995) Structure and expression of a nitrite reductase gene from bean (*Phaseolus vulgaris*) and promoter analysis in transgenic tobacco. Plant Mol Biol 27: 165–177

Schuster C and Mohr H (1990) Appearance of nitrite reductase mRNA in mustard seedling cotyledons is regulated by phytochrome. Planta 181: 327–334

Seith B, Sherman A, Wray JL and Mohr H (1994) Photocontrol of nitrite reductase gene expression in the barley seedling (*Hordeum vulgare* L.). Planta 192: 110–117

Sengupta S, Shaila MS and Rao GR (1997) A novel autophosphorylation mediated regulation of nitrite reductase in *Candida utilis*. FEBS Letters 416: 51–56

Shingles R, Roh MH and McCarty RE (1996) Nitrite transport in chloroplast inner envelope vesicles. I. Direct measurement of proton-linked transport. Plant Physiol 112: 1375–1381

Siddiqi MY, King BJ and Glass DM (1992) Effects of nitrite, chlorate, and chlorite on nitrate uptake and nitrate reductase activity. Plant Physiol 100: 644–650

Siegel LM, Wilkerson JO (1989) Structure and function of spinach ferredoxin-nitrite reductase. In: Wray JL, Kinghorn

JR (eds) Molecular and Genetic Aspects of Nitrate Assimilation, pp 263–283. Oxford Science Publications, Oxford

Sinclair J (1987) Changes in spinach thylakoid activity due to nitrite ions. Photosynth Res 12: 255–263

Sivasankar S and Oaks A (1996) Nitrate assimilation in higher plants: the effects of metabolites and light. Plant Physiol Biochem 34: 609–620

Sivasankar S, Rothstein S and Oaks A (1997) Regulation of the accumulation and reduction of nitrate by nitrogen and carbon metabolites in maize seedlings. Plant Physiol 114: 583–589

Siverio JM, González V, Mendoza-Riquel A, Pérez MD and González G (1993) Reversible inactivation and binding to mitochondria of nitrate reductase by heat shock in the yeast *Hansenula anomala*. FEBS Letters 318: 153–156

Stöhr C (1998) Plasma membrane-bound nitrate reductase in algae and higher plants. In: Asard H, Bérczi A and Caubergs RJ (eds) Plasma Membrane Redox Systems and Their Role in Biological Stress and Disease, pp 103–119. Kluwer Academic Publishers, Dordrecht

Stöhr C (1999) Relationship of nitrate supply with growth rate, plasma membrane-bound and cytosolic nitrate reductase, and tissue nitrate content in tobacco plants. Plant Cell Environ 22: 169–177

Stöhr C and Mäck G (2001) Diurnal changes in nitrogen metabolism of tobacco roots. J Exp Bot 53: 1–7

Stöhr C and Ullrich WR (1997) A succinate-oxidising nitrate reductase is located at the plasma membrane of plant roots. Planta 203: 129–132

Stöhr C, Tischner R and Ward MR (1993) Characterization of the plasma-membrane-bound nitrate reductase in *Chlorella saccharophila* (Krüger) Nadson. Planta 191: 79–85

Stöhr C, Glogau U, Mätschke M and Tischner R (1995a) Evidence for the involvement of plasma-membrane-bound nitrate reductase in signal transduction during blue-light stimulation of nitrate uptake in *Chlorella saccharophila*. Planta 197: 613–618

Stöhr C, Schuler F and Tischner R (1995b) Glycosyl-phosphatidylinositol-anchored proteins exist in the plasma membrane of *Chlorella saccharophila* (Krüger) Nadson: Plasma membrane-bound nitrate reductase as an example. Planta 196: 284–287

Stöhr C, Strube F, Marx G, Ullrich WR and Rockel P (2001) A plasma-membrane-bound enzyme of tobacco roots catalyzes the formation of nitric oxide from nitrite. Planta 212: 835–841

Truong HN, Vaucheret H, Quilleré I, Morot-Gaudry JF and Caboche M (1994) Utilisation de la transgénèse pour l'analyse du métabolisme du nitrate. C R Soc Biol 188: 140–149

Van Camp W, Van Montagu M and Inzé D (1998) H_2O_2 and NO: Redox signals in disease resistance. Trends Plant Sci 3: 330–334

Vaucheret H, Kronenberger J, Lépingle A, Vilaine F, Boutin JP and Caboche M (1992) Inhibition of tobacco nitrite reductase activity by expression of antisense RNA. Plant J 2: 559–569

Vincentz M, Moureaux T, Leydecker MT, Vaucheret H and Caboche M (1993) Regulation of nitrate and nitrite reductase expression of *Nicotiana plumbaginifolia* leaves by nitrogen and carbon metabolites. Plant J 3: 315–324

Wang R, Guegler K, LaBrie ST and Crawford NM (2000) Genomic analysis of a nutrient response in *Arabidopsis* reveals diverse expression patterns and novel metabolic and potential regulatory genes induced by nitrate. Plant Cell 12: 1491–1509

Ward MP, Abbeton MT, Forde BG, Sherman A, Thomas WTB and Wray JL (1995) The *Nir1* in barley is tightly linked to the nitrite reductase apoprotein gene *Nii*. Mol Gen Genet 247: 579–582

Wienkoop S, Ullrich WR and Stöhr C (1999) Kinetic characterization of succinate-dependent plasma membrane-bound nitrate reductase in tobacco roots. Physiol Plant 105: 609–614

Wienkoop S, Schlichting R, Ullrich WR and Stöhr C (2001) Inverse diurnal regulation of the expression of two nitrate reductase transcripts in tobacco roots. Protoplasma 217: 15–19

Wildt J, Kley D, Rockel A, Rockel P and Segschneider HJ (1997) Emission of NO from higher plant species. J Geophys Res 102: 5919–5927

Wray JL (1993) Molecular biology, genetics and regulation of nitrite reduction in higher plants. Physiol Plant 89: 607–612

Wright DP, Huppe HC and Turpin DH (1997) In vivo and in vitro studies of glucose-6-phosphate dehydrogenase from barley root plastids in relation to reductant supply for NO_2^- assimilation. Plant Physiol 114: 1413–1419

Yamasaki H and Sakihama Y (2000) Simultaneous production of nitric oxide and peroxynitrite by plant nitrate reductase: In vitro evidence for the NR-dependent formation of active nitrogen species. FEBS Lett 468: 89–92

Yamasaki H, Sakihama Y and Takahashi S (1999) An alternative pathway for nitric oxide production: New features of an old enzyme. Trends Plant Sci 4: 128–129

Zhang Z, Naughton D, Winyard PG, Benjamin N, Blake DR and Symons MC (1998) Generation of nitric oxide by a nitrite reductase activity of xanthine oxidase: A potential pathway for nitric oxide formation in the absence of nitric oxide synthase activity. Biochem Biophys Res Commun 249: 767–772

Chapter 5

What Limits Nitrate Reduction in Leaves?

Werner M. Kaiser*, Maria Stoimenova and Hui-Min Man
Universität Würzburg, Julius-von-Sachs-Institut für Biowissenschaften, Lehrstuhl für Molekulare Pflanzenphysiologie und Biophysik, Julius-von-Sachs-Platz 2, D-97082, Würzburg, Germany

Summary 63
I. Introduction 64
II. Nitrate Reduction and Nitrate Reductase Activity in Photosynthesizing Leaves 64
III. Nitrate Reduction after Artificial Activation of Nitrate Reductase 65
IV. Is Cytosolic Nitrate Concentration Rate-Limiting? 66
V. Is Nitrate Reduction Limited by NAD(P)H? 68
VI. Conclusions 68
Acknowledgments 70
References 70

Summary

The observation that even drastic over- or underexpression of nitrate reductase (NR) has little effect on biomass production suggests that nitrate reduction in situ and extractable NR activity are not strictly coupled. Rates of nitrate reduction in detached spinach leaves are often, but not always, much lower than NR activity measured in leaf extracts under substrate (nitrate and NADH) saturation. This discrepancy between in vivo and in vitro rates is absent when leaves are illuminated for up to 2 h in high CO_2, becomes obvious when leaves are illuminated in air, and is extremely high when leaves are kept in the dark and NR is artificially activated by anoxia or other treatments. Feeding nitrate through the leaf petiole, which increases the leaf nitrate content, improves nitrate reduction rates in the light (in air) only after several hours. Literature data on cytosolic nitrate concentrations, and measurements of nitrate leakage from leaf discs into nitrate-free solutions, suggest that that cytosolic nitrate is usually not limiting for nitrate reduction in situ. Rather, reductant (NADH) concentration appears to be the limiting factor whenever photosynthesis is suboptimal or absent. This may explain in part why over- or underexpression of NR in transgenic plants has surprisingly little effect on vegetative growth.

*Author for correspondence, email: kaiser@botanik.uni-wuerzburg.de

Christine H. Foyer and Graham Noctor (eds): Photosynthetic Nitrogen Assimilation and Associated Carbon and Respiratory Metabolism, pp. 63–70. © 2002 Kluwer Academic Publishers.

I. Introduction

Plants may take up more nitrate than is immediately reduced, and store any surplus transiently in the vacuole. Nitrate reduction in leaves is usually low in the dark and high in the light. Accordingly, nitrate pools in leaves (at least of herbaceous plants) often increase during the night and decrease during the day. These diurnal pool changes are more obvious the lower the nitrate supply is (Man et al., 1999).

For some time it was believed that NR was itself the most limiting factor in N assimilation. Surprisingly, tobacco plants expressing NR under the control of the CaMV 35S promoter did not grow faster than wild-types, although they had somewhat lower leaf nitrate contents (Vincentz and Caboche, 1991; Foyer et al., 1993; Quillère et al., 1994). Furthermore, *Nicotiana plumbaginifolia* plants lacking post-translational inactivation of NR in the dark did not reduce more nitrate in the dark than wild-type plants (Lejay et al., 1997). On the other hand, mutants with strongly reduced NR activity in the leaves showed little phenotypic response until a large part of the wild-type NR was suppressed (Vaucheret et al., 1990; Crawford et al., 1992), partly because decreased levels of NR protein appear to be compensated by post-translational modulation and modified NR turnover (Scheible et al., 1997). In spite of such complex compensatory regulation, it appears that the rate of nitrate reduction in vivo is not always identical to NR activity in vitro, even when NR is extracted and measured in the presence of divalent cations and protein phosphatase (PP2A) inhibitors, which freeze NR activity at the level believed to exist in vivo.

The question is, therefore: to what extent do NR measurements in vitro, at substrate (NAD(P)H and nitrate) saturation and at optimal pH, reflect nitrate reduction rates in vivo? That question has been under discussion now for more than three decades. The last ten years have provided new insight into the regulatory properties of NR and therefore it seems justified to ask the above question once again. Here, we compare NR activity (as NRact and NRmax, see below) in leaf extracts with nitrate reduction rates of whole leaves under a variety of conditions which are known to drastically change the NR activation state (also compare Kaiser et al., 2000). For that purpose we have used spinach leaves, which give high and stable NR activity in crude extracts. In all the experiments described below, in vitro NR activity was measured at substrate saturation (5 mM nitrate plus 0.2 mM NADH).

Abbreviations: AICAR – 5-aminoimidazole-4-carboxamide1-β-D- ribofuranoside; FW – fresh weight; NR – nitrate reductase; NRact – nitrate reductase activity measured in the presence of divalent cations; NRmax – nitrate reductase activity measured in the presence of EDTA and absence of divalent cations; ZMP – 5-aminoimidazole-4-carboxamide ribonucleotide

II. Nitrate Reduction and Nitrate Reductase Activity in Photosynthesizing Leaves

The catalytic activity of NR is rapidly modulated by reversible phosphorylation on a serine residue (Ser 543 in spinach). P-NR binds a 14-3-3 dimer and becomes totally inactive in the presence of divalent cations. Dephosphorylation, or chelation of divalent cations, releases 14-3-3 and activates NR. Partial inactivation occurs in conditions such as darkness or after removal of CO_2 (for a recent review, see Kaiser et al., 1999). In leaves from spinach grown with good nitrate fertilization, maximum NR activity (+ EDTA, = NRmax) was about 20 μmoles g^{-1} FW h^{-1} (Fig. 1). NR activity measured in the presence of free Mg^{2+}(= NRact) varied between 50% and 80% of NRmax under good photosynthetic conditions (for further experimental details, see Kaiser et al., 2000). If NRact reflects the NR activity in situ, these values suggest that in the leaf, nitrate should be reduced at a rate of 10 to 16 μmoles g^{-1} FW h^{-1}.

A simple way to measure the short term rate of nitrate reduction in situ is to follow nitrate concentration in detached leaves with their petioles in nitrate-free solution. As there is practically no nitrate released from the petiole (data not shown), the decrease of the nitrate content reflects the rate of nitrate reduction, irrespective of the products formed. In detached spinach leaves illuminated in air, the rate of nitrate reduction in situ was considerably lower than NRact (Fig. 1), and decreased within a few hours even further, although NRact and NRmax remained constant.

As NR expression and activation state are responsive to photosynthesis, nitrate reduction and NR activity were also measured at very high ambient CO_2 in order to avoid any limitation of photosynthesis by stomatal closure (Fig. 1). Unexpectedly, in the light under 5% CO_2, leaves (petiole in water) reduced their stored nitrate at a higher rate than in air and, during the first two hours, almost precisely at the rate

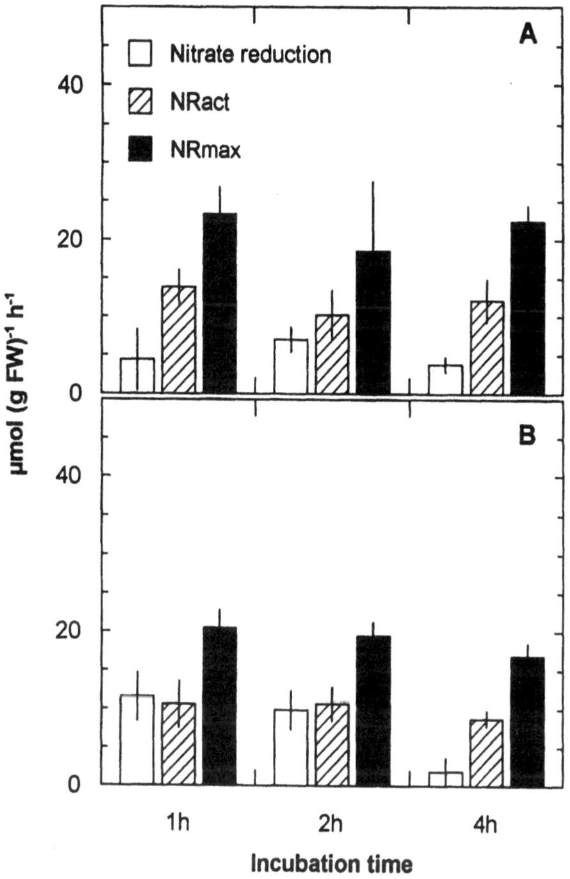

Fig. 1. Nitrate reduction in detached leaves and NRact and NRmax in leaf extracts. Spinach leaves harvested two h into the light period were placed with their petioles in distilled water. Leaves were illuminated at an air temperature of 22 °C at an irradiance of 300 μE m^2 s^{-1}, in air (A) or in 5% CO_2 (B). At the times indicated, leaves were removed and one half of the leaf blade was used for nitrate determination whereas the other half was used for preparing a leaf extract for NR determination. Mean inital nitrate content of the leaves was 62.5 ± 3.5 (μmol g^{-1}FW), mean NRact was 12.3 ± 3.5 and mean NRmax was 21.5 ± 5 (μmol g^{-1}FW h^{-1}). Four independent experiments were carried out with four parallel samples (three leaves each) at each time point (n = 16). Data are modified from Kaiser et al. (2000).

predicted by the in vitro NR assay (NRact). NR activity in the extract was very similar to that in extracts from leaves illuminated in air. As in air, rates of nitrate reduction in situ declined sharply after 4 h in the light, though NRact remained constant (Fig. 1).

III. Nitrate Reduction after Artificial Activation of Nitrate Reductase

NR activity in leaves is not only modulated by light and CO_2, but also by treatments like anoxia, by cellular acidification, by inhibitors or uncouplers of mitochondrial respiration, or by feeding the membrane-permeant 5′-AMP-analog AICAR (5-aminoimidazole-4-carboxamide1-β-D-ribo-furanoside) (for review, see Kaiser et al., 1999). We were interested in examining how such artificial modulation of NR would affect nitrate reduction in vivo.

In the experiment depicted in Fig. 2, NR was activated in the dark to the light level by flushing leaves with nitrogen. Unexpectedly, even though NRact was increased, little nitrate was reduced under these conditions (Fig. 2B). Even this low nitrate reduction, however, led to the accumulation of some nitrite, because nitrite reduction in the chloroplasts was practically zero, as judged from the lack of reduction of added nitrite (data not shown). Similar results were obtained by feeding AICAR in the dark to detached leaves (Fig. 2C). AICAR penetrates the cell membrane and is phosphorylated inside to the 5′-AMP analog 5-aminoimidazole-4-carboxamide ribonucleotide (ZMP). Both 5′-AMP and ZMP promote NR activation in vitro (Kaiser and Huber, 1994; Huber and Kaiser, 1996). As under anoxia, NRact was strongly activated (Fig. 2C), but in situ nitrate reduction remained extremely low, indicating limitation by substrate availability. Here again, nitrite was accumulated, but to a lesser extent than under anoxia. Similar results have been obtained with other NR activating treatments, for instance cellular acidification or feeding uncouplers or inhibitors of mitochondrial electron transport (data not shown).

There are several possibilities to explain the discrepancy between nitrate reduction rates in situ and NRact in vitro:

a) cytosolic nitrate concentrations are too low,

b) cytosolic NADH concentration is too low,

c) in vitro activation state of NR is higher than in situ, e.g. because the 14-3-3-P-NR complex is partly dissociated due to a strong dilution of the cytosol during extraction and reaction,

d) NR operates in vivo under suboptimal pH conditions.

Possibility (c) can be excluded from our consideration. Measurements of NRact under a wide range of dilutions of the leaf sap gave no significant difference

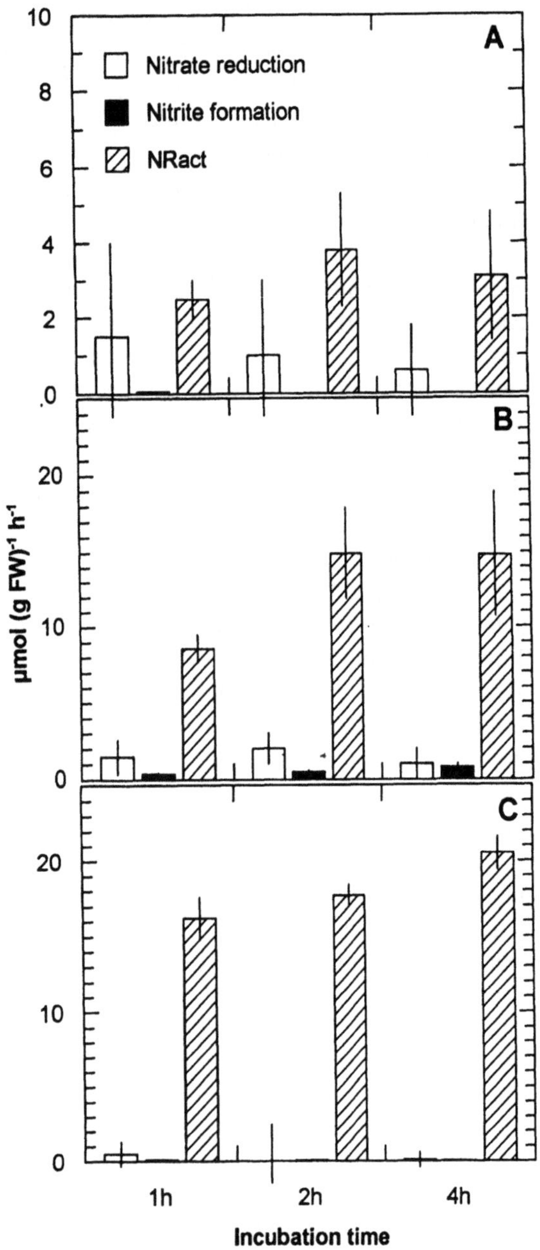

Fig. 2. Nitrate reduction, nitrite formation and NRact in detached spinach leaves after artificial activation of NR. Leaves were harvested as in Fig. 1, and were incubated in the dark in air (A) or in nitrogen (B) with their petiole in water, or in air with the petiole in 20 mM AICAR (C). Mean initial nitrate content was 31.5 ± 7.6 (μmol g^{-1} FW). Mean initial NRact was 10.7 ± 3.5 and NRmax was 22.8 ± 4.1 μmol g^{-1} FW h^{-1} (n = 12). Other conditions as in Fig. 1. Data are modified from Kaiser et al. (2000).

in NRact or in the activation state. Also, addition of 14-3-3 proteins to the reaction mixture resulted in no change of the in vitro NR activity (Kaiser et al., 2000).

The pH-response profile of spinach NR (at substrate saturation and with 10 mM Mg^{2+}) indicated almost constant activity between pH 6.5 and pH 7.5 (Kandlbinder et al., 2000). Accordingly, possibility (d) would require the cytosolic pH to drop below pH 6.5. In tobacco roots, cytosolic pH values reached pH 6.5 only after 3 h of anoxia, as determined by ^{31}P-NMR (unpublished), and the same minimum cytosolic pH was found in anoxic pea leaves (Bligny et al. 1997). It is, therefore, improbable that the low rate of nitrate reduction in situ (under anoxia) resulted from a decreased cytosolic pH. Up to now, there is no indication that unknown compounds (metabolites, proteins) in spinach leaves inhibit NR in vivo. Thus, we are left with the possibility that in vivo either one or both of the two substrates was below saturation and decreased even further with longer illumination time.

IV. Is Cytosolic Nitrate Concentration Rate-Limiting?

This question has been asked frequently over the last 30 years, though with contradictory answers (e.g. Ferrari et al., 1973; Beevers and Hageman, 1980; King et al., 1992). K_M-values of NR are 50 μM for nitrate and 7 μM for NADH, and these values are not affected by inactivation of NR (Kaiser and Spill, 1991). Thus, if nitrate reduction in vivo were nitrate limited, cytosolic nitrate would have to be ≤ 50 μM. Published values for cytosolic nitrate in leaves from nitrate fertilized plants strongly argue against such low cytosolic nitrate. Martinoia et al. (1986, 1987) estimated extravacuolar nitrate concentrations (supposed to reflect mainly cytosolic nitrate) in barley leaves of about 4 to 7 mM. By comparing nitrate contents in pea and spinach leaves freshly harvested in the light period, and in rapidly isolated chloroplasts, a cytosolic nitrate concentration of 3 to 10 mM was found: this concentration appeared to be homeostatically controlled as it remained constant even at very variable nitrate concentrations (5 to 100 mM) in the whole leaf tissue (Schröppel-Meier and Kaiser, 1988; Speer and Kaiser, 1991). More direct nitrate measurements with triple-barreled microelectrodes so far exist only for root epidermal or cortex cells. Here, cytosolic nitrate concentrations were 2 to 4 mM and once again appeared rather constant at variable external concentrations (Miller and Smith, 1996 and refs therein; Van der Leij et al., 1998), thus confirming conclusions drawn by Speer and Kaiser (1991).

Hence, it seems rather improbable that cytosolic nitrate would ever come close to the K_M of NR, at least in well fertilized plants, and nitrate efflux from the vacuole appears to be fast enough to maintain the cytosolic nitrate concentration above values saturating NR, as suggested previously for roots (King et al. 1992).

Nevertheless, we examined the effects of high nitrate feeding on rates of nitrate reduction in vivo. Detached spinach leaves, initially containing about 44 μmol g^{-1} FW nitrate, were fed through their petioles with a high nitrate concentration (30 mM). Nitrate uptake, nitrate content and NR activity were followed in the light (in air) over a 4 h period (Fig. 3). The external nitrate concentration was sufficient to cause a continuous increase in the leaf nitrate content over the experimental period. During the first two hours, in vivo nitrate reduction was again only 50 % of NRact. However, instead of declining thereafter, as in Fig. 1, in vivo rates increased and after 4 h they had come close to NRact.

If cytosolic nitrate was 5 mM at the beginning of the illumination period, and if the volume of the cytosol occupied 10 % of the total leaf water volume, the total amount of nitrate in the cytosol would be 0.5 μmol g^{-1} FW. With NRact = 10 μmol g^{-1}FW h^{-1}, cytosolic nitrate would be completely consumed within 3 min. However, nitrate reduction proceeded for 2 h at a rate of about 10 μmoles g^{-1} FW h^{-1} in high CO_2 or about 5 μmoles g^{-1} FW h^{-1} in air (Fig. 1). Obviously, nitrate efflux from the vacuole was high enough to support these rates for some time without nitrate feeding. Only after some hours in high CO_2 did in vivo rates of nitrate reduction decline sharply, whereas NRact remained almost unchanged. As this late decline was at least partly prevented by nitrate feeding (Fig. 3), cytosolic nitrate apparently became rate limiting only after part of the stored nitrate had been consumed. However, even after 2 h in the light (in air) there was actually enough nitrate (about 40 μmol g^{-1} FW) left over to feed nitrate reduction for a longer time. Nitrate export from the vacuole depends on a continuous production of exchangeable anions (such as malate) or on nitrate/cation co-transport, which may have ceased after 2 h for as yet unknown reasons.

In leaf discs floating on buffer solution in the dark under anoxia, nitrate feeding also increased the (low) rates of nitrite formation (Table 1). Again, this may indicate a limitation by cytosolic nitrate, as concluded decades ago from similar experiments (e.g. Ferrari et

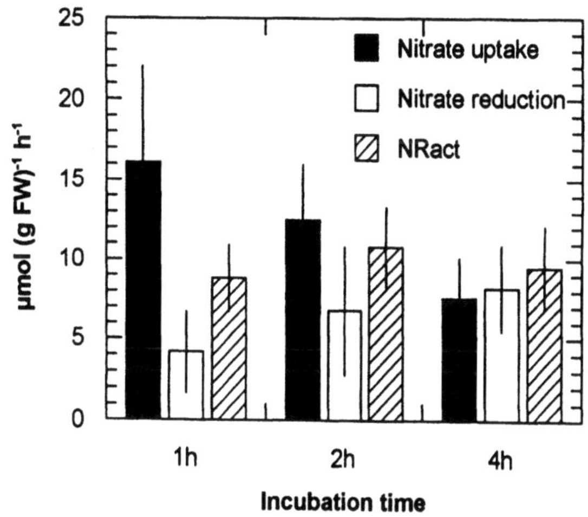

Fig. 3. Effect of nitrate feeding on nitrate reduction, nitrate uptake and NRact of spinach leaves illuminated in air. Each leaf was placed separately in an Eppendorf cup containing 1.5 ml of 30 mM KNO_3 solution. Uptake of the nitrate solution was followed by weighing the cups before and after removal of leaves at the indicated times. Initial nitrate content was 43.5 ± 4.8, which increased to 60.6 ± 7.8 (μmol g^{-1}FW) after 4 h. Nitrate reduction in situ was calculated as the difference between the mean increase in the leaf nitrate content and the mean amount of nitrate taken up with the solution. Data are means from three independent experiments with 4 parallel samples each (n = 12). Other conditions as in Fig. 1.

al. 1973). However, leaf discs floating on a buffer solution may lose nitrate by leakage, thereby decreasing cytosolic and vacuolar nitrate concentrations. We therefore measured the release of nitrate and other anions from discs into a solution containing only 0.1 mM $CaSO_4$ and 40 mM glycinebetaine as osmoticum (Fig. 4). In Fig 4A, nitrate and chloride release (leakage) was measured with discs prepared from leaves with very different initial nitrate content, but a similar chloride content (compare legend). The nitrate content of one part of the leaves had been strongly decreased by illumination (4 h) in 5% CO_2. Initial anion release rates from discs were high, probably indicating efflux from the apoplast. After one hour, leakage was almost linear for the subsequent 3 hours. These nitrate leakage rates were roughly proportional to the initial nitrate content (Fig. 4A). Initial chloride contents were equal in both leaf treatments, and accordingly chloride release rates were identical. Interestingly, nitrate (but not chloride) release was usually somewhat more rapid in the dark than in the light (Fig. 5). This would indicate that cytosolic nitrate in the dark was somewhat higher than in the light, but certainly not lower. Also, nitrate leakage in the dark under nitrogen was about the

Table 1. Effect of high external nitrate concentrations on NR activity and nitrite production rate in leaf discs floating on buffer solution under anoxia in the dark.

Conditions	NRact (μmol g^{-1} FW h^{-1})	NRmax (μmol g^{-1} FW h^{-1})	NO_2^- accumulated (μmol g^{-1} FW h^{-1})
t = 0	4.0 ± 1.6	16.7 ± 3.6	
control 2h	10.2 ±2.2	21.0 ± 5.3	0.60 ± 0.15
control 4 h	15.3 ± 3.0	21.4 ± 4.6	0.80 ± 0.20
+ 90 mM KNO_3, 2h	8.9 ± 2.5	20.9 ± 4.4	0.90 ± 0.20
+ 90 mM KNO_3, 4h	15.3 ± 3.4	21.3 ± 4.5	1.05 ± 0.20

At t = 0, NR activity and nitrite contents were determined in freshly harvested leaves from plants which had been kept in the dark (air) for 1 h. Control = no nitrate added. Data are means from two independent experiments with 4 parallel samples each (n = 8).

same as in light + 5% CO_2 (Fig. 5B). Although these simple experiments do not take into account a possible reabsorption of released nitrate or possible changes in membrane potential, they can be taken as an indication that cytosolic nitrate concentrations are not responsible for the very different nitrate reduction rates observed in light or in the dark, or in the dark under anoxia, where NR was fully active yet nitrate reduction was very low.

V. Is Nitrate Reduction Limited by NAD(P)H?

Why was initial nitrate reduction in situ stimulated by high CO_2 when NRact was approximately the same as in air? At 5% CO_2, the photosynthesis of spinach leaves is at a maximum, as indicated by their rates of oxygen evolution (data not shown). Therefore, more reducing equivalents may be available in the cytosol due to a faster export of triose phosphates from the chloroplast, leading to higher nitrate reduction rates than in air (compare Fig. 1). Unfortunately, it seems almost impossible to measure cytosolic NADH concentrations directly. Heineke et al. (1991) calculated a cytosolic NADH concentration of about 0.7 μM NADH in spinach leaves. Even if NADH were 10-fold higher than the estimated concentration, it would only just reach the K_M for NADH of NR (7 μM: Kaiser and Spill, 1991), suggesting a limitation of NR by NADH, as frequently proposed previously (Abrol et al., 1983, and literature cited therein).

As shown above, nitrate reduction in anoxic leaves in the dark was very low, although NR was highly active. Anoxic cells usually produce ethanol and lactic acid at the expense of NADH in order to maintain glycolytic flux. In tobacco roots, lactic acid and ethanol formation were hardly detectable in air, but increased within 2 h of anoxia to an NADH consumption rate equivalent to about 3 μmol g^{-1} FW (Stoimenova and Kaiser, unpublished). Rates of lactate and ethanol formation in spinach leaves have not yet been determined. In any case, high fermentation rates would not necessarily indicate that cytosolic NADH was sufficient to saturate NR, since K_M values of plant alcohol dehydrogenase for NADH are around 0.08 μM, and thus almost two orders of magnitude lower than the K_M (NADH) of NR.

In order to examine further a possible limitation of nitrate reduction by reducing equivalents, we fed leaf discs floating on buffer solution under anoxia with reduced methylviologen or benzylviologen, which act as artificial electron donors to NR in vitro, and followed nitrite formation under anoxia in the dark. However, in none of these experiments was anoxic nitrite formation stimulated by reduced viologen dyes (data not shown). This indicates either that reductant was not limiting or, more probably, that viologens were oxidized by side reactions in situ.

VI. Conclusions

At least for spinach leaves, which usually give high and stable enzymatic activities in crude extracts, the widely used determination of NR activity in vitro may lead to a considerable overestimation of nitrate reduction rates in vivo. This is especially obvious under conditions where NR is artificially activated.

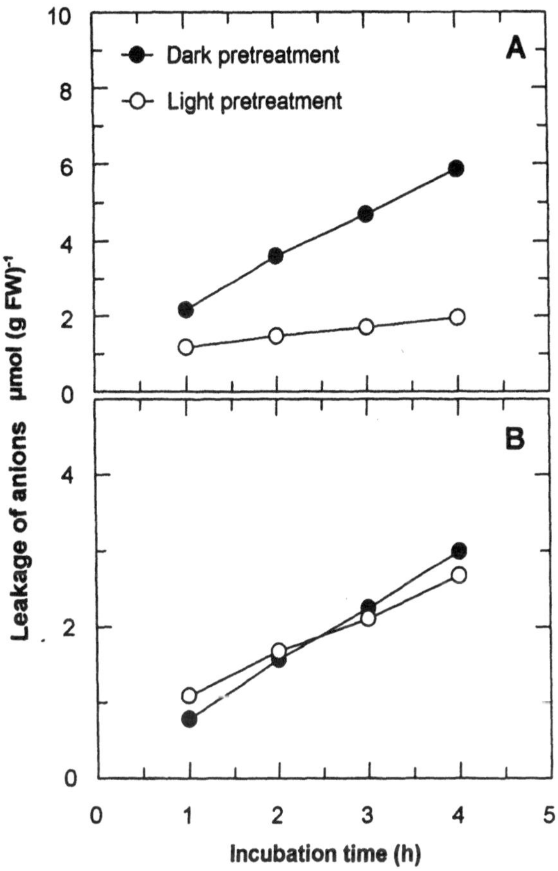

Fig. 4. Leakage of nitrate (A) and chloride (B) from leaf discs floating on a solution containing 0.1 mM $CaSO_4$ and 40 mM glycinebetaine. Ten leaves were harvested at the end of the dark period. One half of the leaves was kept in the dark (closed circles), the other half was illuminated in 5% CO_2, petiole in water (open circles). After 4 h, discs (6 mm diameter) were punched out with a cork borer and 30 discs, corresponding to about 240 mg FW, were collected in 2.5 ml of the above solution in small glass beakers. The discs were washed once by replacing the medium and were then continuously flushed with air under gentle shaking. At the times given in the figure, 0.5 ml aliquots of the medium was removed and immediately subjected to anion chromatography with suppressed conductivity detection (Biotronik IC 1000, Maintal, Germany). The medium was brought back to the original volume by adding 0.5 ml of fresh incubation medium immediately after sampling. Anions were also determined in extracts from discs collected at the beginning of the experiment. Initial nitrate content of dark-pretreated leaves was 34.9, and of light-pretreated leaves 12.3 (μmol g^{-1} FW). Respective chloride contents were 7.6 in both treatments. Each point represents the mean of two separate samples.

Apparently, NR in vivo works at sub-optimal substrate concentrations, except when photosynthesis is operating at maximum rates. Under sub-maximal photosynthetic conditions, and especially in the dark,

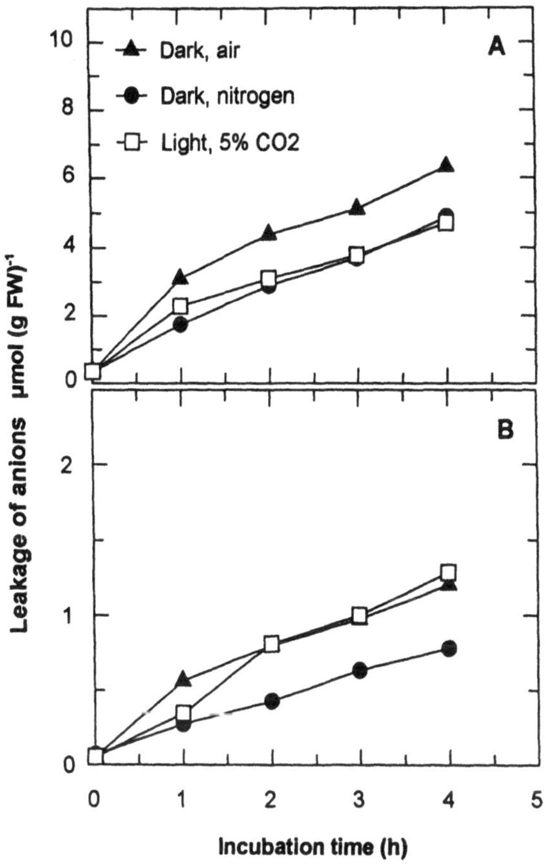

Fig. 5. Leakage of nitrate (A) and chloride (B) from leaf discs incubated in the dark (air or nitrogen) or in the light (5% CO_2). Initial nitrate content was 48.8 ± 8 and chloride content was 3.4 ± 0.5 (μmol g^{-1} FW). Other conditions as in Fig. 4.

NADH, rather than cytosolic nitrate, appears to be the principal factor which limits nitrate reduction in situ. After prolonged illumination, however, cytosolic nitrate may also drop below the level saturating for NR, even when the leaves still contain nitrate well into the millimolar range. Thus, high and continuous photosynthesis rates may be required not only to maintain cytosolic NADH, but also to support a continuously rapid export of nitrate from the vacuole under conditions where nitrate import from the apoplast ceases.

Expression of NR is affected by photosynthesis, as is the post-translational modulation of NR. It seems that the amount and activation state of NR are regulated in such a way that NR activity at substrate saturation is somewhat in excess of the rate in situ, which may enable plants to respond immediately to increased reductant availability.

Acknowledgments

This work was supported in part by the Deutsche Forschungs-Gemeinschaft, Sonderforschungsbereich 251, and by the Graduiertenkolleg 'Pflanze im Spannungsfeld...'. The skilled technical assistance of M. Lesch and E. Wirth is gratefully acknowledged.

References

Abrol YP, Sawhney SK and Naik MS (1983) Light and dark assimilation of nitrate in plants. Plant Cell Environ 6: 595–599

Beevers L and Hageman RH (1980) Nitrate and nitrite reduction. In: Stumpf PK and Conn EE (eds) The Biochemistry of Plants, Vol 5, pp 115—168. Academic Press, New York

Bligny R, Gout E, Kaiser WM, Heber U, Walker D and Douce R (1997) pH regulation in acid stressed leaves of pea plants grown in the presence of nitrate or ammonium salts: Studies involving ^{31}P-NMR spectroscopy and chlorophyll fluorescence. Biochim Biophys Acta 1320: 142–152

Crawford NM (1995) Nitrate: Nutrient and signal for plant growth. Plant Cell 7: 859–868

Crawford NM, Wilkinson JQ and LaBrie ST (1992) Metabolic control of nitrate reduction in *Arabidopsis thaliana*. Aust J Plant Physiol 19: 377–385

Ferrari TE, Yoder OC and Filner P (1973) Anaerobic nitrate production by plant cells and tissues: evidence for two nitrate pools. Plant Physiol 51: 423–431

Foyer CH, Lefebvre JC, Provot M, Vincentz M and Vaucheret H (1993) Modulation of nitrogen and carbon metabolism in transformed *Nicotiana plumbaginifolia* mutant E23 lines expressing either increased or decreased nitrate reductase activity. Aspects Appl Biol 34: 137–145

Heineke D, Riens B, Grosse H, Hoferichter P, Peter U, Flügge UI and Heldt HW (1991) Redox transfer across the inner chloroplast membrane. Plant Physiol 95: 1131–1137

Huber SC and Kaiser WM (1996) 5-Aminoimidazole-4-carboxyamide riboside activates nitrate reductase in darkened spinach and pea leaves. Physiol Plant 98: 833–837

Kaiser WM and Huber SC (1994) Modulation of nitrate reductase in vivo and in vitro: Effects of phosphoprotein phosphatase inhibitors, free Mg^{2+} and 5′-AMP. Planta 193: 358–364

Kaiser WM and Spill D (1991) Rapid modulation of spinach leaf nitrate reductase by photosynthesis. II. In vitro modulation by ATP and AMP. Plant Physiol 96: 368–375

Kaiser WM, Weiner H and Huber SC (1999) Nitrate reductase in higher plants: A case study for transduction of environmental stimuli into control of catalytic activity. Physiol Plant 105: 385–390

Kaiser WM, Kandlbinder A, Stoimenova M, Glaab J (2000) Discrepancy between nitrate reduction in intact leaves and nitrate reductase activity in leaf extracts: What limits nitrate reduction in situ? Planta 210: 801–807

Kandlbinder A, Weiner H and Kaiser WM (2000) Nitrate reductases from leaves of Ricinus (*Ricinus communis* L.) and spinach (*Spinacia oleracea* L.) have different regulatory properties. J Exp Bot 51: 1099–1105

King BJ, Siddiqi MY and Glass ADM (1992) Studies of the uptake of nitrate in barley. V. Estimation of root cytoplasmic nitrate concentration using nitrate reductase activity—implications for nitrate influx. Plant Physiol 99: 1582–1589

Kronzucker HJ, Siddiqi MY, Glass ADM and Kirk GJD (1999) Nitrate and ammonium synergism in rice. A subcellular flux analysis. Plant Physiol 119: 1041–1045

Lejay L, Quillere I, Roux Y, Tillard P, Cliquet JB, Meyer C, Morot-Gaudry JF and Gojon A (1997) Abolition of posttranscriptional regulation of nitrate reductase partially prevents the decrease in leaf NO_3^- reduction when photosynthesis is inhibited by CO_2 deprivation, but not in darkness. Plant Physiol 115: 623–630

Martinoia E, Schramm MJ, Kaiser G, Kaiser WM and Heber U (1986) Transport of anions in isolated barley vacuoles. I. Permeability to anions and evidence for a Cl^- uptake system. Plant Physiol 80: 895–901

Martinoia E, Schramm MJ, Flügge UI and Kaiser G (1987) Intracellular distribution of organic and inorganic anions in mesophyll cells: Transport mechanisms in the tonoplast. In: Marin B (ed) Plant Cell Vacuoles—Their Importance in Solute Compartmentation in Cells and Their Applications in Plant Biotechnology, pp 407—416. Plenum Press, New York

Miller AJ and Smith SJ (1996) Nitrate transport and compartmentation in cereal root cells. J Exp Bot 47: 843–854

Quillère I, Dufosse C, Roux Y, Foyer CH, Caboche M and Morot-Gaudry JF (1994) The effects of deregulation of NR gene expression on growth and nitrogen metabolism of *Nicotiana plumbaginifolia* plants. J Exp Bot 45: 1205–1211

Scheible WR, Gonzales-Fontes A, Morcuende R, Lauerer M, Geiger M, Glaab J, Gojon A, Schulze ED and Stitt M (1997) Tobacco mutants with a decreased number of functional *nia* genes compensate by modifying the diurnal regulation of transcription, post-translational modification and turnover of nitrate reductase. Planta 203: 304–319

Schröppel-Meier G and Kaiser WM (1988) Ion homeostasis in chloroplasts under salinity and mineral deficiency. I. Solute concentrations in leaves and chloroplasts from spinach plants grown under NaCl or $NaNO_3$ salinity. Plant Physiol 87: 822–827

Speer M and Kaiser WM (1991) Ion relations of symplastic and apoplastic space in leaves from *Spinacia oleracea* L. and *Pisum sativum* L. under salinity. Plant Physiol 97: 990–997

Speer M and Kaiser WM (1994) Replacement of nitrate by ammonium as N-source increases salt sensitivity of pea plants. II. Inter- and intracellular solute compartmentation in leaflets. Plant Cell Environ 17: 1223–1231

Van der Leij M, Smith SJ and Miller AJ (1998) Remobilisation of vacuolar stored nitrate in barley root cells. Planta 205: 64–72

Vaucheret H, Chabaud M, Kronenberger J and Caboche M (1990) Functional complementation of tobacco and *Nicotiana plumbaginifolia* nitrate reductase deficient mutants by transformation with the wild-type alleles of the tobacco structural genes. Mol Gen Genet 220: 468–474

Vincentz M and Caboche M (1991) Constitutive expression of nitrate reductase allows normal growth and development of *Nicotiana plumbaginifolia* plants. EMBO J 10: 1027–1035

Chapter 6

The Biochemistry, Molecular Biology, and Genetic Manipulation of Primary Ammonia Assimilation

Bertrand Hirel*
Unité de Nutrition Azotée des Plantes, INRA,
Route de St Cyr, 78026 Versailles, Cedex, France

Peter J. Lea
Department of Biological Sciences, Lancaster University, Lancaster LA1 4YQ, U.K.

Summary 71
I. Introduction: Glutamine Synthetase and Glutamate Synthase, Two Enzymes at the Crossroads Between Carbon and Nitrogen Metabolism 72
II. Glutamine Synthetase 72
A. Plastidic Glutamine Synthetase 74
B. Cytosolic Glutamine Synthetase 76
III. Glutamate Synthase 79
A. Ferredoxin-dependent Glutamate Synthase 79
B. NADH-dependent Glutamate Synthase 83
C. Production of 2-Oxoglutarate for Glutamate Synthase Activity 84
IV. Glutamate Dehydrogenase 85
References 86

Summary

Ammonia is assimilated in the leaves of higher plants by the combined action of chloroplastic glutamine synthetase (GS2) and glutamate synthase (GOGAT). Glutamine synthetase (GS1) is also present in the cytosol of plant cells, in particular in the vascular system, and exists in a number of isoenzymic forms. There are also two distinct forms of GOGAT, which may use either ferredoxin (Fd) or NADH as a source of reductant, the Fd-dependent form being predominant in leaves. In this article, the latest information is presented on the structure, properties and gene regulation of the various forms of both GS and GOGAT. The results of studies which have attempted to modify the activities of the enzymes by genetic manipulation, have been used to identify the roles played by GS and GOGAT in plant metabolism. The role of a third enzyme, glutamate dehydrogenase (GDH), in the deamination of glutamate and production of ammonia, is also discussed.

*Author for correspondence, Email: hirel@versailles.inra.fr

Christine H. Foyer and Graham Noctor (eds): Photosynthetic Nitrogen Assimilation and Associated Carbon and Respiratory Metabolism, pp. 71–92.

I. Introduction: Glutamine Synthetase and Glutamate Synthase, Two Enzymes at the Crossroads Between Carbon and Nitrogen Metabolism

The reduced form of inorganic nitrogen (N) ultimately available to plants for assimilation is ammonia, which is predominantly present as the ammonium ion NH_4^+. Consequently, the rate of ammonia assimilation is likely to be important for plant growth. Ammonia is produced in all plant organs and tissues through a variety of catabolic or anabolic processes, as well as being taken up directly as ammonium ions from the soil, by the roots. For example, ammonia may be generated through nitrate reduction in roots and shoots, through the fixation of atmospheric nitrogen by root nodules, by photorespiring leaves and through the phenylpropanoid pathway. Ammonia may also be released for reassimilation by sink tissues such as young developing or reproductive organs, from N transport compounds and through the breakdown of nitrogenous compounds, including proteins or N containing metabolites (Woodall et al., 1996; Lea and Ireland, 1999) (Fig. 1).

The discovery of the major role of the enzyme couple, glutamine synthetase (GS)/glutamate synthase (GOGAT), in ammonia assimilation in higher plants (Miflin and Lea, 1980) has led to a large number of studies on the mechanisms controlling the tissue- or organ-specific expression of these two proteins, as well as the environmental factors influencing their activity. In higher plants GS and GOGAT are represented by a number of isoenzymes distributed in the cytosol and in the chloroplast. Their relative activities in a given organ or tissue appear to be tightly linked to specific roles in primary N assimilation, ammonia recycling during photorespiration or N remobilization. In particular, in chlorophyllous tissues, ammonia assimilation and recycling are largely dependent on both photosynthetic and photorespiratory metabolism, the former providing carbon (C) skeletons necessary for amino acid biosynthesis (Fig. 1).

In higher plants, as compared to bacteria (Lee et al., 1999), yeast (Beck and Hall, 1999) or algae, very few studies have been carried out on co-regulated gene expression of enzymes involved in N assimilation (including nitrate reduction and ammonia assimilation) and C metabolism. It has been proposed that signals derived from nitrate interact with signals generated further downstream in N and C metabolism (Stitt, 1999). When nitrate is provided to the plant, C is diverted from carbohydrate synthesis to provide the organic acids necessary for the synthesis of glutamine (Gln) and glutamate (Glu) via the GS-GOGAT cycle. Moreover, it seems that signals derived from nitrate and signals derived from the subsequent reactions involved in ammonia assimilation and amino acid biosynthesis interact to coordinate C and N metabolism. There are strong indications that organic acids, amino acids and carbohydrates are some of the primary effectors (Lancien et al., 1999; Oliveira and Coruzzi, 1999; Ferrario et al., 2000) controlling N uptake and assimilation (Lejay et al., 1999) and their coordination with C metabolism. The reaction catalysed by GOGAT may be one of the checkpoints in this coordination: in transgenic plants with reduced enzyme activity, several pathways of amino acid biosynthesis are modified following the accumulation of ammonia, Gln and 2-oxoglutarate (2-OG), which may implicate these compounds as signaling molecules (Ferrario et al., 2000). However, the mechanisms through which these signals are sensed and transmitted remain poorly understood (Hirose and Yamaya, 1999; Sueyoshi et al., 1999).

The recent discovery of homologs of the bacterial PII protein capable of sensing the ratio of Gln to 2-OG in prokaryotes suggests that PII may be one of the components of a complex signal transduction network involved in perceiving plant C/N metabolic status (Hsieh et al., 1998). The identification of a cytokinin-inducible gene that is also sensitive to nitrate application, which possesses similarities to the bacterial signaling system, indicates that common hormonal and nutritional regulatory mechanisms may also function in a cooperative manner in higher plants (Sakakibara et al., 1998). However, it remains to be determined (for example, through the use of knockout mutants) how the loss of these proteins affects the C/N sensing in higher plants.

II. Glutamine Synthetase

GS (EC 6.3.1.2) catalyses the ATP-dependent conversion of Glu to Gln, utilizing ammonia as a

Abbreviations: Asp – aspartate; BSC – bundle sheath cells; Fd – ferredoxin; GDH – glutamate dehydrogenase; Gln – glutamine; Glu – glutamate; GOGAT – glutamate synthase; GS – glutamine synthetase; GS1 – cytosolic glutamine synthetase; GS2 – chloroplastic glutamine synthetase; MC – mesophyll cells; 2-OG – 2-oxoglutarate; PPT – phosphinothricin

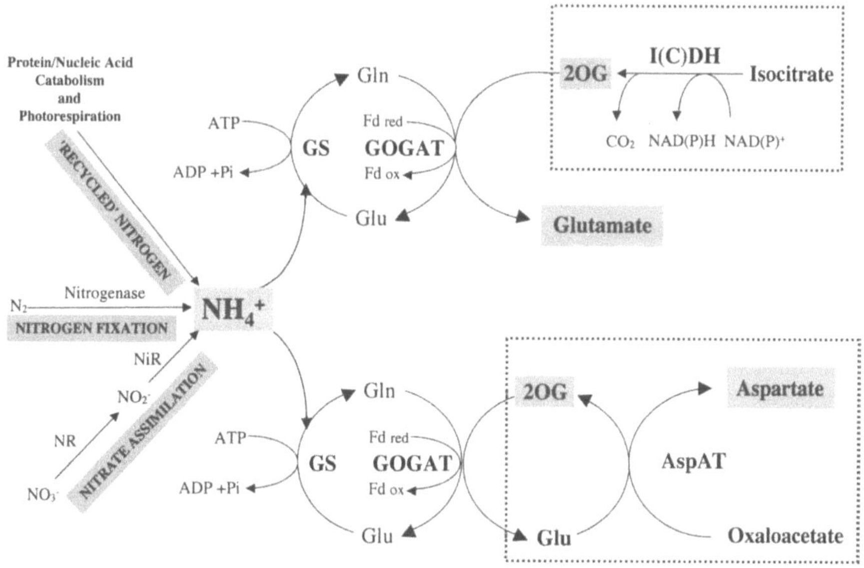

Fig. 1. Scheme showing the origin of ammonia/ammonium production and assimilation by the glutamine synthetase / glutamate synthase (GS/GOGAT) pathway. 2-Oxoglutarate (2OG) may be produced either by isocitrate dehydrogenase (I(C)DH) or aspartate aminotransferase (AspAT). Reproduced from Gálvez et al. (1999) and generously supplied by Michael Hodges. Fd, ferredoxin; Gln, glutamine; Glu, glutamate; NiR, nitrite reductase; NR, nitrate reductase.

substrate. Two major isoforms exist: cytosolic GS (GS1), occurring in the cytosol of leaves and non-photosynthetic organs, and chloroplastic GS (GS2), present only in the chloroplasts of photosynthetic tissues and the plastids of roots or etiolated plants (Cren and Hirel, 1999).

The two isoenzymes were originally identified using a combination of ion-exchange chromatography combined with subcellular fractionation of leaf or root extracts (McNally and Hirel, 1983). Although GS1 and GS2 are different proteins, immunochemical experiments showed that they possess common antigenic determinants and that these antigenic sites are similar between a large number of plant species (Hirel et al., 1984). Using antibodies raised against either GS1 or GS2, immunocytochemical experiments showed that plastidic GS2 is located exclusively in chlorophyllous tissues, where it is associated with the stroma matrix (Botella et al., 1988b). In some species, such as legumes or barley, GS2 has been found to be associated with the plastids in roots (Peat and Tobin, 1996) and root nodules (Brangeon et al., 1989). Both photonic and electronic immunocytochemistry also allowed the localization of GS1 at the cellular and subcellular level. It was found that GS1 is located predominantly in the cytosol of roots, root nodules (Brangeon et al., 1989; Peat and Tobin, 1996) and floral organs (Dubois et al., 1996), and moreover that in shoots and roots of C_3 plants it is localized in the vascular tissue, a high proportion of the protein being concentrated in the phloem companion cells (Peat and Tobin, 1996; Dubois et al., 1996; Sakurai et al., 1996). The situation appears to be different in C_4 plants, since a large proportion of GS protein was found in the cytosol of both mesophyll and bundle sheath cells (Becker et al., 1993). A unique situation was found in pine seedlings, in which GS was exclusively localized in the cytosol even though chloroplasts were fully differentiated in the seedlings studied (García-Gutiérrez et al., 1998).

Anti-GS antisera have been used to perform quantitative estimations of the relative amount of GS1 and GS2 subunits in different plant species and tissues (Becker et al., 1992, 2000; Woodall et al., 1996). This approach showed that the relative proportions of the cytosolic and plastidic GS may vary between different organs of the same plant or between different plant species, depending on their photosynthetic type (C_3 or C_4), natural habitat (McNally and Hirel, 1983; McNally et al., 1983) or whether they are woody species (Woodall et al., 1996; García-Gutiérrez et al., 1998). Originally, using ion-exchange chromatography, four main groups of plants were defined according to their GS isoenzyme composition (McNally et al., 1983). Group A consisted of plants with only GS1 activity, the achlorophyllous non-photosynthetic parasites. Species in group B and C were mainly represented by C_3 plants and contained only or predominantly GS2 activity. In group D, composed of C_4 plants or tropical legumes, approximately equal amounts of GS1 and GS2 were detected in leaf protein extracts. Subsequently, it was found that GS2 was absent from woody plants such as pine (García-Gutiérrez et al., 1998), whereas in others such as *Trientalis europaea,* the order of GS1 and GS2 elution from ion-exchange columns was the opposite to that of other species examined, for which GS1 always eluted first (Parry et al., 2000). The physiological significance of the different distribution of GS1 and GS2 in C_3 and C_4 plants remains largely unexplained, but seems to be tightly linked to the photosynthetic metabolism of plants originating from tropical regions.

A. Plastidic Glutamine Synthetase

In all plant species studied, GS2 is encoded by one nuclear gene per haploid genome. This gene encodes a polypeptide exhibiting a molecular mass of 43 to 45 kDa (depending on the plant species examined), which combines to form an octameric complex that is the native GS enzyme (Forde and Cullimore, 1989). In all the GS2 subunits, an N-terminal signal peptide of almost 50 amino acids is found, which targets the protein to the chloroplastic compartment (Lightfoot et al., 1988). In addition, a conserved region of 16 amino acids, characteristic of the GS2 protein, is present at the C-terminal part of the subunit. In some plants, such as tobacco, plastidic GS subunits may also be represented by several polypeptides differing in their charge (Hirel et al., 1984; Lara et al., 1984) or size (Valpuesta et al., 1989), whereas in pea (Tingey et al., 1987), only a single polypeptide could be detected after isoelectric separation or SDS gel electrophoresis. Glycosylation of GS2 has also been reported (Nato et al., 1984; Miranda-Ham and Loyola-Vargas, 1992), though the significance of this remains unclear: it may be involved in the turnover of the protein during senescence (Miranda-Ham and Loyola-Vargas, 1992).

Basic enzymatic studies led to the proposal that GS2 activity is modulated by light through changes in pH, Mg^{2+} concentration and adenylate nucleotide concentration (Hirel et al., 1983). It was then hypothesized that these mechanisms may be a way of controlling the flux of ammonia in the chloroplast to allow a fine tuning between N and C assimilation during the day/night transition. In addition, the high level of GS2 activity in the chloroplast when compared to GOGAT activity has led to the suggestion that due to a high affinity for ammonia (K_m in the μM range), the enzyme could be involved in ammonia detoxification within the different cellular compartments (Givan, 1979). In particular, it has been well established that in C_3 plants massive amounts of ammonia are released in the mitochondria during photorespiration, leading to the hypothesis that one of the two GS isoenzymes may be involved in reassimilating the excess of photorespiratory ammonia. Since significant amount of cytosolic GS are present in a number of plant species, Keys et al. (1978) proposed that, due its localization close to the mitochondria, the enzyme may be directly involved in this process. These authors used an in vitro reconstituted system composed of isolated mitochondria supplemented with purified GS to demonstrate that ammonia released during the decarboxylation of glycine could be reassimilated by GS. The photorespiratory N cycle was then proposed, in which GS in the cytosol and ferredoxin-dependent GOGAT (Fd-GOGAT) in the chloroplast recycle in a cooperative manner the ammonia released during the photorespiratory process.

A subsequent survey of the different GS isoform complement in a selection of higher and lower plants clearly demonstrated that several species, regardless of their classification or their ecological habit, did not contain any leaf cytosolic GS activity, at least in the mesophyll cells (McNally et al., 1983). Following this new finding, the role of GS1 during photorespiration was extensively debated. The matter was resolved when barley mutants lacking plastidic GS activity were isolated due to their inability to survive in air, thus demonstrating that GS2 was necessary for reassimilation of photorespiratory ammonia (Blackwell et al., 1988a,b). Analysis of mutants with reduced Fd-GOGAT activity (Leegood et al., 1995) confirmed that not only chloroplastic ammonia reassimilation but also generation of Glu through the GS/GOGAT cycle, was required to overcome the toxic build-up of metabolites derived from photorespiration. It was therefore presumed that the level of both GS2 and Fd-GOGAT gene expression in illuminated leaves would be regulated primarily with respect to the high rate of photorespiration to avoid the detrimental accumulation and/or depletion of ammonium, Gln or Glu. Experiments were therefore conducted to determine whether suppression of photorespiration led to down-regulation of GS2 and Fd-GOGAT expression (Edwards and Coruzzi, 1989; Cock et al., 1991). Contradictory results were obtained depending of the level of atmospheric CO_2 used to inhibit photorespiration (Migge et al., 1997). However, using a moderate increase in CO_2 concentration (by 300 rather than 2000-4000 $\mu L.L^{-1}$), Migge et al. (1997) did not observe any effect on the expression of either GS2 or Fd-GOGAT.

Since the regulation of chloroplastic ammonia assimilation and reassimilation was shown to be directly related to photorespiration and thus photosynthesis, it was thought that as for many chloroplast proteins, light may be involved in the regulation of GS2 expression and activity. Indeed, in C_3 and C_4 plants, it was found that light plays a fundamental role in the regulation of GS2 both at the transcriptional and post-transcriptional levels (Ireland and Lea, 1999). Following illumination of etiolated leaves and cotyledons, an increase in both GS2 transcript and protein have been observed in most C_3 or C_4 species examined. This increase was found to be more rapid during a transition from dark to light than during the illumination of etiolated leaves, since in the first instance chloroplasts are already fully differentiated (Hirel et al., 1982; Edwards and Coruzzi, 1989; Galvez et al., 1990). The influence of light-dependent factors on GS2 expression was confirmed when etiolated plants were exposed to different wavelengths of the spectrum. Experiments with white, red, far-red or blue light showed that both phytochrome and the blue-light photoreceptor are involved in the positive response to light (Edwards and Coruzzi, 1989; Becker et al., 1992; Migge et al., 1998). More detailed studies on *Pinus sylvestris* demonstrated that light regulation of GS2 expression occurs coarsely at the transcriptional level and more finely at the post-translational level (Elmlinger et al., 1994), and involves modifications in subunit composition, as has also been shown in tomato seedlings (Migge et al., 1998). However, the biological role of the post-translational modification of GS2 subunit composition is still unknown.

Compared to the large number of other studies describing the light perception and the subsequent signal transduction pathway regulating the expression of genes encoding proteins and enzymes implicated in the photosynthetic process (Bowler and Chua, 1994), very little is known about the mechanism controlling *GS2* gene transcription, likely because additional environmental and developmental factors are also involved (see below). In order to identify light-responsive elements in the *GS2* promoter, transgenic plants expressing promoter-reporter gene fusion constructs have been produced. A 323 bp promoter fragment from pea *GS2* contains *cis*-acting elements responsible for the light-regulation of the *GUS* reporter gene in the leaf mesophyll cells of mature transgenic tobacco or *Arabidopsis thaliana* (Tjaden et al., 1995). However, since a basal level of *GUS* expression was detected in etiolated cotyledons, it was suggested that promoter elements other than the light-responsive one may be involved in *GS2* gene expression in non-photosynthetic tissues. Similarly, it was shown that a 460 bp fragment of the *Phaseolus vulgaris GS2* promoter was sufficient for light-regulation and specific photosynthetic tissue expression of the *GUS* reporter gene in transgenic tobacco (Cock et al., 1992).

In conjunction with light, metabolites such as nitrate, ammonia, amino acids or carbohydrates may also play a regulatory role in controlling the production of GS2 in leaves (Mäck, 1995; Migge et al., 1996). In the presence of an N source and illumination with red or far-red light, etiolated tomato seedlings synthesize two types of GS2 polypeptides while only one is detected in the presence of ammonium. Thus, specific wavelengths (via phytochrome), and also nitrate, can modify the GS2 subunit composition of tomato at the post-translational level (Migge et al., 1998). In the majority of plant species examined so far, ammonia does not seem to have any effect on chloroplastic GS activity. However, in both rice and tobacco leaves, chloroplastic *GS2* gene transcription is enhanced following the addition of ammonia to the growth medium (Kozaki et al., 1992; Lancien et al., 1999). In barley plants supplemented with ammonia, an increase in GS2 corresponding to a change in the subunit composition of the native holoenzymes has also been observed (Mäck, 1995).

Light may also exert an effect indirectly through changes in C metabolites derived from photosynthesis. For example, in dark-adapted *A. thaliana* seedlings, sucrose enhances *GS2* expression, thus mimicking the effect of light. This result suggests that light exerts an indirect effect on *GS2* gene expression and that an efficient photosynthetic activity producing sucrose and/or another metabolizable sugar is required to control *GS2* gene expression (Melo-

Oliveira et al., 1996). In addition, Oliveira and Coruzzi (1999) have shown that *GS2* gene expression and activity are controlled by the relative abundance of C skeletons versus amino acids

There is increasing evidence suggesting that in addition to light and metabolites, the functionality of the plastids is prerequisite for optimal GS2 activity. It is well known that temperature is an important environmental factor controlling the expression of several genes involved in the photosynthetic process. In pea and barley plants grown at 15 °C instead of 25 °C, a 50% reduction in GS2 activity was observed after two days, while the activity of GS1 was unaffected (Woodall et al., 1996), again indicating that an optimal photosynthetic activity is required to attain full GS activity in the chloroplast. Similarly, when tomato plants were infected by the pathogen *Pseudomonas syringae* or treated with the GS inhibitor, phosphinothricin (PPT), Pérez-Garcia et al. (1998) observed a rapid leaf chlorosis. Following these two treatments, a decrease of both *GS2* gene expression and protein content, concomitant with an increase in *GS1* expression, was observed when plants were exposed to light. In contrast, in non-photosynthetic conditions, these modifications were not observed, leading to the conclusion that light-dependent factors are involved in controlling the expression of the two GS isoenzymes (Pérez-Garcia et al., 1998). In particular, the authors hypothesized that the decrease in chloroplastic GS following PPT treatment is the result of chloroplast degeneration due to a down-regulation of photosynthetic genes by the GS inhibitor. A similar situation seems to occur during natural senescence, when a rapid decrease in chloroplastic GS activity is associated with the degeneration of chloroplasts and the concomitant loss of photosynthetic functions (Kamachi et al., 1991; Masclaux et al., 2000).

Plastidic GS activity may also exercise significant control over apoplastic NH_4^+ concentrations in photosynthetic tissues. The temperature-mediated displacement of the chemical equilibrium between gaseous and aqueous ammonia may greatly influence the emission of gaseous ammonia from leaves, provoking serious negative environmental impacts associated with acidification and eutrophication and a loss of up to 5% of the shoot N content (see Schjoerring et al. (2000) for a review). Although far less well documented, leaf developmental stage may be an important parameter influencing final GS2 activity. Mäck and Tischner (1994) proposed that a progressive modification of the holoenzyme structure from an octameric form to a tetrameric form may be a means of controlling both the enzyme activity and stability during leaf ontogeny. This process may be related to a specific function of GS2 at certain stages of plant development, as revealed by the positive effect of GS2 overexpression on the growth of young tobacco seedlings (Migge et al., 2000). In these plants, the significant increase in biomass production was attributed to a more efficient incorporation of ammonium into organic molecules thus increasing the relative amounts of some amino acids such as Gln, Glu and Asp. Although this increase had no repercussions for plant soluble protein content, it was hypothesized that some unknown metabolic adjustments, possibly involving other N containing molecules such as polyamines, may be responsible for the positive effect on plant growth.

B. Cytosolic Glutamine Synthetase

Like GS2, GS1 is an octameric protein, but with smaller subunits, ranging from 38 to 41 kDa depending on the species. In some species, such as tobacco (Dubois et al., 1996), tomato (Becker et al., 1992) sugar beet (Brechlin et al., 1999), or pine (Cantón et al., 1999), leaf GS1 is composed of a single type of subunit of similar size whereas in others, such as soybean (Hirel et al., 1987), two subunits of different size have been identified as components of the holoenzyme. Additional experiments using 2-D gel analysis indicated that in French bean, a single GS1 subunit may be composed of two polypeptides of different charge (Lara et al., 1984). However, the significance of these differences, in terms of enzyme activity and physiological function, remains unknown. To investigate further the relationship between the structure and the function of the various GS1 polypeptides, preliminary studies using in vitro mutagenesis were undertaken (Clemente and Márquez, 1999), allowing the identification of key amino acid residues important for the catalytic properties of the enzyme. In addition, Carvalho et al. (1997) showed that two different GS1 polypeptides synthesized in vitro are able to self-assemble. However, further work is required to establish whether the kinetic properties of the enzymes produced in vitro are physiologically relevant.

Although there is normally only one gene encoding GS2, studies on a wide range of species have shown that GS1 is encoded by a complex multigene family which varies from three to six genes. In pea, three *GS1* genes are expressed in leaves, predominantly in the phloem cells. Two of the genes, *GS3A* and *GS3B*, have high sequence identity in both coding (99%) and noncoding (96%) regions and the two polypeptides differ by only three amino acids (Walker and Coruzzi, 1989; Walker et al., 1995). The genes encoding GS1 in maize have been the subject of intense study by two independent research groups. Of the four *GS1* cDNA clones isolated by Sakakibara et al. (1992), *GS1a* and *GS1b* were strongly expressed in etiolated maize leaves and exhibited a small increase during greening, whereas *GS1c* and *GS1d* mRNAs were barely detectable in etiolated leaves. The expression of *GS1c* decreased during greening, while *GS1d* increased. Five *GS1* genes have been isolated from maize by Li et al. (1993). Three of these ($GS_{1\text{-}2}$, $GS_{1\text{-}3}$ and $GS_{1\text{-}4}$) were expressed in mature leaves, seedling shoots and stems. $GS_{1\text{-}5}$, and to a much lesser extent $GS_{1\text{-}1}$, were expressed in the seedling shoot and stem, but not in the leaves. In the amphidiploid *Brassica napus*, formed by the fusion of two *Brassica* species, there was evidence that at least four *GS1* and two *GS2* genes were expressed, these being derived from both parents (Ochs et al., 1999). Phylogenetic analyses of plant GS genes have been carried out by Doyle (1991) and Biesiadka and Legocki (1997). Following the isolation of two different forms of GS in the chloroplasts of *Trientalis europaea*, a further analysis was carried out by Parry et al. (2000). Clear divisions were identified between *GS1* and *GS2* and between monocot and dicot gene sequences. The genes encoding GS in plants are termed type II and are similar in all eukaryotes, but are distinct from those found in prokaryotes, which are designated type I. Recently, Mathis et al. (2000) have demonstrated that type I genes are also present in plants and may represent a separate small gene family that is expressed in a number of different organs.

Compared to roots and root nodules (Ireland and Lea, 1999), relatively few studies have focused on *GS1* gene expression in leaves and the subsequent synthesis of the corresponding polypeptides. In many C_3 plants a gene encoding GS1, that is constitutively expressed in roots, is induced in leaves after the onset of leaf senescence (Ochs et al., 1999; Brugière et al., 2000). In C_4 plants, *GS1* gene expression appears to be generally constitutive in both roots and shoots since high levels of GS1 are always present in both organs regardless of the developmental stage of the plant (Sakakibara et al., 1992; Li et al., 1993). It is generally found that in the leaf vascular tissue one or two members of the cytosolic GS multigene family are constitutively expressed in C_3 and C_4 graminaceous and non-graminaceous plants (Li et al., 1993; Dubois et al., 1997). Interestingly, in tomato, a plant which possesses only a single GS1 subunit, treatment with PPT, or infection by the plant pathogen *Pseudomonas syringae,* induced the synthesis of a novel GS1 subunit of slightly lower molecular weight (Pérez-Garcia et al., 1998). This induction was attributed to the accumulation of ammonia under stress conditions, leading to the hypothesis that at least one member of the GS1 multigene family encoding a specific GS polypeptide is induced for an optimal enzyme activity adapted to physiological stress conditions. This had led to the suggestion that this kind of stress-adaptive mechanism, during which both GS1 gene expression and activity are induced, also occurs in senescing green tissues (Kamachi et al., 1991; Pearson and Ji, 1994). During senescence, most chloroplastic assimilatory functions, including primary ammonia assimilation, are progressively reduced and replaced by metabolism allowing the remobilization of protein N in the cytosol (Feller and Fisher, 1994; Brugière et al., 2000; Masclaux et al., 2000).

A similar shift seems to occur when plants are subjected to either water stress (Bauer et al., 1997) or pathogen infection (Pérez-Garcia et al., 1995), suggesting that common molecular control mechanisms may be involved in enhancing GS1 gene and protein expression in stressed leaves. It is still a matter for discussion whether these common regulatory mechanisms are part of a general signaling network controlling the various responses associated with leaf senescence (physiological or stress-induced), or whether they can be triggered by specific metabolic changes associated with leaf ageing. However, the use of transgenic plants overexpressing a heterologous gene encoding GS1 in the leaf cytosol of plants where the native gene is not normally expressed, demonstrated that leaf N remobilization can be prematurely induced (Vincent et al., 1997). This result suggests that metabolic signals may trigger the induction of genes involved in leaf N remobilization, whether ammonia assimilation in the cytosol is naturally induced in senescing leaves (Masclaux et al., 2000) or forced by overexpressing GS in the leaf cytosol (Hirel et al., 1992).

The occurrence of cytosolic GS protein in the phloem (Sakurai et al., 1996) has also led to speculation concerning its role during N transport and mobilization. It is still a matter of discussion whether the phloem-specific GS isoenzyme plays a non-overlapping role compared to the other GS isoenzymes expressed in roots and leaves. Recent work by Brugière et al. (1999) suggests that it does: using transgenic tobacco plants impaired in GS1 activity in the phloem, it was demonstrated that the enzyme plays in important role in proline production, particularly under conditions of water shortage. In some plants, however, vascular GS1 seems to function in conjunction with the rest of the ammonia assimilatory pathway. In rice, for example, vascular GS1 is the only GS1 detected, even in senescing leaves, and it has been proposed that the enzyme plays a major role during leaf N remobilization for grain-filling (Sakurai et al., 1996). Once again, it seems that species-specific adaptive mechanisms exist whereby cytosolic ammonia assimilation may be either turned on, enhanced or maintained during leaf development.

Despite the few species-specific characteristics in terms of leaf GS1 localization and mode of expression, the current consensus is that leaf cytosolic Gln synthesis is associated with the process of N remobilization of C_3 plants, rather than with photosynthesis and photorespiration. If so, an interesting question arises concerning the origin of the C skeletons for cytosolic Gln synthesis. A possible metabolic pathway has been proposed that involves the transamination of amino acids released following protein hydrolysis, which would thereby contribute to the pools of pyruvate, acetyl CoA or 2-OG (Buchanan-Wollaston, 1997). However, this hypothesis requires further experimentation to assess the role of either glutamate dehydrogenase (GDH) (Robinson et al., 1992; Masclaux et al., 2000) or NADH-GOGAT (Yamaya et al., 1992) in providing Glu for the reaction catalysed by GS1.

The apparent compartmentation of N remobilization and transport in the cytosol of either leaf mesophyll or leaf vascular tissue does not seem to be so evident in C_4 plants where approximately equal amounts of GS1 and GS2 are present (McNally et al., 1983; Becker et al., 1993). An extreme case was found in pine seedlings, in which GS2 was not induced after transfer from dark to light, despite apparent high photosynthetic and photorespiratory capacity (García-Gutiérrez et al., 1998). Due to the lack of physiological studies using either transgenic plants or mutants deficient in leaf mesophyll GS1 activity, no firm hypotheses have been proposed to assign a role for GS1 in either C_4 plants or gymnosperms. In C_4 plants, N assimilation is divided between two distinct photosynthetic cell types, mesophyll cells (MC) and bundle sheath cells (BSC), nitrate reduction occurring in MC and photorespiratory ammonia reassimilation in BSC. Since GS1 was found to be present in both cell types, Becker et al. (2000) suggested that in BSC, GS1 in conjunction with GDH could function in the generation of Gln for the transport of reduced N to the phloem, whereas GS1 in MC may contribute to the efficient utilization and recycling of N, characteristics of C_4 plants (Oaks, 1994).

In gymnosperms, it was hypothesized that the predominance of GS1 regardless of the photosynthetic capacities of the seedlings was the result of adaptation to the etiolation response during germination and/or to darkened habitats (García-Gutiérrez et al., 1998). The reaction catalysed by GS1 may also be an important factor influencing plant growth and development in trees as revealed by the significant increase in both soluble protein and height of transgenic poplar overexpressing a gene encoding a pine cytosolic GS (Gallardo et al., 1999). This observation reinforces the current idea that GS1 may be an important checkpoint for plant productivity, if we consider its role in assimilating or recycling ammonia in a variety of anabolic or catabolic processes. The resulting Gln is then used to transport most of the combined N to different organs or cell types during plant growth and development (Harrison et al., 2000).

In conclusion, the possible functions of the different GS isoenzymes can be summarized as follows. Leaf plastidic GS2 plays a ubiquitous role in ammonia assimilation or reassimilation in conjunction with the various metabolic processes associated with the photosynthetic capacity of the leaf. This ubiquity of function may also be explained by the fact that in all higher plant species examined so far, GS2 is encoded by a single gene per haploid genome. Therefore, species-specific adaptive mechanisms such as post-transcriptional modifications, enzyme subunit polymerization, or rate of protein turnover, may have been selected during evolution to fulfil different functions confined to a single organelle, the chloroplast. In contrast, ammonia assimilation or recycling in the cytosol, instead of being restricted to a single sub-cellular compartment, is carried out in different plant parts by multiple isoenzymes, differentially expressed in various organs or tissues, according to both the developmental stage and the physiological status of the plant. From an evolutionary point of view, this may explain why cytosolic GS1 is encoded by a multigene family, each member encoding a single and unique polypeptide. To form the holoenzyme, these polypeptides can assemble into homo-octamers or hetero-octamers, depending on the organ or the physiological status of a given organ or tissue. The exact nature of the molecular mechanisms that control, in a coordinated manner, the various events between *GS1* gene transcription and holoprotein assembly and turnover, is still an enigma. Deciphering the metabolic and developmental signal(s) involved will be one of the main future goals, in order to explain how the different GS1 isoenzymes may be able to control ammonia assimilation in particular and N metabolism in general, for optimal plant growth and development.

III. Glutamate Synthase

GOGAT catalyses the Fd- or NADH-dependent conversion of Gln and 2-OG to two molecules of Glu:

$$\text{Glutamine} + \text{2-oxoglutarate} + 2H^+ + 2e^- \rightarrow \text{2 Glutamate}$$

In the original publication describing the reaction, pea chloroplasts were shown to be able to convert Gln and 2-OG to two molecules of Glu, utilising light as the source of reductant (Lea and Miflin, 1974). Later studies indicated that illuminated chloroplasts were also able to catalyse O_2 evolution, in the presence of 2-OG and ammonia (Anderson and Done, 1977; Anderson and Walker, 1983). This process was attributed to the combined reaction of both GS and GOGAT. The capacity of chloroplasts to convert inorganic N into amino acids can, therefore, be considered a true photosynthetic reaction, in the same way as CO_2 assimilation or nitrite reduction.

A. Ferredoxin-dependent Glutamate Synthase

Fd-GOGAT (EC 1.4.7.1) was first isolated from pea leaves (Lea and Miflin, 1974) and may represent 1% of the total leaf protein (Márquez et al., 1988). The enzyme has been shown to be monomeric with molecular masses of 165 kDa in pea (Wallsgrove et al., 1977) and tomato (Migge et al., 1998), 154 kDa in barley (Márquez et al., 1988), 164 kDa in tobacco (Zehnacker et al., 1992), 168 kDa in pine (García-Guttiérrez et al., 1995), 180 kDa in *A. thaliana* (Suzuki and Rothstein, 1997), 160 kDa in soybean (Turano and Muhitch, 1999). The enzyme in spinach and *Chlamydomonas reinhardtii* has been shown to contain one FMN, one FAD and one [3Fe-4S] cluster per molecule (Hirasawa et al., 1992). However, later studies with spinach indicated that the enzyme did not contain FAD (Hirasawa et al., 1996). The assay of Fd-GOGAT has been greatly facilitated by the use of methyl viologen as a source of reductant, rather than Fd itself, provided that a saturating concentration is employed (Márquez et al., 1988). Using three different monoclonal antibodies raised against the tobacco enzyme, Suzuki et al. (1994) were able to show that Fd and methyl viologen were recognized by the same domain. Using N-bromosuccinimide, Hirasawa et al. (1998) demonstrated that modification of two tryptophan residues in spinach GOGAT prevented the binding of both Fd and methyl viologen to the enzyme protein and caused a severe inhibition of GOGAT activity. An involvement of thioredoxin in the activation of Fd-GOGAT has also been proposed (Lichter and Häberlein, 1998).

The first report of the sequence of a cDNA clone encoding Fd-GOGAT was made by Sakakibara et al. (1991). The maize cDNA was shown to encode a polypeptide of 1616 amino acids, including a chloroplast transit peptide sequence of 97 amino acids. The molecular mass of the mature protein was calculated as 165kDa, in agreement with the value determined by SDS-PAGE. In the sequence of the mature polypeptide, 633 amino acids (42% of the sequence) were shown to be identical to the *E. coli* NADPH-dependent enzyme. The sequence also contained a short region similar to the potential FMN-binding region of yeast flavocytochrome b_2. Only one copy of the gene was detected in maize (Sakakibara et al. 1991). A cDNA clone encoding 70% of the amino acids of tobacco Fd-GOGAT was isolated by Zehnacker et al. (1992). The co-linear amino acid sequences of the tobacco and maize enzymes were 85% homologous. The tobacco sequence also shared a conserved region with the large subunit of the *E. coli* enzyme and again a putative FMN-binding site was detected. Only one copy of the gene was detected in the diploid species

Nicotiana sylvestris, but two copies were present in the amphidiploid *Nicotiana tabacum,* which could account for the presence of two polypeptides of very similar molecular mass (Zehnacker at al., 1992). A specific 1.3 kb cDNA fragment, encoding approximately 30% of the amino terminal portion of mature barley Fd-GOGAT, was amplified, cloned and sequenced by Avila et al. (1993). This sequence was 87% identical to the maize sequence (Sakakibara et al., 1991) at the nucleotide level and 88% identical at the amino acid level, but surprisingly did not overlap with the tobacco sequence (Zehnacker et al., 1992). A putative Gln-binding site, based on similarities of the sequence with *purF*-type amidotransferases, was identified in the amino terminal region of the barley enzyme protein (Avila et al., 1993). A cDNA clone encoding 1483 amino acids has been isolated from spinach. The amino acid sequence was 83% identical to the maize enzyme and was 43.3% identical to the large subunit of the *Azospirillum brasilense* NADPH-dependent enzyme and 39.1% identical to the *E. coli* enzyme. Only one copy of the spinach gene was detected (Nalbantoglu et al., 1994). A cDNA clone encoding the C-terminal third of the protein was isolated from Scots pine, the amino acid sequence of which again showed a high homology with the previously published sequences and also the presence of a putative FMN-binding site (García-Gutiérrez et al., 1995). Partial sequences encoding Fd-GOGAT have now also been characterized from grapevine (Loulakakis and Roubelakis-Angelakis, 1997) and soybean (Turano and Muhitch, 1999).

Suzuki and Rothstein (1997) analyzed in detail the predicted amino acid sequence of a full length cDNA clone isolated from *A. thaliana.* The N-terminal region upstream of Cys 132, which contained a high percentage of basic amino acids and a continuous serine sequence, was identified as the chloroplast transit peptide. The N-terminal domain of the enzyme protein, was shown to contain a Cys132-His340-Asp173 triad, which is similar to that found in *purF*-type glutamine amidotransferases and is presumably the Gln-binding site. The region Leu 1210 to Arg 1267 was identified as the FMN-binding site. In addition, three Cys residues at 1263, 1269 and 1274 were predicted to be involved in the binding of FeS clusters. Additional glycine-rich potential adenylate-binding sites were also identified in the *A. thaliana* amino acid sequence (Suzuki and Rothstein, 1997). Temple et al. (1998) constructed a phylogenetic tree based on the amino acid sequences of regions common to all eubacterial and eukaryotic GOGAT proteins. With the exception of the *Synechocystis sp. gltB* gene product, all of the Fd-GOGAT proteins clustered together. The analysis indicated that the eukaryotic and bacterial enzymes are closely related and that the genes are probably derived from the eubacterial precursors of chloroplasts, consistent with an endosymbiotic origin of chloroplasts. Support for this conclusion is provided by the finding that the Fd-GOGAT gene isolated from the red alga *Antithamnion* sp. is encoded in the plastid genome (Valentin et al., 1993).

Mutants lacking Fd-GOGAT have been isolated in *A. thaliana* (Somerville and Ogren, 1980) and barley (Kendall et al., 1986). The mutants accumulated very high concentrations of Gln when grown in air and exhibited major changes in amino acid metabolism (Blackwell et al., 1988a,b; Häusler at al., 1996). These mutants were identified via their requirement for growth in elevated CO_2 and by the development of severe stress symptoms in normal air, due to an inability to carry out photorespiration (Leegood et al., 1995). These mutants are discussed further in Chapter 8 (Keys and Leegood).

The characteristics of the mutants lacking GOGAT activity, as well as all the early molecular studies, indicated that there was only one gene encoding Fd-GOGAT in higher plants. It therefore came as a great surprise that in a key review article, Lam et al. (1996) proposed that there were in fact two expressed genes in *A. thaliana.* In their definitive study on Fd-GOGAT genes, Coschigano et al. (1998) sequenced thirteen cDNA clones, of which twelve were identical and were designated *GLU1,* while the thirteenth was designated *GLU2.* The two nucleotide sequences were 71% identical and the predicted amino acid sequences had 80% identity, differing primarily at the N and C terminals. Both cDNAs encoded an N-terminal extension that was characteristic of a chloroplast transit peptide. The *GLU1* gene was the major form expressed in the leaves, while *GLU2* was expressed at very low levels in leaves, but more abundantly in roots. The *GLU1* gene mapped to a region of chromosome 5, while *GLU2* mapped to chromosome 2. Coschigano et al. (1998) warned that 'it is likely that other species contain a second gene for Fd-GOGAT, that may have been missed in other cDNA screens, because of low expression levels.' Interestingly, more recent evidence from grapevine (Loulakakis and Roubelakis-Angelakis, 1997), *A. thaliana* (Suzuki and Rothstein, 1997) and soybean

(Turano and Muhitch, 1999) has indicated that these plants may also have two genes encoding Fd-GOGAT.

Light has been shown to cause a large increase in Fd-GOGAT activity in cotyledons and leaves (Wallsgrove et al., 1982; Suzuki et al., 1987; Hecht et al., 1988; Zehnacker at al., 1992; Fernandez-Conde et al., 1995; Pajuelo et al., 1997; Turano and Muhitch, 1999). In sunflower leaves, rhythmic fluctuations in enzyme activity have been detected that were not dependent upon light/dark cycles (Fernandez-Conde et al., 1995). In sugar beet leaves, Fd-GOGAT activity reached a maximum at the end of the dark period and then decreased steadily during the light period to reach 58% of the starting value and even more dramatic reductions in GS activity were detected (Schjoerring et al., 2000).

In maize, although Fd-GOGAT mRNA could be detected in dark-grown leaves, the level increased eight-fold, four days after transfer into the light (Sakakibara et al., 1992a,b). Similar results were obtained with etiolated tobacco leaves that had been exposed to light for two days (Zehnacker et al., 1992). The activity of Fd-GOGAT in tomato seedlings increased four-fold following the transfer to light for one day, accompanied by a similar increase in the enzyme protein and an even more striking change in mRNA abundance (15-fold increase: Becker et al., 1993a). More recently, a number of groups have confirmed that light induces increased abundance of the Fd-GOGAT mRNA and the enzyme protein in a range of plants (Loulakakis and Roubelakis-Angelakis, 1997; Pajuelo et al., 1997; Suzuki and Rothstein, 1997; Turano and Muhitch, 1999). In *A. thaliana*, the level of Fd-GOGAT *GLU1* transcripts was shown to increase dramatically in response to light in as short a time as 3 h, reaching a peak at 24 h, while only a small effect was noted with the *GLU2* transcript. Sucrose was able partially to replace the effect of light on the *GLU1* mRNA, but had no effect on *GLU 2* (Coschigano et al., 1998). Similar stimulatory effects of sucrose, in the absence of light have previously been demonstrated for other genes encoding enzymes of N assimilation, e.g. nitrate reductase, nitrite reductase and chloroplastic GS.

It has been suggested that the induction of enzyme activity is a phytochrome-mediated response (Hecht et al., 1988; Becker et al., 1993a), which may also include a specific blue/UV-A light receptor (Teller et al. 1996). Further work by Migge et al. (1998) has confirmed that both UV-A and UV-B can increase Fd-GOGAT transcripts, protein and activity in etiolated tomato seedlings. In pine and other gymnosperm seedlings, even when grown in the dark, the chloroplasts synthesize chlorophyll and enzymes involved in CO_2 assimilation. In a range of different pine seedlings, there were substantial increases in the levels of Fd-GOGAT activity, polypeptide and mRNA during germination, which were the same in either dark- or light-grown plants (García-Gutiérrez et al., 1995, 1998). It was argued that the ability of the pine seedlings to synthesize Glu in the dark is essential if the full photosynthetic development of the chloroplasts is to take place (Cánovas et al., 1998).

Fd-GOGAT activity has been shown to increase in maize in response to nitrate and ammonium ions (Sakakibara et al., 1992b). More recently, the interaction between light and N sources has been studied in maize leaves (Suzuki at al., 1996). Fd-GOGAT activity and polypeptide increased three- to five-fold following transfer of etiolated seedlings to nitrate or ammonia in the light, but not in the dark. A corresponding five-fold increase in mRNA encoding the enzyme was also detected under the same conditions (Suzuki et al., 1996). In tobacco leaves, a small decrease in the expression of the genes encoding chloroplastic GS and Fd-GOGAT following N starvation was observed, but the effect was much less marked than for both nitrate reductase and nitrite reductase. Both GS and GOGAT mRNA levels were restored by the application of nitrate and Gln, while ammonia or Glu only increased the Fd-GOGAT mRNA (Migge and Becker, 1996). In soybean cotyledons or leaves, there was very little evidence of any change in Fd-GOGAT activity, protein or mRNA, following transfer from zero N to nitrate or ammonium, in either the light or dark (Turano and Muhitch, 1999). Similar results were also obtained with tomato seedlings (Migge et al., 1998). In grapevine cell cultures, nitrate induced a slight stimulatory effect on the level of Fd-GOGAT mRNA, while ammonium ions were inhibitory (Loulakakis and Roubelakis-Angelakis, 1997). In transgenic tobacco plants overexpressing chloroplastic GS, with elevated concentrations of both Glu and Gln, there was no evidence of any change in the expression of Fd-GOGAT (Migge et al., 2000).

The cumulative information on the expression of the Fd-GOGAT genes in leaves and cotyledons clearly indicates that light is by far the major regulatory factor and that the N source plays only a minor role. These findings, together with previous studies of photorespiratory mutants deficient in leaf Fd-GOGAT activity (Somerville and Ogren, 1980; Kendall et al., 1986), strengthen the current consensus that the major role of the enzyme is to reassimilate the ammonia liberated during photorespiration (Keys et al., 1978). However, growing tobacco plants under conditions of elevated CO_2, which should suppress photorespiration, had no effect on Fd-GOGAT mRNA or protein synthesis (Migge et al., 1997). As argued by Stitt and Krapp (1999), it is very difficult to identify molecules that could signal the N status of a plant and hence control gene expression, when there is such a high rate of Gln synthesis from photorespiratory ammonia release, rather than from primary nitrate reduction.

The photorespiratory mutants of barley lacking Fd-GOGAT activity have proved invaluable in the study of the expression of Fd-GOGAT mRNA and protein, with considerable variation being detected amongst the different mutants (Avila et al., 1993). More recently, Suzuki and Rothstein (1997) and Coschigano et al. (1998) have re-examined the expression of Fd-GOGAT genes in the mutant of *A. thaliana* lacking enzyme activity (originally designated *gluS* and now renamed *gls*). The *gls1* mutant allele and the *GLU1* gene mapped to the same local region of chromosome 5. Coschigano et al. (1998) argued that, as the mutant was unable to respond to exogenously supplied inorganic N, the Fd-GOGAT product of *GLU1* is involved in primary N metabolism as well as assimilating the ammonia released during photorespiration. They also proposed that the product of the *GLU2* gene, which has a much higher level of expression in roots, and is not altered in the mutant, is a housekeeping gene used for synthesizing basal levels of Glu. Interestingly, a much earlier study on the N metabolism of mutants of barley lacking leaf Fd-GOGAT had indicated that the root contained significant amounts of enzyme activity and was able to synthesize Glu from exogenously supplied ^{15}N-nitrate (Joy et al., 1992).

Ferrario-Méry et al. (2000) obtained 56 independent primary transformed tobacco lines expressing an antisense construct of Fd-GOGAT. The transformed plants exhibited between 10 and 90% of the normal leaf and root enzyme activity and reductions in NADH-GOGAT activity were also detected in the roots. Plants containing less than 60% of the normal GOGAT activity exhibited severe chlorosis when exposed to air but grew normally at a CO_2 concentration of 4000 μL.L^{-1}. The leaves accumulated Gln, ammonia and 2-OG following exposure to air. The concentrations of soluble Glu, alanine and Asp decreased, while glycine remained constant and a range of other amino acids including serine, the Asp family and aromatic amino acids increased. Ferrario-Méry et al. (2000) argued strongly that the increases in individual amino acids were not due to increased proteolysis, although such a possibility cannot be ruled out. They proposed that the accumulation of ammonia and Gln could instigate pathways of signal transduction that may modulate several pathways of amino acid biosynthesis. Similar evidence has been provided from the changes in amino acid metabolism observed when the biosynthesis of histidine was blocked using specific inhibitors (Guyer at al., 1995).

Following the original discovery of Fd-GOGAT activity in pea chloroplasts (Lea and Miflin, 1974), the leaf enzyme has now been shown to be solely localized in chloroplasts (Wallsgrove et al., 1979; Suzuki and Gadal, 1984). Using immunogold antibody localization techniques in tomato, the enzyme protein was detected in the chloroplast stroma of mesophyll, xylem parenchyma and epidermal cells (Botella et al., 1988a). In maize, Western blot analysis of isolated cells (Becker et al., 2000) and immunofluorescence studies (Becker et al., 1993b), indicated that the Fd-GOGAT protein was predominantly (if not totally) localized in the BSC chloroplasts, confirming earlier activity measurements carried out by Harel et al. (1977). Intact maize leaf BSC have also been shown to convert Gln and 2-OG to Glu at high rates in a light driven reaction (Valle and Heldt, 1992). In rice leaves, Fd-GOGAT activity and protein were shown to be highest in the MC of the fully expanded green leaf blades and were greatly reduced in the leaf sheaths and developing non-green leaf blades (Yamaya et al., 1992).

As indicated previously, Fd-GOGAT is also present in non-photosynthetic tissues. In pea roots, the enzyme is located in the plastids (Emes and Fowler, 1979) and mechanisms have been proposed for the supply of reductant via the oxidative pentose phosphate pathway (Bowsher et al., 1992). Fd-GOGAT has also been shown to be localized in the plastids of rice, maize, bean, barley and pea roots (Suzuki et al., 1981) and the activity and protein were not influenced by the availability of N (Yamaya et al., 1995). In soybean seedlings, increases in Fd-GOGAT activity, protein and mRNA abundance were detected in roots supplied with ammonium nitrate in the dark, although the effects were less obvious in plants grown in the light. An interesting additional finding was the observation that Fd-GOGAT activity in soybean roots increased following the addition of ammonium sulfate, without a corresponding increase in protein or mRNA (Turano and Muhitch, 1999).

In tobacco, the enzyme protein has been isolated from pistils and anthers as well as leaves, but not from roots, corollas or stems (Zehnacker et al., 1992). During the ripening of the tomato fruit, the activity of the enzymes of the photorespiratory cycle decreased dramatically, but a high level of activity of both NADH-GOGAT and Fd-GOGAT was maintained in the red fruit (Gallardo et al., 1993). Plastids isolated from developing tomato fruits were shown to carry out the light-dependent conversion of Gln and 2-OG to Glu, but exogenous glucose-6-phosphate in the dark could only support 18% of the maximum activity (Büker et al., 1998).

B. NADH-dependent Glutamate Synthase

Early reports indicated that NADH-GOGAT (EC 1.4.1.14) was able to use either NADPH or NADH as a coenzyme, but it is now established that the NADH-dependent enzyme is the predominant form in higher plant tissues. It is unlikely that NADH-GOGAT plays a major role in photosynthetic N metabolism, and so the discussion of this enzyme will be relatively brief. In green leaves the activity is low in comparison to the Fd-GOGAT activity (Wallsgrove et al., 1982; Avila et al., 1984, 1987; Hecht et al., 1988) but high levels of NADH-GOGAT activity and protein are present in the non-green and developing leaf blades of rice (Yamaya et al., 1992). Tissue print immunoblots utilizing specific antisera indicated that NADH-GOGAT was located in the large and small vascular bundles of unexpanded rice leaves. Enzyme protein was detected in vascular parenchyma cells (metaxylem and metaphloem parenchyma cells) and mestome sheath cells of the young leaf blade before emergence (Hayakawa et al., 1994). In rice roots, the NADH-GOGAT immunogold-labeling density was high in the plastids of the cells of the epidermis and exodermis, cortical parenchyma and vascular parenchyma (Hayakawa et al., 1999).

NADH-GOGAT has been purified from rice suspension culture cells and shown to be a monomer with a molecular mass of 196 kDa (Hayakawa et al., 1993). Antisera raised against the enzyme protein did not cross-react with the Fd-GOGAT protein and were used for the leaf localization studies described above. cDNA clones encoding NADH-GOGAT have been obtained from *A. thaliana* (Lam et al., 1996) while cDNA and genomic clones have been isolated from rice (Goto et al., 1998). When N-starved rice seedlings were transferred to 1 mM NH_4Cl, the level of the NADH-GOGAT activity and protein increased more than ten-fold in the root within one day (Yamaya et al., 1995; Ishiyama et al., 1998). Increases in NADH-GOGAT mRNA were also detected within 12 hours, following the application of concentrations of ammonium ions as low as 50 μM to rice cell cultures or roots (Hirose et al., 1997), and it was proposed that Gln may act as the signal for the increase in transcription. However, okadaic acid, a potent inhibitor of protein serine/threonine phosphatases, also induced the accumulation of NADH-GOGAT in rice cell cultures, and so the precise signaling mechanism controlling gene expression is still not clear (Hirose and Yamaya, 1999). Light or various N treatments had little effect on NADH-GOGAT activity in cotyledons, leaves or hypocotyls/stems of soybean. However, enzyme activity in the roots increased 14-fold following the addition of ammonium salts to N-starved seedlings, but only seven-fold after addition of KNO_3. Smaller increases in NADH-GOGAT protein and mRNA were also detected following the addition of the various N sources (Turano and Muhitch, 1999).

Early studies indicated that NADH-GOGAT appears to play a major role in legume root nodules, where the activity increases dramatically following the onset of nitrogen fixation (Awonaike et al., 1981). NADH-GOGAT has been purified to homogeneity from alfalfa (*Medicago sativa*) root nodules and shown to be a monomer of approximately 200 kDa. Using antisera raised against the protein, Gregerson et al. (1993) isolated a 7.2 kb cDNA clone that encoded the 240 kDa NADH-GOGAT. Several important regions were identified in the amino acid sequence, which shared significant sequence identity with the maize Fd-GOGAT and the *E. coli* NADPH-GOGAT. The complete gene was shown to be 14 kb long and to be composed of 22 exons interrupted by 21 introns. The Vance laboratory has carried out a series of excellent detailed studies on the expression and localization of NADH-GOGAT in alfalfa (Vance et al., 1995; Temple et al., 1998; Trepp et al., 1999a,b). A detailed discussion of these data is beyond the scope of this chapter.

C. Production of 2-Oxoglutarate for Glutamate Synthase Activity

It is the GOGAT reaction which represents the immediate interface between N and C metabolism. Utilising spinach chloroplasts supplied with ^{14}C and ^{15}N-labelled metabolites, Woo et al. (1987a) established that malate was the key metabolite regulating the entry and exit of the substrates and products of the GS/GOGAT reaction. A two-translocator model was proposed in which malate is the counterion for both the import of 2-OG into the chloroplast via a 2-OG transporter and the export of Glu via a dicarboxylate transporter (Fig. 2), in a cascade-like manner (Flügge et al., 1988). The gene encoding the 2-OG/malate translocator from spinach chloroplasts has now been isolated and the amino acid sequence of the protein determined. The translocator contains a very long (10 kDa) hydrophilic transit peptide with a final molecular mass of 50 kDa. Twelve hydrophobic transmembrane helices were identified that were connected by hydrophilic domains. When the 2-OG/malate translocator was expressed in yeast cells, the substrate specificity and capacities to transport malate, fumarate, succinate and 2-OG were shown to be very similar to those determined for spinach chloroplast membranes (Weber et al., 1995). More recently, the Glu/malate translocator, the amino acid sequence of which shows 50% identity to the 2-OG/malate translocator, has also been cloned from spinach and *Flavaria* species and expressed in yeast. The predominant substrates are Asp, Glu and malate, although there is evidence of overlapping substrate specificities, when compared to the 2-OG/malate transporter (A. Weber, unpublished).

Arabidopsis and barley mutants lacking the chloroplastic 2-OG transporter have been described (Somerville and Ogren, 1983; Wallsgrove et al., 1986) which show similar phenotype to the Fd-GOGAT mutants discussed above (for further discussion, Chapter 8 (Keys and Leegood)). Weber and his colleagues have inserted antisense constructs for both the chloroplastic 2-OG/malate and Glu/malate translocator into tobacco. Plants lacking the 2-OG/malate translocator exhibited stress symptoms and accumulated nitrate, ammonia and glyoxylate with reduced concentrations of amino acids. Somewhat surprisingly, the loss of the Glu/malate translocator had very little effect on the phenotype, indicating that Glu may be carried across the chloroplast membrane by more than one translocator (A Weber, unpublished).

If ammonia is being rapidly recycled, as in photorespiration, then there is little demand for additional 2-OG for Glu synthesis. However, if there is primary nitrate assimilation or ammonia is derived from the metabolism of a transport compound, then there is a requirement for a supply of 2-OG. The two obvious sources of 2-OG are either from the oxidative decarboxylation of isocitrate catalysed by isocitrate dehydrogenase or the transamination of Glu by Asp

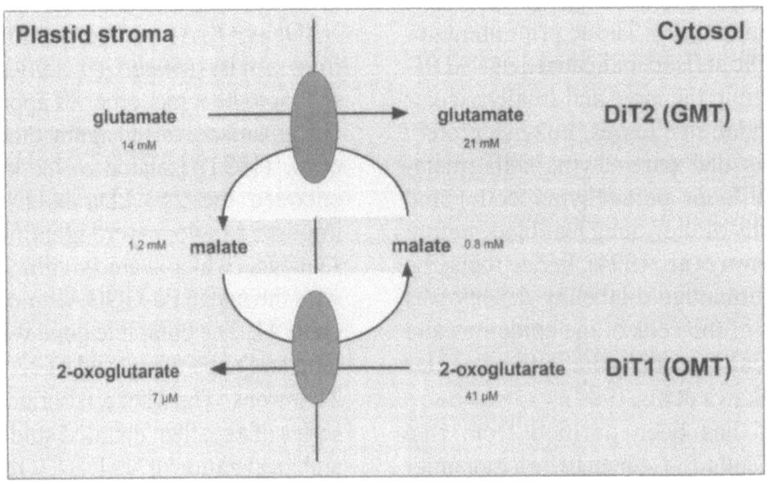

Fig. 2. The two-translocator system for the transport of dicarboxylic acids into and out of the chloroplasts. Generously supplied by Andreas Weber. Metabolite concentrations are taken from Winter et al. (1994). GMT, glutamate-malate exchange transporter; OMT, 2-oxoglutarate-malate exchange transporter.

aminotransferase (Schultz et al., 1998), which requires the input of oxaloacetate (Fig. 1). The major form of isocitrate dehydrogenase in green leaves utilizes NADP as the coenzyme and is localized in the cytosol (Gálvez et al., 1999). In Scots pine seedlings, during chloroplast development, there was a correlation between the mRNA levels of isocitrate dehydrogenase, GS and GOGAT (Palomo et al., 1998). However at later stages of development of the cotyledons and in the hypocotyl, there was no such correlation, indicating a secondary role for 2-OG production, possibly as a substrate for dioxygenases involved in secondary metabolism (Palomo et al., 1998). In transgenic potato plants, in which NADP-isocitrate dehydrogenase had been reduced to 8% of the wild type activity, no changes in growth rate or flowering were noted. In addition, no changes in C or N metabolism were detected, indicating that other sources of 2-OG are available within a plant leaf (Kruse et al., 1998).

Despite low activity and high instability, a mitochondrial form of isocitrate dehydrogenase, which utilizes NAD as a coenzyme, has now been examined in detail in tobacco (Lancien et al., 1998). The addition of both ammonium and nitrate ions was shown to stimulate the synthesis of mRNA encoding NAD-isocitrate dehydrogenase in both the roots and shoots of N-starved tobacco, while NADP-isocitrate dehydrogenase was relatively unaffected by the same treatments (Lancien et al., 1999). It is therefore clear that a range of different enzyme pathways are available for the synthesis of 2-OG which, considering the importance of the metabolite in both C and N metabolism, is not surprising. However, in a very interesting series of experiments, Schjoerring et al. (2000) have recently demonstrated that the 2-OG content of sugar beet leaves fell dramatically during the middle of day to zero, but recovered by the end of the light period. During this daytime inversion, the activity of both GS and Fd-GOGAT decreased steadily. The contradictory observations found in the literature probably reflect the fact that the contribution by different enzymes may vary depending on the tissue, developmental age and specific physiological conditions (Lancien et al., 2000).

IV. Glutamate Dehydrogenase

GDH (EC 1.4.1.2) catalyses the following reversible reaction:

$$\text{Glutamate} + H_2O + NAD^+ \leftrightarrow \text{2-oxoglutarate} + NH_3 + NADH + H^+$$

Thus the enzyme could either play a role in the assimilation of ammonia or be responsible for the deamination of amino acids to liberate ammonia. For the last 25 years plant biochemists have attempted to design experiments to establish the role of GDH in higher plants, the results of which have frequently given rise to further discussion and argument. A considerable amount of evidence has accumulated that indicates that over 95% of the ammonia that is available to plants, is assimilated via the GS/GOGAT pathway (Lea and Ireland, 1999). However, proposals that GDH could operate in the direction of ammonia assimilation have been put forward on a regular basis (Yamaya and Oaks, 1987; Oaks 1994; Melo-Oliveira et al., 1996). Others have argued equally strongly that GDH operates in the direction of Glu deamination (Robinson et al., 1992; Fox et al., 1995; Stewart et al., 1995).

Plant GDH has a very high Km for ammonia, is activated by calcium, and is localized in the mitochondria (Srivastava and Singh 1987; Turano, 1998). The native enzyme protein exists as a hexamer, with subunits ranging from 41-45 kDa, and there is strong evidence that there are at least two different subunits, which can randomly associate to form a range of different isoenzymic forms (Cammaerts and Jacobs, 1985; Loulakakis and Roubelakis-Angelakis, 1996; Bechtold et al., 1998). cDNA clones have been isolated that encode higher plant GDH, including maize (Sakakibara et al., 1995), grapevine (Syntichaki et al., 1996), *A. thaliana* (Melo-Oliveira et al., 1996; Turano et al., 1997) and tomato (Purnell et al., 1997). Considerable similarities with the amino acid sequence obtained from animals, bacteria and algae were noted and putative binding sites for NADH, 2-OG and Glu have been identified. Of the two distinct cDNA clones isolated from *A. thaliana*, *GDH1* and *GDH2* encoded mitochondria-targeted proteins of molecular mass 43 and 42.5 kDa, of which only GDH2 had a putative EF-hand that could be involved in Ca^{2+} binding (Turano et al., 1997). Pavesi et al. (2000) isolated two GDH genes from *Asparagus officinalis* and carried out a phylogenetic analysis which demonstrated that the plant genes were more closely related to those of thermophilic archaebacterial and eubacterial species, rather than eukaryotic fungi.

The addition of ammonium ions to plants almost

invariably causes an increase in measurable GDH activity (Srivastava and Singh, 1987; Ireland and Lea, 1999), and similar effects have also been observed following carbohydrate starvation (Robinson et al., 1992; Athwal et al., 1997). GDH activity has also been shown to increase following the onset of senescence (Srivastava and Singh, 1987; Bechtold et al., 1998), a time when carbohydrate concentrations fall and ammonia increases. In grapevine, Loulakakis and Roubelakis-Angelakis (1992) demonstrated that the ammonia-induced increase in GDH activity was due to the synthesis of the 43kDa subunit. In *A. thaliana,* the expression of both *GDH1* and *GDH2* was stimulated by darkening and ammonia, and the synthesis of *GDH1* mRNA was repressed by light or sucrose (Melo-Oliveria et al., 1996). Turano et al. (1997) carried out a more detailed study on GDH gene expression in *A. thaliana,* in which enzyme activity, subunit composition and mRNA accumulation were determined following changes of N source in both the light and dark. The investigators concluded that although there were similarities in the regulation of the two genes, *GDH1* and *GDH2* were not co-ordinately expressed. Particular differences were noted in the lack of sucrose-mediated repression and larger dark stimulation of *GDH2* as compared to *GDH1* mRNA synthesis.

Taking into account the two reactions catalysed by GDH and the obvious regulation of gene expression and enzyme activity, its is clear that GDH must play an important role at the interface of C and N metabolism. Despite the availability of mutants of maize and *A. thaliana* lacking one of the GDH subunits (Pryor, 1990; Melo-Oliveira et al., 1996), this role is still not clear. Ameziane et al. (2000) transformed tobacco with the *gdhA* gene of *E. coli* encoding a high affinity assimilatory NADPH-dependent GDH, targeted to the cytosol. Over a three year period the *gdhA* transgenic tobacco produced significantly more dry weight than the control plants, particularly during water shortage. Increases in soluble amino acids (in particular proline) and carbohydrates were also detected. In addition, the transgenic plants were less sensitive to PPT. Other studies on transformed plants expressing the assimilatory GDH from *E. coli* or *Chlorella sorokiniana* have also demonstrated increased growth and improved stress tolerance (Schmidt and Miller, 1999; S. J. Temple, unpublished).

References

Ameziane R, Bernhard K and Lightfoot D (2000) Expression of the bacterial *gdhA* gene encoding a NADPH glutamate dehydrogenase gene in tobacco affects plant growth and development. Plant Soil 221: 47–57

Anderson JW and Done J (1997) Polarographic study of ammonia assimilation by isolated chloroplasts. Plant Physiol 60: 354–359

Anderson JW and Walker DA (1983) Ammonia assimilation and O_2 evolution by a reconstituted chloroplast system in the presence of 2-oxoglutarate and glutamate. Planta 159: 247–253

Athwal GS, Pearson J and Laurie S (1997) Regulation of glutamate dehydrogenase activity by manipulation of nucleotide supply in *Daucus carota* suspension cultures. Physiol Plant 101: 503–509

Avila C, Cánovas F, Nunez de Castro I and Valpuesta V (1984) Separation of two forms of glutamate synthase in leaves of tomato (*Lycopersicon esculentum).* Biochem Biophys Res Commun 122: 1125–1130

Avila C, Botella JR, Cánovas FM, de Castro I and Valpuesta V (1987) Different characteristics of the two glutamate synthases in the green leaves of *Lycopersicon esculentum.* Plant Physiol 85: 1036–1039

Avila C, Márquez AJ, Pajuelo P, Cannell ME, Wallsgrove RM and Forde BG (1993) Cloning and sequence analysis of a cDNA for barley ferredoxin-dependent glutamate synthase and molecular analysis of photorespiratory mutants deficient in the enzyme. Planta 189: 475–483

Awonaike KO, Lea PJ and Miflin BJ (1981) The localisation of the enzymes of ammonia assimilation in root nodules of *Phaseolus vulgaris.* Plant Sci Lett 23:189–195

Bauer D, Biehler K, Fock H, Carrayol E, Hirel B, Migge A and Becker T W (1997) A role for cytosolic glutamine synthetase in the remobilization of leaf nitrogen during water stress in tomato. Physiol Plant 99: 241–248

Bechtold U, Pahlich E and Lea PJ (1998) Methionine sulphoximine does not inhibit pea and wheat glutamate dehydrogenase. Phytochemistry 49: 347–354

Beck T and Hall N (1999) The TOR signalling pathway controls nuclear localization of nutrient-regulated transcription factors. Nature 402: 689–692

Becker TW, Caboche M, Carrayol E and Hirel B (1992) Nucleotide sequence of a tobacco cDNA encoding plastidic glutamine synthetase and light-inducibility, organ specificity and diurnal rhythmicity in the expression of the corresponding genes of tobacco and tomato. Plant Mol Biol 19: 367–379

Becker TW, Nef-Campa C, Zehnacker C and Hirel B (1993a) Implication of the phytochrome light regulation of the tomato gene(s) encoding ferredoxin dependent glutamate synthase. Plant Physiol Biochem 31: 725–729

Becker TW, Perrot-Rechenman C, Suzuki A and Hirel B (1993b) Subcellular and immunocytochemical localisation of the enzymes involved in ammonia assimilation in mesophyll and bundle sheath cells of maize. Planta 191: 129–136

Becker TW, Carrayol E and Hirel B (2000) Glutamine synthetase and glutamate dehydrogenase isoforms in maize leaves: Localization, relative proportion and their role in ammonium assimilation or nitrogen transport. Planta 211: 800–806

Biesiadka J and Legocki AB (1997) Evolution of the glutamine synthetase genes in plants. Plant Sci 128: 51–58
Blackwell RD, Murray AJS, Lea PJ and Joy KW (1988a) Photorespiratory amino donors, sucrose synthesis and the induction of CO_2 fixation in barley deficient in glutamine synthetase and/or glutamate synthase. J Exp Bot 39: 845–858
Blackwell RD, Murray AJS, Lea PJ, Kendall AC, Hall NP, Turner JC and Wallsgrove RM (1988b) The value of mutants unable to carry out photorespiration. Photosynth Res 16: 155–176
Botella JR, Verbelen JP and Valpuesta V (1988a) Immunocytolocalization of ferredoxin-GOGAT in the cells of green leaves and cotyledons of *Lycopersicon esculentum*. Plant Physiol 87: 255-257
Botella JR, Verbelen JP and Valpuesta V (1988b) Immunocytolocalization of glutamine synthetase in green leaves and cotyledons of *Lycopersicon esculentum*. Plant Physiol 88: 943–946
Bowler C and Chua NH (1994) Emerging themes of plant signal transduction. Plant Cell 6: 1529–1541
Bowsher CG, Boulton EL, Rose J, Nayagam S and Emes MJ (1992) Reductant for glutamate synthase is generated by the oxidative pentose phosphate pathway in non-photosynthetic root plastids. Plant J 2: 893–898
Brangeon J, Hirel B and Forchion A (1989) Immunogold localisation of glutamine synthetase in soybean leaves, roots and nodules. Protoplasma 151: 88–97
Brugière N, Dubois F, Limami A, Lelandais M, Roux Y, Sangwan R and Hirel B (1999) Glutamine synthetase in the phloem plays a major role in controlling proline production. Plant Cell 11: 1995–2011
Brugière N, Dubois F, Masclaux C, Sangwan R and Hirel B (2000) Immunolocalization of glutamine synthetase in senescing tobacco (*Nicotiana tabacum* L.) leaves suggests that ammonia assimilation is progressively shifted to the mesophyll cytosol. Planta 211: 519–527
Buchanan-Wollaston V (1997) The molecular biology of leaf senescence. J Exp Bot 48: 181–199
Büker M, Schunemann D and Borchert S (1998) Enzymic properties and capacities of developing tomato (*Lycopersicon esculentum* L.) fruit plastids. J Exp Bot 49: 681–691
Cammaerts D and Jacobs M (1985) A study of the role of glutamate dehydrogenase in the nitrogen metabolism of *Arabidopsis thaliana*. Plant Sci Lett 31: 65–73
Cánovas FM, Cantón FR, García-Gutiérrez A, Gallardo F and Crespillo R (1998) Molecular physiology of glutamine and glutamate biosynthesis in developing seedlings of conifers. Physiol Plant 103: 287–294
Cantón FR, Suarez MF, Jose -Estanyol M and Cánovas F (1999) Expression analysis of a cytosolic glutamine synthetase gene in cotyledons of scots pine seedlings: Developmental regulation and spatial distribution of specific transcripts. Plant Mol Biol 40: 623–634
Carvalho H, Sunkel C, Salema R and Cullimore JV (1997) Heteromeric assembly of the cytosolic glutamine synthetase polypeptides of *Medicago truncatula*: Complementation of a *glnA Escherichia coli* mutant with a plant domain-swapped enzyme. Plant Mol Biol 35: 623–632
Clemente MT and Marquez AJ (1999) Site-directed mutagenesis of Glu-297 from the *[alpha]*-polypeptide of *Phaseolus vulgaris* glutamine synthetase alters kinetic and structural properties and confers resistance to L-methionine sulfoximine. Plant Mol Biol 40: 835–845
Cock JM, Brock IW, Watson AT, Swarup R, Morby AP and Cullimore JV (1991) Regulation of glutamine synthetase genes in leaves of *Phaseolus vulgaris*. Plant Mol Biol 17: 761–771
Cock JM, Hemon P and Cullimore JV (1992) Characterization of the gene encoding the plastid-located glutamine synthetase of *Phaseolus vulgaris*: Regulation of beta-glucuronidase gene fusions in transgenic tobacco. Plant Mol Biol 18: 1141–1149
Coschigano KT, Melo-Oliveria R, Lim J and Coruzzi GM (1998) *Arabidopsis gls* mutants and distinct Fd-GOGAT genes: implications for photorespiration and primary nitrogen assimilation. Plant Cell 10: 741–752
Cren M and Hirel B (1999) Glutamine synthetase in higher plants: Regulation of gene and protein expression from the organ to the cell. Plant Cell Physiol 40: 1187–1193
Doyle JJ (1991) Evolution of higher plant glutamine synthetase genes: Tissue specificity as a criterion for predicting orthology. Mol Biol Evol 8: 366–377
Dubois F, Brugière N, Sangwan RS and Hirel B (1996) Localisation of tobacco cytosolic glutamine synthetase enzymes and the corresponding transcripts show organ- and cell-specific patterns of protein synthesis and gene expression. Plant Mol Biol 31: 803–817
Edwards JW and Coruzzi GM (1989) Photorespiration and light act in concert to regulate the expression of the nuclear gene for chloroplast glutamine synthetase. Plant Cell 1: 241–248
Elmlinger MW, Bolle C, Batschauer A, Oelmuller R and Mohr H (1994) Coaction of blue light absorbed by phytochrome in control of glutamine synthetase gene expression in Scots pine (*Pinus sylvestris* L.) seedlings. Planta 192: 189–194
Emes MJ and Fowler MW (1979) The intracellular location of enzymes of nitrogen assimilation in the apices of pea roots. Planta 144: 249—253
Feller U and Fischer A (1994) Nitrogen metabolism in senescing leaves. CRC Crit Rev Plant Sci 13: 241–273
Fernandez-Conde ME, de la Haba P and Maldonado JM (1995) Rhythmic variations of ferredoxin-glutamate synthase activity in sunflower (*Helianthus annuus*) leaves. J Plant Physiol 145: 682-685
Ferrario-Méry S, Suzuki A, Kunz C, Valadier MH, Roux Y, Hirel B and Foyer CH (2000) Modulation of amino acid metabolism in transformed tobacco plants deficient in Fd-GOGAT. Plant and Soil 221: 67–79
Flügge IU, Woo KC and Heldt HW (1988) Characteristics of 2-oxoglutarate and glutamate transport in spinach chloroplasts. Planta 174: 534–541
Forde BG and Cullimore JV (1989) The molecular biology of glutamine synthetase in higher plants. Oxford Surveys of Plant Mol Cell Biol 6: 247–296
Fox GG, Ratcliffe RG, Robinson SA and Stewart GR (1995) Evidence for deamination by glutamate dehydrogenase in higher plants: Commentary. Can J Bot 73: 1112–1115
Gallardo F, Cantón FR, García-Gutiérrez A and Cánovas FM (1993) Changes in photorespiratory enzymes and glutamate synthases in ripening tomatoes. Plant Physiol Biochem 31: 189–196
Gallardo F, Fu J, Cantón FR, Garcia-Gutiérrez A, Cánovas F and Kirby EG (1999) Expression of a conifer glutamine synthetase gene in transgenic poplar. Planta 210: 19–26
Gálvez S, Gallardo F and Cánovas F (1990) The occurrence of

glutamine synthetase isoenzymes in tomato plants is correlated to plastid differentiation. J Plant Physiol 137: 1–4

Gálvez S, Lancien M and Hodges M (1999) Are isocitrate and 2-oxoglutarate involved in the regulation of glutamate synthesis? Trends Plant Sci 4: 484–490

García-Gutiérrez A, Cantón FR, Gallardo F, Sánchez-Jiménez F and Cánovas FM (1995) Expression of ferredoxin-glutamate synthase in dark grown pine seedlings. Plant Mol Biol 27: 115–128

García-Gutiérrez A, Dubois F, Cantón FR, Gallardo F, Sangwan RS and Cánovas FM (1998) Two different modes of early development and nitrogen assimilation in gymnosperm seedlings. Plant J 13: 187–199

Givan CV (1979) Metabolic detoxification of ammonia in higher plants. Phytochemistry 18: 375–382

Goto S, Akagawa T, Kojima S, Hayakawa T and Yamaya T (1998) Organisation and structure of a NADH-dependent glutamate synthase gene from rice plants. Biochim Biophys Acta 1387: 293–308

Gregerson RG, Miller SS, Twary SN, Gantt JS and Vance CP (1993) Molecular characterisation of NADH-glutamate synthase from alfalfa nodules. Plant Cell 5: 215–226

Guyer D, Patton D and Ward E (1995) Evidence for cross pathway regulation of metabolic gene expression in plants. Proc Natl Acad Sci USA 92: 4997–5000

Harel E, Lea PJ and Miflin BJ (1977) The localisation of enzymes of nitrogen assimilation in maize leaves and their activities during greening. Planta 134: 195-203

Harrison J, Brugière N, Phillipson B, Ferrario-Méry S, Becker T, Limami A and Hirel B (2000) Manipulating the pathway of ammonia assimilation through genetic manipulation and breeding. Consequences on plant physiology and development. Plant Soil 221: 81–93

Häusler RE, Bailey KJ, Lea PJ and Leegood RC (1996) Control of photosynthesis in barley mutants with reduced activities of glutamine synthetase and glutamate synthase. III. Aspects of glyoxylate metabolism and effects of glyoxylate on the activation state of ribulose-1,5-bisphosphate carboxylase-oxygenase. Planta 200: 388–396

Hayakawa T, Yamaya T, Mae T and Ojima K (1993) Changes in content of two glutamate synthase proteins in spikelets of rice (*Oryza sativa*) plants during ripening. Plant Physiol 101: 1257–1262

Hayakawa T, Nakamura T, Hattori F, Mae T, Ojima K and Yamaya T (1994) Cellular localisation of NADH-dependent glutamate synthase protein in vascular bundles of unexpanded leaf blades and young grains of rice plants. Planta 193: 455–460

Hayakawa T, Hopkins L, Peat LJ, Yamaya T and Tobin AK (1999) Quantitative intercellular localisation of NADH-dependent glutamate synthase protein in different types of root cells in rice plants. Plant Physiol 119: 409–416

Hecht U, Oelmuller R, Schmidt S and Mohr H (1988) Action of light, nitrate and ammonium on the levels of NADH- and ferredoxin-dependent glutamate synthases in the cotyledons of mustard seedlings. Planta 175: 130–138

Hirasawa M, Robertson DE, Ameyibor E, Johnson MK and Knaff DB (1992) Oxidation-reduction properties of the ferredoxin linked glutamate synthase from spinach leaf. Biochim Biophys Acta 1100: 105–108

Hirasawa M, Hurley JK, Salamon Z, Tollin G and Knaff DB (1996) Oxidation-reduction and transient kinetic studies of spinach ferredoxin dependent glutamate synthase. Arch Biochem Biophys 330: 209–215

Hirasawa M, Hurley JK, Salamon Z, Tollin G, Markley JL, Cheng H, Xia B and Knaff DB (1998) The role of aromatic and acidic amino acids in the electron transfer reaction catalysed by spinach ferredoxin dependent glutamate synthase. Biochim Biophys Acta 1363: 134–146

Hirel B and Gadal P (1982) Glutamine synthetase isoforms in leaves of a C_4 plant: *Sorghum vulgare* L. Physiol Plant 54: 69–74

Hirel B, Lavergne D, McNally SF and Gadal P (1983) Some regulatory properties of glutamine synthetase isoforms in a C_4 plant *Pennisetum americanum* L. (Leeke). Plant Sci 111: 413–422

Hirel B, McNally SF, Gadal P, Sumar N and Stewart GR (1984) Cytosolic glutamine synthetase in higher plants. A comparative immunological study. Eur J Biochem 138: 63–66

Hirel B, Bouet C, King B, Layzell D, Jacobs F and Verma DPS (1987) Glutamine synthetase genes are regulated by ammonia provided externally or by symbiotic nitrogen fixation. EMBO J 6: 1167–1171

Hirel B, Marsolier MC, Hoarau A, Hoarau J, Brangeon J, Schafer R and Verma DPS (1992) Forcing expression of a soybean root glutamine synthetase gene in tobacco leaves induces a native gene encoding cytosolic enzyme. Plant Mol Biol 20: 207–218

Hirose N and Yamaya T (1999) Okadaic acid mimics nitrogen-stimulated transcription of the NADH-glutamate synthase in rice cell cultures. Plant Physiol 121: 805–812

Hirose N, Hayakawa T and Yamaya T (1997) Inducible accumulation of mRNA for NADH-dependent glutamate synthase in rice roots in response to ammonium ions. Plant Cell Physiol 38: 1295–1297

Hsieh MH, Lam HM, Van der Loo F and Coruzzi G (1998) A PII-like protein in *Arabidopsis:* Putative role in nitrogen sensing. Proc Natl Acad Sci USA 95: 13965–13970

Ireland RJ and Lea PJ (1999) The enzymes of glutamine, glutamate, asparagine and aspartate metabolism. In: Singh BK (ed) Plant Amino Acids, pp 49–109. Marcel Dekker, New York

Ishiyama K, Hayakawa T and Yamaya T (1998) Expression of NADH-dependent glutamate synthase protein in epidermis and exodermis of rice roots in response to the supply of nitrogen. Planta 204: 288–294

Joy KW, Blackwell RD and Lea PJ (1992) Assimilation of nitrogen in mutants lacking enzymes of the glutamate synthase cycle. J Exp Bot 43: 139–145

Kamachi K, Yamaya T, Mae T and Ojima K (1991) A role for glutamine synthetase in the remobilization of leaf nitrogen during natural senescence in rice leaves. Plant Physiol 96: 411–417

Kendall AC, Wallsgrove RM, Hall NP, Turner JC and Lea PJ (1986) Carbon and nitrogen metabolism in barley (*Hordeum vulgare*) mutants lacking ferredoxin-dependent glutamate synthase. Planta 168: 316–323

Keys AJ, Bird IF, Cornelius MJ, Lea PJ, Wallsgrove RM and Miflin BJ (1978) Photorespiratory nitrogen cycle. Nature 275: 741–743

Kozaki A, Sakamoto A and Takeba G (1992) The promoter of the gene for plastidic glutamine synthetase (GS2) from rice is developmentally regulated and exhibits substrate-induced expression in transgenic tobacco plants. Plant Cell Physiol 33: 233–238

Kruse A, Fiew S, Heineke D and Müller-Röber B (1998) Antisense inhibition of cytosolic NADP-dependent isocitrate dehydrogenase in transgenic potato plants. Planta 205: 82–91

Lam HM, Coschigano KT, Oliveira IC, Melo-Oliveira R and Coruzzi GM (1996) The molecular genetics of nitrogen assimilation into amino acids in higher plants. Annu Rev Plant Physiol Plant Mol Biol 47: 569–593

Lancien M, Gadal P and Hodges M (1998) Molecular characterisation of higher plant NAD-dependent isocitrate dehydrogenase: evidence for a heteromeric structure by complementation of yeast mutants. Plant J 16: 325–333

Lancien M, Ferrario-Méry S, Roux Y, Bismuth E, Masclaux C, Hirel B, Gadal P and Hodges M (1999) Simultaneous expression of NAD-dependent isocitrate dehydrogenase and other Krebs cycle genes after nitrate resupply to short-term nitrogen-starved tobacco. Plant Physiol 120: 717–725

Lancien M, Gadal P and Hodges M (2000) Enzyme redundancy and the importance of 2-oxoglutarate in higher plant ammonium assimilation. Plant Physiol 123: 817–824

Lara M, Porta H, Padilla J, Folch J and Sanchez F (1984) Heterogeneity of glutamine synthetase polypeptides in *Phaseolus vulgaris.* Plant Physiol 76: 1019–1023

Lea PJ and Ireland RJ (1999) Nitrogen metabolism in higher plants. In: Singh BK (ed) Plant Amino Acids, pp. 1–47. Marcel Dekker, New York

Lea PJ and Miflin BJ (1974) An alternative route for nitrogen assimilation in higher plants. Nature 251: 680–685

Lee HM, Vazquez-Bermudez MF and Tandeau de Marsac N (1999) The global nitrogen regulator NtcA regulates transcription of the signal transducer PII (GlnB) and influences its phosphorylation level in response to nitrogen and carbon supply in the cyanobacterium *Synechococcus sp.* strain PCC 7942. J Bacteriol 181: 2697–2702

Leegood RC, Lea PJ, Adcock MD and Häusler, R.E. (1995) The regulation and control of photorespiration. J Exp Bot 46: 1397–1414

Lejay L, Tillard P, Lepetit M, Olive FD, Filleur S, Daniel-Vedele F and Gojon A (1999) Molecular and functional regulation of two NO_3^- uptake systems by N-and C-status of *Arabidopsis* plants. Plant J 18: 509–519

Li MG, Villemur R, Hussey PJ, Silflow CD, Gantt JS and Snustad DP (1993) Differential expression of six glutamine synthetase genes in *Zea mays.* Plant Mol Biol 23: 401–407

Lichter A and Häberlein I (1998) A light dependent redox signal participates in the regulation of ammonia fixation in chloroplasts of higher plants. J Plant Physiol 153: 83–90

Lightfoot DA, Nicola KG and Cullimore JV (1988) The chloroplast-located glutamine synthetase of *Phaseolus vulgaris* L.: Nucleotide sequence, expression in different organs and uptake into isolated chloroplasts. Plant Mol Biol 11: 191–202

Loulakakis KA and Roubelakis-Angelakis KA (1992) Ammonium induced increase in NADH-glutamate dehydrogenase activity is caused by de novo synthesis of the α-subunit. Planta 187: 322–327

Loulakakis KA and Roubelakis-Angelakis KA (1996) The seven NAD(H)-glutamate dehydrogenase isoenzymes exhibit similar anabolic activities. Physiol Plant 96: 29–35

Loulakakis KA and Roubelakis-Angelakis KA (1997) Molecular cloning and characterisation of cDNAs coding for ferredoxin-dependent glutamate synthase from *Vitis vinifera.* Physiol Plant 101: 220–228

Mäck G (1995) Organ-specific changes in the activity and subunit composition of glutamine synthetase isoforms of barley (*Hordeum vulgare* L.) after growth on different level of NH_4^+. Planta 196: 231–238

Mäck G and Tischner R (1994) Activity of the tetramer and octamer of glutamine synthetase isoforms during primary leaf ontogeny of sugar beet (*Beta vulgaris* L.). Planta 194: 353–359

Márquez AJ, Avila C, Forde BG and Wallsgrove RM (1988) Ferredoxin-glutamate synthase from barley leaves: Rapid purification and partial characterization. Plant Physiol Biochem 26: 645–651

Masclaux C, Valadier MH, Brugière N, Morot-Gaudry JF and Hirel B (2000) Characterization of the sink/source transition in tobacco (*Nicotiana tabacum* L.) shoots in relation to nitrogen management and leaf senescence. Planta 211: 510–518

Mathis R, Gamas P, Meyer Y and Cullimore JV (2000) The presence of GSI-like genes in higher plants: Support for the paralagous evolution of GSI and GSII genes. J Mol Evol 50: 116–122

McNally SF and Hirel B (1983) Glutamine synthetase isoforms in higher plants. Physiol Vég 21: 761–774

McNally SF, Hirel B, Gadal P, Mann F and Stewart GR (1983) Glutamine synthetases of higher plants. Plant Physiol 72: 22–25

Melo-Oliveira R, Cinha-Oliveira I and Coruzzi GM (1996) *Arabidopsis* mutant analysis and gene regulation define a non-redundant role for glutamate dehydrogenase in nitrogen assimilation. Proc Natl Acad Sci USA 96: 4718–4723

Miflin BJ and Lea PJ (1980) Ammonium assimilation. In: BJ Miflin (ed) The Biochemistry of Plants. Amino Acids and Derivatives, pp 169–202. Academic Press, New York

Migge A and Becker TW (1996) In tobacco leaves, the genes encoding the nitrate-reducing or ammonium-assimilating enzymes, are regulated differentially by external nitrogen sources. Planta 200: 213–220

Migge A, Meya G, Carrayol E, Hirel B and Becker TW (1996) Regulation of the subunit composition of tomato plastidic glutamine synthetase by light and the nitrogen source. Planta 200: 213–220

Migge A, Carrayol E, Kunz C, Hirel B, Fock H and Becker T (1997) The expression of the tobacco genes encoding plastidic glutamine synthetase or ferredoxin-dependent glutamate synthase does not depend on the rate of nitrate reduction, and is unaffected by suppression of photorespiration. J Exp Bot 48: 1175–1184

Migge A, Carrayol E, Hirel B, Lohmann M, Meya G and Becker TW (1998) Influence of UV-A or UV-B light and of nitrogen source on the induction of ferredoxin-dependent glutamate synthase in etiolated tomato cotyledons. Plant Physiol Biochem 36: 789–797

Migge A, Carrayol E, Hirel B and Becker TW (2000) Leaf-specific overexpression of plastidic glutamine synthetase stimulates the growth of transgenic tobacco seedlings. Planta 210: 252–260

Miranda-Ham ML and Loyola-Vargas VM (1992) Purification and characterization of glutamine synthetase from leaves of *Cataranthus roseus.* Plant Physiol Biochem 30: 585–592

Nalbantoglu B, Hirasawa M, Moomaw C, Nguyen H, Knaff DB and Allen R (1994) Cloning and sequencing of the gene encoding spinach ferredoxin-dependent glutamate synthase. Biochim Biophys Acta 1183: 557–561

Nato F, Hirel B, Nato A and Gadal P (1984) Chloroplastic glutamine synthetase from tobacco leaves. A glycosylated protein. FEBS Lett 175: 443–446

Oaks A (1994) Primary nitrogen assimilation in higher plants and its regulation. Can J Bot 62:739–750

Ochs G, Schock G, Trishler M, Kosemund K and Wild A (1999) Complexity and expression of the glutamine synthetase gene family in the amphidiploid crop *Brassica napus*. Plant Mol Biol 39: 395–405

Oliveira IC and Coruzzi GM (1999) Carbon and amino acids reciprocally modulate the expression of glutamine synthetase in *Arabidopsis*. Plant Physiol 121: 301–309

Pajuelo P, Pajuelo E, Forde BG and Marquéz AJ (1997) Regulation of the expression of ferredoxin glutamate synthase in barley. Planta 203: 517–525

Palomo J, Gallardo F, Suarez MF and Cánovas FM (1998) Purification and characterisation of $NADP^+$-linked isocitrate dehydrogenase from Scots pine. Evidence for different physiological roles of the enzyme in primary development. Plant Physiol 118: 617–626

Parry G, Woodall J, Nuotio S and Pearson J (2000) Glutamine synthetase isoforms in *Trientalis europea*: A biochemical and molecular approach. Plant Soil 221: 39–45

Pavesi A, Ficarelli A, Tassi F and Restivo FM (2000) Cloning of two glutamate dehydrogenase cDNAs from *Asparagus officinalis:* Sequence analysis and evolutionary implications. Genome 43: 306–316

Pearson J and Ji YM (1994) Seasonal variation of leaf glutamine synthetase isoforms in temperate deciduous trees strongly suggests different functions of the enzyme. Plant Cell Environ 17: 1331–1337

Peat LJ and Tobin A (1996) The effect of nitrogen nutrition on the cellular localization of glutamine synthetase isoforms in barley roots. Plant Physiol 111: 1109–1117

Pérez-Garcia A, Cánovas FM, Gallardo F, Hirel B and de Vicente A (1995) Differential expression of glutamine synthetase isoforms in detached leaflets infected with *Pseudomonas syringae* pv.tomato. Mol Plant Microbe In 8: 96–103

Pérez-Garcia A, De Vicente A, Cantón FR, Cazorla FM, Codina JC, García-Gutiérrez A and Cánovas FM (1998) Light-dependent changes of tomato glutamine synthetase in response to *Pseudomonas syringae* infection or phosphinothricin treatment. Physiol Plant 102: 377–384

Pryor AJ (1990) A maize glutamic dehydrogenase null mutant is cold temperature sensitive. Maydica 35: 367–372

Purnell MP, Stewart GR and Botella JR (1997) Cloning and characterisation of glutamate dehydrogenase cDNA from tomato (*Lycopersicon esculentum L.*). Gene 186: 249–254

Robinson SA, Phillips R and Stewart GR (1992) Regulation of glutamate dehydrogenase activity in relation to carbon limitation and protein catabolism in carrot suspension cultures. Plant Physiol 98: 1190–1195

Sakakibara H, Watanabe M, Hase T and Sugiyama T (1991) Molecular cloning and characterization of complementary DNA encoding for ferredoxin-dependent glutamate synthase in maize leaf. J Biol Chem 266: 2028–2035

Sakakibara H, Kawabata S, Takahashi H, Hase T and Sugiyama T (1992a) Molecular cloning of the family of glutamine synthetase genes from maize: Expression of genes for glutamine synthetase and ferredoxin-dependent glutamate synthase in photosynthetic and non-photosynthetic tissues. Plant Cell Physiol 33: 49–58

Sakakibara H, Kawabata S, Hase T and Sugiyama T (1992b) Differential effects of nitrate and light on the expression of glutamine synthetases and ferredoxin-dependent glutamate synthase in maize. Plant Cell Physiol 33: 1193–1198

Sakakibara H, Fujii K and Sugiyama T (1995) Isolation and characterisation of a cDNA that encodes maize glutamate dehydrogenase. Plant Cell Physiol 36: 789–797

Sakakibara H, Suzuki M, Takei K, Deji A, Taniguchi M and Sugiyama T (1999) A response-regulator homologue possibly involved in nitrogen signal transduction mediated by cytokinin in maize. Plant J 14: 337–344

Sakurai N, Hayakawa T, Nakamura T and Yamaya T (1996) Changes in the cellular localisation of cytosolic glutamine synthetase protein in vascular bundles of rice leaves at various stages of development. Planta 200: 306–311

Schjoerring JK, Husted S, Mäck G, Höier-Nielsen K, Finnemann J and Mattsson M (2000) Physiological regulation of plant-atmosphere ammonia exchange. Plant Soil 221: 95–102

Schmidt RR and Miller P (1999) Novel polypeptides and polynucleotides relating to the alpha and beta-subunits of glutamate dehydrogenase and methods of use. US patent 5879941

Schultz CJ, Hsu M, Miesak B and Coruzzi GM (1998) *Arabidopsis* mutants define an in vivo role for isoenzymes of aspartate aminotransferase in plant nitrogen assimilation. Genetics 149: 491–499

Somerville CR and Ogren WL (1980) The inhibition of photosynthesis in *Arabidopsis* mutants lacking glutamate synthase activity. Nature 286: 257–259

Somerville SC and Ogren WL (1983) An *Arabidopsis thaliana* mutant defective in chloroplastic dicarboxylate transport. Proc Natl Acad Sci USA 80: 1290–1294

Srivastava HS and Singh RP (1987) Role and regulation of L-glutamate dehydrogenase activity in higher plants. Phytochemistry 26: 597–610

Stewart GR, Shatilov, VR, Turnbull MH, Robinson SA and Goodall R (1995) Evidence that glutamate dehydrogenase plays a role in oxidative deamination of glutamate in seedlings of *Zea mays.* Aust J Plant Physiol 22: 805–809

Stitt M (1999) Nitrate regulation of metabolism. Curr Opin Plant Biol 2: 178–186

Stitt M and Krapp A (1999) The interaction between elevated carbon dioxide and nitrogen nutrition: The physiological and molecular background. Plant Cell Environ 22: 583–621

Sueyoshi K, Mitsuyama T, Sugimoto T, Kleinhofs A, Warner RL and Oji Y (1999) Effect of inhibitors for signalling components on the expression of the genes for nitrate reductase and nitrite reductase in excised barley leaves. Soil Sci Plant Nutr 45: 1015–1019

Suzuki A and Gadal P (1984) Glutamate synthase: Physicochemical and functional properties of different forms in higher plants and other organisms. Physiol Vég 22: 471–486

Suzuki A and Rothstein S (1997) Structure and regulation of ferredoxin dependent glutamate synthase from *Arabidopsis thaliana*—cloning of cDNA, expression in different tissues of wild type and *gltS* mutant strains and light induction. Eur J Biochem 243: 708–718

Suzuki A, Gadal P and Oaks A (1981) Intracellular distribution of enzymes associated with nitrogen assimilation in roots. Planta 151: 457–461

Suzuki A, Audet C and Oaks A (1987) Influence of light in the ferredoxin-dependent glutamate synthase reaction in maize leaves. Plant Physiol 84: 578–581

Suzuki A, Vergnet C, Morot-Gaudry J-F, Zehnacker C and Grosclaude J (1994) Immunological characterisation of ferredoxin and methyl viologen interacting domains of glutamate synthase using monoclonal antibodies. Plant Physiol Biochem 32: 619–626

Suzuki A, Burkhart W and Rothstein S (1996) Nitrogen effects on the induction of ferredoxin dependent glutamate synthase and its messenger RNA, in maize leaves under the light. Plant Sci 114: 83–91

Syntichaki KM, Loulakakis KA and Roubelakis-Angelakis KA (1996) The amino acid sequence similarity of plant glutamate dehydrogenase to the extremophilic archeal enzyme conforms to its stress-related function. Gene 168: 87–92

Teller S, Schmidt KH and Appenroth KJ (1996) Ferredoxin dependent, but not NADH-dependent glutamate synthase is regulated by phytochrome and a blue/UV-A light receptor in turions of *Spirodela polyrhiza*. Plant Physiol Biochem 34: 713–719

Temple SJ, Vance CP and Gantt JS (1998) Glutamate synthase and nitrogen assimilation. Trends Plant Sci 3: 51–56

Tingey SV, Walker E and Coruzzi GM (1987) Glutamine synthetase genes of pea encode distinct polypeptides which are differentially expressed in leaves, roots and nodules. EMBO J 6: 1–9

Tjaden G, Edwards JW and Coruzzi GM (1995) *Cis* elements and *trans*-acting factors affecting regulation of a nonphotosynthetic light-regulated gene for chloroplast glutamine synthetase. Plant Physiol 108: 1109–1117

Trepp GB, van de Mortel M, Yoshioka H, Miller SS, Samac DA, Gantt JS and Vance CP (1999a) NADH-glutamate synthase in alfalfa roots. Genetic regulation and cellular expression. Plant Physiol 119: 817–828

Trepp GB, Temple SJ, Bucciarelli B, Shi LF and Vance CP (1999b) Expression map for genes involved in nitrogen and carbon metabolism in alfalfa root nodules. Mol Plant Microbe In 12: 526–535

Turano FJ (1998) Characterisation of mitochondrial glutamate dehydrogenase from dark-grown soybean seedlings. Physiol Plant 104: 337–354

Turano FJ and Muhitch MJ (1999) Differential accumulation of ferredoxin- and NADH-dependent glutamate synthase activities, peptides and transcripts in developing soybean seedlings in response to light, nitrogen and nodulation. Physiol Plant 107: 407—418

Turano FJ, Thakkar SS, Fang T and Weisemann JM (1997) Characterisation and expression of NAD(H) dependent glutamate dehydrogenase genes in *Arabidopsis*. Plant Physiol 113: 1329–1341

Valentin K, Kostrzewa M and Zetsche K (1993) Glutamate synthase is plastid-encoded in a red alga: Implications for the evolution of glutamate synthases. Plant Mol Biol 23: 77–85

Valle EM and Heldt HW (1992) Synthesis of glutamine and glutamate by intact bundle sheath cells of maize (*Zea mays*). Planta 187: 221-223

Valpuesta V, Perez-Rodriguez MJ, Quesada MA and Botella JR (1989) Two polypeptides from green tomato leaves recognized by antibodies against chloroplastic glutamine synthetase. Plant Physiol Biochem 27: 963–966

Vance CP, Miller SS, Gregerson RG, Samac DA, Robinson DL and Gantt JS (1995) Alfalfa NADH-dependent glutamate synthase: Structure of the gene and importance in symbiotic nitrogen fixation. Plant J 8: 345–358

Vincent R, Fraisier V, Chaillou S, Limami MA, Deléens E, Phillipson B, Douat C, Boutin JP and Hirel B (1997) Overexpression of a soybean gene encoding cytosolic glutamine synthetase in shoots of transgenic *Lotus corniculatus* L. plants triggers changes in ammonium assimilation and plant development. Planta 201: 424–433

Walker EL and Coruzzi GM (1989) Developmentally regulated expression of the gene family for cytosolic glutamine synthetase in *Pisum sativum*. Plant Physiol 91: 702–708

Walker EL, Weeden NF, Taylor CB, Green P and Coruzzi GM (1995) Molecular evolution of duplicate copies of genes encoding cytosolic glutamine synthetase in *Pisum sativum*. Plant Mol Biol 29: 1111–1115

Wallsgrove RM, Harel E, Lea PJ and Miflin BJ (1977) Studies on glutamate synthase from the leaves of higher plants. J Exp Bot 28: 588–596

Wallsgrove RM, Lea PJ and Miflin BJ (1979) Distribution of enzymes of nitrogen metabolism within pea leaf cells. Plant Physiol 63: 232–236

Wallsgrove RM, Lea PJ and Miflin BJ (1982) The development of NAD(P)H-dependent and ferredoxin dependent glutamate synthase in greening barley and pea leaves. Planta 153: 473–476

Wallsgrove RM, Kendall AC, Hall NP, Turner JC and Lea PJ (1986) Carbon and nitrogen metabolism in a barley (*Hordeum vulgare*) mutant with impaired chloroplast dicarboxylate transport. Planta 168: 324–329

Weber A, Menzlaff E, Arbinger B, Gutensohn M, Eckerson C and Flügge U-I (1995) The 2-oxoglutarate/malate translocator of chloroplast envelope membranes: Molecular cloning of a transporter containing a 12-helix motif and expression of a functional protein in yeast cells. Biochemistry 34: 2621–2627

Winter H, Robinson G and Heldt HW (1994) Subcellular volumes and metabolite concentrations in spinach leaves. Planta 193: 530–535

Woo KC, Boyle FA, Flügge IU and Heldt HW (1987) ^{15}N-ammonia assimilation, 2-oxoglutarate transport and glutamate export in spinach chloroplasts in the presence of dicarboxylates in the light. Plant Physiol 85: 621–625

Woodall J, Boxall JG, Forde BG and Pearson J (1996) Changing perspectives in plant nitrogen metabolism: The central role of glutamine synthetase. Sci Prog 79: 1–26

Yamaya T and Oaks A (1987) Synthesis of glutamate by mitochondria—an anaplerotic function for glutamate dehydrogenase. Physiol Plant 70: 749–756

Yamaya T, Hayakawa T, Tanasawa K, Kamachi K, Mae T and Ojima K (1992) Tissue distribution of glutamate synthase and glutamine synthetase in rice leaves. Occurrence of NADH-dependent glutamate synthase protein and activity in the unexpanded non-green leaf blades. Plant Physiol 100: 1427–1432

Yamaya T, Tanno H, Hirose N, Watanabe S and Hayakawa T (1995) A supply of nitrogen causes an increase in the level of

NADH-dependent glutamate synthase protein and the activity of the enzyme, in the roots of rice. Plant Cell Physiol 36: 1197–1204

Zehnacker C, Becker TW, Suzuki A, Caboche M and Hirel B (1992) Purification and properties of tobacco ferredoxin-dependent glutamate synthase and isolation of corresponding cDNA clones. Light-inducibility and organ-specificity of gene transcription and protein expression. Planta 187: 266–274

Chapter 7

Regulation of Ammonium Assimilation in Cyanobacteria

Francisco J. Florencio* and José C. Reyes
Instituto de Bioquímica Vegetal y Fotosíntesis. Centro de Investigaciones Cientificas Isla de la Cartuja. Universidad de Sevilla-CSIC. Av. Américo Vespucio s/n, E-41092 Sevilla, Spain

Summary 93
I. Introduction 94
II. Ammonium Uptake 94
III. The Glutamine Synthetase/Glutamate Synthase Pathway 96
A. Two Types of Glutamine Synthetase in Cyanobacteria 96
B. Glutamate Synthase: Two Enzymes for Two Redox Carriers 98
C. Isocitrate Dehydrogenase Provides 2-Oxoglutarate for Ammonia Assimilation 100
D. Glutamate Dehydrogenase: To Aminate or to Deaminate?—That is the Question 102
IV. Regulation of Ammonium Assimilation 103
A. Global Nitrogen Control by NtcA 103
B. Post-Transcriptional Regulation of GSI by Protein Protein Interaction 105
C. How do Cyanobacteria Sense Nitrogen? 108
V. Future Perspectives 109
Acknowledgments 109
References 109

Summary

Ammonia assimilation constitutes a central part of the cyanobacterial metabolism closely linked to photosynthesis. Ammonium taken up directly from the medium by specific permeases, or resulting from the metabolization of alternative nitrogen sources, is incorporated into carbon skeletons by the sequential action of two enzymes: glutamine synthetase (GS) and glutamate synthase (GOGAT). Two types of GS (GSI and GSIII) and two types of GOGAT (ferredoxin-GOGAT and NADH-GOGAT) have been described in cyanobacteria. Carbon skeletons required for ammonium assimilation are supplied in the form of 2-oxoglutarate, which is synthesized by isocitrate dehydrogenase (ICDH). Glutamate dehydrogenase (GDH) is also present in some cyanobacteria, but its role in ammonium assimilation seems to be limited to specific growth conditions. Regulation of the GS-GOGAT pathway is essential for the carbon/nitrogen balance in cyanobacteria. Both the level of GS protein and GS activity are finely controlled by different environmental conditions, such as nitrogen and carbon availability. The transcription factor NtcA increases the expression of ammonium permease, ICDH, GSI and GSIII under conditions of nitrogen limitation. Furthermore, in the cyanobacterium *Synechocystis* sp. PCC 6803, NtcA represses the synthesis of two inhibitory polypeptides (IF7 and IF17) that inactivate GSI by protein-protein direct interaction.

*Author for correspondence, Email: floren@cica.es

Christine H. Foyer and Graham Noctor (eds): Photosynthetic Nitrogen Assimilation and Associated Carbon and Respiratory Metabolism, pp. 93–113.

I. Introduction

Ammonium is the form of nitrogen (N) incorporated into carbon (C) skeletons by plants, fungi and bacteria in a process known as ammonium assimilation. Since N is a constituent of most biomolecules, the control of the rate of ammonium assimilation is an important task that organisms have to accomplish in order to maintain their C/N homeostasis and growth rates.

Ammonium is the most reduced inorganic form of N. However, N is commonly present in the environment in less reduced forms such as nitrate, nitrite, and N_2 or in organic compounds such as urea, amino acids, etc. Inorganic compounds of N have to be reduced to ammonium before their incorporation into C skeletons in processes that require reducing equivalents and energy (Chapters 3–5). In a similar way, organic compounds require metabolization to yield free ammonium. Therefore, from the point of view of energetic economy, ammonium is the preferred N source. This is an almost universal rule for microorganisms and plants and, in the presence of ammonium, many different regulatory mechanisms are devoted to the repression and/or the inhibition of proteins involved in the utilization of alternative N sources. N sources other than ammonium are typically alluded to as poor N sources.

Cyanobacteria are photosynthetic prokaryotes that carry out oxygenic photosynthesis like plants. Most cyanobacteria can use nitrate, nitrite or ammonium ions as N sources and some strains are also able to fix N_2 or to utilize urea, cyanate or some amino acids. The processes of nitrate and nitrite reduction in cyanobacteria have been reviewed previously (Guerrero and Lara, 1987; Flores and Herrero, 1994; Flores et al., 1999). Different aspects of N_2 fixation in cyanobacteria have also been reviewed in Flores and Herrero (1994); Wolk et al. (1994); Böhme (1998); Haselkorn (1998); Mulholland and Capone (2000) and finally the assimilation of organic N sources has been reviewed by Flores and Herrero (1994). Ammonium, taken up directly from the medium by specific permeases, or resulting from the metabolization of alternative N sources, is incorporated into C skeletons mainly through the sequential operation of two enzymes, glutamine synthetase (GS) and glutamate synthase (GOGAT), in a cycle commonly known as the GS-GOGAT pathway (Fig. 1). The reaction catalyzed by GS involves the ATP-dependent amidation of Glu to yield Gln (Purich, 1998). GOGAT then catalyzes the reductive transfer of the amide group from Gln to 2-oxoglutarate (2-OG) to yield two molecules of Glu. The C skeleton required for ammonium assimilation is 2-OG, which is synthesized by isocitrate dehydrogenase (ICDH), an enzyme of the tricarboxylic acid cycle. Direct amination of 2-OG catalyzed by glutamate dehydrogenase (GDH) also takes place in some cyanobacteria. However, assimilation of ammonium through GDH seems to be quantitatively of low importance under normal growth conditions (see below). Nitrogen atoms contained in Glu and Gln are then distributed to a number of N-containing metabolites such as amino acids, purines, pyrimidines, porphyrins and amino sugars. Therefore, the GS-GOGAT pathway represents the connecting step between C and N metabolism and requires two direct photosynthetic products, ATP and reducing power (Fig. 1). This central position in metabolism makes the pathway susceptible to regulation by different environmental conditions, such as N and C availability, and by photosynthetic growth conditions. A landmark in the field was the identification of the DNA-binding protein, NtcA, a transcription factor that plays a central role in the regulation of GS. NtcA also controls the expression of genes involved in the utilization of N sources other than ammonium.

In this Chapter we will discuss the advances made in recent years in our understanding of ammonium transport and assimilation and the regulation of these processes in cyanobacteria.

Abbreviations: 2OG –2-oxoglutarate; CAP – catabolite activator protein; CRP – cAMP receptior protein; DON – 6-diazo-5-oxo-L-norleucine; Fd – ferredoxin; GDH – glutamate dehydrogenase; Gln – glutamine; Glu – glutamate; GOGAT –glutamate synthase; GS – glutamine synthetase; ICDH – isocitrate dehydrogenase; IF – inactivating factor; MSX – L-methionine-DL-sulphoximine; TCA – tricarboxylic acid; WT – wild-type

II. Ammonium Uptake

Ammonium solutions always contain ammonia (pKa[ammonium/ammonia], 9.25), which can diffuse through biological membranes (Kleiner, 1981). However, the concentration of free ammonium in aquatic environments is usually extremely low, which probably provoked the evolution of ammonium transport systems that concentrate ammonium inside the cell. Ammonium transport has been characterized in a number of cyanobacteria using [^{14}C]methyl-ammonium as a probe. Thus, ammonium effectively

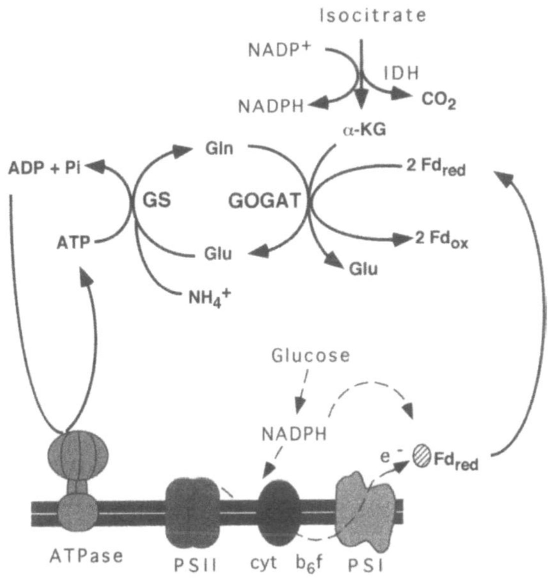

Fig. 1. The GS-GOGAT pathway showing its relationship to photosynthetic metabolism. Abbreviations: Cyt b_6f, cytochrome b_6f complex; Fd_{ox}, oxidized ferredoxin; Fd_{red}, reduced ferredoxin; GS, glutamine synthetase; GOGAT, glutamate synthase; IDH, isocitrate dehydrogenase; α-KG, α-ketoglutarate (= 2-oxoglutarate); PSII, photosystem II; PSI, photosystem I.

inhibits methylammonium uptake in cyanobacteria, as in many other organisms, supporting the idea of ammonium being the natural substrate of the methylammonium uptake systems (but see Soupene et al. (1998)). Pioneering studies using this method in cyanobacteria were carried out in the unicellular strain *Anacystis* R2 (*Synechococcus* sp. PCC 7942) and in the filamentous strain *Anabaena variabilis* (Boussiba et al., 1984; Rai et al., 1984). Both strains exhibit a biphasic kinetic with a rapid high-affinity phase, related to the entry of methylammonium and a slower phase associated with its metabolization via GS. This second slower phase is abolished by incubating the cells with L-methionine-DL-sulphoximine (MSX), a specific inhibitor of GS, supporting the idea that GS is able to catalyze the synthesis of γ-glutamylmethylamide using Glu and methylammonium as substrates. This has been further characterized by thin layer chromatography in the cyanobacterium *Synechocystis* sp. PCC 6803 (Montesinos et al., 1998) and in enteric bacteria (Soupene et al., 1998).

Three genes encoding methylammonium/ammonium permeases (*amt1*, *amt2* and *amt3*) have recently been characterized in the cyanobacterium *Synechocystis* sp. PCC 6803 (Montesinos et al., 1998). In addition, one *amt1* homologue has also recently been characterized in *Synechococcus* sp. PCC 7942 (Vázquez-Bermúdez, 2000). Cyanobacterial Amt permeases show between 37 and 27% sequence identity with methylammonium/ammonium permeases (MEP) from plants, yeast and other bacteria. Characterized MEPs are highly hydrophobic polypeptides that bear 12 putative membranes spanning regions. The energetic and molecular mechanisms responsible for the transport of the ammonium species across the membrane are unknown. While most work suggests that ammonium transport is an active process that concentrates NH_4^+ inside the cell, Kustu and coworkers reported that enteric bacteria AmtB permease increases the rate of equilibration of the uncharged species (NH_3) across the membrane (Soupene et al., 1998). These data are clearly in contradiction with previous results from the Barnes laboratory which support a mechanism for methylammonium/ammonium accumulation which requires K^+ antiport and is driven by the electrochemical K^+ gradient (Jayakumar et al., 1985). In cyanobacteria methylammonium accumulation seems to be also an energy-requiring process, sensitive to uncouplers and ATPase inhibitors. Inhibition by K^+ and by triphenylmethylphosphonium indicates that the accumulation is dependent on the membrane potential (Boussiba et al., 1984; Rai et al., 1984). Interestingly, in *Synechocystis* sp. PCC 6803, the *amt3* gene might be cotranscribed with an ORF that encodes a putative potassium channel protein (Montesinos et al., 1998).

Mutagenesis of the three *amt* permeases of *Synechocystis* sp. PCC 6803, has shown that under the conditions tested, *amt1* is responsible for more than 95% of the [^{14}C]methylammonium uptake and that the products of the other two ORFs contribute very little to the uptake activity. Expression of the three genes is induced in the absence of ammonium, and especially under conditions of N deprivation, *amt1* being expressed at higher levels than the other two *amt* genes (Montesinos et al., 1998). These results confirm previous data suggesting that [^{14}C]methylammonium uptake activity is repressed by ammonium (Boussiba et al., 1984; Rai et al., 1984). The fact that Amt protein expression is induced in the absence of its substrate points to a role of the Amt permeases under conditions of very low concentration of ammonium which are the typical conditions in natural habitats. Under conditions of ammonium availability, diffusion of ammonia through the cytoplasmic membrane is probably enough to support growth at alkaline or neutral pH. Thus, *amt* single mutants or *amt1/amt3* and *amt1/amt2* double mutants grow normally using ammonium as N source. However, a triple mutant is not yet available and therefore it is not yet formally demonstrated that diffusion alone is able to allow ammonium dependent growth. Recent experiments carried out in a *Synechococcus* sp. PCC 7942 *amt1* mutant point to a role of Amt permeases in recovering NH_3 that diffuses out of the cell when growing in N sources other than ammonium (Vázquez-Bermúdez, 2000).

The structure of the *amt1* promoter follows the canonical features of the NtcA-dependent promoters. In addition, NtcA protein binds to a fragment containing the *amt1* putative regulatory region. These data suggest that *amt1* belongs to the NtcA regulon (see below). The genes *amt2* and *amt3* are induced in a similar way to *amt1*. Whether these two genes are under the control of NtcA has not been established.

Since the product of *amt1* is responsible for more than 95% of the ammonium uptake activity, the reason why *Synechocystis* sp. PCC 6803 requires three *amt* genes remains a mystery. Maybe *amt2* and *amt3* are induced under certain unknown environmental conditions, where they could be responsible for a higher percentage of the uptake activity. Three different *amt* genes closely related to *amt1* and less similar to *amt2* or *amt3* are also present in *Anabaena* sp. PCC 7120. However, only one *amt* gene (closer to *amt1* than to *amt2* or *amt3*) is present in the *Prochlorococcus* genome. As will be discussed later in this review, the existence of more than one gene whose products have the same or similar function is relatively common for genes involved in the ammonium assimilation pathways in cyanobacteria.

III. The Glutamine Synthetase/Glutamate Synthase Pathway

There are two pathways for 2-OG amination: directly through GDH or by the sequential action of two enzymes, GS and GOGAT. Activity of all three enzymes was detected in several cyanobacterial strains in the early seventies (Dharmawardene et al., 1973; Neilson and Doudoroff, 1973; Lea and Miflin, 1975). However, metabolic labeling in several fixing strains using [^{13}N]N_2 and $^{13}NH_4^+$, demonstrated that the first labeled organic compound is Gln followed by Glu (Wolk et al., 1976; Meeks et al., 1977; Meeks et al., 1978). These results, together with the strong inhibition of ammonium assimilation produced by MSX (Stewart and Rowell, 1975), a specific inhibitor of GS, clearly point to the GS-GOGAT cycle as the major pathway of ammonium assimilation in cyanobacteria. This conclusion is further supported by the fact that a *Synechocystis* sp. PCC 6803 GDH deficient mutant strain is not significantly impaired in ammonium assimilation under normal growth conditions (Chávez et al., 1999).

In this section, we will first describe recent findings concerning the enzymes that constitute the GS-GOGAT pathway and their structural genes. Then we shall analyze the role of ICDH in the synthesis of 2-OG and, finally, we will try to shed light on the possible function of GDH in N metabolism in cyanobacteria.

A. Two Types of Glutamine Synthetase in Cyanobacteria

GS converts Glu and ammonium to Gln in the presence of divalent cations (generally Mg^{2+} or Mn^{2+}) using the energy of ATP hydrolysis. Three different types of GS have been found so far. Most prokaryotes have a dodecameric GS (known as GS type I, GSI),

Table 1. Physicochemical characteristics of the three types of glutamine synthetase

	GSI	GSII	GSIII
Molecular Mass			
native	600 kDa	350 kDa	500 kDa
subunit	50 kDa	40 kDa	80 kDa
AA number of the subunit	About 500	About 400	About 700
Subunits	12	8	6
K_m (ammonium)	0.05 to 0.2 mM	0.2 mM	0.2 mM
Inhibitors	MSX DL-phosphinothricin	MSX DL-phosphinothricin	MSX DL-phosphinothricin
Organisms	Prokaryotes	Eukaryotes *Rhizobium* *Agrobacterium* *Streptomyces* *Frankia*	Cyanobacteria Bacteroidaceae Clostridiaceae Diatoms ?

AA, amino acids

composed of 12 identical subunits (M_r, about 50,000) arranged in two superimposed hexagonal rings. Eukaryotic GS (GS type II, GSII) is an octameric enzyme with subunits of M_r, about 40,000. GSI and GSII should not be confused with GS1 and GS2 from plants (Chapter 6, Hirel and Lea), which both belong to the GSII type. Members of the family *Rhizobiaceae* and certain Actinomycetales harbor both a GSI and a GSII-like enzyme (Merrick and Edwards, 1995; Eisenberg et al., 2000). A third type of GS (GS type III, GSIII), composed of six identical subunits (M_r, about 75,000) was initially identified in *Bacteroides fragilis*, and later in several other bacteria (Woods and Reid, 1993; Reyes and Florencio, 1994; Crespo et al., 1998). These three types of GS are quite different in amino acid sequence. However, five domains of homology among all known GSs can be identified (see alignment in Reyes and Florencio (1994)). Structural studies on the *Salmonella typhimurium* GSI using X-ray crystallography show that these five domains contain amino acids that form part of the catalytic site or that are involved in the binding of two divalent metal ions that are constituents of the enzyme. The 3-D structure also shows that each active site is formed at the interface between the C-terminal domain of one subunit and the N-terminal domain of an adjacent subunit within a hexameric ring (reviewed in Eisenberg et al. (2000)). Some molecular and kinetic properties of the different types of GS are summarized in Table 1.

GSI has been purified from a number of cyanobacterial strains (*Anabaena* sp. PCC 7120, *Anabaena azollae*, *Anacystis nidulans*, *Synechocystis* sp. PCC 6803, *Calothrix* sp. PCC 7601 and *Phormidium laminosum)* (Stacey et al., 1977; Orr et al., 1981; Florencio and Ramos, 1985; Blanco et al., 1989; Mérida et al., 1990; Crespo et al., 1999). All cyanobacterial GSIs were similar to each other in size and subunit composition and also similar to other prokaryotic GSIs. K_m for the different substrates of GSIs ranges from 20 to 170 μM for ammonium, from 0.35 to 5 mM for Glu, and from 0.3 to 0.7 mM for ATP. Structural genes for GSI (denoted *glnA*) have also been cloned and sequenced from several cyanobacteria (Fisher et al., 1981; Elmorjani et al., 1992; Wagner et al., 1993; Reyes and Florencio, 1995a; Crespo et al., 1999). Cyanobacterial *glnA*-encoded polypeptides show more than 75% amino acid sequence identity among them and about 50% amino acid identity with respect to enterobacterial GSIs. Site-directed mutagenesis of Asp-51 from the *Anabaena azollae* GSI suggests that, as previously stated for the enterobacterial GSI Asp-50, this residue may be involved in ammonium binding (Crespo et al., 1999).

Analysis of a *Synechococcus* sp. PCC 7002 *glnA* mutant strain revealed the surprising result that although no *glnA* mRNA could be detected in the mutant cells, they retained about 60% of wild-type (WT) Gln biosynthetic activity (Wagner et al., 1993). We had observed that a *glnA* mutant strain of *Synechocystis* sp. PCC 6803 was not a Gln auxotroph, and presented a low level of GS activity. These results led us to investigate the existence of a second gene encoding GS in this organism. This alternative GS encoding gene (named *glnN*) was cloned by complementation of a *glnA Escherichia coli* mutant auxotroph of Gln (Reyes and Florencio, 1994). The *glnN* gene encodes a type III GS, homologous to GSIIIs from *Bacteroides fragilis* (44% identity) and *Butyrivibrio fibrisolvens* (41%). Heterologous Southern and western blotting suggest that GSIII is present in many other non-nitrogen fixing cyanobacteria but not in nitrogen fixers (Reyes and Florencio, 1994; García-Domínguez et al., 1997). In fact, *glnN* genes have been recently cloned from two other cyanobacteria; *Pseudanabaena* sp. PCC 6903 and *Synechococcus* sp. PCC 7942 (Crespo et al., 1998; Sauer et al., 2000). The case of *Pseudanabaena* sp. PCC 6903 is particularly interesting, because this strain lacks GSI (and *glnA* gene) and has only GSIII. Therefore, cyanobacteria can be classified into three categories with respect to the type of GS present: cyanobacteria that only have GSI (like *Anabaena* sp. PCC 7120 and several other N_2 fixers), those that only contain GSIII (like *Pseudanabaena* sp. PCC 6903) or those that have both (such as *Synechocystis* sp. PCC 6803 and *Synechococcus* sp. PCC 7942).

Recombinant *Synechocystis* sp. PCC 6803 GSIII expressed in *E. coli* has been purified (García-Domínguez et al., 1997). Biosynthetic activity of GSIII requires the same substrates and cofactors as GSI and GSII enzymes. Apparent K_m values for ATP, Glu and ammonium are also similar to those of the *Synechocystis* sp. PCC 6803 GSI. However, optimum pH was about 8.25 in contrast to the neutral optimum pH of GSI. The physiological significance of this difference remains unknown.

GSIII has been found in different species from very different taxonomic groups: *Bacteroidaceae* (*Bacteroides fragilis* and *Prevotella melaninogenica*), *Clostridiaceae* (*Ruminococcus flavefaciens* and *Butyrivibrio fibrisolvens*), Deioncoccales (*Deionococcus radiodurans*) and cyanobacteria. The origin of GSIII and its phylogenetic relationship with GSI and GSII are unknown. The fact that GSIII is present in phylogenetically unrelated taxonomic groups suggests that GSIII was present in a putative common ancestor and that it later was lost in certain taxa, probably due to the redundancy caused by GSI. The reason why GSIII instead of GSI has prevailed in some taxa is unknown. More interesting is the possibility that GSIII exists in algae. Recently, partial cDNA sequences that may code for GSIII from four different diatoms and from a *Chlorophyceae* have been deposited in the databases (AF251001 to AF251004, AB016770). This is the first evidence of the existence of GSIII in eukaryotes. Whether GSIII is encoded in the chloroplast or in the nuclear genome is not yet known. However, the high homology (more than 80% identity) between the cyanobacterial and the algal sequences suggests that algal genes encoding GSIII come from the endosymbiotic cyanobacteria that gave rise to the chloroplast.

Synechocystis sp. PCC 6803 *glnN* null mutants grow normally under all the conditions tested. However, a *glnA/glnN* double mutant is not viable, even in the presence of Gln in the culture medium (Reyes and Florencio, 1994). Since *Synechocystis* sp. PCC 6803 exhibits GDH activity, which could support ammonium assimilation, and since Gln can be taken up by the cells (Labarre et al., 1987; Flores and Muro-Pastor, 1990), the reason why it is not possible to segregate a GS deficient strain is an unsolved question.

The role of GSIII in cyanobacterial N metabolism is another subject that requires further investigation. In *Synechocystis* sp. PCC 6803 cells growing on nitrate, GSI is responsible for 97% of the total GS activity, while the *glnN* product (GSIII) accounts for only about 3%. However, after 24 h of N deprivation, the activity corresponding to the GSIII represents about 20% of the total GS activity. Induction of GSIII specifically under conditions of N deficiency is also observed in several different cyanobacterium strains where GSIII coexists with GSI (Reyes and Florencio, 1994; García-Domínguez et al., 1997). This pattern of expression, together with the lack of GSIII in cyanobacteria able to fix N_2, suggests that the presence of GSIII gives a selective advantage when a combined N source is not present, for cyanobacteria that are unable to fix N_2. This has been recently demonstrated with a *glnN* mutant of *Synechococcus* sp. PCC 7942. Thus, *glnN* mutant cells present a low recovery rate after long periods of N deficiency (Sauer et al., 2000). It is worth noting that GSIII kinetic properties do not seem to suggest that this enzyme is more efficient in the assimilation of ammonium than GSI under N deficiency conditions (García-Domínguez et al., 1997).

B. Glutamate Synthase: Two Enzymes for Two Redox Carriers

In photosynthetic organisms, as described in Chapter 6 (Hirel and Lea), two types of GOGAT

(Fd-GOGAT and NADH-GOGAT) synthesize Glu by the transfer of the Gln amide group to the C skeleton 2-OG, in a reductive step that involves two electrons. A third type of GOGAT using NADPH for reduction is present in non-photosynthetic bacteria (Temple et al., 1998). NADPH-GOGAT has been extensively characterized in *E. coli* and *Azospirillum brasilense* (recently reviewed by Vanoni and Curti (1999)), and is composed of two different subunits. The large one, named α subunit, has a molecular mass about of 150 kDa and contains a flavin, FMN and an iron-sulfur cluster [3Fe-4S] and its structure has recently been elucidated (Binda et al., 2000). The small subunit, named β subunit, of about 50 kDa, contains a FAD and two iron-sulfur clusters of type [4Fe-4S], probably localized at the surface of interaction between both subunits.

Fd-GOGAT has been well characterized from different photosynthetic sources, including higher plants, green algae and cyanobacteria (Galván et al., 1984; Knaff and Hirasawa, 1991; Lam et al., 1995; Navarro et al., 1995). The enzyme is a monomer of about 170 kDa molecular mass, containing as prosthetic groups the flavin FMN and the [3Fe-4S] cluster, and is therefore similar to the bacterial α subunit both with respect to size and prosthetic groups (Garcia et al., 1977; Hirasawa and Tamura, 1984; Márquez et al., 1986; Knaff et al., 1991; Sakakibara et al., 1991; Marqués et al., 1992a; Hirasawa et al., 1996; Navarro et al., 2000).

NADH-GOGAT has been purified and characterized from *Medicago sativa* and yeast and it is also a monomer of about 200 kDa. It contains the same prosthetic groups as the bacterial α subunit in its N-terminal domain and an additional iron-sulfur cluster and a flavin (probably FAD) in its C-terminal domain. NADH-GOGAT C-terminal domain is similar to the bacterial β-subunit (Gregerson et al., 1993; Cogoni et al., 1995). In cyanobacteria, NADH-GOGAT is composed by two different subunits: a large one of 160 kDa with homology to Fd-GOGAT and the bacterial α subunit and a small one of 60 kDa homologous to the β subunit of bacterial NADPH-GOGAT (Okuhara et al., 1999). It is interesting to note that cyanobacterial and higher plant NADH-GOGAT show a high degree of homology. Since NADH-GOGAT of higher plants is a single polypeptide and the cyanobacterial enzyme is composed by two different polypeptides it has been suggested that the plant enzyme is probably the fusion of these two polypeptides, which were present in the primitive cyanobacteria that gave rise to the chloroplast (Fig. 2) (Lam et al., 1995; Temple et al., 1998; Okuhara et al., 1999).

The gene encoding Fd-GOGAT (*glsF*, formerly named *gltS*) has been cloned from higher plants, and sequenced (Sakakibara et al., 1991; Nalbantoglu et al., 1994; Lam et al., 1995). Sequences have also been obtained from the red algae *Porphyra purpurea* and *Antithamnium* sp. (Reith and Munholland, 1993; Valentin et al., 1993) and from the cyanobacteria *Synechocystis* sp. PCC 6803, *Plectonema boryanum* and *Anabaena* sp. PCC 7120 (Navarro et al., 1995; Okuhara et al., 1999; Martín-Figueroa et al., 2000). Detailed analysis of the Fd-GOGAT sequences available reveals several well-conserved regions assigned to the iron-sulfur cluster with the cysteine motif (CX_5CX_4C), the FMN binding domain and the Gln-amide transferase domain, as well as an amino acid signature present only in the Fd-dependent enzymes (Vanoni and Curti, 1999).

Although some cyanobacteria exhibit GDH activity, it is clear that GOGAT is the obligatory pathway for Glu formation, since mutants lacking

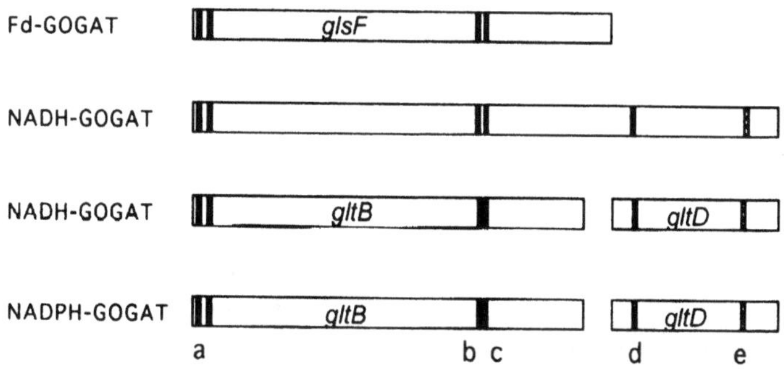

Fig. 2. Schematic representation of the different glutamate synthases. Different domains are marked. a, glutamine amide transferase domain; b, FMN binding-domain; c, 3Fe-4S cluster, d, FAD binding domain; e, NADH or NADPH binding site. For gene identification, see text.

this enzyme activity have not been obtained. Only in cyanobacteria containing both GOGATs could mutants lacking the Fd-dependent enzyme be obtained, such as *P. boryanum* and *Synechocystis* sp. PCC 6803 (Navarro et al., 1995; Okuhara et al., 1999). The *Synechocystis glsF* mutant is able to grow photosynthetically, using different N sources, without differences in their growth rates or chlorophyll content. In contrast, a *P. boryanum* mutant lacking Fd-GOGAT (*glsF* mutant), but not the corresponding NADH-GOGAT mutant, shows a phenotype of N deficiency at high light intensity and 2% CO_2, suggesting a primary role for Fd-GOGAT in N assimilation at high photosynthetic growth rates that cannot be supported by NADH-GOGAT (Okuhara et al., 1999). In those cyanobacteria where NADH-GOGAT is present, the protein is encoded by the *gltB* and *gltD* genes (large and small subunit, respectively). In the case of *P. boryanum* both genes are organized close together, as an operon, but not in *Synechocystis* sp. PCC 6803, where the genes are located far away from each other (Kaneko et al., 1996; Okuhara et al., 1999). *gltB* and *gltD* mutants obtained in *P. boryanum* and *Synechocystis* sp. PCC 6803 do not show differences in growth as compared to the WT strains, suggesting an accessory role for NADH-GOGAT in N assimilation in cyanobacteria (Okuhara et al., 1999; E. Martín-Figueroa and F. J. Florencio, unpublished).

To date all cyanobacteria studied contain Fd-GOGAT. This is particularly interesting in the case of the N_2-fixing heterocyst forming cyanobacteria, such as *Anabaena* sp. PCC 7120, since it is important to know if the GS-GOGAT pathway operates in their specialized cells, the heterocysts. Recent studies combining analysis by immunoblotting using Fd-GOGAT antibodies and analysis of *glsF* transcript abundance have demonstrated that Fd-GOGAT is absent from the heterocysts, while GS is very abundant. These data indicate that the N_2 fixed by the nitrogenase is incorporated into Glu by GS, and Gln or another amino acid has to be exported to the vegetative cells, where Gln can be used by Fd-GOGAT to synthesize Glu (Martín-Figueroa et al., 2000). Furthermore, kinetic analysis using Fd from the heterocysts or from vegetative cells also indicated that heterocystous Fd was unable to serve as an efficient electron donor for Fd-GOGAT (Schmitz et al., 1996).

Studies concerning GOGAT gene expression are scarce. Our data indicate that the amount of the *glsF* transcript in *Anabaena* sp. PCC 7120 is similar whether the cells have been grown on nitrate, ammonium or N_2 as N source (Martín-Figueroa et al., 2000). In addition, data on the amount of enzyme and enzyme activity in several cyanobacteria, also suggest that Fd-GOGAT is not subject to fluctuation depending on the N source or the growth light intensity. These data substantiate the idea that the control of N flux is exerted at the level of GS. No data are available about NADH-GOGAT gene expression in cyanobacteria.

Phylogenetic analysis of the GOGAT genes clearly indicates that cyanobacterial genes *glsF* and *gltB* are probably the result of a gene duplication, and that both Fd-GOGAT and NADH-GOGAT are the ancestors of the corresponding higher plant enzymes (Temple et al., 1998). Taking this into account, the primitive cyanobacterium that gave rise to the chloroplast should have contained genes for both enzymatic activities. While in red algae the *glsF* gene is still encoded in the chloroplast, in higher plants both GOGAT genes are nuclear, although both enzymes are localized in the plastid (Reith and Munholland, 1993; Valentin et al., 1993; Temple et al., 1998). That higher plant NADH-GOGAT seems to be the result of the fusion of the *gltB* and *gltD* genes is supported by the genomic structure of *gltB-gltD* in *P. boryanum* where *gltD* is only 106 bp downstream of *gltB* (Okuhara et al., 1999). The reason why two different GOGATs are present in some cyanobacteria is unknown, but a probable hypothesis can be related to the electron source available for reduction. It could be speculated that those cyanobacteria able to growth heterotrophically could more easily use a pyridine nucleotide like NADH, instead of Fd, which requires a lower redox potential for its reduction (midpoint potentials (eV) for NADH and Fd: –0.32 and –0.42, respectively).

C. Isocitrate Dehydrogenase Provides 2-Oxoglutarate for Ammonia Assimilation

Isocitrate dehydrogenase (ICDH) carries out the oxidative decarboxylation of isocitrate to yield 2-OG with the concomitant reduction of a pyridine nucleotide. 2-OG is not only important because it is one of the substrates of the GS-GOGAT cycle, but also because it is a key metabolite with regulatory functions that will be discussed later. Although 2-OG can be synthesized by Glu-dependent transaminases, the main enzyme involved in its synthesis in cyanobacteria is ICDH.

Cyanobacteria have an incomplete TCA cycle

lacking the 2-OG dehydrogenase enzyme complex (Stanier and Cohen-Bazire, 1977). Therefore, in cyanobacteria, since 2-OG produced in the ICDH reaction cannot be further oxidized, it directly enters the GS-GOGAT cycle. Indeed, decrease of ICDH activity leads to a depletion of the intracellular Glu pool (Vega-Palas and Florencio, unpublished). This fact connects the ICDH reaction to biosynthetic N metabolism and rules out a role of ICDH in energy production as in other organisms. Although NADP-dependent and NAD-dependent ICDHs have been described in prokaryotes, most bacteria have only the NADP-linked enzyme. Cyanobacterial ICDH is strictly dependent on NADP and no NAD-ICDH activity has been reported so far (Friga and Farkas, 1981; Muro-Pastor and Florencio, 1992; Muro-Pastor and Florencio, 1994). NADP-ICDH has been purified from the unicellular cyanobacteria *Synechocystis* sp. PCC 6803 and *Anacystis nidulans* (Friga and Farkas, 1981; Muro-Pastor and Florencio, 1992) and from the filamentous strains *Anabaena* sp. PCC 7120 and *Phormidium laminosum* (Muro-Pastor and Florencio, 1994; Pardo et al., 1999). The enzyme is composed of two identical subunits ($M_r \approx 55{,}000$) and shows kinetic and physicochemical parameters similar to the *E. coli* NADP-ICDH. *E. coli* NADP-ICDH is inactivated by phosphorylation when acetate is present as a C source, in order to increase the flow of isocitrate to the glyoxylate cycle (Laporte and Koshland, 1982). This pathway is a bypass of the TCA cycle, present in plants and in different bacterial groups, which allows synthesis of glucose using acetate as the sole C source. Such phosphorylation of NADP-ICDH has not been found in *Synechocystis* sp. PCC 6803 NADP-ICDH, in agreement with the lack of glyoxylate cycle in this cyanobacterium (M. I. Muro-Pastor and F. J. Florencio, unpublished). This may explain why most cyanobacteria are unable to use acetate as C source in heterotrophic growth (Stal and Moezelaar, 1997).

The *icd* genes (genes encoding ICDH) from *Synechocystis* sp. PCC 6803 and *Anabaena* sp. PCC 7120 have been cloned and sequenced (Muro-Pastor and Florencio, 1994; Muro-Pastor et al., 1996). Both NADP-ICDHs show a high amino acid sequence similarity with the NADP-ICDH from other prokaryotes such as *E. coli* and *Vibrio* sp. (about 55% amino acid identity) and *B. subtilis* (52.5% amino acid identity). The most significant difference between these bacterial NADP-ICDH sequences is the presence of an insertion of 44 amino acid residues in the cyanobacterial proteins. This extra stretch (amino acid residues 286 to 329) is conserved in the three cyanobacterial sequences analyzed (from *Synechocystis* sp. PCC 6803, *Anabaena* sp. PCC 7120 and *Prochlorococcus marinus*) and seems to be an exclusive characteristic of NADP-ICDHs from cyanobacteria. The predicted secondary structure for this region is an α-helix, located within the small α/β domain described in the *E. coli* enzyme, but its function in the cyanobacterial enzymes is unknown.

Attempts to completely segregate *Synechocystis* sp. PCC 6803 or *Anabaena* sp. PCC 7120 *icd* mutants have been unsuccessful, indicating that *icd* is an essential gene for these cyanobacteria (Muro-Pastor and Florencio, 1994; Muro-Pastor et al., 1996). Growth analysis of the non-segregated *Anabaena* sp. PCC 7120 *icd* mutant proved that ICDH is required especially for diazotrophic growth (growth sustained by N_2 fixation). Addition of 2-OG or proline (which is easily converted to Glu) to the medium partially rescued the growth defect but did not permit complete segregation (Muro-Pastor and Florencio, 1994). In this respect it is worth noting that ICDH is an abundant enzyme in the heterocysts (Martín-Figueroa et al., 2000). The strict requirement for ICDH in diazotrophic growth has two possible explanations: first, the requirement for 2-OG might be higher on N free-medium, due to the fact that ammonium assimilation is restricted to heterocysts under these conditions. Second, a role of ICDH as an electron donor to nitrogenase has also been proposed in different heterocystous cyanobacteria (Karni and Tel-Or, 1983; Bothe and Neuer, 1988).

Interestingly, expression of the *icd* gene is controlled by the N source in *Synechocystis* sp. PCC 6803 and in *Anabaena* sp. PCC 7120 (Muro-Pastor et al., 1996). Thus, levels of ICDH activity and *icd* transcript increase three- to five-fold under conditions of N deficiency which is in good agreement with the increase in the pool of 2-OG observed under these conditions (Mérida et al., 1991). Transcription of the *Synechocystis* sp. PCC 6803 *icd* gene is activated by the transcription factor NtcA (Muro-Pastor et al., 1996) (see below).

What is the physiological significance of the increase in ICDH activity under conditions of N stress? Most organisms tend to coordinate their C and N metabolism in order to maintain the C/N balance. Why would the concentration of C skeletons be increased when N is scarce? One possibility is that 2-OG plays a role in N stress signaling (Section

IV.C). Nitrogen starvation provokes an increase in the level of 2-OG that triggers the increase in the level of ICDH activity, which in turn synthesizes more 2-OG. This positive feedback mechanism maintains higher and higher levels of the signaling molecule 2-OG. The second possibility is a kinetic reason. When the ammonium concentration is low, the increase in the concentration of the other substrates of the GS-GOGAT pathway will facilitate the reaction. Both possibilities are not mutually exclusive and may both be correct.

D. Glutamate Dehydrogenase: To Aminate or to Deaminate?—That is the Question

A number of cyanobacterial strains present GDH activity (NADP-GDH) (Neilson and Doudoroff, 1973; Chávez, 1992). As previously mentioned, metabolic labeling experiments, together with the dramatic effect provoked by the GS inhibitor MSX on ammonium assimilation, indicate that GDH plays a minor role in ammonium assimilation in cyanobacteria.

NADP-GDH has been purified from *Synechocystis* sp. PCC 6803 and from *Phormidium laminosum* (Florencio et al., 1987; Martinez-Bilbao et al., 1988). The enzyme is a hexamer of identical subunits with a molecular weight of about 300,000 (Chávez, 1992). The apparent K_m value for ammonium is between 1 and 3 mM, which argues against a role of this enzyme in primary N assimilation. However, cyanobacterial GDH catalyzes the amination of 2-OG preferentially over the reverse deaminating reaction, suggesting a preference for Glu synthesis instead of Glu catabolism. The *gdhA* gene of *Synechocystis* sp. PCC 6803, coding for NADP-GDH, was cloned by complementation of an *E. coli gdhA* mutant (Chávez et al., 1995). In such a genetic background, the *Synechocystis* sp. PCC 6803 NADP-GDH is able to operate as the only ammonium-assimilating enzyme. A NADP-GDH homologue ORF appears also in the genome of *Anabaena* sp. PCC 7120 but not in the genome of *Prochlorococcus marinus*. The amino acid sequences deduced from the *gdhA* genes show high identity with GDHs from archaebacteria (42–47%), some Gram-positive bacteria (40–44%), plants (40–42%) and mammals (37%). In contrast, cyanobacterial GDHs are much less similar to enterobacterial and fungal NADP-GDHs. A minor NAD-dependent activity has also been detected in *Synechocystis* sp. PCC 6803 (Chávez and Candau, 1991) but no putative gene coding for an independent NAD-GDH has been identified in the sequenced genome of this cyanobacterium (Kaneko et al., 1996). This activity could be ascribed to a secondary activity of another enzyme or to a member of a new family of GDHs.

Levels of *Synechocystis* NADP-GDH activity or *gdhA* mRNA are not affected by the N source. However, evolution of NADP-GDH during the growth curve follows a complex pattern. NADP-GDH activity level is high during the first 24 h of growth, dropping abruptly after 48 h. Then, the activity increases progressively, reaching the maximum at the late stage of growth (Chávez, 1992). The level of *gdhA* mRNA follows a parallel pattern reaching maximum levels very early in the exponential phase and close to the stationary phase (J. M. Lucena and P. Candau, personal communication). Analysis of the promoter using a reporter gene strongly suggests that *gdhA* mRNA amount is controlled at the level of transcription (Chávez et al., 1995). The recent characterization of a *Synechocystis* sp. PCC 6803 *gdhA* mutant lacking NADP-GDH gives some clues about the role of this enzyme (Chávez et al., 1999). Cells lacking NADP-GDH grow normally in the exponential growth phase but show a significantly decreased content of phycobiliproteins, antenna Photosystem II pigments in cyanobacteria. Since phycobiliproteins are degraded under conditions of N stress, the amount of these proteins can be taken as an indicator of the nutritional state of the cells (Collier and Grossman, 1994). The reduction of the level of phycobiliproteins in the exponential phase suggests that *gdhA* mutants are slightly N-stressed, which in turn indicates an aminating role of the enzyme. This notion is corroborated by the observation that the growth of the *gdhA* mutant is impaired in the late stage of the culture. Competition experiments between the WT and the null mutant confirmed that the presence of NADP-GDH confers a selective advantage on *Synechocystis* sp. PCC 6803 in late stages of growth. Competition experiments carried out with *E. coli* GDH-deficient mutants suggest that GDH is used in Glu synthesis when the cells are limited in energy (and C), while the GS-GOGAT pathway is used when the cells are not under energy limitation (Helling, 1994, 1998). The reason for this could be that the GDH enzyme does not use ATP, as does the GS-GOGAT pathway. A similar interpretation can explain the behavior of *Synechocystis* sp. PCC 6803 *gdhA* mutants in late stages of growth.

In cultures close to the stationary phase, autoshading produced by high cell density decreases photosynthesis and thus causes energy limitation. Under these conditions NADP-GDH may contribute significantly to ammonium assimilation. Interestingly, GS specific activity decreases strongly in the stationary phase of *Synechocystis* sp. PCC 6803 cultures (J. C. Reyes and F. J. Florencio, unpublished). The exact role of NADP-GDH in early exponential phase is not explained by this hypothesis and remains to be established. Further work is required to identify the *cis* and *trans* regulatory elements that control the striking pattern of expression of the *gdhA* gene during the growth curve.

IV. Regulation of Ammonium Assimilation

We have noted above that the GS-GOGAT pathway represents the connecting step between C and N metabolism. Because of this, it is not surprising that both the activity and the synthesis of the first enzyme of the pathway, GS, are tightly regulated in many organisms including cyanobacteria. In contrast, the level of GOGAT activity and level of expression of *gltB*, *gltD* and *glsS* genes seem not to be affected by N availability in cyanobacteria. In most bacterial systems studied, the control of GS activity responds to C and N signals. In the presence of abundant C sources, N deficiency results in a high level of GS activity. In contrast, when a N rich source is present, GS activity is down-regulated.

The term 'nitrogen control' designates the regulatory circuits that control the utilization of the different N sources (ammonium, nitrate, nitrite, N_2, urea, etc) in coordination with the level of GS synthesis and activity. In cyanobacteria, the main element shown to be responsible for N control is the transcription factor NtcA.

A. Global Nitrogen Control by NtcA

NtcA belongs to the cAMP receptor protein (CRP) family of bacterial DNA-binding proteins and it has been shown to activate transcription of a number of promoters in the absence of ammonium. The *ntcA* gene was first isolated in the laboratory of Flores and Herrero by complementation of a pleiotropic mutant of *Synechococcus* sp. PCC 7942. Mutant cells lacking *ntcA* were unable to grow using nitrate as N source and showed low nitrate reductase, nitrite reductase, GS and methylammonium transport activities in the absence of ammonium (Vega-Palas et al., 1990, 1992). A biochemical approach in the laboratory of Golden led to the identification of a DNA-binding protein, termed VF1, able to bind to several N regulated promoters (P*glnA*, P*nifHDK*, P*xis*) in *Anabaena* sp. PCC 7120 (Chastain et al., 1990). Cloning of the gene encoding VF1 (*bifA* gene) by an in vivo transcriptional interference method demonstrated that *Anabaena* sp. PCC 7120 VF1 (BifA) is the same protein as *Anabaena* sp. PCC 7120 NtcA (Frias et al., 1993; Wei et al., 1993). *ntcA* genes have been cloned from several cyanobacteria and the deduced amino acid sequences show more than 63% identity (Frias et al., 1993).

NtcA is composed of 222 to 225 amino acids and contains a DNA-binding helix-turn-helix motif in the carboxyl terminus (Vega-Palas et al., 1992). Recombinant *Anabaena* sp. PCC 7120 NtcA protein expressed in *E. coli* is a dimer in solution (Wisen et al., 1999). DNAse I footprinting experiments together with alignment of NtcA binding sites indicate that NtcA binds the consensus palindromic sequence GTAN$_8$TAC centered at around –41.5 nucleotides with respect to the transcription start point (Luque et al., 1994). Jiang et al. (2000) have reported an extended consensus site (NGTa-N$_{7,8,9}$-tACA) based on in vitro selection of DNA-binding motifs from a random library, using the *Anabaena* sp. PCC7120 NtcA protein. Figure 3 shows the distribution of natural NtcA-binding sites in NtcA-dependent promoters. NtcA-dependent promoters also present a canonical sigma–70 *E. coli*-like –10 box. Therefore, the structure of the NtcA-activated promoters is similar to the class II CRP-dependent promoters. In Class I CRP-dependent promoters, the DNA-binding site for CRP is located upstream of the site for RNA polymerase, at about –61.5 (Ebright, 1993; Busby and Ebright, 1997). No NtcA-dependent promoters have been identified with these characteristics.

Regulatory regions upstream of *glnA* genes are often quite complex, presenting NtcA-dependent and NtcA-independent overlapping promoters. These overlapping promoters can determine several transcription start points (as is the case in *Anabaena* sp. PCC 7120 and *Calothrix* sp. PCC 7601 *glnA* genes (Tumer et al., 1983; Elmorjani et al., 1992) or only one transcription start point (*Synechocystis* sp. PCC 6803 and *Synechococcus* sp. PCC 7942 (Luque et al., 1994; Reyes et al., 1997). In general, *glnA* genes are transcribed at basal levels from NtcA-

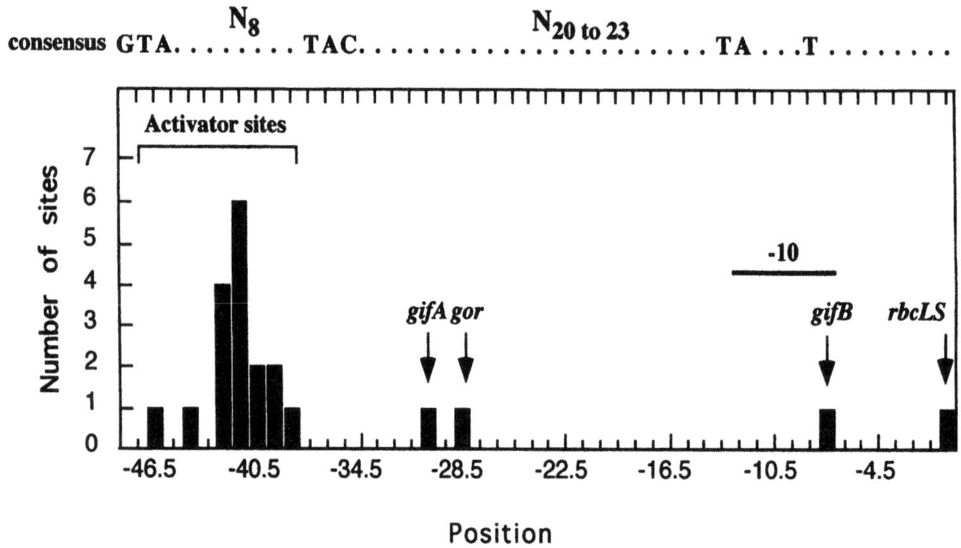

Fig. 3. Histogram showing the position of NtcA binding sites in NtcA-dependent promoters. The x-axis gives nucleotide number with respect to the transcription start point. The middle point of the sequence palindrome is considered as the position of the DNA binding site. The consensus NtcA binding site is shown above the figure. Activator sites are grouped around position –41.5. For gene identification, see text.

independent promoters, in the presence of ammonium. NtcA-dependent *glnA* promoters are induced in cells that use nitrate as N source, but maximal level of expression is usually reached in the absence of N source (Tumer et al., 1983; Elmorjani et al., 1992; Cohen-Kupiec et al., 1993, 1995; Wagner et al., 1993; Frias et al., 1994, 1995; Reyes et al., 1997). As previously mentioned nitrate can be considered a poor N source for cyanobacteria since it has to be reduced to ammonium before its incorporation to the GS-GOGAT pathway. Therefore, nitrate-growing cells are partially N-limited in good agreement with the fact that *glnA* genes present a medium level of induction. How this gradation of levels of activation is accomplished is not understood, but is probably related to the mechanism that controls NtcA activity. Recombinant *Synechocystis* sp. PCC 6803 or *Synechococcus* sp. PCC 7942 NtcA purified from *E. coli* binds the respective *glnA* promoters with a dissociation constant (K_d) of 2.5 to 3.5 × 10^{-8} M (Reyes et al., 1997; Vázquez-Bermúdez, 2000). In contrast, the *E. coli* catabolite activator protein (CAP)-cAMP complex binds to the *lac* promoter with an affinity three orders of magnitude higher (K_d = 3.8 × 10^{-11} M) (Takahashi et al., 1989). How can this low binding affinity of NtcA be explained? The existence of two NtcA conformations in vivo, one with high binding affinity for its DNA target under conditions of N limitation and another with low binding affinity under conditions of N excess, is an attractive possibility. Increase in the NtcA binding affinity in vivo could be mediated by binding of an allosteric modulator, covalent modification or interaction with other regulatory proteins. To date, NtcA has not been shown to be modified in response to the N status of the cells. A redox regulation of NtcA has been postulated based on the increase in DNA binding in vitro, in the presence of the reducing agent dithiothreitol (Jiang et al., 1997). Relevance of this putative redox control in vivo is unknown. Another possibility is that NtcA binds constitutively to its target sequence and that transcription activation activity is modulated by N status. However, a repressor role of NtcA by direct interference with the RNA polymerase binding site, which will be discussed in the next section, implies that NtcA binding activity has to be regulated.

In addition to the *glnA* genes, NtcA activates transcription of a number of cyanobacterial genes under conditions of N limitation. Known genes that are under the control of NtcA in *Synechococcus* sp. PCC 7942, *Synechocystis* sp. PCC 6803 and *Anabaena* sp. PCC 7120 are summarized in Table 2. NtcA-activated genes could be classified into two subgroups based on their pattern of expression: i) genes that are induced in a medium that contains nitrate as N source, and ii) genes that are significantly upregulated only in the absence of an N source. The first category corresponds to genes that are induced under conditions of partial N limitation. In this

Table 2. Identified genes that belong to the NtcA regulon from different cyanobacteria

Strain	NtcA regulon
Synechoccocus sp. PCC 7942	*glnA*[1], *ntcA*[1], *nir operon*[1], *amt1*[2], *glnB*[3], *nirB*[4]
Synechocystis sp. PCC 6803	*glnA*[5], *glnN*[5], *icd*[6], *glnB*[7], *amt1*[8], *gifA*[9], *gifB*[9]
Anabaena sp. PCC 7120	*glnA*[10], *nifHDK*[10], *sixA*[10], *rbcLS*[10], *ntcA*[11], *nir operon*[12], *hetC*[13], *petH*[14], *gor*[16]

[1](Luque et al., 1994), [2](Vázquez-Bermúdez, 2000), [3](Lee et al., 1999), [4](Maeda et al., 1998), [5](Reyes et al., 1997), [6](Muro-Pastor et al., 1996), [7](Garcia-Dominguez and Florencio, 1997), [8] (Montesinos et al., 1998), [9](García-Domínguez et al., 2000), [10](Chastain et al., 1990), [11](Ramasubramanian et al., 1996), [12](Frias et al., 1997), [13](Muro-Pastor et al., 1999), [14](Valladares et al., 1999), [15](Jiang et al., 1995)

category are *glnA* genes and *nir* operons from several cyanobacteria. The second category corresponds to genes that are induced only upon severe N limitation, and includes *glnB*, *amt1* and *icd* genes from several cyanobacteria, *glnN* from *Synechocystis* sp PCC 6803 and genes involved in N_2 fixation and heterocyst differentiation from *Anabaena* sp. PCC 7120 (e.g. *hetC*, *petH* and *nifHDK)*. How is this hierarchy of promoter induction established at the molecular level? The most attractive hypothesis is that NtcA recognizes different DNA-binding motifs with differential affinities, the consensus sequence $GTAN_8TAC$ being the one to which NtcA displays highest affinity. This hypothesis predicts that genes that respond to partial N limitation have NtcA binding sites close to the consensus. In contrast, genes that only respond to strong N limitation harbor non-consensus NtcA binding sites. Some experimental data confirm this hypothesis. For example, *glnA* gene promoters that are induced under conditions of partial N limitation present consensus NtcA binding sites (see, for example, Luque et al. (1994); Reyes et al. (1997)). The *Synechocystis* sp. PCC 6803 *icd* gene, which is slightly induced in the presence of nitrate, but strongly induced in the absence of N source, presents a non-consensus NtcA binding site (**$GTAN_8TGC$**) (Muro-Pastor et al., 1996). The *Anabaena* sp. PCC 7120 *nifH* gene and *Synechocystis* sp. PCC 6803 *glnN* gene which are only expressed under strong N deficiency, present NtcA binding sites even more distant to the consensus ($ACTN_8$**TAC** and **GTA**N_8GTC respectively) (Chastain et al., 1990; Reyes et al., 1997).

B. Post-Transcriptional Regulation of GSI by Protein-Protein Interaction

It became evident in the early 1980s that cyanobacterial GSI was not subjected to covalent modification by adenylylation (Fisher et al., 1981), the classical system for GSI activity control, much studied in enteric bacteria and present also in many other bacterial groups (Merrick and Edwards, 1995). A number of studies were then devoted to the elucidation of the regulation of GSI activity by feedback inhibitors (Sawhney and Nicholas, 1978; Stacey et al., 1979; McMaster et al., 1980; Orr and Haselkorn, 1981; Tuli and Thomas, 1981) and divalent cation availability (Ip et al., 1983), following the regulation model of the *Bacillus* GSI (Deuel and Prusiner, 1974). Most of the cyanobacterial GSIs are inhibited in vitro by Ser, Ala, and ADP at concentrations between 1 and 5 mM (Stacey et al., 1977; Sawhney and Nicholas, 1978; McMaster et al., 1980; Orr and Haselkorn, 1981; Florencio and Ramos, 1985; Blanco et al., 1989; Mérida et al., 1990). Although a combined effect of several amino acids and nucleotides has been proposed as a regulatory mechanism, the role of these feedback inhibitors in vivo has not been demonstrated in cyanobacteria.

In the cyanobacterium *Synechocystis* sp. PCC 6803, addition of ammonium to cells growing on nitrate provokes a quick drop of GSI activity within 30 min after ammonium upshift (Mérida et al., 1991). This decrease in GSI activity occurs without reduction of the level of GSI protein and can be reversed upon removal of ammonium from the medium. These data suggested that *Synechocystis* sp. PCC 6803 GSI was inactivated in vivo by a reversible mechanism. The fact that inactive GSI could be reactivated in vitro by increasing the pH or the ionic strength of the buffer suggested that GSI inactivation was provoked by the direct binding of a metabolite or a polypeptide (Mérida et al., 1991). The inactive GSI monomer could be cross-linked to two small polypeptides, which strongly supported the hypothesis of the existence of two small inhibitory peptides which were designed IF (Inactivating Factors) (Reyes and Florencio, 1995b). Finally, co-purification of the inactive GSI with two polypeptides of 7 and 17 kDa, allowed the molecular identification of the two inactivating factors, IF7 and IF17. Direct binding of IF7 or IF17 to the GSI yields an inactive GSI-IFs complex (García-Domínguez et al., 1999). IF7 and IF17 are homologous proteins encoded by two unlinked genes, *gifA* and *gifB*, respectively. The Δ*gifA* and Δ*gifB Synechocystis* sp. PCC 6803 mutant strains are severely impaired in GSI inactivation and the double mutant Δ*gifA*/Δ*gifB* is completely deficient in GSI inactivation. Expression of both genes is maximal in the presence of ammonium, when GSI is inactivated. Analysis of the *gifA* and *gifB* promoters (PgifA and PgifB) has revealed the existence of NtcA-binding sites at −8.5 and −30.5 bp upstream of *gifB* and *gifA* transcription start-points, respectively (García-Domínguez et al., 1999; García-Domínguez et al., 2000). Certain activators of the CAP family of transcription factors can also mediate repression. This has been clearly characterized for several promoters controlled by CAP or the fumarate and nitrate reduction (FNR) transcriptional regulator (Collado-Vides et al., 1991; Kolb et al., 1993). In these cases transcription factor binding sites overlap the RNA polymerase-binding sites between −40 and +20. The position of the NtcA binding sites in PgifA and PgifB strongly suggested a repressive role of NtcA in the regulation of both promoters. This repressive role has been confirmed by the constitutive expression of *gifA* and *gifB* genes in an NtcA mutant (García-Domínguez et al., 2000). A role for NtcA as a repressor has also been hypothesized based on the presence of NtcA binding sites in typically repressive positions in the *gor* and the *rbcLS* promoters (Chastain et al., 1990; Jiang et al., 1995). However, NtcA-dependent repression of these two promoters has not been demonstrated in vivo. A comparison between the positions of NtcA-repressor sites and NtcA-activator sites is presented in Fig. 3.

Recombinant IF7 and IF17 produced in *E. coli* are able to inactivate GSI in vitro, suggesting that both factors can interact with GSI without further modification (García-Domínguez et al., 1999).

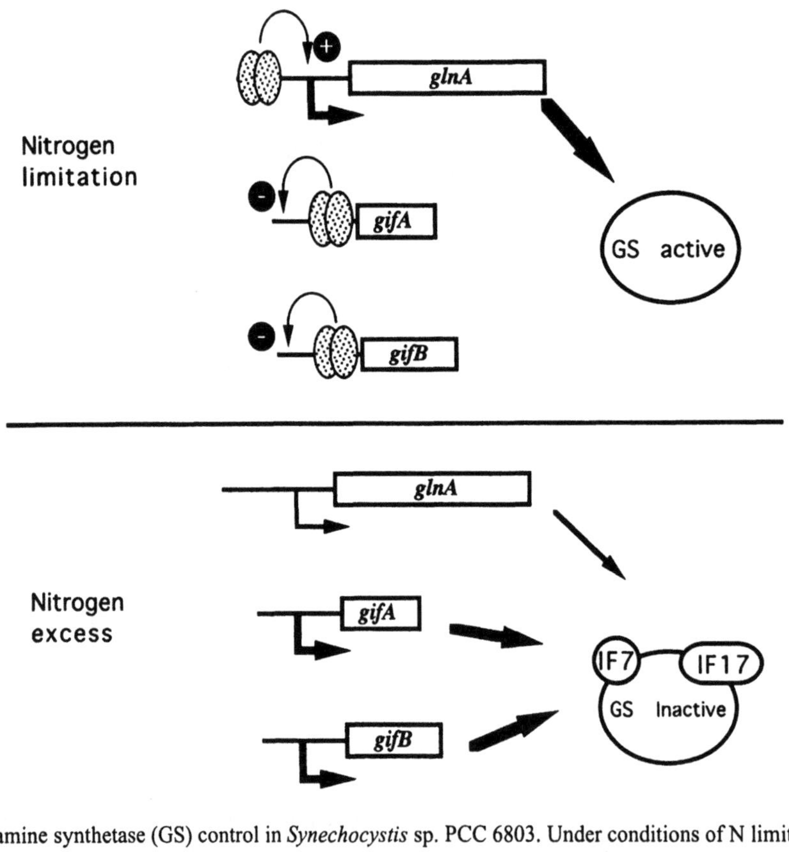

Fig. 4. Model of glutamine synthetase (GS) control in *Synechocystis* sp. PCC 6803. Under conditions of N limitation the transcription factor NtcA represses *gifA* and *gifB* promoters and activates the *glnA* promoter. This situation gives active GS. Under conditions of N excess *gifA* and *gifB* are derepressed while *glnA* is expressed at basal levels. Under these conditions IF7 and IF17 are synthesized and bind to the GS which thus becomes inactive.

Therefore, formation of GSI-IF complexes seems to be determined only by the intracellular concentration of IFs. Our current model for the regulation of GSI in *Synechocystis* sp. PCC 6803 is as follows (Fig. 4): Under N deficiency NtcA in its active form activates transcription of *glnA* (about four-fold induction) and represses *gifA* and *gifB* genes. Under these conditions GSI activity is high. Under conditions of N excess, NtcA is in an inactive form and is unable either to induce *glnA* or repress *gif* genes, and derepression of *gifA* and *gifB* inactivates GSI. This situation is also characterized by a basal level of transcription of the *glnA* gene. Under these conditions, therefore, GSI activity is low.

IF-GSI binding stoichiometry, as well as the mechanism by which GSI activity is inhibited, remain unknown. Preliminary results suggest that access of the substrates to the GSI active site is not blocked by the binding of IF to GSI (Reyes and Florencio, unpublished).

One important question remains to be addressed: Ammonium-dependent *Synechocystis* sp. PCC 6803 GSI inactivation is a reversible process. Thus, GSI is fully reactivated within ten minutes after removing ammonium from the culture medium. GSI can also be reactivated in vitro, in cell-free extracts by treatment with alkaline phosphatase (Mérida et al., 1991). These data indicate that some other elements involved in control of IF-GSI interaction remain to be identified.

ORFs that show amino acid sequence similarity to IF7 are present in other cyanobacteria such as *Anabaena* sp. PCC 7120 or *Anabaena azollae*, suggesting that a system of GSI activity control similar to the one described in *Synechocystis* sp. PCC 6803 may be widespread among cyanobacteria. Ammonium-promoted down-regulation of GS from other cyanobacterial strains has been reported and the extent as well as the kinetics of inactivation vary between the species (Rowell et al., 1977, 1979; Tuli and Thomas, 1980, 1981). In some strains long-term reduction of GS activity upon ammonium upshift could simply be a consequence of decreased transcription of *glnA* genes. Therefore, whether the IF based system of GS control is widely distributed among cyanobacteria requires further investigation.

What is the physiological significance of the GS inactivation system in cyanobacteria? The fact that diverse short-term GSI inactivation systems are present in many different prokaryotic groups suggests that these mechanisms play an important role in bacterial physiology. Bacteria have to respond quickly to dramatic changes in their surroundings. Under conditions of N limitation, GS activity is very high. However, since ammonium is absent, the intracellular concentration of Gln is very low and Glu is the most abundant amino acid. Seconds after ammonium upshift, the pool of Glu decreases dramatically, while the Gln level increases reciprocally (about 30- to 60-fold increase) (Rowell et al., 1977; Flores et al., 1980; Mérida et al., 1991b; Tapia et al., 1996). This indicates that the efficiency of the GS-GOGAT pathway in the assimilation of ammonium increases 30- to 60-fold in the presence of ammonium. In order to maintain the homeostasis of internal amino acid pools and the C/N balance, levels of activity of the GS-GOGAT pathway need to be readjusted. This regulatory process is carried out in the short term by inactivating GS and in the long term by regulating the expression of *glnA*, *glnN* and *icd* genes. In fact, 30 min after a shift in ammonium concentration, the amino acid pool size is restored. If this is the role of the GSI inactivation mechanism, *gif* mutants of *Synechocystis* sp. PCC 6803 should be unable to restore amino acid pools after ammonium upshift. Thus, 16 h after ammonium addition to nitrate growing *Synechocystis* sp. PCC 6803 cells, the Gln pool is about 100-fold higher in the Δ*gifA*/Δ*gifB* strain than in the WT strain (Muro-Pastor et al., 2001).

Synechocystis sp. PCC 6803 and *Synechococcus* sp. PCC 6301 GSI are also inactivated when the cultures are transferred to darkness, emphasizing the connection between N assimilation and photosynthesis (Marqués et al., 1992b; Reyes et al., 1995). Recent experiments in *Synechocystis* sp. PCC 6803 indicate that the molecular mechanism by which GSI is inactivated in the dark is the same as that which operates in ammonium-mediated GSI inactivation (Reyes et al., 1995; M. García-Domínguez and F. J. Florencio, unpublished). In fact, transcription of *gifA* and *gifB* genes from *Synechocystis* sp. PCC 6803 is upregulated in the dark. Two observations suggest that it is not the presence or the absence of light per se that controls the GS activity. First, GSI inactivation can be also reproduced by DCMU in the light. Interestingly DBMIB did not have the same effect. Second, dark- and DCMU-mediated inactivation of GSI can be prevented by the presence of glucose in the culture medium (Reyes et al., 1995; M. García-Domínguez and F. J. Florencio, unpublished). These data suggest that C metabolism and/or the redox state of the cell are involved in control of *gif* genes. Whether this regulation operates through NtcA remains to be investigated.

C. How do Cyanobacteria Sense Nitrogen?

How cells perceive N limitation is a basic question in microbiology that is far from being solved even in enteric bacteria, the most studied model system. Thus, although it has been postulated that transcription and covalent modification of enterobacterial GS are controlled by the ratio of Gln and 2-OG, in vivo evidence for this is limited (Ninfa et al., 2000). A rigorous and exhaustive study from the Kustu laboratory strongly suggests that N limitation is perceived by enteric bacteria as a decrease in the intracellular concentration of the Gln pool (Ikeda et al., 1996). This was deduced from the inverse correlation between N-limited growth and the intracellular concentration of Gln. This correlation is far less obvious in *Bacillus subtilis*, suggesting that other metabolites are probably involved in N sensing (Hu et al., 1999). Unfortunately, the above works did not report intracellular 2-OG concentrations.

In cyanobacteria, ammonium sensing requires its metabolization through the GS-GOGAT pathway. Thus, ammonium-promoted repression of nitrate or N_2 utilization does not occur in the presence of inhibitors of the GS-GOGAT pathway (Stewart and Rowell, 1975; Herrero et al., 1981). In fact, inhibition of the GS-GOGAT pathway with MSX or DON, or a *glnA* null mutation, makes NtcA-dependent genes insensitive to ammonium (Suzuki et al., 1993; Muro-Pastor et al., 2001). This suggests that some metabolites related to the GS-GOGAT pathway are involved in N signaling.

In *Synechocystis* sp. PCC 6803 N starvation attenuates the ammonium-mediated derepression of *gifA* and *gifB*, and the consequent inactivation of GSI (Mérida et al., 1991b; García-Domínguez et al., 2000). The degree of inactivation is inversely related to the incubation time in the absence of N source, suggesting that a metabolite that is accumulated in the absence of N is responsible for the attenuation. This fact has been interpreted as an interaction between C and N signals and suggests that GS regulation is modulated through the internal balance between C-N compounds and C compounds. An important regulatory metabolite could be 2-OG, which is accumulated under conditions of N starvation (see below and Section III.C) (Mérida et al., 1991b; Tapia et al., 1996a,b). Both inhibition of GS by MSX and inhibition of GOGAT by DON are perceived as N limitation by the cell. Both inhibitors cause an increase in the intracellular 2-OG pool but have opposite effects on the Gln pool. While inhibition of GS provokes a dramatic decrease in Gln, inhibition of GOGAT leads to a 10- to 15-fold increase in the Gln pool (Mérida et al., 1991b). Data presented hitherto indicate a correlation between the intracellular pool of 2-OG and the condition of N starvation. However, these data do not demonstrate that 2-OG is the signaling molecule that transmits information on the C/N status (Mérida et al., 1991b; Tapia et al., 1996a,b). Further experiments, using genetic tools instead of chemical inhibitors, may enable identification and evaluation of signaling metabolites.

From the above discussion a key question remains concerning signaling: Which protein senses the putative signaling molecules? The discovery of a cyanobacterial protein homologous to the enterobacterial PII protein was a promising advance (Harrison et al., 1990). In enteric bacteria the PII protein is modified by uridylylation through the action of the uridylyltransferase enzyme, a bifunctional enzyme able to carry out both uridylyl-transfer and removal, depending on the concentration of Gln and 2-OG. The PII protein is involved in controlling both the transcription and the activity of the enterobacterial GSI (Ninfa et al., 2000). Cyanobacterial PII (encoded by the *glnB* gene) is modified by phosphorylation according to cell N status (Forchhammer and Tandeau de Marsac, 1994, 1995). Cyanobacterial and enterobacterial PII both bind 2-OG and ATP. Analysis of a *Synechococcus* sp. PCC 7942 *glnB* null mutant demonstrated that PII is involved in control of the ammonium-mediated inhibition of nitrate uptake (Forchhammer and Tandeau de Marsac, 1995; Lee et al., 1998). In this mutant strain regulation of the expression of the *nir* operon, which is controlled by *ntcA*, is not affected but induction of the *glnN* gene by N starvation is attenuated (Sauer et al., 2000). Furthermore, a *Synechocystis* sp. PCC 6803 mutant strain harboring a PII (S54A) protein that cannot be phosphorylated shows normal regulation of *glnA* gene and inactivation of GSI (García-Domínguez and Florencio, unpublished). Therefore, there is no conclusive evidence regarding a putative role of the PII protein in the control of the GS-GOGAT pathway in cyanobacteria.

V. Future Perspectives

In our opinion future research is likely to develop within three major fields. First, structural studies of some of the proteins discussed above will be of great interest. Fd- and NADH-dependent GOGATs are especially interesting proteins with several prosthetic groups involved in intermolecular and intramolecular electron transfer. Elucidation of the GOGAT 3-D structure will open the door for future structure-function studies assisted by directed mutagenesis. Second, the *Synechocystis* sp. PCC 6803 system for GSI inactivation is a model for GS control previously not found in other organisms. Whether this system of GSI control is present in other cyanobacteria and in other prokaryotes is an interesting question. The stoichiometry of the IF-GSI complexes, the IF-GSI interaction sites, and the mechanism of inactivation and reactivation, are subjects that remain to be investigated. Finally, the questions of the regulation of the activity of the transcription factor NtcA, and how the N status is sensed and signaled, are related subjects that require intense study before we can paint a more complete picture of the regulatory pathways that control ammonium assimilation in cyanobacteria.

Acknowledgments

We thank Marika Lindahl, María Isabel Muro, and Mário García-Domínguez for critical reading of the manuscript. Work in the authors' laboratory was supported by grants PB94-1444, PB97-0732 from the Ministerio de Educación y Ciencia, by Junta de Andalucía (group CV1-0112) and by European Union Project CI1-CT94-0053.

References

Blanco F, Alana A, Llama MJ and Serra JL (1989) Purification and properties of glutamine synthetase from the non-N_2-fixing cyanobacterium *Phormidium laminosum*. J Bacteriol 171: 1158–1165

Binda C, Bossi R, Wakatsuki S, Artz S, Coda A, Curti B, Vanoni MA and Mattevi A (2000) Cross-talk and ammonia channeling between active centres in the unexpected domain arrangement of glutamate synthase. Struct Fold Des 8: 1299–1308

Böhme H (1998) Regulation of nitrogen fixation in heterocyst-forming cyanobacteria. Trends Plant Sci 3: 346–351

Bothe H and Neuer G (1988) Electron donation to nitrogenase in heterocysts. Methods Enzymol 167: 469–501

Boussiba S, Dilling W and Gibson J (1984) Methylammonium transport in *Anacystis nidulans* R-2. J Bacteriol 160: 204–210

Busby S and Ebright RH (1997) Transcription activation at class II CAP-dependent promoters. Mol Microbiol 23: 853–859

Chastain CJ, Brusca JS, Ramasubramaniam TS, Wei T-F and Golden JW (1990) A sequence-specific DNA-binding factor (VF1) from *Anabaena* sp strain PCC7120 vegetative cells binds to three adjacent sites in the *xisA* upstream region. J Bacteriol 172: 5044–5051

Chávez S (1992) Glutamato deshidrogenasas en cianobacterias. Ph.D thesis, Universidad de Sevilla, Sevilla, Spain

Chávez S and Candau P (1991) An NAD-specific glutamate dehydrogenase from cyanobacteria. Identification and properties. FEBS Lett 285: 35–38

Chávez S, Reyes JC, Chauvat F, Florencio FJ and Candau P (1995) The NADP-glutamate dehydrogenase of the cyanobacterium *Synechocystis* 6803: Cloning, transcriptional analysis and disruption of the *gdhA* gene. Plant Mol Biol 28: 173–188

Chávez S, Lucena JM, Reyes JC, Florencio FJ and Candau P (1999) The presence of glutamate dehydrogenase is a selective advantage for the cyanobacterium *Synechocystis* sp. strain PCC 6803 under nonexponential growth conditions. J Bacteriol 181: 808–813

Cogoni C, Valenzuela L, González-Halphen D, Olivera H, Macino G, Ballario P and González A (1995) *Saccharomyces cerevisiae* has a single glutamate synthase gene coding for a plant-like high-molecular-weight polypeptide. J Bacteriol 177: 792–798

Cohen-Kupiec R, Gurevitz M and Zilberstein A (1993) Expression of *glnA* in the cyanobacterium *Synechococcus* sp. strain PCC 7942 is initiated from a single *nif*-like promoter under various nitrogen conditions. J Bacteriol 175: 7727–7731

Cohen-Kupiec R, Zilberstein A and Gurevitz M (1995) Characterization of cis elements that regulate the expression of *glnA* in *Synechococcus* sp. strain PCC 7942. J Bacteriol 177: 2222–2226

Collado-Vides J, Magasanik B and Gralla JD (1991) Control site location and transcriptional regulation in *Escherichia coli*. Microbiol Rev 55: 371–394

Collier JL and Grossman AR (1994) A small polypeptide triggers complete degradation of light-harvesting phycobiliproteins in nutrient-deprived cyanobacteria. EMBO J 13: 1039–1047

Crespo JL, García-Domínguez M and Florencio FJ (1998) Nitrogen control of the *glnN* gene that codes for GS type III, the only glutamine synthetase in the cyanobacterium *Pseudanabaena* sp. PCC 6903. Mol Microbiol 30: 1101–1112

Crespo JL, Guerrero MG and Florencio FJ (1999) Mutational analysis of Asp51 of *Anabaena azollae* glutamine synthetase. D51E mutation confers resistance to the active site inhibitors L- methionine-DL-sulfoximine and phosphinothricin. Eur J Biochem 266: 1202–1209

Deuel TF and Prusiner S (1974) Regulation of glutamine synthetase from *Bacillus subtilis* by divalent cations, feedback inhibitors, and L-glutamine. J Biol Chem 249: 257–264

Dharmawardene MW, Haystead A and Stewart WD (1973) Glutamine synthetase of the nitrogen-fixing alga *Anabaena cylindrica*. Arch Mikrobiol 90: 281–295

Ebright RH (1993) Transcription activation at Class I CAP-dependent promoters. Mol Microbiol 8: 797–802

Eisenberg D, Gill HS, Pfluegl GM and Rotstein SH (2000) Structure-function relationships of glutamine synthetases. Biochim Biophys Acta 1477: 122–145

Elmorjani K, Liotenberg S, Houmard J and Tandeau de Marsac N (1992) Molecular characterization of the gene encoding glutamine synthetase in the cyanobacterium *Calothrix* sp. PCC 7601. Biochem Biophys Res Commun 189: 1296–1302

Fisher R, Tuli R and Haselkorn R (1981) A cloned cyanobacterial gene for glutamine synthetase functions in *Escherichia coli*, but the enzyme is not adenylylated. Proc Natl Acad Sci USA 78: 3393–3397

Florencio FJ and Ramos JL (1985) Purification and characterization of glutamine synthetase from the unicellular cyanobacterium *Anacystis nidulans*. Biochim Biophys Acta 838: 39–48

Florencio FJ, Marqués S and Candau P (1987) Identification and characterization of a glutamate dehydrogenase in the unicellular cyanobacterium *Synechocystis* sp. PCC 6803. FEBS Lett 233: 37–41

Flores E and Herrero A (1994) Assimilatory nitrogen metabolism and its regulation. In: Bryant A (ed) The Molecular Biology of Cyanobacteria, pp 487–517. Kluwer Academic Publishers, Dordrecht

Flores E and Muro-Pastor AM (1990) Mutation and kinetic analysis of basic amino acid transport in the cyanobacterium *Synechocystis* sp. PCC 6803. Arch Microbiol 154: 521–527

Flores E, Guerrero MG and Losada M (1980) Short-term ammonium inhibition of the nitrate utilization by *Anacystis nidulans* and other cyanobacteria. Arch Microbiol 128: 137–144

Flores E, Muro-Pastor AM and Herrero A (1999) Cyanobacterial nitrogen assimilation genes and NtcA-dependent control of gene expression. In: Peschek GA, Löffelhardt W and Schmetterer G (eds) The phototrophic prokaryotes, pp 463–477. Kluwer Academic/Plenum Publishers, New York

Forchhammer K and Tandeau de Marsac N (1994) The PII protein in the cyanobacterium *Synechococcus* sp. strain PCC 7942 is modified by serine phosphorylation and signals the cellular N-status. J Bacteriol 176: 84–91

Forchhammer K and Tandeau de Marsac N (1995a) Functional analysis of the phosphoprotein PII (*glnB* gene product) in the cyanobacterium *Synechococcus* sp. strain PCC 7942. J Bacteriol 177: 2033–2040

Forchhammer K and Tandeau de Marsac N (1995b) Phosphorylation of the PII protein (*glnB* gene product) in the cyanobacterium *Synechococcus* sp. strain PCC 7942: Analysis of in vitro kinase activity. J Bacteriol 177: 5812–5817

Frias JE, Merida A, Herrero A, Martin-Nieto J and Flores E (1993) General distribution of the nitrogen control gene *ntcA* in cyanobacteria. J Bacteriol 175: 5710–5713

Frias JE, Flores E and Herrero A (1994) Requirement of the regulatory protein NtcA for the expression of nitrogen assimilation and heterocyst development genes in the cyanobacterium *Anabaena* sp. PCC 7120. Mol Microbiol 14: 823–832

Frias JE, Flores E and Herrero A (1997) Nitrate assimilation gene cluster from the heterocyst-forming cyanobacterium *Anabaena* sp. strain PCC 7120. J Bacteriol 179: 477–486

Friga GM and Farkas GL (1981) Isolation and properties of an isocitrate dehydrogenase from *Anacystis nidulans*. Arch Microbiol 129: 331–334

Galván F, Márquez AJ and Vega JM (1984) Purification and molecular properties of ferredoxin-glutamate synthase from *Chlamydomonas reinhardtii*. Planta 162: 180–187

Garcia E, Bancroft S, Rhee SG and Kustu S (1977) The product of a newly identified gene, *glnF*, is required for synthesis of glutamine synthetase in *Salmonella*. Proc Natl Acad Sci USA 74: 1662–1666

García-Domínguez M and Florencio FJ (1997) Nitrogen availability and electron transport control the expression of *glnB* gene (encoding PII protein) in the cyanobacterium *Synechocystis* sp. PCC 6803. Plant Mol Biol 35: 723–734

García-Domínguez M, Reyes JC and Florencio FJ (1997) Purification and characterization of a new type of glutamine synthetase from cyanobacteria. Eur J Biochem 244: 258–264

García-Domínguez M, Reyes JC and Florencio FJ (1999) Glutamine synthetase inactivation by protein-protein interaction. Proc Natl Acad Sci USA 96: 7161–7166

García-Domínguez M, Reyes JC and Florencio FJ (2000) NtcA represses transcription of *gifA* and *gifB*, genes that encode inhibitors of glutamine synthetase type I from *Synechocystis* sp. PCC 6803. Mol Microbiol 35: 1192–1201

Gregerson RG, Miller SS, Twary SN, Gantt JS and Vance CP (1993) Molecular characterization of NADH-dependent glutamate synthase from alfalfa nodules. Plant Cell 5: 215–226

Guerrero MG and Lara C (1987) Assimilation of inorganic nitrogen. In: Fay P and Van Baalen C (eds) The Cyanobacteria, pp 163–185. Elsevier Science Publishers, New York

Harrison MA, Keen JN, Findlay JB and Allen JF (1990) Modification of a *glnB*-like gene product by photosynthetic electron transport in the cyanobacterium *Synechococcus* 6301. FEBS Lett 264: 25–28

Haselkorn R (1998) How cyanobacteria count to 10. Science 282: 891–892

Helling RB (1994) Why does *Escherichia coli* have two primary pathways for synthesis of glutamate? J Bacteriol 176: 4664–4668

Helling RB (1998) Pathway choice in glutamate synthesis in *Escherichia coli*. J Bacteriol 180: 4571–4575

Herrero A, Flores E and Guerrero MG (1981) Regulation of nitrate reductase levels in the cyanobacteria *Anacystis nidulans*, *Anabaena* sp. strain 7119, and *Nostoc* sp. strain 6719. J Bacteriol 145: 175–180

Hirasawa M and Tamura G (1984) Flavin and iron-sulfur containing ferredoxin-linked glutamate synthase from spinach leaves. J Biochem 95: 983–994

Hirasawa M, Hurley JK, Salamon Z, Tollin G and Knaff DB (1996) Oxidation-reduction and transient kinetic studies of spinach ferredoxin-dependent glutamate synthase. Arch Biochem Biophys 330: 209–215

Hu P, Leighton T, Ishkhanova G and Kustu S (1999) Sensing of nitrogen limitation by *Bacillus subtilis*: Comparison to enteric bacteria. J Bacteriol 181: 5042–5050

Ikeda TP, Shauger AE and Kustu S (1996) *Salmonella typhimurium* apparently perceives external nitrogen limitation as internal glutamine limitation. J Mol Biol 259: 589–607

Ip SM, Rowell P and Stewart WD (1983) The role of specific cations in regulation of cyanobacterial glutamine synthetase. Biochem Biophys Res Commun 114: 206–213

Jayakumar A, Epstein W and Barnes EM, Jr. (1985) Characterization of ammonium (methylammonium)/potassium antiport in *Escherichia coli*. J Biol Chem 260: 7528–7532

Jiang F, Hellman U, Sroga GE, Bergman B and Mannervik B (1995) Cloning, sequencing, and regulation of the glutathione reductase gene from the cyanobacterium *Anabaena* PCC 7120. J Biol Chem 270: 22882–22889

Jiang F, Mannervik B and Bergman B (1997) Evidence for redox regulation of the transcription factor NtcA, acting both as an activator and a repressor, in the cyanobacterium *Anabaena* PCC 7120. Biochem J 327: 513–517

Jiang F, Wisen S, Widersten M, Bergman B and Mannervik B (2000) Examination of the transcription factor NtcA-binding motif by in vitro selection of DNA sequences from a random library. J Mol Biol 301: 783–793

Kaneko T, Sato S, Kotani H, Tanaka A, Asamizu E, Nakamura Y, Miyajima N, Hirosawa M, Sugiura M, Sasamoto S, Kimura T, Hosouchi T, Matsuno A, Muraki A, Nakazaki N, Naruo K, Okumura S, Shimpo S, Takeuchi C, Wada T, Watanabe A, Yamada M, Yasuda M and Tabata S (1996) Sequence analysis of the genome of the unicellular cyanobacterium *Synechocystis* sp. strain PCC6803. II. Sequence determination of the entire genome and assignment of potential protein-coding regions. DNA Res 3: 109–136

Karni L and Tel-Or E (1983) Isocitrate dehydrogenase as a potential electron donor to nitrogenase of *Nostoc muscorum*. In: Papageorgiou GC and Packer L (eds) Photosynthetic Prokaryotes, pp 257–264. Elsevier Biomedical, New York

Kleiner D (1981) The transport of NH_3 and NH_4^+ across biological membranes. Biochim Biophys Acta 639: 41–52

Knaff DB and Hirasawa M (1991) Ferredoxin-dependent chloroplast enzymes. Biochim Biophys Acta 1056: 93–125

Knaff DB, Hirasawa M, Ameyibor E, Fu W and Johnson MK (1991) Spectroscopic evidence for a [3Fe-4S] cluster in spinach glutamate synthase. J Biol Chem 266: 15080–15084

Kolb A, Busby S, Buc H, Garges S and Adhya S (1993) Transcriptional regulation by cAMP and its receptor protein. Annu Rev Biochem 62: 749–795

Labarre J, Thuriaux P and Chauvat F (1987) Genetic analysis of amino acid transport in the facultatively heterotrophic cyanobacterium *Synechocystis* sp. strain 6803. J Bacteriol 169: 4668–4673

Lam HM, Coschigano K, Schultz C, Melo-Oliveira R, Tjaden G, Oliveira I, Ngai N, Hsieh MH and Coruzzi G (1995) Use of *Arabidopsis* mutants and genes to study amide amino acid biosynthesis. Plant Cell 7: 887–898

LaPorte D and Koshland DE Jr (1982) A protein with kinase and phosphatase activities involved in regulation of tricarboxylic acid cycle. Nature 300: 458–460

Lea PJ and Miflin BJ (1975) Glutamate synthase in blue-green algae. Biochem Soc Trans 3: 381–384

Lee HM, Flores E, Herrero A, Houmard J and Tandeau de Marsac N (1998) A role for the signal transduction protein PII in the control of nitrate/nitrite uptake in a cyanobacterium. FEBS Lett 427: 291–295

Lee HM, Vazquez-Bermudez MF and Tandeau de Marsac N (1999) The global nitrogen regulator NtcA regulates transcription of the signal transducer PII (GlnB) and influences its phosphorylation level in response to nitrogen and carbon supplies in the Cyanobacterium *Synechococcus* sp. strain PCC 7942. J Bacteriol 181: 2697–2702

Luque I, Flores E and Herrero A (1994) Molecular mechanism for the operation of nitrogen control in cyanobacteria. EMBO J 13: 2862–2869

Maeda S, Kawaguchi Y, Ohe TA and Omata T (1998) cis-Acting sequences required for NtcB-dependent, nitrite-responsive positive regulation of the nitrate assimilation operon in the cyanobacterium *Synechococcus* sp. strain PCC 7942. J Bacteriol 180: 4080–4088

Marqués S, Florencio FJ and Candau P (1992a) Purification and characterization of the ferredoxin-glutamate synthase from the unicellular cyanobacterium *Synechococcus* sp. PCC 6301. Eur J Biochem 206: 69–77

Marqués S, Candau P and Florencio FJ (1992b) Light-mediated regulation of glutamine synthetase activity in the unicellular cyanobacterium *Synechococcus* sp. PCC 6301. Planta 187: 247–253

Márquez AJ, Gotor C, Romero LC, Galván F and Vega JM (1986) Ferredoxin-glutamate synthase from *Chlamydomonas reinhardtii*. Prosthetic groups and preliminary studies of mechanism. Internat J Biochem 18: 531–535

Martín-Figueroa E, Navarro F and Florencio FJ (2000) The GS-GOGAT pathway is not operative in the heterocysts. Cloning and expression of *glsF* gene from the cyanobacterium *Anabaena* sp. PCC 7120. FEBS Lett 476: 282–286

Martinez-Bilbao M, Martinez A, Urkijo I, Llama MJ and Serra JL (1988) Induction, isolation, and some properties of the NADPH-dependent glutamate dehydrogenase from the nonheterocystous cyanobacterium *Phormidium laminosum*. J Bacteriol 170: 4897–4902

McMaster BJ, Danton MS, Storch TA and Dunham VL (1980) Regulation of glutamine synthetase in the blue-green alga *Anabaena* flos- aquae. Biochem Biophys Res Commun 96: 975–983

Meeks JC, Wolk CP, Thomas J, Lockau W, Shaffer PW, Austin SM, Chien WS and Galonsky A (1977) The pathways of assimilation of $^{13}NH_4^+$ by the cyanobacterium, *Anabaena cylindrica*. J Biol Chem 252: 7894–7900

Meeks JC, Wolk CP, Lockau W, Schilling N, Shaffer PW and Chien WS (1978) Pathways of assimilation of $[^{13}N]N_2$ and $^{13}NH_4^+$ by cyanobacteria with and without heterocysts. J Bacteriol 134: 125–130

Mérida A, Leurentop L, Candau P and Florencio FJ (1990) Purification and properties of glutamine synthetases from the cyanobacteria *Synechocystis* sp. strain PCC 6803 and *Calothrix* sp. strain PCC 7601. J Bacteriol 172: 4732–4735

Mérida A, Candau P and Florencio FJ (1991a) In vitro reactivation of in vivo ammonium-inactivated glutamine synthetase from *Synechocystis* sp. PCC 6803. Biochem Biophys Res Commun 181: 780–786

Mérida A, Candau P and Florencio FJ (1991b) Regulation of glutamine synthetase activity in the unicellular cyanobacterium *Synechocystis* sp. strain PCC 6803 by the nitrogen source: Effect of ammonium. J Bacteriol 173: 4095–4100

Merrick MJ and Edwards RA (1995) Nitrogen control in bacteria. Microbiol Rev 59: 604–622

Montesinos ML, Muro-Pastor AM, Herrero A and Flores E (1998) Ammonium/methylammonium permeases of a Cyanobacterium. Identification and analysis of three nitrogen-regulated *amt* genes in *Synechocystis* sp. PCC 6803. J Biol Chem 273: 31463–31470

Mulholland MR and Capone DG (2000) The nitrogen physiology of the marine N_2-fixing cyanobacteria *Trichodesmium* spp. Trends Plant Sci 5: 148–153

Muro-Pastor AM, Valladares A, Flores E and Herrero A (1999) The hetC gene is a direct target of the NtcA transcriptional regulator in cyanobacterial heterocyst development. J Bacteriol

181: 6664–6669

Muro-Pastor MI and Florencio FJ (1992) Purification and properties of NADP-isocitrate dehydrogenase from the unicellular cyanobacterium *Synechocystis* sp. PCC 6803. Eur J Biochem 203: 99–105

Muro-Pastor MI and Florencio FJ (1994) NADP(+)-isocitrate dehydrogenase from the cyanobacterium *Anabaena* sp. strain PCC 7120: Purification and characterization of the enzyme and cloning, sequencing, and disruption of the *icd* gene. J Bacteriol 176: 2718–2726

Muro-Pastor MI, Reyes JC and Florencio FJ (1996) The $NADP^+$-isocitrate dehydrogenase gene (*icd*) is nitrogen regulated in cyanobacteria. J Bacteriol 178: 4070–4076

Muro-Pastor MI, Reyes JC and Florencio FJ (2001) Cyanobacteria perceive nitrogen status by sensing intracellular 2-oxoglutarate levels. J Biol Chem 276: 38320–38328

Nalbantoglu B, Hirasawa M, Moomaw C, Nguyen H, Knaff DB and Allen R (1994) Cloning and sequencing of the gene encoding spinach ferredoxin-dependent glutamate synthase. Biochim Biophys Acta 1183: 557–561

Navarro F, Chávez S, Candau P and Florencio FJ (1995) Existence of two ferredoxin-glutamate synthases in the cyanobacterium *Synechocystis* sp. PCC 6803. Isolation and insertional inactivation of *gltB* and *gltS* genes. Plant Mol Biol 27: 753–767

Navarro F, Martín-Figueroa E, Candau P and Florencio FJ (2000) Ferredoxin-dependent iron-sulfur flavoprotein glutamate synthase (GlsF) from the cyanobacterium *Synechocystis* sp. PCC 6803. Expression and Assembly in *Escherichia coli*. Arch Biochem Biophys 379: 267–276

Neilson AH and Doudoroff M (1973) Ammonia assimilation in blue-green algae. Arch Mikrobiol 89: 15–22

Ninfa AJ, Jiang P, Atkinson MR and Peliska JA (2000) Integration of antagonistic signals in the regulation of nitrogen assimilation in *Escherichia coli*. In: Stafman, ER and Chock, PB (eds) Current Topic in Cellular Regulation, Vol 36, pp 31–75. Academic Press, New York

Okuhara H, Matsumura T, Fujita Y and Hase T (1999) Cloning and inactivation of genes encoding ferredoxin- and NADH-dependent glutamate synthases in the cyanobacterium *Plectonema boryanum*. Imbalances in nitrogen and carbon assimilations caused by deficiency of the ferredoxin-dependent enzyme. Plant Physiol 120: 33–42

Orr J and Haselkorn R (1981) Kinetic and inhibition studies of glutamine synthetase from the cyanobacterium *Anabaena* 7120. J Biol Chem 256: 13099–13104

Orr J, Keefer LM, Keim P, Nguyen TD, Wellems T, Heinrikson RL and Haselkorn R (1981) Purification, physical characterization, and NH_2-terminal sequence of glutamine synthetase from the cyanobacterium *Anabaena* 7120. J Biol Chem 256: 13091–13098

Pardo MA, Llama MJ and Serra JL (1999) Purification, properties and enhanced expression under nitrogen starvation of the $NADP^+$-isocitrate dehydrogenase from the cyanobacterium *Phormidium laminosum*. Biochim Biophys Acta 1431: 87–96

Purich DL (1998) Advances in the enzymology of glutamine synthesis. In: Purich DL (ed) Amino Acid Metabolism, pp 9–42. John Wiley & Sons, New York

Rai AN, Rowell P and Stewart WDP (1984) Evidence for an ammonium transport system in free-living and symbiotic cyanobacteria. Arch Microbiol 137: 241–246

Ramasubramanian TS, Wei TF, Oldham AK and Golden JW (1996) Transcription of the *Anabaena* sp. strain PCC 7120 *ntcA* gene: multiple transcripts and NtcA binding. J Bacteriol 178: 922–926

Reith M and Munholland J (1993) A high resolution gene map of the chloroplast genome of the red alga *Porphyra purpurea*. Plant Cell 5: 465–475

Reyes JC and Florencio FJ (1994) A new type of glutamine synthetase in cyanobacteria: The protein encoded by the glnN gene supports nitrogen assimilation in *Synechocystis* sp. strain PCC 6803. J Bacteriol 176: 1260–1267

Reyes JC and Florencio FJ (1995a) Electron transport controls transcription of the glutamine synthetase gene (*glnA*) from the cyanobacterium *Synechocystis* sp. PCC 6803. Plant Mol Biol 27: 789–799

Reyes JC and Florencio FJ (1995b) A novel mechanism of glutamine synthetase inactivation by ammonium in the cyanobacterium *Synechocystis* sp. PCC 6803. Involvement of an inactivating protein. FEBS Lett 367: 45–48

Reyes JC, Crespo JL, García-Domínguez M and Florencio FJ (1995) Electron transport controls glutamine synthetase activity in the facultative heterotrophic cyanobacterium *Synechocystis* sp. PCC6803. Plant Physiol 109: 899–905

Reyes JC, Muro-Pastor MI and Florencio FJ (1997) Transcription of glutamine synthetase genes (*glnA* and *glnN*) from the cyanobacterium *Synechocystis* sp. strain PCC 6803 is differently regulated in response to nitrogen availability. J Bacteriol 179: 2678–2689

Rowell P, Enticott S and Stewart WDP (1977) Glutamine synthetase and nitrogenase activity in the blue-green alga *Anabaena cylindrica*. New Phytol 79: 41–54

Rowell P, Sampaio MJAM, Ladha JK and Stewart WDP (1979) Alteration of cyanobacterial glutamine synthetase activity in vivo in response to light and NH_4^+. Arch Microbiol 120: 195–200

Sakakibara H, Watanabe M, Hase T and Sugiyama T (1991) Molecular cloning and characterization of complementary DNA encoding for ferredoxin-dependent glutamate synthase in maize leaf. J Biol Chem 266: 2028–2035

Sauer J, Dirmeier U and Forchhammer K (2000) The *Synechococcus* strain PCC 7942 *glnN* product (glutamine synthetase III) helps recovery from prolonged nitrogen chlorosis. J Bacteriol 182: 5615–5619

Sawhney SK and Nicholas DJ (1978) Effects of amino acids, adenine nucleotides and inorganic pyrophosphate on glutamine synthetase from *Anabaena cylindrica*. Biochim Biophys Acta 527: 485–496

Schmitz S, Navarro F, Kutzki CK, Florencio FJ and Böhme H (1996) Glutamate 94 of [2Fe-2S]-ferredoxins is important for efficient electron transfer in the 1:1 complex formed with ferredoxin-glutamate synthase (GltS) from *Synechocystis* sp. PCC 6803. Biochim Biophys Acta 1277: 135–140

Soupene E, He L, Yan D and Kustu S (1998) Ammonia acquisition in enteric bacteria: Physiological role of the ammonium/methylammonium transport B (AmtB) protein. Proc Natl Acad Sci USA 95: 7030–7034

Southern JA, Parker JR and Woods DR (1986) Expression and purification of glutamine synthetase cloned from *Bacteroides fragilis*. J Gen Microbiol 132: 2827–2835

Stacey G, Tabita FR and Van Baalen C (1977) Nitrogen and ammonia assimilation in the cyanobacteria: purification of glutamine synthetase from *Anabaena* sp. strain CA. J Bacteriol 132: 596–603

Stacey G, Bottomley PJ, Van Baalen C and Tabita FR (1979) Control of heterocyst and nitrogenase synthesis in cyanobacteria. J Bacteriol 137: 321–326

Stal LJ and Moezelaar R (1997) Fermentation in cyanobacteria. FEMS Microbiol Rev 21: 179–211

Stanier RY and Cohen-Bazire G (1977) Phototrophic prokaryotes: The cyanobacteria. Annu Rev Microbiol 31: 225–274

Stewart WD and Rowell P (1975) Effects of L-methionine-DL-sulphoximine on the assimilation of newly fixed NH_3, acetylene reduction and heterocyst production in *Anabaena cylindrica*. Biochem Biophys Res Commun 65: 846–856

Suzuki I, Sugiyama T and Omata T (1993) Primary structure and transcriptional regulation of the gene for nitrite reductase from the cyanobacterium *Synechococcus* PCC 7942. Plant Cell Physiol 34: 1311–1320

Takahashi M, Blazy B, Baudras A and Hillen W (1989) Ligand-modulated binding of a gene regulatory protein to DNA. Quantitative analysis of cyclic-AMP induced binding of CRP from *Escherichia coli* to non-specific and specific DNA targets. J Mol Biol 207: 783–796

Tapia MI, Llama MJ and Serra JL (1996a) Regulation of nitrate assimilation in the cyanobacterium *Phormidium laminosum*. Planta 198: 24–30

Tapia MI, Ochoa de Alda JAG, Llama MJ and Serra JL (1996b) Changes in intracellular amino acids and organic acids induced by nitrogen starvation and nitrate or ammonium resupply in the cyanobacterium *Phormidium laminosum*. Planta 198: 526–531

Temple SJ, Vance CP and Gantt JS (1998) Glutamate synthase and nitrogen assimilation. Trends Plant Sci 3: 51–56

Tuli R and Thomas J (1980) Regulation of glutamine synthetase in the blue-green alga *Anabaena* L-31. Biochim Biophys Acta 613: 526–533

Tuli R and Thomas J (1981) In vivo regulation of glutamine synthetase by ammonium in the cyanobacterium *Anabaena* L-31. Arch Biochem Biophys 206: 181–189

Tumer NE, Robinson SJ and Haselkorn R (1983) Different promoters for the *Anabaena* glutamine synthetase during growth using molecular or fixed nitrogen. Nature 306: 337–342

Valentin K, Kostrzewa M and Zetsche K (1993) Glutamate synthase is plastid-encoded in a red alga: Implications for the evolution of glutamate synthases. Plant Mol Biol 23: 77–85

Valladares A, Muro-Pastor AM, Fillat MF, Herrero A and Flores E (1999) Constitutive and nitrogen-regulated promoters of the *petH* gene encoding ferredoxin:NADP$^+$ reductase in the heterocyst-forming cyanobacterium *Anabaena* sp. FEBS Lett 449: 159–164

Vanoni MA and Curti B (1999) Glutamate synthase: A complex iron-sulfur flavoprotein. Cell Mol Life Sci 55: 617–638

Vázquez-Bermúdez MF (2000) Estudios sobre el control de la asimilación del nitrógeno en la cianobacteria *Synechococcus* sp. mediado por el regulador transcripcional NtcA. Ph.D. thesis, Universidad de Sevilla, Sevilla, Spain

Vega-Palas MA, Madueno F, Herrero A and Flores E (1990) Identification and cloning of a regulatory gene for nitrogen assimilation in the cyanobacterium *Synechococcus* sp. strain PCC 7942. J Bacteriol 172: 643–647

Vega-Palas MA, Flores E and Herrero A (1992) NtcA, a global nitrogen regulator from the cyanobacterium *Synechococcus* that belongs to the Crp family of bacterial regulators. Mol Microbiol 6: 1853–1859

Wagner SJ, Thomas SP, Kaufman RI, Nixon BT and Stevens SE, Jr. (1993) The glnA gene of the cyanobacterium *Agmenellum quadruplicatum* PR-6 is nonessential for ammonium assimilation. J Bacteriol 175: 604–612

Wei TF, Ramasubramanian TS, Pu F and Golden JW (1993) *Anabaena* sp. strain PCC 7120 *bifA* gene encoding a sequence-specific DNA-binding protein cloned by in vivo transcriptional interference selection. J Bacteriol 175: 4025–4035

Wisen S, Jiang F, Bergman B and Mannervik B (1999) Expression and purification of the transcription factor NtcA from the cyanobacterium *Anabaena* PCC 7120. Protein Expr Purif 17: 351–357

Wolk CP, Thomas J, Shaffer PW, Austin SM and Galonsky A (1976) Pathway of nitrogen metabolism after fixation of ^{13}N-labeled nitrogen gas by the cyanobacterium, *Anabaena cylindrica*. J Biol Chem 251: 5027–5034

Wolk CP, Ernst A and Elhai J (1994) Heterocyst metabolism and development. In: Bryant A (ed) The Molecular Biology of Cyanobacteria, pp 769–823. Kluwer Academic Publishers, Dordrecht

Woods DR and Reid SJ (1993) Recent developments on the regulation and structure of glutamine synthetase enzymes from selected bacterial groups. FEMS Microbiol Rev 11: 273–283

Chapter 8

Photorespiratory Carbon and Nitrogen Cycling: Evidence from Studies of Mutant and Transgenic Plants

Alfred J. Keys*
IACR-Rothamsted, Harpenden, Hertfordshire AL5 2JQ, U.K.

Richard C. Leegood
Robert Hill Institute and Department of Animal and Plant Sciences, University of Sheffield, Sheffield, S10 2TN, U.K.

Summary 115
I. Introduction 116
A. Physiological and Biochemical Background 116
B. Selection of Photorespiratory Mutants 116
C. The Value of Mutant and Transgenic Plants for Understanding Photorespiration 118
II. Entry of Carbon into the Photorespiratory Pathway 119
III. Recycling of Carbon to the Reductive Pentose Phosphate Pathway 120
A. Mutants Impaired in the Conversion of Glycine to Serine 120
B. Mutants Lacking Hydroxypyruvate Reductase 121
C. Alternative Pathways for Photorespiratory Carbon Recycling 122
D. Intracellular Transport of Photorespiratory Metabolites 124
IV. Recycling of Nitrogen Associated with Photorespiration 124
A. Serine-Glyoxylate Aminotransferase 124
B. Recycling of Ammonia 125
1. Glutamine synthetase 125
2. Glutamate synthase 126
V. Feedback from Photorespiration on Other Processes 127
A. Feedback on the Reductive Pentose Phosphate Pathway 127
B. Feedback on Gene Expression 128
VI. Role of Photorespiration During Stress 129
Conclusions 130
References 130

Summary

Photorespiratory mutants represent the most complete set of mutants for any metabolic pathway in plants. The photorespiratory pathway is also a prime example of the integration and co-ordination of carbon and nitrogen (N) metabolism. Studies of mutant and transgenic plants with lesions in photorespiratory metabolism have confirmed its cyclic nature, its origin in the reductive pentose phosphate pathway, and the associated N cycling. They have led to new insights into the nature of these processes and aspects of their regulation and control. Unlike most pathways in plants, the specific isozymes in chloroplasts, mitochondria and peroxisomes involved

*Author for correspondence, email: alfred.keys@bbsrc.ac.uk

Christine H. Foyer and Graham Noctor (eds): Photosynthetic Nitrogen Assimilation and Associated Carbon and Respiratory Metabolism, pp. 115–134.

in photorespiratory metabolism have been unequivocally identified. In the case of some mutations, isozymes not specifically involved in photorespiratory metabolism provide an alternative route and so by-pass the lesion. We discuss how, in mutant plants in which the recycling of N is defective, the normal photorespiratory pathway involving glycine decarboxylation may be partially by-passed by a modified form of photorespiration involving glyoxylate decarboxylation. Apart from metabolic feedback, we also discuss how mutants have been used to study the regulation of gene expression and the role of photorespiration during light and drought stress.

I. Introduction

A. Physiological and Biochemical Background

Photorespiration is the production of CO_2 and uptake of O_2 by metabolism in the light that differs from respiration involving the Krebs cycle in its response to O_2. Krebs cycle (night-time) respiration is saturated in 2% O_2, whereas photorespiration is not saturated at 20% O_2. Photorespiration is virtually stopped by elevating CO_2 to three or four times the atmospheric concentration.

Photorespiratory metabolism is a cyclic process in which carbon (C) is removed from ribulose 1,5-bisphosphate (RuBP) in the reductive pentose phosphate (RPP) pathway and returned as phosphoglycerate and CO_2 (Chollet and Ogren, 1975; Lorimer and Andrews, 1981). Associated with the metabolic pathway of photorespiration is a release and re-fixation of ammonia in the photorespiratory nitrogen (N) cycle (Keys et al., 1978). The involvement of common intermediates in the recycling of the C and N means a great degree of interdependence and integration of the two cycles. The metabolic pathways (Fig. 1) are initiated by the oxygenation of RuBP catalyzed by Rubisco to produce glycolate 2-P. This compound is dephosphorylated in the chloroplast by phosphoglycolate phosphatase (PGP) and the glycolic acid formed is oxidized in the peroxisome by glycolate oxidase. The resulting glyoxylic acid is aminated by aminotransferases in the peroxisome, mainly Ser-glyoxylate (SGAT) and glutamate-glyoxylate aminotransferase (GGAT), to form Gly. In the mitochondria, two molecules of Gly are converted to one molecule each of CO_2, ammonia and Ser catalyzed by a complex involving four proteins called Gly decarboxylase (GDC) together with Ser hydroxymethyltransferase (SHMT). GDC is a mitochondrial multi-enzyme complex catalyzing the conversion of Gly, NAD and tetrahydrofolate (THF) to CO_2, NH_3, NADH and N^5,N^{10}-methylene THF (Oliver, 1994; Douce and Neuburger, 1999; Douce and Heldt, 2000). The N comprising the amino group of Ser is recycled in the amination of glyoxylate earlier in the pathway, catalyzed by SGAT, producing hydroxypyruvate and Gly. Hydroxypyruvate is reduced by NADH to glycerate, catalyzed by hydroxypyruvate reductase (NADH-HPR) in the peroxisomes and the glycerate is phosphorylated in the chloroplasts to glycerate 3-P to rejoin the C reduction cycle. The ammonia produced in the conversion of Gly to Ser is recovered by conversion to Glu by the combined operation of chloroplastic glutamine synthetase (GS2) and ferredoxin-dependent glutamate synthase (Fd-GOGAT). Glutamate directly or indirectly donates its amino group to glyoxylate to complete the recycling of N. The recycling of C and N requires energy for the refixation of ammonia, the phosphorylation of glycerate, and especially for the reduction of the resulting glycerate 3-P to the level of RuBP, the intermediate initially wasted in the initiation of photorespiration through the oxygenase activity of Rubisco. The entire process thus results in the liberation of a quarter of the C initially present in glycolate 2-P as CO_2, and of equal amounts of NH_3.

Abbreviations: AGAT – alanine-glyoxylate aminotransferase; Ala – alanine; Ci – intercellular CO_2 concentration; Fd-GOGAT – ferredoxin-dependent glutamate synthase; FW – fresh weight; GDC – glycine decarboxylase; GGAT – glutamate-glyoxylate aminotransferase; Gln – glutamine; Glu – glutamate; Gly – glycine; GS1 – cytosolic glutamine synthetase; GS2 – chloroplastic glutamine synthetase; HPR – hydroxypyruvate reductase; *nia* – nitrate reductase coding sequence; 2-OG – 2-oxoglutarate; PEPc – phosphoenolpyruvate carboxylase; PGP – phosphoglycolate phosphatase; RPP – reductive pentose phosphate (RPP pathway = Calvin cycle); Rubisco – ribulose 1,5-bisphosphate carboxylase-oxygenase; RuBP – ribulose 1,5-bisphosphate; Ser – serine; SGAT – serine-glyoxylate aminotransferase; SHMT – serine hydroxymethyltransferase; THF – tetrahydrofolate

B. Selection of Photorespiratory Mutants

During the past two decades our knowledge and understanding of photorespiratory metabolism has been refined by the selection and study of mutant plants lacking the specific isozymes involved. Mutant plants have been selected on the principle that a

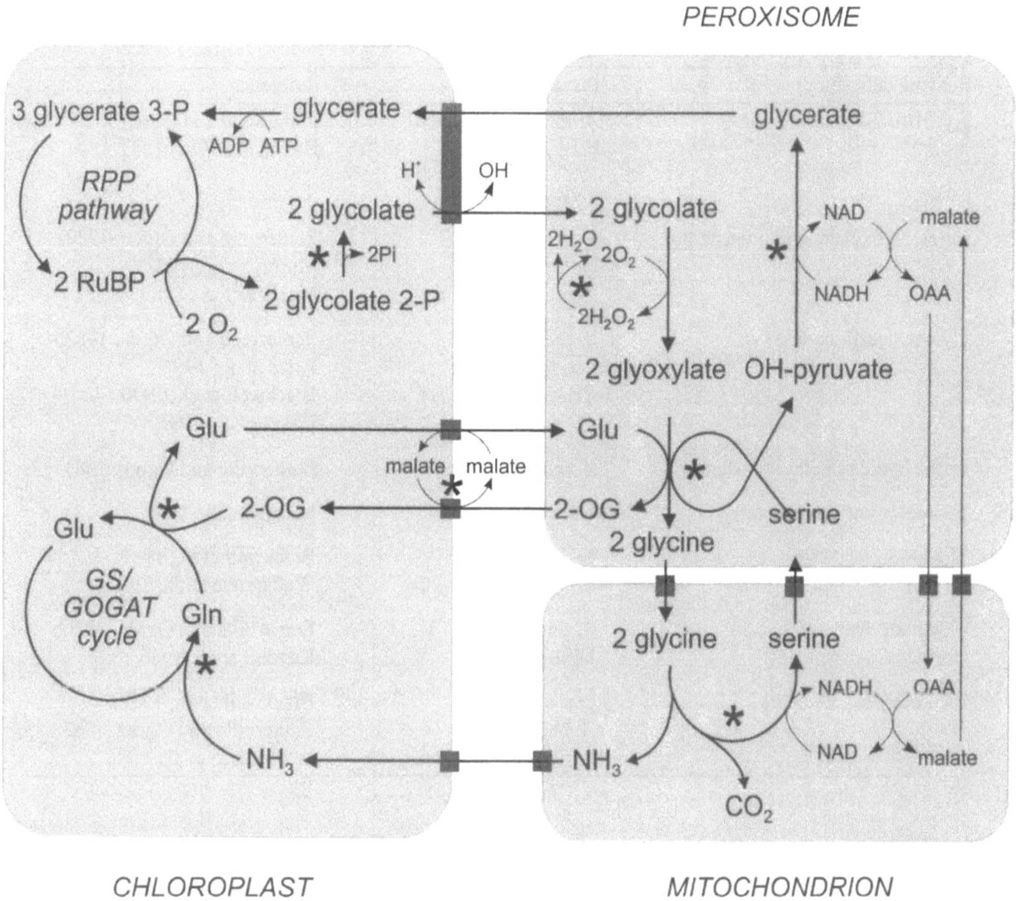

Fig. 1. Photorespiratory carbon and nitrogen metabolism. Oxygenation of two molecules of RuBP yields two molecules of glycerate-3-P (C3) and two molecules of glycolate 2-P (C2). Glycolate 2-P is salvaged by converting it into glycerate 3-P in the photorespiratory cycle. This can then re-enter the RPP pathway. One molecule each of CO_2 and NH_3 are liberated in glycine decarboxylation which utilizes two molecules of glycine. Note that the carboxylation reaction of Rubisco is not shown and that one carbon is released, hence only 9 carbons in glycerate 3-P are available to regenerate RuBP. NH_3 is reassimilated in the chloroplast by GS2 and Fd-GOGAT in the GS/GOGAT cycle. Steps for which one or more mutants are available are indicated by asterisks. Metabolite transporters are indicated by solid rectangles, though in certain cases these are speculative, as for ammonia transport between organelles and for glycine and serine transport across the mitochondrial membrane. Glu = glutamate, Gln = glutamine, 2-OG = 2-oxoglutarate, OAA = oxaloacetate, OH-pyruvate = hydroxypyruvate.

lesion in the photorespiratory pathway will disadvantage C_3 plants for growth in ambient air but that it will not be disadvantageous when they are grown in elevated CO_2 to competitively decrease oxygenation of RuBP. Thus, although photorespiratory intermediates may be used for other metabolic processes, such as protein and peptide synthesis (Ongun and Stocking, 1965; Madore and Grodzinsky, 1984; Noctor et al., 1998), and photorespiration plays a role in stress protection (Section VI), photorespiration does not appear to be an essential pathway for the growth and normal development of plants. This concept was the basis for the selection of mutants with defects in enzymes involved in photorespiratory metabolism (Somerville and Ogren, 1979). Seed (M_1) of *Arabidopsis thaliana* was treated with a chemical mutagen, ethyl methane sulfonate, germinated and grown to maturity to produce seed (M_2). The M_2 seed was germinated and grown in air enriched with CO_2 to suppress photorespiration. Plants that did not thrive were discarded and the rest transferred to grow in normal air. Those plants then showing signs of stress were returned to grow in CO_2-enriched air, in which plants with defects in photorespiratory metabolism recovered. Such mutants occurred at a frequency of

Table 1. Mutants of plants lacking enzymes of the photorespiratory carbon and nitrogen cycle. For further details of individual lines see references and the reviews by Blackwell et al. (1988a) and Lea and Forde (1994).

Enzyme deficiency	Plant	Reference
glycolate-2-P phosphatase	*Arabidopsis thaliana*	Somerville and Ogren,1979
	barley	Hall et al., 1987
catalase	barley	Kendall et al., 1983
serine:glyoxylate aminotransferase	*A. thaliana*	Somerville and Ogren, 1980a
	barley	Murray et al., 1987
	Nicotiana sylvestris	McHale et al., 1989
glycine decarboxylase	*A. thaliana*	Somerville and Ogren, 1982b
	barley	Artus et al., 1994
	Amaranthus edulis	Blackwell et al., 1990
		Dever et al., 1995
serine hydroxymethyltransferase	*A. thaliana*	Somerville and Ogren, 1981
hydroxypyruvate reductase	barley	Murray et al., 1989
glutamine synthetase	barley	Blackwell et al., 1987
		Wallsgrove et al., 1987
glutamate synthase (ferredoxin-dependent)	*A. thaliana*	Somerville and Ogren, 1980b
	barley	Kendall et al., 1986
dicarboxylate transport	pea	Blackwell et al., 1988b
	A. thaliana	Somerville and Ogren, 1983
	barley	Wallsgrove et al., 1986

about 1 per 1000 plants screened. Table 1 summarizes the photorespiratory mutants that have been generated in the C_3 plants, *Arabidopsis*, pea, tobacco, barley and in the C_4 plant, *Amaranthus edulis*. Despite the presence of a CO_2-concentrating mechanism in C_4 plants, the residual photorespiratory metabolism is not insignificant and conditional lethal photorespiratory mutants have been selected in *Amaranthus edulis* by a similar screen to that used to select photorespiratory mutants of C_3 species (Dever et al., 1995). Where photorespiratory mutants have been subjected to genetic analysis the mutations have proved to be the result of effects on single recessive nuclear genes. In the homozygous state, mutants do not express significant amounts of catalytic activity of the isozyme involved in photorespiratory metabolism. In addition a number of lines that appear to have altered rates of photorespiration have been isolated in tobacco (Zelitch and Day, 1968), including mutants with increased catalase, that have lower apparent rates of photorespiration (Zelitch, 1992). Tobacco plants have also been selected for improved growth in low CO_2. However, these show changes in leaf structure and respiration, rather than changes in rate of photorespiration (Delgado et al., 1993). A mutant of *Chlamydomonas reinhardtii* deficient in GS2 has also been characterized (López-Siles et al., 1999).

C. The Value of Mutant and Transgenic Plants for Understanding Photorespiration

This chapter is concerned with the extension of our knowledge of photorespiratory C and N cycling through the characterization and use of these mutants and transgenic plants, but it largely avoids topics that have already been reviewed by Blackwell et al. (1988a) and Lea and Forde (1994). Characterization and study of the mutants produced in these programs have confirmed the main features of a pathway of photorespiration suggested by the biochemical and physiological studies, indicated the isoforms of enzymes involved and drawn attention to the importance of membrane translocators. It has also provided evidence for amounts of C and N processed through the various steps in the metabolic pathways. Thus by observing the rate of accumulation of intermediates before the lesion, and the decrease in intermediates after the lesion, when mutants are transferred from non-photorespiratory to photorespiratory conditions, an indication of the flux in the pathway can be obtained. Analysis of photorespiratory mutants indicates that the following alterations could lead to reduced rates of photosynthesis: (i) an impairment of the recycling of the C in the photorespiratory pathway resulting in a depletion of RPP pathway intermediates; (ii) an impairment of photorespiratory N reassimilation leading to a decline in the N status of the leaf and a reduction in the amount of photosynthetic proteins; (iii) accumulation of photorespiratory metabolites feeding back on RPP pathway activity.

Photorespiratory mutants have also proved valuable in studying the control of photorespiratory metabolism by using heterozygous plants. While, in the long term, the homozygous photorespiratory mutants are not viable at ambient CO_2 concentrations, heterozygotes can be grown in air. Heterozygous mutants of GS2, Fd-GOGAT, GDC and SGAT have been used to study the control exerted by photorespiratory enzymes on photosynthetic and photorespiratory metabolism (Häusler et al., 1994a,b, 1996; Wingler et al., 1997, 1999a,b,c, 2000). The control strength of any enzyme on the flux through a pathway can be quantified in the form of a flux control coefficient (C^J_E, where J = flux and E = the enzyme concerned). It expresses the fractional change in flux which occurs when the activity of an enzyme is changed by a fractional amount (Kacser, 1987). The sum of all flux control coefficients in a pathway is 1.0. Normally flux control coefficients for individual enzymes are rather less than 1.0, as a result of control being shared by the enzymes in a pathway, unless one-sided limitations are imposed. Thus control by photorespiratory enzymes increases when the flux through that pathway increases, as in low CO_2 and high light. Stitt and his colleagues have demonstrated elegantly how flux control coefficients change with different environmental conditions in transgenic tobacco plants that have less Rubisco (Stitt and Schulze, 1994). It should be noted that 'rate-limiting steps' cannot be determined by the method of Kozaki and Takeba (1996).

II. Entry of Carbon into the Photorespiratory Pathway

It is the properties of Rubisco that ultimately determine the rate at which C enters the photorespiratory pathway. The rate increases with temperature and light intensity and is competitively determined by the relative concentrations of CO_2 and O_2 in the chloroplast stroma. The concentration of CO_2 in the chloroplast stroma depends on the boundary layer, stomatal and liquid phase conductances, temperature, and on the rate of assimilation of CO_2. At 25 °C, in good light, it can be predicted from the properties of Rubisco that the amount of C entering the photorespiratory pathway in a C_3 herbaceous plant in ambient air is of a similar magnitude to net C assimilation. At lower temperatures the amount is less.

One of the initial aims of selecting photorespiratory mutants was to isolate a mutant Rubisco that lacked the oxygenase activity, since this is the only way in which photorespiration can be beneficially decreased (Somerville and Ogren, 1980a). Somerville and Ogren (1980a) have described attempts to obtain revertants of an SGAT-deficient *Arabidopsis* mutant with decreased oxygenase activity of Rubisco. M_8 seed of a mutant line was re-mutated with ethyl methane sulfonate, germinated and grown to maturity in non-photorespiratory conditions. 500,000 of the seeds produced (M_2) were germinated and grown in continuous light in air. Only seven plants with wild-type characteristics were found and all had restored activities of SGAT activity. Thus the reversions all derived from the original lesion. Similar attempts to obtain reversions using a PGP-deficient mutant, and with a double mutant containing lesions in both PGP and SGAT, yielded no survivors with decreased Rubisco oxygenase activities (Somerville and Ogren, 1982a). Knowledge gained subsequently of the structure of Rubisco proteins, and the catalytic mechanisms of carboxylation and oxygenation (Harpel and Hartman, 1994; Cleland et al., 1998), and studies by in vitro mutagenesis (Bainbridge et al., 1995), showed that there is essentially no chance that a single (point) mutation can produce a major decrease in the oxygenase activity relative to the carboxylase activity of Rubisco.

Studies of the PGP-deficient mutants of *Arabidopsis* (Somerville and Ogren, 1979) and barley (Hall et al., 1987) confirm the initial steps in entry of C into the photorespiratory pathway and give an indication of rate. Firstly, plants transferred in the light from non-photorespiratory conditions to air containing $^{14}CO_2$ accumulated ^{14}C in glycolate 2-P but little, compared to a wild-type control, in glycolate, Gly and Ser. When plants were treated with 2-hydroxy-3-butynoic acid, an inhibitor of glycolate oxidase, under non-photorespiratory conditions before exposure to $^{14}CO_2$ in photorespiratory conditions, wild-type *Arabidopsis* plants accumulated glycolate while the mutant did not (Somerville and Ogren, 1979). This result confirmed that the glycolic acid in photorespiratory metabolism came from glycolate 2-P and not from any other source. Since PGP is a chloroplast enzyme and Rubisco in vitro produces glycolate 2-P from RuBP in the presence of oxygen, it is likely that the oxygenase activity is the sole initial reaction of the photorespiratory pathway. Since the mutant plant produced little CO_2 that could be attributed to photorespiration (evolution of CO_2 into CO_2-free air in the light or in a post-illumination burst) glycolate 2-P is clearly an intermediate of the pathway responsible for photorespiration. Heterozygous plants with 50% of the wild-type activity of PGP had gas-exchange characteristics in air which were indistinguishable from the wild-type, showing that this enzyme exerts no control on CO_2 assimilation in the wild-type in air (Hall et al., 1987).

The proportion of the total ^{14}C assimilated from $^{14}CO_2$ into glycolate 2-P during photosynthesis in photorespiratory conditions was less than might be predicted from the properties of Rubisco. In the *Arabidopsis* mutant, 19% of the total ^{14}C assimilated was found in glycolate 2-P after 2 min in $^{14}CO_2$ (Somerville and Ogren, 1979); in the barley mutant the corresponding value was 26% after 5 min photosynthesis (Hall et al., 1987). If metabolism of glycolate 2-P were completely blocked, nearer to 50% of the ^{14}C would be expected to accumulate in glycolate 2-P to be consistent with a rate of photorespiratory CO_2 release that is 25% of net photosynthesis. One factor will be that the oxygenase activity of Rubisco converts C1 and C2 of RuBP into glycolate 2-P but these two carbons do not initially receive the new C entering the RPP pathway (Bassham, 1964). In the mutant this situation may be exaggerated by altered concentrations of intermediates of the reduction cycle because glycolate 2-P inhibits triose phosphate isomerase (Anderson, 1981). This would change the rate of randomization of isotope among the C atoms of the phosphorylated sugars in the cycle. Some glycolate 2-P is also dephosphorylated by non-specific phosphatases so that essentially the block is not total. Thus traces of ^{14}C were found (Somerville and Ogren, 1979) in glycolate, Gly and Ser in the *Arabidopsis* mutant following photosynthesis in the presence of $^{14}CO_2$.

Although no glycolate oxidase mutant was recovered in any species by the specific selection procedure, the selection of a catalase mutant of barley (Kendall et al., 1983) supports the evidence that glycolate oxidase is involved in photorespiratory metabolism. In the oxidation of glycolate by glycolate oxidase, hydrogen peroxide is produced in the peroxisome. It is clear that oxidative damage in the catalase mutant leads to death of the plants in photorespiratory conditions. Interestingly, the mutant makes large quantities of the antioxidant glutathione when placed in mild photorespiratory conditions (Smith et al., 1984). Transgenic tobacco plants with less glycolate oxidase activity have been shown to be more sensitive to photoinhibition in high light, but no effect on electron transport was observed until glycolate oxidase activity was reduced below 40% of the wild-type, implying a low control coefficient for this enzyme in air (Yamaguchi and Nishimura, 2000).

III. Recycling of Carbon to the Reductive Pentose Phosphate Pathway

A. Mutants Impaired in the Conversion of Glycine to Serine

Mutants with lesions in the GDC complex of proteins perhaps give the best means of assessing flux in the photorespiratory pathway because it is at this step that photorespired CO_2 is normally generated. An *Arabidopsis* mutant (Somerville and Ogren, 1982b) lacking GDC activity accumulated 45% of the total ^{14}C derived from photosynthesis in Gly after 10 min in air containing $^{14}CO_2$. This would be consistent with a flux through the photorespiratory pathway giving a photorespiration rate approaching 25% the rate of net photosynthesis. When $^{14}CO_2$ was replaced by $^{12}CO_2$ and photosynthesis allowed to continue the proportion of ^{14}C in Gly rose to 50% in 5 to 10 min and then remained constant for a further 10 min. This suggests that, in this mutant, metabolism of photorespiratory Gly was completely blocked. Consistent with this conclusion was the absence of GDC activity in mitochondria isolated from leaves of this mutant. A barley mutant lacking the H protein and with reduced P protein in the GDC complex (Wingler et al., 1997) accumulated 66 % of ^{14}C taken up by leaves in Gly after 5 min photosynthesis in $^{14}CO_2$ (Blackwell et al., 1990), consistent with an even higher flux of C through the photorespiratory pathway. The accumulation of Gly measured by amino acid analysis (Blackwell et al., 1990) suggests a much lower flux than the ^{14}C accumulation in Gly. There is evidence (Blackwell et al., 1990) of some decarboxylation of [^{14}C] Gly added to detached leaves and of some Gly-bicarbonate exchange catalyzed by extracts, so the barley mutant may have residual GDC activity, perhaps not in the mesophyll cells. The more likely explanation lies in the accumulation of amino groups sequestered in Gly. This causes a decline in amino donors (amino acids that can transaminate glyoxylate) and results in accumulation of glyoxylate that is then decarboxylated via an alternative pathway (Section III.C; Somerville and Ogren, 1981).

With certain of the mutants selected by the screen devised by Somerville and Ogren (1979), it is clear that the lesions are lethal in photorespiratory conditions because they prevent C released into photorespiratory metabolism from being recycled into the RPP pathway. A particular example is the SHMT mutant of *A. thaliana*. Mutants of *Arabidopsis* lacking the mitochondrial SHMT activity, which catalyzes the step following Gly decarboxylation, were isolated by Somerville and Ogren (1981). Like the GDC mutants, the SHMT mutant accumulates Gly. From the proportion of ^{14}C in Gly (47–48%), following photosynthesis in photorespiratory conditions in $^{14}CO_2$, a rate of photorespiration close to 25% of net assimilation can be deduced. One of the mutants released CO_2 into CO_2-free air in the light at some 30% the rate shown by the wild type. This release was oxygen-dependent in a manner similar to CO_2 release in photorespiration. It was assumed therefore to be from decarboxylation of glyoxylate that accumulated because of a shortage of amino donors. Evolution of CO_2 by the mutant was abolished if the leaves were supplied with $(NH_4)_2SO_4$, which also partly prevented the decline in net photosynthesis under photorespiratory conditions. Somerville and Somerville (1983) measured the rate of Gly accumulation in leaves of one of the SHMT mutants supplemented with 30 mM Ser plus 30 mM NH_4Cl during photosynthesis in 2, 21 and 50% O_2 in air. Gly accumulated at 0.08, 0.53 and 1.52 μmol.μmol^{-1} CO_2 assimilated. Converting these values to rates of photorespiratory CO_2 release, were the pathway not blocked, gives 4, 36 and 316% of net photosynthesis. The amounts of Ser and NH_4Cl were sufficient to maintain the rate of photosynthesis at that of wild-type leaves treated similarly, even in 50% O_2, for some 20 min. Because the supplemented mutant leaves did not produce CO_2, the internal CO_2 concentration may be less than in the intact wild-type leaf so that the oxygenase activity of Rubisco in the stroma would be stimulated. Hence photorespiration rates based on Gly accumulation in these mutants could be an overestimation of rates in wild-type plants under the same conditions (Somerville and Somerville, 1983). Somerville and Ogren (1983) also showed that the rate of Gly accumulation is decreased with increased CO_2. This is entirely consistent with lack of mitochondrial SHMT blocking completely photorespiratory metabolism and the recycling of C. With this mutant, adequate exogenous supplies of intermediates in the metabolic pathway after the block not only prevent temporarily the decrease in photosynthesis in photorespiratory conditions, but also prolonged the accumulation of Gly (Somerville and Somerville, 1983) and prevented CO_2 release from glyoxylate. In contrast, Ser supplied via the transpiration stream to a GDC mutant and a putative Gly transport mutant (Blackwell et al., 1990) only partly restored the rate of photosynthesis; this was almost certainly because uptake by this means is too slow. The photosynthetic rate of an SGAT mutant could not be restored by supplying hydroxypyruvate, glycerate, Glu or ammonium sulfate in the transpiration stream (Murray et al., 1987).

Using mutants of barley deficient in GDC, Wingler et al. (1997) concluded that this enzyme has no control over CO_2 assimilation under normal growth conditions, but that appreciable control becomes apparent under conditions leading to higher rates of photorespiration, with C^J_{GDC} increasing to 0.34 in the wild-type in low CO_2 and high light.

B. Mutants Lacking Hydroxypyruvate Reductase

Carbon from the CO_2 released in photorespiration in the mitochondria is partly directly refixed in the chloroplasts without escaping to the outside of leaves (Loreto et al., 1999). The remaining C taken out of the reduction cycle as glycolate 2-P is returned as glyceric acid to be phosphorylated in the chloroplast by glycerate kinase. No glycerate kinase mutant has been identified, but the properties of a barley mutant with a lesion in NADH-dependent HPR have been described (Murray et al., 1989). Net photosynthesis in air was decreased by only 25% in this mutant in which the NADH-HPR activity was only 5% of that in the wild type. Clearly this enzyme has a low control coefficient in air, although it is likely to increase under more photorespiratory conditions. The most notable effects of the mutation were decreased conversion of Ser added to the leaf to sucrose, an increased accumulation of ^{14}C in Ser following photosynthesis in air containing $^{14}CO_2$, and decreased synthesis of glycerate 3-P. After 5 min in air containing $^{14}CO_2$, following 60 min in 1% O_2 and 340 ppm CO_2, ^{14}C in Ser was 30% of the total assimilated compared to 7.7% in the wild-type barley. This difference is less than would be expected if the flux of C into glycolate were equal to net photosynthesis. Because some 25% of the added [^{14}C] Ser was still converted to sucrose by leaves of the mutant, it is concluded that alternative enzymes are responsible for reduction of some hydroxypyruvate. There is an NADPH-dependent hydroxypyruvate reductase (Givan and Kleczkowski, 1992) that probably converts glyoxylate to glycolate, perhaps counteracting any accumulation of glyoxylate in the cytosol, and a glyoxylate reductase that is largely cytosolic and that is NADPH-dependent (Givan et al., 1988; Kleczkowski et al., 1988, 1990). It has a similar activity to the NADPH-HPR.

Amino acid changes during 180 min after leaves were transferred to air showed that the amount of Ser in the mutant increased steadily at a mean rate of 125 $nmol.min^{-1}.g^{-1}FW$ compared to 12 nmol for the wild-type control. The corresponding rates of net photosynthesis were about 4.5 and 6 $\mu mol.min^{-1}.g^{-1}FW$. Thus the accumulation of C in Ser in this mutant indicates an amount of C in the photorespiratory cycle of less than 10% of net photosynthesis.

C. Alternative Pathways for Photorespiratory Carbon Recycling

Studies of photorespiratory mutants have revealed possible alternative pathways for C recycling in photorespiration. The fact that very little glyoxylate accumulates in illuminated leaves of mutants completely lacking a number of photorespiratory enzymes suggests that glyoxylate generated in the photorespiratory pathway can be metabolized in reactions other than transamination (Fig. 2). Thus the SGAT and GDC mutants showed very low accumulation of glyoxylate (Chastain and Ogren, 1989; Wingler et al., 1999b), in contrast to the accumulation of Gly that occurs in mutants lacking GDC (Somerville and Ogren, 1983). This means either that glyoxylate accumulation feeds back on Rubisco or glycolate oxidase to decrease the production and metabolism of glycolate or that glyoxylate is metabolized via an alternative route. Another piece of evidence is that the non-enzymic reaction of glyoxylate with H_2O_2 to generate formate and CO_2, which occurs in vitro and in isolated peroxisomes (Chang and Huang, 1981), also occurs in vivo (Grodzinski and Butt, 1976). Thus an SHMT-deficient mutant of *Arabidopsis thaliana*, unable to metabolize Gly, was shown to convert glyoxylate to CO_2 once all the amino donors were depleted (Somerville and Ogren, 1981). However, as discussed in Section III.A, if ammonia and Ser were supplied, then photorespiratory CO_2 evolution was prevented, implying that the preferred route of glyoxylate metabolism is amination rather than conversion to CO_2 and formate (Somerville and Somerville, 1983).

The decarboxylation of glyoxylate to formate is believed to proceed non-enzymically due to direct oxidation of glyoxylate by H_2O_2 (Fig. 2), although Häusler et al. (1996) suggested that it might be regulated. However, it has been suggested that non-enzymic decarboxylation of glyoxylate is unlikely under normal circumstances, since H_2O_2 is rapidly destroyed by catalase (Walton, 1982). A catalase-deficient mutant of barley, in which the capacity for H_2O_2 removal is decreased, did not show increased CO_2 production, suggesting that the availability of H_2O_2 plays a minor role in regulating the fate of glyoxylate (Kendall et al., 1983). In contrast, a tobacco mutant with increased catalase showed decreased photorespiration. Mutants of tobacco with 40% more catalase had rates of CO_2 assimilation which were 9% higher than the wild-type at 30 °C and 21% higher at 38 °C (higher rates of photorespiration relative to photosynthesis obtain at higher temperatures), indicating significant control of the rate of CO_2 assimilation by catalase in the wild-type (Zelitch, 1989). It has been proposed that decreased H_2O_2 in these plants would lead to less non-oxidative decarboxylation of glyoxylate and hydroxypyruvate under highly photorespiratory conditions and thus greater net CO_2 fixation (Zelitch, 1989, 1992).

Assuming that glyoxylate is decarboxylated to formate in some of the photorespiratory mutants, the fate of any formate generated from glyoxylate is less clear. Halliwell (1973) showed that spinach beet contained a formyl-THF synthase activity that could convert formate and Gly to Ser. A similar conversion occurs in chloroplasts (Shingles et al., 1984). Formate

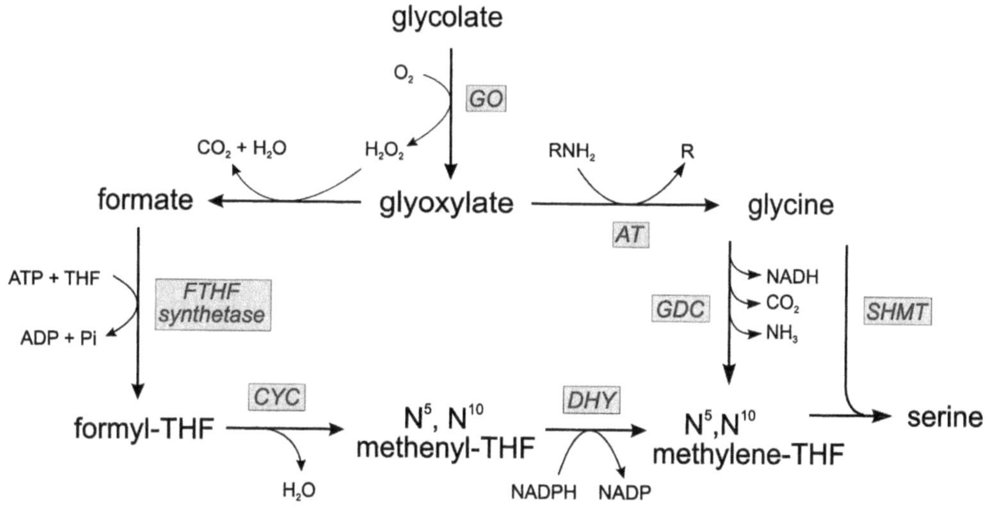

Fig. 2. Alternative pathways of glyoxylate metabolism. Glyoxylate is generated from glycolate by the action of glycolate oxidase (GO). In the normal photorespiratory pathway, glyoxylate is transminated to glycine, catalyzed by a range of transaminases (AT). Glycine is converted to N^5,N^{10}-methylene-THF via glycine decarboxylase (GDC) and serine is formed by serine hydroxymethyltransferase (SHMT) from N^5,N^{10}-methylene-THF and glycine, shown on the right-hand side. In a possible alternative pathway, shown on the left-hand side, glyoxylate is oxidatively converted to formate. Formate is then converted into N^5,N^{10}-methylene-THF (C1-THF synthase pathway) by 10-formyl-THF synthetase (FTHF synthetase) and N^5,N^{10}-methenyl-THF cyclohydrolase (CYC, DHY).

is activated by reacting with THF in the C1-THF synthase pathway, which provides C_1 units for the synthesis of purines, thymidylate, methionine and formylmethionyl-tRNA (Cossins and Chen, 1997). The enzymes involved in the C1-THF synthase pathway in plants are a monofunctional N^{10}-formyl-THF synthetase (Nour and Rabinowitz, 1991, 1992) and a bifunctional N^5,N^{10}-methylene-THF dehydrogenase:N^5,N^{10}-methenyl-THF cyclohydrolase (Kirk et al., 1995). The reactions catalyzed by these enzymes result in the formation of N^5,N^{10}-methylene-THF from formate. Since the methylene group of N^5,N^{10}-methylene-THF is incorporated into Ser in the SHMT reaction, formate, instead of the N^5,N^{10}-methylene-THF produced in the GDC reaction, can be used as an alternative substrate for the formation of Ser (Gifford and Cossins 1982a,b; Prabhu et al., 1996). Together with the C1-THF synthase/SHMT pathway, the oxidative decarboxylation of glyoxylate to formate could, therefore, form a GDC-independent bypass to the normal photorespiratory pathway, as shown for *Euglena gracilis* (Yokota et al., 1985). However, formate could also be converted to CO_2 by an NAD-formate dehydrogenase in the mitochondria (Halliwell, 1974). Formate can also be oxidized to CO_2 in peroxisomes (Leek et al., 1972) or chloroplasts (Zelitch, 1972).

Häusler et al. (1996) suggested that such an alternative pathway of glyoxylate metabolism could be a mechanism by which the loss of N as NH_3 is reduced in heterozygous barley mutants with reduced activities of GS2 (Section IV.B.1). These showed changes in both oxalate (another possible product of glyoxylate metabolism) and formate that mirrored changes in ammonia. The possibility of such a bypass to glyoxylate transamination operating in higher plants was also studied in homozygous GDC mutants of barley and in the C_4 plant, *Amaranthus edulis* (Wingler et al., 1999b). In contrast to wild-type plants, the mutants showed a light-dependent accumulation of glyoxylate and formate, which was suppressed in high (0.7%) CO_2. After growth in air, the activity and amount of N^{10}-formyl-THF synthetase were increased in the mutants compared to the wild types. A similar induction of N^{10}-formyl-THF synthetase occurred when leaves were incubated with Gly under illumination, but not in the dark. In addition, the barley mutant was capable of incorporating [^{14}C]formate and [2-^{14}C]glycolate into Ser. Since the GDC activity in the mutant (1% of wild-type activity; Wingler et al., 1997) was too low to support the rate of Ser formation from glycolate, the formation of Ser must have occurred via a GDC-independent pathway. Together, these results indicate that the mutants are able to bypass the normal photorespiratory pathway by oxidative decarboxylation of glyoxylate and formation of Ser from formate, thereby partially compensating for the lack of GDC activity, although it must be emphasized that this is not a route with a high capacity.

D. Intracellular Transport of Photorespiratory Metabolites

Transport processes in the photorespiratory pathway are indicated in Fig. 1. Numerous transporters exist to shuttle photorespiratory metabolites, to support transamination, and to supply or export the reductant generated or consumed in the photorespiratory pathway. During photorespiration, the re-assimilation of ammonia in the chloroplast depends on the recycling of 2-OG produced during the transamination between Glu and glyoxylate in the peroxisome. This means transport through the chloroplast envelope of 2-OG into the chloroplast and Glu out again (Fig. 1). In spinach chloroplasts, dicarboxylate transport involves the exchange of dicarboxylic acids such as malate, succinate, 2-OG, aspartate and Glu. This is catalyzed by two separate processes, involving a 2-OG translocator, which exchanges 2-OG for succinate, fumarate and malate, but not Glu, and a general dicarboxylate translocator, which can exchange Glu for malate (Woo et al., 1987; Flügge et al., 1988; Yu and Woo, 1992). There is also evidence for a separate Gln translocator, which also translocates Glu, but no other dicarboxylic acids. Current knowledge of the molecular biology of these transporters is described in Chapter 6 of this volume.

A mutant of *A. thaliana* lacking the chloroplast 2-OG translocator has been of great value in distinguishing the different transport processes associated with the photorespiratory pathway (Somerville and Ogren, 1983; Somerville and Somerville, 1985). Mutants in both barley (Wallsgrove et al., 1986) and *Arabidopsis* (Somerville and Ogren, 1983) were recovered with characteristics similar to the GOGAT mutants in that ammonia and Gln increased while Glu, Ala, Gly and Ser decreased. These mutants had wild-type activities of GOGAT and GS, and were less sensitive to air than the GOGAT mutants. In the *Arabidopsis* mutant a deficiency in a 42 kDa envelope protein was shown (Somerville and Somerville, 1985). This was suggested to be a component of a dicarboxylate transporter. Studies of chloroplasts of the *Arabidopsis* mutant also showed decreased uptake of 2-OG, aspartate, Glu and malate. The barley mutant transferred to air showed an initial rate of Gln accumulation corresponding to a rate of photorespiration of some 50% the rate of photosynthesis (Wallsgrove et al., 1986). Wallsgrove et al. (1986) suggest that the lower sensitivity of the dicarboxylate translocator mutants, compared to GOGAT mutants, is because, as 2-OG builds up, passive diffusion into the chloroplasts overcomes the limitation on transport. Thus chloroplasts from the mutant showed good rates of oxygen evolution with added 2-OG at 50 mM, but not at 1 mM, in the presence of 20 mM Gln; with chloroplasts from wild-type barley 1 mM 2-OG was adequate.

In the mitochondria, Gly oxidation occurs at extremely high rates during photorespiration, suggesting that both Gly and Ser might be actively transported (Oliver, 1987) although there is also evidence for passive movement of Gly (Day and Wiskich, 1980; Shingles et al., 1984). It may be that Ser must be rapidly removed from the mitochondria in order to allow the continuous production of Ser by SHMT, the equilibrium value for this reaction being unfavorable to Ser formation (Besson et al., 1993). An oxaloacetate carrier (Ebbighausen et al., 1985; Zoglowek et al., 1988) enables the shuttling of malate and oxaloacetate, catalyzing the transfer of reducing equivalents from the mitochondria to the peroxisomes for the reduction of hydroxypyruvate. Although no mutant in any of these mitochondrial transport processes has been identified, Blackwell et al. (1990) suggested that a Gly-accumulating mutant of barley might be deficient in mitochondrial Gly transport.

IV. Recycling of Nitrogen Associated with Photorespiration

A. Serine-Glyoxylate Aminotransferase

In the mitochondria, the GDC complex and SHMT convert two molecules of Gly to one of Ser and one of ammonia. Several mutants have been isolated in which the enzyme SGAT is missing from the peroxisome. This is the enzyme that can be regarded as re-cycling half of the N in photorespiratory metabolism. In two mutants of *Arabidopsis* without SGAT activity (Somerville and Ogren, 1980a), the amount of ^{14}C accumulating in Gly and Ser upon transfer to air containing $^{14}CO_2$ for 10 min was approximately double that in wild type plants treated similarly, totaling some 43 % of the ^{14}C assimilated. This is consistent with the direct involvement of SGAT in photorespiratory metabolism and with a considerable flux of C. Mutants of barley (Murray et al., 1987) and *Nicotiana sylvestris* (McHale, 1989) lacking SGAT also accumulate Ser. In both barley and *Arabidopsis* the SGAT lesion caused a rapid decrease in photosynthesis when plants were transferred from non-photorespiratory to photorespiratory conditions and eventually photorespiratory conditions were lethal. Shortage of amino donors is assumed to become a major problem as N accumulates in Ser (Murray et al., 1987).

Havir and McHale (1988) showed that a reduction in the activity of SGAT by 50% in plants of *Nicotiana sylvestris* had no detectable effect on the pattern of metabolism of [2-^{14}C] glycolate, implying no major perturbations of metabolism, but fluxes were not directly measured, nor were the environmental conditions altered so as to stimulate the rate of photorespiration. No evidence was found for compensating increases in other aminotransferase activities. Wingler et al. (1999a) showed that in heterozygous barley with 45–60% of wild type activities of SGAT, a reduction in SGAT resulted in the accumulation of Ser and, to a lesser extent, Gly, indicating that the flux through the photorespiratory pathway was restricted. However rates of photosynthesis were not affected by the reduction in SGAT activity even in low CO_2 and high light.

B. Recycling of Ammonia

Ammonia is generated by the GDC system and is largely recycled by the GS/GOGAT system in the chloroplasts at the expense of photosynthetic energy. Little ammonia escapes to the external atmosphere from healthy leaves of wild-type plants (Schjoerring et al., 1993) because of its extremely high solubility in aqueous phases and the efficiency of its re-assimilation. Thus leaves have an ammonia compensation point which is close to the K_m (NH_3) of GS (Farquhar et al., 1980). However, even in leaves of wild-type plants, there is some accumulation of ammonia in the light followed by reassimilation in the dark, while the accumulation and loss of ammonia is exacerbated in heterozygous plants with lower activities of GS2 (Häusler et al., 1994a; Mattsson et al., 1997), and especially in mutants which lack the chloroplastic GS2 (Blackwell et al., 1987). Much of the ammonia in the leaf would be expected to accumulate in the acidic vacuolar compartment or the apoplast as NH_4^+. This indicates the necessity for NH_3 transport within plant cells. A high affinity ammonia transporter (K_i (NH_4^+) for [^{14}C]methylamine uptake = 10 μM) was identified in *Arabidopsis* (Ninnemann et al., 1994) and, in tomato leaves, the mRNAs of two ammonia transporters declined at elevated CO_2 and both were diurnally regulated, one being expressed after the onset of light and one in darkness. Both may be involved in the retrieval of photorespiratory ammonia (von Wirén et al., 2000).

1. Glutamine synthetase

The recycling of photorespiratory ammonia was shown to depend on GS (Wallsgrove et al., 1980) and the recovery of eight allelic GS mutants of barley that would not grow in photorespiratory conditions (Wallsgrove et al., 1987) showed that the lesion was in the chloroplast isozyme (GS2) and that cytosolic GS (GS1) was not primarily involved. This situation has been made clearer by the subsequent finding that GS1 is associated with the phloem rather than the mesophyll (Edwards et al., 1990; Chapter 6, Hirel and Lea). The rate of accumulation of ammonia in the mutant leaf when transferred to photorespiratory conditions from photosynthesis in 1% O_2, 350 ppm CO_2 should equal 25% of the rate of C flux into the photorespiratory pathway. Ammonia accumulated in the first 30 min after the transfer was 50 μmol.g^{-1} FW while the mean rate of net photosynthesis was equivalent to approximately 4 μmol.min^{-1}.g^{-1} FW. Wallsgrove et al. (1987) claim this rate of ammonia accumulation, approximately 40% of net photosynthesis, to be an underestimate of photorespiration because some ammonia would be reassimilated by the non-chloroplast isozyme of GS which was not affected by the mutation. An alternative view is that this rate of ammonia release is on the high side for 20 °C and a moderate light intensity (400 μmol quanta.m^{-2}.s^{-1}) and that the extra ammonia might arise from increased nitrate reduction, perhaps because of the decrease in Gln or an increase in Glu in the tissue.

Häusler et al. (1994a) have shown that, in heterozygous barley mutants, a decrease in GS2 resulted in a decrease in leaf protein and Rubisco. This probably results from a limitation on ammonia re-assimilation which leads to the accumulation of ammonia and results in some loss of NH_3 from the plant (Mattsson et al., 1997). This results in a decrease in amino acid pools in the light, although these are partially restored during darkness, presumably by re-assimilation of the remaining ammonia and/or by mobilization from other parts of the plant or by assimilation of inorganic N. In the heterozygous barley plants, ammonia increased gradually down to 66% of wild-type GS2 activity. However, with a further decrease in GS2 activity, ammonia contents decreased, suggesting an inhibition of its generation. This could only come about by a temporary inhibition of photorespiration, perhaps by engagement of an alternative pathway of glyoxylate metabolism (Section III.C), since the ability to reassimilate ammonia via GS2 was less. Total amino acid contents showed an inverse relationship to the ammonia contents in that they decreased gradually in the wild-type to the 66% GS2 mutant, but exhibited a slight upward trend at GS2 activities below 55%. A rise in the activities of AGAT and GGAT and a fall in SGAT and Ser suggested that transamination of glyoxylate by Glu and Ala may assume more importance than by Ser as GS2 declined (Blackwell et al., 1988b; Häusler et al., 1996).

Häusler et al. (1994b) investigated the relationship between the quantum efficiency of CO_2 assimilation and the quantum efficiency of Photosystem II (ΦPSII) (Genty et al., 1989) in heterozygous barley plants with less GS2. The ratio ΦPSII/ΦCO_2 is a measure of the electron requirement for CO_2 assimilation and independent of changes in the absolute rate of CO_2 fixation caused by changes in Rubisco in the mutants. The most striking feature was a decrease in the electron requirement for CO_2 assimilation in the heterozygous GS2 mutants. At low intercellular CO_2 concentrations (Ci, particularly below 100 ppm), the average ΦPSII/ΦCO_2 ratio was reduced in the GS-mutant compared to the wild-type and was diminished by about 45% at the CO_2 compensation point, despite the fact that the 47% GS2 mutant had rates of CO_2 assimilation over a wide range of Ci and irradiance which were comparable to the wild-type. In moderate light and ambient CO_2 there were apparently no differences in the electron requirement per CO_2 assimilated for the whole range of GS2 mutants, but increasing temperatures or irradiance, which favor the oxygenation of RuBP and hence the flux through the photorespiratory pathway, also led to a decrease in the electron requirement in the GS2 mutant. Häusler et al. (1994b) discussed several reasons why the electron requirement for CO_2 assimilation might be reduced. First, it could be that the decrease in electron transport represents that normally used to assimilate ammonia via GS2 and Fd-GOGAT. However, even at the CO_2 compensation point, assimilation of all the ammonia generated in photorespiration would only account for about 15% of the rate of electron transport. Second, there could be an inhibition of photo-respiratory CO_2 release. An inhibition of Gly decarboxylation would decrease CO_2 release and decrease the electron requirement for net CO_2 fixation and for the re-cycling of intermediates back into the RPP pathway and for the re-assimilation of ammonia. It would also reduce the loss of ammonia during photorespiration, as indicated by the measurements of ammonia and amino acids, discussed above. In the short-term this could be advantageous, particularly under N-limited conditions. The decrease in amino acids would limit amino donors for the transaminases and lead to alternative pathways for the metabolism of glyoxylate (e.g. to formate, oxalate or other organic acids). Thus once ammonia accumulated and amino acids decreased, photorespiratory loss of further NH_3 would tend to be curtailed. Third, there could be an additional carboxylation process which utilizes less photosynthetic energy, for example, activation of PEPc.

The data obtained on changes in fluxes also allowed an analysis of the control of CO_2 assimilation and electron transport by GS2. The control exercised by GS2 in the wild-type as well as the plants with 50% GS2 depended strongly upon the environmental conditions, and it increased as rates of photo-respiration increased, rising to C^J_{GS} = 0.31 in the wild-type in high light and low CO_2 (Häusler et al., 1994b). The data did not provide any evidence that post-translational modifications of the activity of GS2 are able to compensate for decreases in GS2 activity, in contrast to the regulation by phos-phorylation of GS1 (Finnemann and Schjoerring, 2000).

2. Glutamate synthase

Mutants lacking chloroplastic Fd-GOGAT have been generated in both *Arabidopsis* and barley. Three allelic mutants of *Arabidopsis* had less than 5% the activity of Fd-GOGAT of the wild type (Somerville and Ogren, 1980b). *Arabidopsis* contains two Fd-GOGAT genes, *glu*1 (or *gls1*) and *glu*2. *Glu*1 has the major role in photorespiration, but also primary N assimilation in leaves, whereas *glu2* functions in the roots (Coschigano et al., 1998).

A characteristic of the Fd-GOGAT mutants is a steady accumulation of NH_3 upon transfer to photorespiratory conditions (Somerville and Ogren, 1980b; Blackwell et al., 1988b; Kendall et al., 1986). This is also observed in transgenic tobacco deficient in Fd-GOGAT (Ferrario-Méry et al., 2000). This is because Gln synthesis becomes rapidly limited by a shortage of Glu that is normally regenerated by Fd-GOGAT and because photorespiration is not entirely suppressed (Joy et al., 1992). From the rate of increase in NH_3 in the leaf upon transfer from darkness into photorespiratory conditions of 50% O_2 with 357 ppm CO_2, balance N_2, a rate of photorespiration of some 25% of net photosynthesis can be deduced. Under these conditions photosynthesis rose to a maximum after 5 min and then declined to some 12% of the rate of the wild-type control. The estimate of 25% of net photosynthesis is lower would be expected under these conditions. However, because the barley mutants contain wild-type activities of NADH-GOGAT, and suffer rapid damage, it is probably not possible to consider questions of flux into photorespiratory metabolism. Although there was no change in the relative amounts of SGAT, GGAT and AGAT in barley mutants (Häusler et al., 1996), transgenic tobacco showed a decline in Ala with decreasing Fd-GOGAT, suggesting its increased utilization as an amino donor (Ferrario-Méry et al., 2000). As with the SGAT, GDC and SHMT mutants, decarboxylation of glyoxylate may be an alternative route of C flow.

In heterozygous barley plants with reduced Fd-GOGAT activity, a 30–40% decrease in the contents of total amino acids and a decrease in leaf protein and Rubisco was apparent under conditions of enhanced photorespiratory flux and in the dark (Häusler et al., 1994b). In transgenic tobacco, only a 20% reduction in Fd-GOGAT caused accumulation of Gln and 2-OG (Ferrario-Méry et al., 2000). In contrast to the GS2 mutants, there was apparently no difference in the $\Phi PSII/\Phi CO_2$ ratios in both Fd-GOGAT mutants compared to the wild-type under any condition. However, there was an inhibition of CO_2 assimilation in the Fd-GOGAT mutants under conditions of *low* photorespiration (at high Ci) (Häusler et al., 1994b). There are a range of possible explanations, some involving modifications of transport across the chloroplast envelope. An inhibition of CO_2 assimilation at high Ci might occur if the re-entry of glycerate, an intermediate of the photorespiratory C pathway, were restricted (Harley and Sharkey, 1991). The data would also be consistent with the operation of a triose-P/glycerate 3-P shuttle between the stroma and the cytosol, in order to balance the ATP supply. It could also be that changes in the Glu pool could affect the operation of the dicarboxylate translocator on the chloroplast envelope and so interfere with the exchange of redox equivalents with the cytosol.

V. Feedback from Photorespiration on Other Processes

A. Feedback on the Reductive Pentose Phosphate Pathway

Work on photorespiratory mutants of *Arabidopsis* has shown a photorespiration-induced decrease in the activation state of Rubisco (Chastain and Ogren, 1985). A deactivation of Rubisco occurred under photorespiratory conditions in mutants with lesions either in GDC or further along the photorespiratory pathway. Deactivation of Rubisco also occurred in protoplasts treated with a GDC inhibitor (Chastain and Ogren, 1985; Créach and Stewart, 1982) and in leaves with diminished GS activity (Wendler et al., 1992). The data suggest that photorespiratory metabolites preceding GDC (glycolate, glyoxylate or Gly) may cause a decrease in the activation state of Rubisco. Of these three metabolites, only glyoxylate brought about an inhibition of CO_2 fixation (Oliver and Zelitch, 1977; Lawyer et al., 1983; Chastain and Ogren, 1989) and a decrease of Rubisco activation state in isolated chloroplasts (Chastain and Ogren, 1989). It was shown that glyoxylate accumulated in the mutants in which Rubisco deactivation occurred. Häusler et al. (1996) have shown that, in photorespiratory mutants with reduced activities of GS2, the activation state of Rubisco was strongly inversely correlated with the leaf content of glyoxylate, further suggesting that the amount of glyoxylate might control the activity of Rubisco. The mechanism by which glyoxylate might regulate the activation state of Rubisco is unclear. Glyoxylate at unphysiologically high concentrations can inhibit Rubisco by formation of a Schiff base with a lysyl residue within the catalytic site (Cook et al., 1985). However, it seems more probable that inhibition occurs through some effect on the Rubisco activase system (Campbell and

Ogren, 1990). Rubisco activase was itself discovered by the isolation of a mutant of *Arabidopsis* that grew well only in elevated CO_2 concentrations (Somerville et al., 1982).

This then raises the question of whether or not Rubisco in the chloroplast encounters sufficient concentrations of glyoxylate in vivo to bring about inhibition. The concentration of glyoxylate in the whole leaf ranges between 10 and 50 $\mu mol.m^{-2}$ (Chastain and Ogren, 1989; Häusler et al., 1996). If all this glyoxylate were contained within the stroma (assuming 25 $\mu l.mg^{-1}$ chlorophyll), its concentration would be between 1 and 5 mM. Since glyoxylate is one of the metabolites which is considered to be channeled within the peroxisomal matrix (Heupel and Heldt, 1994) glyoxylate could be considerably less concentrated in the stroma. However, the concentrations of glyoxylate needed to inhibit Rubisco activation are less than 100 μM (Chastain and Ogren, 1989).

Other photorespiratory metabolites possibly involved in feedback on the RPP pathway include glycerate (Schimkat et al., 1990) and glycolate 2-P. In mutants lacking PGP (Somerville and Ogren, 1979), photosynthetic CO_2 assimilation was very rapidly inhibited upon exposure to air. There was a large decrease in the amount of RuBP (Chastain and Ogren, 1989) but the activation state of Rubisco was maintained (Chastain and Ogren, 1985). Since glycolate 2-P accumulates in these mutants under photorespiratory conditions and is a potent inhibitor of triose-P isomerase (Anderson, 1981), it was inferred that inhibition of triose-P isomerase inhibited the regeneration of RuBP.

One intriguing effect is the extreme sensitivity of Fd-GOGAT mutants to photorespiratory conditions. The Fd-GOGAT mutants of barley are among the most sensitive of the photorespiratory mutants studied, with yellow-brown lesions appearing within 4 h in air. During 10 min photosynthesis in $^{14}CO_2$ in air after transfer from a CO_2-enriched atmosphere more than 15% of the ^{14}C assimilated appeared in gluconate 6-P. The appearance of considerable amounts of gluconate 6-P (Wallsgrove et al., 1987) as a product of photosynthesis following initial transfer to photorespiratory conditions suggests that active oxygen species are formed, leading to the activation of glucose-6-phosphate dehydrogenase under oxidizing conditions via the ferredoxin-thioredoxin regulatory system.

B. Feedback on Gene Expression

Expression of most of the photorespiratory enzymes, i.e. glycolate oxidase, catalase, HPR, SGAT, P-, H- and T-proteins of the GDC complex, and SHMT, is induced by light (Raman and Oliver, 1997; McClung et al., 2000). The enzyme whose expression has been most intensively analyzed is NADH-dependent HPR. Induction of the expression of the HPR gene in cucumber by light involves a phytochrome-dependent component (Bertoni and Becker, 1993). In the dark, expression of the HPR gene can be induced by cytokinin (Chen and Leisner, 1985; Andersen et al., 1996). It has also been suggested that photorespiratory metabolites have an effect on the expression of the HPR gene. Although there was no effect of high CO_2 on HPR activity in pea (Thibaud et al., 1995), when photorespiration in cucumber plants was suppressed in high CO_2, the HPR mRNA decreased (Bertoni and Becker, 1996). The reduced expression of the HPR gene observed in high CO_2 (Bertoni and Becker, 1996) could be due to sugar-mediated changes in gene expression (Wingler et al., 1998). However, the increase in sugar contents observed in drought-stressed barley (Wingler et al., 1999c) led to an increase in HPR protein in the leaves (Wingler et al., 2000). This was also the case in SGAT and GDC heterozygous barley mutants, suggesting that either a general drought-related signal or a metabolite formed in the photorespiratory pathway before the GDC reaction (e.g. glycolate) could act as the signal.

In barley mutants with reduced activities of GDC, it has been shown that the amount of P-protein was reduced in plants that had a content of H-protein that was lower than 60% of wild-type contents, while the amounts of T- and L-proteins were normal (Blackwell et al., 1990; Wingler et al., 1997). This indicates that the mutation in this GDC mutant is probably in a gene encoding H-protein and that the synthesis of P-protein is also regulated downwards, when the formation of functional GDC complexes is limited by the availability of H-protein. Very small amounts of H-protein (about 1% of wild-type) were detected (Wingler et al., 2000). There was no difference in the content of H-protein in the roots of the GDC mutant compared to the wild type. This suggests that, in addition to the photorespiratory gene for H-protein, barley, like other plants, contains a second gene which is constitutively expressed in roots and leaves. The housekeeping function of this minor isoform of GDC appears to be Gly catabolism associated with

C1 metabolism (Mouillon et al., 1999).

In contrast to their cytosolic counterparts, the expression of GS2 and Fd-GOGAT is not strongly regulated by N supply but is highly responsive to light (Hecht et al., 1988; Migge et al., 1996; Migge and Becker, 1996). The involvement of photorespiratory signals in the regulation of the expression of GS2 has been suggested by Edwards and Coruzzi (1989). In their work, suppression of photorespiration in 2% CO_2 led to a decrease in the GS2 mRNA in pea. Similarly in bean, longer term exposure of plants to 4% CO_2 led to a lower expression of GS2 than in plants grown in air, although there was no short-term effect on the GS2 mRNA, when *Phaseolus vulgaris* plants grown in high CO_2 were transferred into air (Cock et al., 1991). On the other hand, growth of *Arabidopsis* or tobacco plants at 0.3% CO_2, which is probably high enough to suppress photorespiration almost completely, did not affect the amount of GS2 or Fd-GOGAT mRNA compared to plants grown in air (Beckmann et al., 1997; Migge et al., 1997). Similarly suppression of photorespiration in an SHMT-deficient mutant of *Arabidopsis* did not result in changes in GS2, but did result in an increase in SHMT transcripts, perhaps as a result of negative feedback from a photorespiratory metabolite (Beckmann et al., 1997). Beckmann et al. (1997) have argued that this discrepancy with earlier results may arise from the fact that exposure to very high levels of CO_2 (2–4%) may involve acclimation of C and N metabolism rather than a simple suppression of photorespiration. However, von Wirén et al. (2000) observed repression of GS2 and SHMT transcripts in tomato at 800 ppm as compared with 400 ppm CO_2. An involvement of photorespiratory metabolites in regulating the expression of GS2, Fd-GOGAT and other enzymes therefore remains an open question.

Dzuibany et al. (1998) have used the Fd-GOGAT deficient mutant of *Arabidopsis* to test the hypothesis that Gln regulates the amount of the transcript of the nitrate reductase gene, *nia*2 (Vincentz et al., 1993). Their results indicate that endogenously accumulated Gln in the mutant does not influence *nia*2 transcript abundance, and that exogenously applied Gln probably affects nitrate uptake.

VI. Role of Photorespiration During Stress

Despite enormous research effort over more than 40 years, it is not yet agreed whether photorespiration has an essential function in plants. However, under conditions of high light it is a significant mechanism by which plants dissipate excessive energy. The efficient consumption of ATP and reductant by photorespiration allows it to protect against photoinhibition (Osmond, 1981). Kozaki and Takeba (1996) have utilized transgenic tobacco under- and over-expressing GS2 to study responses to light stress. Although the claim was made that these plants had altered rates of photorespiration, this is unlikely because events downstream of Rubisco, such as the activity of GS2 are unlikely to affect the rate of oxygenation, unless it leads to a very large change in the activation state of Rubisco (Section V.A). Since photorespiration was estimated by measuring the post-illumination CO_2 burst, it seems more likely that metabolite pools (e.g. Gly) were altered in the transgenics. However, GS overexpressors do show altered N metabolism, with large changes in ammonia and amino acids (e.g. Migge et al., 2000). Leaves of plants over-expressing GS did appear to be less susceptible to photoinhibition (estimated by changes in Fv/Fm) and chlorophyll loss was less than in wild-types or in plants with less GS (Kozaki and Takeba, 1996). Similarly, Hoshida et al. (2000), suggest that transgenic rice over-expressing GS2 had an increased capacity for photorespiration and an increased tolerance to salt and chilling stress. In both of these examples, a more detailed evaluation is needed. Yamaguchi and Nishimura (2000) have shown that photoinhibition, estimated as a decrease in Fv/Fm following illumination at 500 $\mu mol.m^{-2}s^{-1}$, was enhanced in transgenic tobacco plants expressing less than about 40% of the wild-type glycolate oxidase.

In drought-stressed leaves, as during light stress, the importance of mechanisms protecting the photosynthetic apparatus is increased because CO_2 assimilation is decreased, resulting in a reduced electron requirement for photosynthesis. Under conditions of mild to moderate drought stress, the decline in photosynthesis mainly results from lower Ci caused by stomatal closure (Kaiser, 1987; Lal et al., 1996; Sánchez-Rodríguez et al., 1999) rather than damage to the photosynthetic apparatus (Cornic, 2000). Under these conditions, activities of photosynthetic enzymes do not decrease (Sharkey and Seemann, 1989; Lal et al., 1996; Sánchez-Rodríguez et al., 1999; Wingler et al., 1999c). In the long-term, however, drought stress has been shown to result in lower fructose-1,6-bisphosphatase activities in *Casuarina equisetifolia* (Sánchez-Rodríguez et al., 1999), and to a decline in the amounts of sedoheptulose-1,7-bisphosphatase and NADP-dependent glyceraldehyde-3-phosphate dehydrogenase proteins in barley (Wingler et al., 1999c). The amounts of photorespiratory enzyme proteins (proteins of the GDC complex, GS2, SGAT) were not affected by drought stress, while the amount of NADH-HPR increased (Wingler et al., 1999c). In combination, the decline in Ci and sustained activities of Rubisco and photorespiratory enzymes are likely to result in increased rates of photorespiration, not only relative to photosynthesis, but also in absolute terms. Therefore, photorespiration could serve as an important means to maintain electron flow.

Wingler et al. (1999c) utilized heterozygous barley mutants which contained approx. 50% of wild-type activities of the photorespiratory enzymes, GS2, GDC and SGAT, to study the role of photorespiration during drought stress. These mutants have normal rates of photosynthesis in moderate light and in ambient CO_2. In low CO_2, on the other hand, photosynthesis is reduced in the GS2 and GDC mutants. The rationale behind the study was that if photorespiration were increased in dehydrated leaves, photosynthesis should decline to a greater extent in the mutants than in the wild-type with increasing drought stress, and the control exerted by the photorespiratory enzymes on photosynthesis should increase. In well-watered plants, reduced activities of GS2, GDC or SGAT did not affect photosynthesis. With decreasing water potential, rates of CO_2 assimilation declined almost linearly in the wild-type. In the mutants with reduced activities of photorespiratory enzymes, this decline was accelerated, resulting in lower rates of CO_2 assimilation at moderate drought stress. The control exerted by photorespiratory enzymes on photosynthesis was, therefore, increased in moderately drought-stressed leaves. However, under severe drought stress, the rates of CO_2 assimilation were equally low in the wild-type and in the mutants. Together with the lower rates of photosynthesis, the calculated values for the oxygenase reaction of Rubisco indicated that during moderate drought stress (when the calculation of Ci was probably still valid) photorespiration was increased (Wingler et al., 2000). This was also indicated by an increase in Gly contents in drought-stressed leaves of the GDC mutant (Wingler et al., 1999c).

The lower rates of photosynthesis in the heterozygous mutants were accompanied by decreased quantum efficiencies of PSII electron transport. This decreased electron consumption in photosynthesis and photorespiration in the mutants did not lead to a decline in Fv/Fm, which would have indicated chronic photoinhibition. Instead, energy dissipation by non-photochemical quenching increased. In the SGAT and GDC mutants, this was accompanied by a strong increase in the formation of zeaxanthin. As shown by Brestic et al. (1995) and Demmig-Adams et al. (1988), xanthophyll-cycle dependent energy dissipation seems to be an important mechanism for protecting against the deleterious effect of light in drought-stressed leaves.

Conclusions

Photorespiratory mutants represent the most complete set for any metabolic pathway in plants. Consequently, photorespiration is one of the most clearly defined pathways in plant metabolism. Photorespiratory mutants still have considerable potential to increase our understanding of the processes of the C-N interactions involved in photorespiration, both at the level of the regulation of gene expression and at the level of regulation and control of metabolism, as well as increasing our understanding of plant responses to stress. Transformation of photorespiratory mutants also offers the exciting possibility of complementing the lesions with mutant forms of enzymes.

References

Andersen BR, Jin G, Chen R, Ertl JR and Chen C-M (1996) Transcriptional regulation of hydroxypyruvate reductase gene expression by cytokinin in etiolated pumpkin cotyledons. Planta 198: 1–5

Anderson LE (1981) Chloroplast and cytoplasmic enzymes. II. Pea leaf triose phosphate isomerase. Biochim Biophys Acta 235: 237–244

Artus NN, Naito S and Somerville CR (1994) A mutant of *Arabidopsis thaliana* that defines a new locus for glycine decarboxylation. Plant Cell Physiol 35: 879–885

Bainbridge G, Madgwick P, Parmar S, Mitchell R, Paul M, Pitts J, Keys AJ and Parry MAJ (1995) Engineering Rubisco to change its catalytic properties. J Exp Bot 46: 1269–1276

Bassham JA (1964) Kinetic studies of the photosynthetic carbon reduction cycle. Annu Rev Plant Physiol 15: 101–120

Beckmann K, Dzuibany C, Biehler K, Fock H, Hell R, Migge A and Becker TW (1997) Photosynthesis and fluorescence quenching, and mRNA levels of GS2 or of mitochondrial Ser hydroxymethyl transferase (SHMT) in the leaves of the wild-type and of the SHMT-deficient stm mutant of *Arabidopsis thaliana* in relation to the rate of photorespiration. Planta 202: 379–386

Bertoni GP and Becker WM (1993) Effects of light fluence and wavelength on expression of the gene encoding cucumber hydroxypyruvate reductase. Plant Physiol 103: 933–941

Bertoni GP and Becker WM (1996) Expression of the cucumber hydroxypyruvate reductase gene is down-regulated by elevated CO_2. Plant Physiol 112: 599–605

Besson V, Rébeillé F, Neuburger M, Douce R and Cossins EA (1993) Effects of tetrahydrofolate polyglutamates on the kinetic parameters of serine hydroxymethyltransferase and glycine decarboxylase from pea leaf mitochondria. Biochem J 292: 425–430

Blackwell RD, Murray AJS and Lea PJ (1987) Inhibition of photosynthesis in barley with decreased levels of chloroplastic glutamine synthetase activity. J Exp Bot 38: 1799–1809

Blackwell RD, Murray AJS, Lea PJ, Kendall AC, Hall NP, Turner JC and Wallsgrove RM (1988a) The value of mutants unable to carry out photorespiration. Photosynth Res 16: 155–176

Blackwell RD, Murray AJS, Lea PJ and Joy KW (1988b) Photorespiratory amino donors, sucrose synthesis and the induction of CO_2 fixation in barley deficient in glutamine synthetase and glutamate synthase. J Exp Bot 39: 845–858

Blackwell RD, Murray AJS and Lea PJ (1990) Photorespiratory mutants of the mitochondrial conversion of glycine to serine. Plant Physiol 94: 1316–1322

Brestic M, Cornic G, Fryer MJ and Baker NR (1995) Does photorespiration protect the photosynthetic apparatus in French bean leaves from photoinhibition during drought stress? Planta 196: 450–457

Campbell WJ and Ogren WL (1990) Glyoxylate inhibition of ribulose bisphosphate carboxylase/oxygenase activation in intact, lysed and reconstituted chloroplasts. Photosynth Res 23: 257–268

Chang C-C and Huang AHC (1981) Metabolism of glycolate in isolated spinach leaf peroxisomes. Kinetics of glyoxylate, oxalate, carbon dioxide, and glycine formation. Plant Physiol 67: 1003–1006

Chastain CJ and Ogren WL (1985) Photorespiration-induced reduction of ribulose bisphosphate carboxylase activation level. Plant Physiol 77: 851–856

Chastain CJ and Ogren WL (1989) Glyoxylate inhibition of ribulosebisphosphate carboxylase/oxygenase activation state in vivo. Plant Cell Physiol 30: 937–944

Chen C-M and Leisner SM (1985) Cytokinin-modulated gene expression in excised pumpkin cotyledons. Plant Physiol 77: 99–103

Chollet R and Ogren WL (1975) Regulation of photorespiration in C_3 and C_4 species. Bot Rev 41:137–179

Cleland WW, Andrews TJ, Gutteridge S, Hartman FC and Lorimer GH (1998) The mechanism of Rubisco: The carbamate as general base. Chem Rev 98: 549–561

Cock JM, Brock IW, Watson AT, Swarup R, Morby AP and Cullimore JV (1991) Regulation of glutamine synthetase genes in leaves of *Phaseolus vulgaris*. Plant Mol Biol 17: 761–771

Cook CM, Mulligan RM and Tolbert NE (1985) Inhibition and stimulation of ribulose-1,5-bisphosphate carboxylase by glyoxylate. Arch Biochem Biophys 240: 392–401

Cornic G (2000) Drought stress inhibits photosynthesis by decreasing stomatal aperture—not by affecting ATP synthesis. Trends Plant Sci 5: 187–188

Coschigano KT, Melo-Oliveira R, Lim J and Coruzzi GM (1998) *Arabidopsis gls* mutants and distinct Fd-GOGAT genes: Implications for photorespiration and primary nitrogen assimilation. Plant Cell 10: 741–752

Cossins EA and Chen L (1997) Folates and one-carbon metabolism in plants and fungi. Phytochemistry 45: 437–452

Créach E and Stewart CR (1982) Effects of aminoacetonitrile on net photosynthesis, ribulose-1,5-bisphosphate levels, and glycolate pathway intermediates. Plant Physiol 70: 1444–1448

Day DA and Wiskich JT (1980) Glycine transport by pea leaf mitochondria. FEBS Lett 112: 191–194

Delgado E, Parry MAJ, Lawlor DW, Keys AJ and Medrano H (1993) Photosynthesis, ribulose-1,5-bisphosphate carboxylase and leaf characteristics of *Nicotiana tabacum* L. genotypes selected by survival at low CO_2 concentrations. J Exp Bot 44: 1–7

Demmig-Adams B, Winter K, Krüger A and Czygan F-C (1988) Zeaxanthin and the heat dissipation of excess light energy in *Nerium oleander* exposed to a combination of high light and water stress. Plant Physiol 87: 17–24

Dever LV, Blackwell RD, Fullwood NJ, Lacuesta M, Leegood RC, Onek LA, Pearson M and Lea PJ (1995) The isolation and characterization of mutants of the C_4 photosynthetic pathway. J Exp Bot 46: 1363–1376

Douce R and Heldt HW (2000) Photorespiration. In: Leegood RC, Sharkey TD and von Caemmerer S (eds) Photosynthesis: Physiology and Metabolism, pp 115–136. Kluwer Academic Publishers, Dordrecht

Douce R and Neuburger M (1999) Biochemical dissection of photorespiration. Curr Opin Plant Biol 2: 214–222

Dzuibany C, Haupt S, Fock H, Biehler K, Migge A and Becker TW (1998) Regulation of nitrate reductase transcript levels by glutamine accumulating in the leaves of a ferredoxin-dependent glutamate synthase-deficient *gluS* mutant in *Arabidopsis thaliana*, and by glutamine provided via the roots. Planta 206: 515–522

Ebbighausen H, Chen J and Heldt HW (1985) Oxaloacetate translocator in plant mitochondria. Biochim Biophys Acta 810: 184–199

Edwards JW and Coruzzi GM (1989) Photorespiration and light act in concert to regulate the expression of the nuclear gene for chloroplast glutamine synthetase. Plant Cell 1: 241–248

Edwards JW, Walker EL and Coruzzi GM (1990) Cell-specific expression in transgenic plants reveals non-overlapping roles for chloroplast and cytosolic glutamine synthetase. Proc Natl Acad Sci USA 87: 3459–3463

Farquhar GD, Firth PM, Wetselaar R and Weir B (1980) On the gaseous exchange of ammonia between leaves and the environment: Determination of the ammonia compensation point. Plant Physiol 66: 710–714

Ferrario-Méry S, Suzuki A, Kunz C, Valadier MH, Roux Y, Hirel B and Foyer CH (2000) Modulation of amino acid metabolism in transformed tobacco plants deficient in Fd-GOGAT. Plant Soil 221: 67–79

Finnemann J and Schjoerring JK (2000) Post-translational regulation of cytosolic glutamine synthetase by reversible phosphorylation and 14-3-3 protein interaction. Plant J 24: 171–181

Flügge UI, Woo KC and Heldt HW (1988) Characteristics of 2-OG and glutamate transport in spinach chloroplasts. Studies with a double-silicone-layer centrifugation technique and in liposomes. Planta 174: 534–541

Genty B, Briantais JM and Baker NR (1989) The relationship between the quantum yield of photosynthetic transport and quenching of chlorophyll fluorescence. Biochim Biophys Acta 990: 87–92

Gifford DJ and Cossins EA (1982a) The nature of formate metabolism in greening barley leaves. Phytochemistry 21: 1479–1484

Gifford DJ and Cossins EA (1982b) Relationships between glycollate and formate metabolism in greening barley leaves. Phytochemistry 21: 1485–1490

Givan CV and Kleczkowski LA (1992) The enzymic reduction of glyoxylate and hydroxypyruvate in leaves of higher plants. Plant Physiol 100: 552–556

Givan CV, Tsutakawa S, Hodgson JM, David N and Randall DD (1988) Glyoxylate reductase activity in pea leaf protoplasts. Nucleotide specificity and subcellular location. J Plant Physiol 132: 593–599

Grodzinski B and Butt VS (1976) Hydrogen peroxide production and release of carbon dioxide during glycolate oxidation in leaf peroxisomes. Planta 128: 225–231

Hall NP, Kendall AC, Lea PJ, Turner JC and Wallsgrove RM (1987) Characteristics of a photorespiratory mutant of barley (*Hordeum vulgare* L.) deficient in phosphoglycollate phosphatase. Photosynth Res 11: 89–96

Halliwell B (1973) The role of formate in photorespiration. Biochem Soc Trans 1: 1147–1150

Halliwell B (1974) Oxidation of formate by peroxisomes and mitochondria from spinach leaves. Biochem J 138: 77–85

Harley PC and Sharkey TD (1991) An improved model of C_3 photosynthesis at high CO_2: Reversed O_2 sensitivity explained by lack of glycerate reentry into chloroplasts. Photosynth Res 27: 169–178

Harpel MR and Hartman FC (1994) Structure, function, regulation and assembly of ribulose-1,5-bisphosphate carboxylase/oxygenase. Annu Rev Biochem 63: 197–234

Häusler RE, Blackwell RD, Lea PJ and Leegood RC (1994a) Control of photosynthesis in barley leaves with reduced activities of glutamine synthetase or glutamate synthase. I. Plant characteristics and changes in nitrate, ammonium and amino acids. Planta 194: 406–417

Häusler RE, Lea PJ and Leegood RC (1994b) Control of photosynthesis in barley leaves with reduced activities of glutamine synthetase or glutamate synthase. II. Control of electron transport and CO_2 assimilation. Planta 194: 418–435

Häusler RE, Bailey KJ, Lea PJ and Leegood RC (1996) Control of photosynthesis in barley leaves with reduced activities of glutamine synthetase or glutamate synthase. III. Aspects of glyoxylate metabolism and the effects of glyoxylate on the activation state of ribulose-1,5-bisphosphate carboxylase-oxygenase. Planta 200: 388–396

Havir EA and McHale NA (1988) A mutant of *Nicotiana sylvestris* lacking serine glyoxylate aminotransferase: Substrate specificity of the enzyme and fate of [2-^{14}C] glycolate in plants with genetically altered enzyme levels. Plant Physiol 87: 806–808

Hecht U, Oelmüller R and Mohr H (1988) Action of light, nitrate and ammonium on the levels of NADH- and ferredoxin-dependent glutamate synthases in the cotyledons of mustard seedlings. Planta 195: 130–138

Heupel R and Heldt HW (1994) Protein organisation in the matrix of leaf peroxisomes. Eur J Biochem 220: 165–172

Hoshida H, Tanaka Y, Hibino T, Hayashi Y, Tanaka A, Takabe T and Takabe T (2000) Enhanced tolerance to salt stress in transgenic rice that overexpresses chloroplast glutamine synthetase. Plant Mol Biol 43: 103–111

Joy KW, Blackwell RD and Lea PJ (1992) Assimilation of nitrogen in mutants lacking enzymes of the glutamate synthase cycle. J Exp Bot 43:139–146

Kacser H (1987) Control of metabolism. In: Davies DD (ed) The Biochemistry of Plants: Biochemistry of Metabolism. Vol 11, pp 39–67. Academic Press, London

Kaiser WM (1987) Effects of water deficit on photosynthetic capacity. Physiol Plant 71: 142–149

Kendall AC, Keys AJ, Turner JC, Lea PJ and Miflin BJ (1983) The isolation of a catalase deficient mutant of barley (*Hordeum vulgare*). Planta 159: 505–511

Kendall AC, Wallsgrove RM, Hall NP, Turner JC and Lea PJ (1986) Carbon and nitrogen assimilation in barley (*Hordeum vulgare* L.) mutants lacking ferredoxin-dependent glutamate synthase. Planta 168: 316 –323

Keys AJ, Bird IF, Cornelius MJ, Lea PJ, Wallsgrove RM and Miflin BJ (1978) Photorespiratory nitrogen cycle. Nature 275: 741–743

Kirk CD, Chen L, Imeson H and Cossins EA (1995) A 5,10-methylenetetrahydrofolate dehydrogenase:5,10-methenyltetrahydrofolate cyclohydrolase protein from *Pisum sativum*. Phytochemistry 39: 1309–1317

Kleczkowski LA, Givan CV, Hodgson JM and Randall DD (1988) Subcellular location of NADPH-dependent hydroxypyruvate reductase activity in leaf protoplasts of *Pisum sativum* L. and its role in photorespiratory metabolism. Plant Physiol 88: 1182–1185

Kleczkowski LA, Edwards GE, Blackwell RD, Lea PJ and Givan CV (1990) Enzymology of the reduction of hydroxypyruvate and glyoxylate in a mutant of barley lacking peroxisomal hydroxypyruvate reductase. Plant Physiol 94: 819–825

Kozaki A and Takeba G (1996) Photorespiration protects C_3 plants from photooxidation. Nature 384: 557–560

Lal A, Ku MSB and Edwards GE (1996) Analysis of inhibition of photosynthesis due to water stress in the C_3 species *Hordeum vulgare* and *Vicia faba*: Electron transport, CO_2 fixation and carboxylation capacity. Photosynth Res 49: 57–69

Lawyer AL, Cornwell KL, Gee SL and Bassham JA (1983) Glyoxylate and glutamate effects on photosynthetic carbon metabolism in isolated chloroplasts and mesophyll cells of spinach. Plant Physiol 72: 420–425

Lea PJ and Forde BG (1994) The use of mutants and transgenic plants to study amino acid metabolism. Plant Cell Environ 17: 541–556

Leek AE, Halliwell B and Butt VS (1972) Oxidation of formate and oxalate in peroxisomal preparations from leaves of spinach beet (*Beta vulgaris* L.). Biochim Biophys Acta 286: 299–311

Lopez-Siles FJ, Cardenas J and Franco AR (1999) Biochemical and genetic analysis of a *Chlamydomonas reinhardtii* mutant devoid of chloroplastic glutamine synthetase activity. Planta 207: 436–441

Loreto F, Delphine S and di Marco G (1999) Estimation of photorespiratory carbon dioxide recycling during photosynthesis. Aust J Plant Physiol 26: 733–736

Lorimer GH and Andrews TJ (1981) The C_3 chemo- and photorespiratory carbon oxidative cycle. In: Stumpf PK and Conn EE (eds) The Biochemistry of Plants, Vol 8, pp 329–374. Academic Press, London

Madore M and Grodzinski B (1984) Effect of oxygen concentration on ^{14}C–photoassimilate transport from leaves of *Salvia splendens* L. Plant Physiol 76: 782–786

Mattsson M, Häusler RE, Leegood RC, Lea PJ and Schjoerring JK (1997) Leaf-atmosphere ammonia exchange in barley mutants with reduced activities of glutamine synthetase. Plant Physiol 114: 1307–1312

McClung CR, Hsu M, Painter JE, Gagne JM, Karlsberg SD and Salomé PA (2000) Integrated temporal regulation of the photorespiratory pathway. Circadian regulation of two *Arabidopsis* genes encoding serine hydroxymethyltransferase. Plant Physiol 123: 381–391

McHale NA, Havir EA and Zelitch I (1989) Photorespiratory toxicity in autotrophic cell cultures of a mutant of *Nicotiana sylvestris* lacking serine-glyoxylate aminotransferase activity. Planta 179: 67–72

Migge A and Becker TW (1996) In tobacco leaves, the genes encoding the nitrate reducing or the ammonium-assimilating enzymes are regulated differently by external nitrogen-sources. Plant Physiol Biochem 34: 665–671

Migge A, Meya G, Carrayol E, Hirel B and Becker TW (1996) Regulation of the subunit composition of tomato plastidic glutamine synthetase by light and the nitrogen source. Planta 200: 213–220

Migge A, Carrayol E, Kunz C, Hirel B, Fock H and Becker T (1997) The expression of the tobacco genes encoding plastidic glutamine synthetase or ferredoxin-dependent glutamate synthase does not depend on the rate of nitrate reduction, and is unaffected by suppression of photorespiration. J Exp Bot 48: 1175–1184

Migge A, Carrayol E, Hirel B and Becker B (2000) Leaf-specific overexpression of plastidic glutamine synthetase stimulates the growth of transgenic tobacco seedlings. Planta 210: 252–260

Mouillon J-M, Aubert S, Bourguignon J, Gout E, Douce R and Rébeillé F (1999) Glycine and serine catabolism in non-photosynthetic higher plant cells: Their role in C1 metabolism. Plant J 20: 197–205

Murray AJS, Blackwell RD, Joy KW and Lea PJ (1987) Photorespiratory N donors, aminotransferase specificity and photosynthesis in a mutant barley deficient in serine:glyoxylate aminotransferase activity. Planta 172: 106–113

Murray AJS, Blackwell RD and Lea PJ (1989) Metabolism in a mutant of barley lacking NADH-dependent hydroxypyruvate reductase, an important photorespiratory enzyme. Plant Physiol 91: 395–400

Ninnemann O, Jauniaux J-C and Frommer WB (1994) Identification of a high affinity NH_4^+ transporter from plants. EMBO J 13: 3464–3471

Noctor G, Arisi AC-M, Jouanin L, Kunert KJ, Rennenberg H and Foyer CH (1998) Glutathione: Biosynthesis, metabolism and relationship to stress tolerance explored in transformed plants. J Exp Bot 49: 623–647

Nour JM and Rabinowitz JC (1991) Isolation, characterization, and structural organization of 10-formyltetrahydrofolate synthetase from spinach leaves. J Biol Chem 266: 18363–18369

Nour JM and Rabinowitz JC (1992) Isolation and sequencing of the cDNA for spinach 10-formyltetrahydrofolate synthetase. Comparisons with the yeast, mammalian, and bacterial proteins. J Biol Chem 23: 16292–16296

Oliver DJ (1987) Glycine uptake by pea leaf mitochondria: A proposed model for the mechanism of glycine-serine exchange. In: Moore AL, Beechey RB eds. Plant Mitochondria. Structural, Functional and Physiological Aspects. New York: Plenum Press, 219–222

Oliver DJ (1994) The glycine decarboxylase complex from plant mitochondria. Annu Rev Plant Physiol Plant Mol Biol 45: 323–327

Oliver DJ and Zelitch I (1977) Increasing photosynthesis by inhibiting photorespiration with glyoxylate. Science 196: 1450–1451

Ongun A and Stocking CR (1965) Effects of light on the incorporation of serine into the carbohydrates of chloroplast and non-chloroplast fractions of tobacco leaves. Plant Physiol 40: 819–824

Osmond CB (1981) Photorespiration and photoinhibition. Some implications for the energetics of photosynthesis. Biochim Biophys Acta 639: 77–98

Prabhu V, Chatson KB, Abrams GD and King J (1996) ^{13}C nuclear magnetic resonance detection of interactions of serine hydroxymethyltransferase with C1-tetrahydrofolate synthase and glycine decarboxylase complex activities in *Arabidopsis*. Plant Physiol 112: 207–216

Raman R and Oliver DJ (1997) Light-dependent control of photorespiratory gene expression. In: Pessarakli M (ed) Handbook of Photosynthesis, pp 381–389. Marcel Dekker, New York

Sánchez-Rodríguez J, Pérez P and Martínez-Carrasco R (1999) Photosynthesis, carbohydrate levels and chlorophyll-fluorescence estimated intercellular CO_2 in water-stressed *Casuarina equisetifolia* Forst. & Forst. Plant Cell Environ 22: 867–873

Schimkat D, Heineke D and Heldt HW (1990) Regulation of sedoheptulose-1,7-bisphosphatase by sedoheptulose-7-phosphate and glycerate, and of fructose-1,6-bisphosphatase by glycerate in spinach chloroplasts. Planta 181: 97–103

Schjoerring JK, Kyllingsbaek A, Mortensen JV and Byskov-Nielsen S (1993) Field investigations of ammonia exchange between barley plants and the atmosphere. I. Concentration profiles and flux densities of ammonia. Plant Cell Environ 16: 161–167

Sharkey TD and Seemann JR (1989) Mild water stress effects on carbon-reduction-cycle intermediates, ribulose bisphosphate carboxylase activity, and spatial homogeneity of photosynthesis in intact leaves. Plant Physiol 89: 1060–1065

Shingles R, Woodrow L and Grodzinski B (1984) Effects of glycolate pathway intermediates on glycine decarboxylation and serine synthesis in pea (*Pisum sativum* L.). Plant Physiol 74: 705–710

Smith IK, Kendall AC, Keys AJ, Turner JC and Lea PJ (1984) Increased levels of glutathione in a catalase-deficient mutant of barley (*Hordeum vulgare* L). Plant Sci Lett 37: 29–33

Somerville CR and Ogren WL (1979) A phosphoglycolate phosphatase-deficient mutant of *Arabidopsis*. Nature 280: 833–836

Somerville CR and Ogren WL (1980a) Photorespiration mutants

of *Arabidopsis thaliana* deficient in serine-glyoxylate aminotransferase activity. Proc Natl Acad Sci USA 77: 2684–2687

Somerville CR and Ogren WL (1980b) Inhibition of photosynthesis in *Arabidopsis* mutants lacking leaf glutamate synthase activity. Nature 286: 257–289

Somerville CR and Ogren WL (1981) Photorespiratory deficient mutants of *Arabidopsis thaliana* lacking mitochondrial serine transhydroxymethylase activity. Plant Physiol 67: 666–671

Somerville CR and Ogren WL (1982a) Genetic modification of photorespiration. Trends Biochem Sci 7: 171–174

Somerville CR and Ogren WL (1982b) Mutants of the cruciferous plant *Arabidopsis thaliana* lacking glycine decarboxylase activity. Biochem J 202: 373–380

Somerville SC and Ogren WL (1983) An *Arabidopsis thaliana* mutant defective in chloroplast dicarboxylate transport. Proc Natl Acad Sci USA 80: 1290–1294

Somerville SC and Somerville CR (1983) The effect of oxygen and carbon dioxide on photorespiratory flux determined from glycine accumulation in a mutant of *Arabidopsis thaliana.* J Exp Bot 34: 415–424

Somerville SC and Somerville CR (1985) A mutant of *Arabidopsis* deficient in chloroplast dicarboxylate transport is missing an envelope protein. Plant Sci Lett 37: 217–220

Somerville CR, Portis AR and Ogren WL (1982) A mutant of *Arabidopsis thaliana* which lacks activation of RuBP carboxylase in vivo. Plant Physiol 70: 381–387

Stitt M and Schulze ED (1994) Does Rubisco control the rate of photosynthesis and plant growth? An exercise in molecular ecophysiology. Plant Cell Environ 17: 465–487

Thibaud M-C, Cortez N, Rivière H and Betsche T (1995) Photorespiration and related enzymes in pea (*Pisum sativum*) grown in high CO_2. J Plant Physiol 146: 596–603

Vincentz M, Moureaux T, Leydecker M-T, Vaucheret H and Caboche M (1993) Regulation of nitrate and nitrite reductase expression in *Nicotiana plumbaginifolia* leaves by nitrogen and carbon metabolites. Plant J 3: 315–324

von Wirén N, Lauter F-R, Ninnemann O, Gillissen B, Walch-Liu P, Engels C, Jost W and Frommer WB (2000) Differential regulation of three functional ammonium transporter genes by nitrogen in root hairs and by light in leaves of tomato. Plant J 21: 167–175

Wallsgrove RM, Keys AJ, Bird IF, Cornelius MJ, Lea PJ and Miflin BJ (1980) The location of glutamine synthetase in leaf cells and its role in the reassimilation of ammonia release in photorespiration. J Exp Bot 31: 1005–1017

Wallsgrove RM, Kendall AC, Hall NP, Turner JC and Lea PJ (1986) Carbon and nitrogen metabolism in a barley (*Hordeum vulgare*) mutant with impaired chloroplast dicarboxylate transport. Planta 168: 324–329.

Wallsgrove RM, Turner JC, Hall NP, Kendall AC and Bright SWJ (1987) Barley mutants lacking chloroplast glutamine synthetase. Biochemical and genetic analysis. Plant Physiol 83: 155–158

Walton NJ (1982) Glyoxylate decarboxylation during glycolate oxidation by pea leaf extracts: Significance of glyoxylate and extract concentrations. Planta 155: 218–224

Wendler C, Putzer A and Wild A (1992) Effect of glufosinate (phosphinothricin) and inhibitors of photorespiration on photosynthesis and ribulose-1,5-bisphosphate carboxylase activity. J Plant Physiol 139: 666–671

Wingler A, Lea PJ and Leegood RC (1997) Control of photosynthesis in barley plants with reduced activities of glycine decarboxylase. Planta 202: 171–178

Wingler A, von Schaewen A, Leegood RC, Lea PJ and Quick WP (1998) Regulation of leaf senescence by cytokinin, sugars and light. Plant Physiol 116: 329–335

Wingler A, Ann VJ, Lea PJ and Leegood RC (1999a) Serine-glyoxylate aminotransferase exerts no control on photosynthesis. J Exp Bot 50: 719–722

Wingler A, Lea PJ and Leegood RC (1999b) Photorespiratory metabolism of glyoxylate and formate in glycine accumulating mutants of barley and *Amaranthus edulis*. Planta 207:518–526

Wingler A, Quick WP, Bungard RA, Bailey KJ, Lea PJ and Leegood RC (1999c) The role of photorespiration during drought stress: an analysis utilising barley mutants with reduced activities of photorespiratory enzymes. Plant Cell Environ 22: 361–373

Wingler A, Lea PJ, Quick WP and Leegood RC (2000) Photorespiration: Metabolic pathways and their role in stress protection. Phil Trans R Soc Lond B 355: 1517–1529

Woo KC, Flügge UI and Heldt HW (1987) A two translocator model for the transport of 2-oxoglutarate and glutamate in chloroplasts during ammonia assimilation in the light. Plant Physiol 84: 624–632

Yamaguchi K and Nishimura M (2000) Reduction to below threshold levels of glycolate oxidase activities in transgenic tobacco enhances photoinhibition during irradiation. Plant Cell Physiol 41: 1397–1406

Yokota A, Komura H and Kitaoka S (1985) Different metabolic fate of two carbons of glycolate in its conversion to serine in *Euglena gracilis*. Arch Biochem Biophys 242: 498–506

Yu J-W and Woo KC (1992) Ammonia assimilation and metabolite transport in isolated chloroplasts. I. Kinetic measurement of 2-oxoglutarate and malate uptake via the 2-oxoglutarate translocator in oat and spinach chloroplasts. Aust J Plant Physiol 19: 653–658

Zelitch I (1972) The photooxidation of glyoxylate by envelope-free spinach chloroplasts and its relation to photorespiration. Arch Biochem 150: 698–707

Zelitch I (1989) Selection and characterization of tobacco plants with novel O_2-resistant photosynthesis. Plant Physiol 90: 1457–1464

Zelitch I (1992) Factors affecting expression of enhanced catalase activity in a tobacco mutant with O_2-resistant photosynthesis. Plant Physiol 98: 1330–1335

Zelitch I and Day PR (1968) Variation in photorespiration. The effect of genetic differences in photorespiration on net photosynthesis in tobacco. Plant Physiol 43: 1838–1844

Zoglowek C, Krömer S and Heldt HW (1988) Oxaloacetate and malate transport of plant mitochondria. Plant Physiol 87: 109–115

Chapter 9

The Regulation of Plant Phosphoenolpyruvate Carboxylase by Reversible Phosphorylation

Jean Vidal*, Nadia Bakrim and Michael Hodges
Institut de Biotechnologie des Plantes, UMR CNRS 8618, Université de Paris-Sud, 91405 Orsay Cedex, France

Summary 135
I. Introduction 136
II. Properties of Phosphoenolpyruvate Carboxylase 136
III. The Enzyme's Physiological Context 137
IV. Reversible Modulation in vivo by a Regulatory Phosphorylation Cycle 139
A. Phosphoenolpyruvate Carboxylase as a Target for Phosphorylation 139
B. Identification of the Phosphoenolpyruvate Carboxylase Protein Kinase 140
C. The Transduction Cascade 141
1. Alkalization of the Cytosol in C_4 Mesophyll Cells 141
2. Phosphoinositide-Specific Phospholipase C and Inositol-1,4,5-Trisphosphate 142
3. Calcium and Upstream Calcium-Dependent Protein Kinase(s) 143
4. A Similar Cascade in Crassulacean Acid Metabolism Plants? 143
V. Significance of Regulatory Phosphorylation of the Photosynthetic Isoform 144
VI. Regulatory Phosphorylation of the C_3 Form: Importance in Anaplerosis 145
VII. Conclusions and Perspectives 148
References 148

Summary

Phosphoenolpyruvate carboxylase (PEPc) is a multifaceted enzyme that serves different physiological functions in plants. In C_3 plants, an important role is in the anaplerotic supply of carbon skeletons for biosynthetic functions such as amino acid synthesis, whereas C_4 and crassulacean acid metabolism (CAM) species also have a specific, highly active isoform that catalyses primary CO_2 fixation in the photosynthesis pathway. More effort has been concentrated to date on the regulation of the latter, photosynthetic form of PEPc. It has long been known that this form of the enzyme is subject to allosteric control by opposing photosynthesis-related metabolites in the cytosol of the mesophyll cells. The discovery of a phosphorylation process acting on photosynthetic PEPc revitalized interest in this enzyme and the ensuing wealth of data has highlighted signaling mechanisms acting in the regulation of plant metabolism. In C_4 plants, the cascade depends upon a cross-talk between the two neighboring photosynthetic cell types, involves classical second messengers like pH, phosphoinositide-specific phospholipase C, inositol-1,4,5-trisphosphate and calcium, leading to up-regulation of the activity of a Ca^{2+}-independent, C_4 PEPc-specific protein-serine/threonine kinase (PEPcK), which finally phosphorylates PEPc. The final activity of C_4 PEPc and the resulting carbon flux to bundle sheath cells are dependent on the mutual interaction between metabolite and covalent control mechanisms acting on this enzyme. Recent results have suggested that a similar regulatory circuit is operative at night in mesophyll cells of CAM leaves. It has become clear that the anaplerotic PEPc which is found in all plant types, is also regulated

*Author for correspondence, email: jean.vidal@ibp.u-psud.fr

Christine H. Foyer and Graham Noctor (eds): Photosynthetic Nitrogen Assimilation and Associated Carbon and Respiratory Metabolism, pp. 135–150.

by a PEPcK and that phosphorylation of PEPc in C_3 plant leaves functions in the coordination of carbon and nitrogen assimilation. We discuss the extent to which parallels can be drawn between the regulation of the different isoforms of PEPc.

I. Introduction

Phosphoenolpyruvate carboxylase (EC 4.1.1.31, PEPc) catalyzes the exergonic β-carboxylation of phosphoenolpyruvate (PEP) by HCO_3^- ($\Delta G = -7$ kcal mol^{-1}) in the presence of a divalent cation, generally Mg^{2+}. The reaction proceeds through a stepwise mechanism involving the reversible, rate-limiting formation of carboxyphosphate and the enolate of pyruvate. Carboxyphosphate is split into inorganic phosphate and free CO_2 within the active site, and the produced CO_2 then reacts with the enolate species to form oxaloacetate (OAA) (Chollet et al., 1996). PEPc is a widely distributed enzyme in plants, green algae and micro-organisms but absent in yeast and animals (Andreo et al., 1987). In higher plants, it catalyses a pivotal reaction related to such important processes as C_4 and crassulacean acid metabolism (CAM) photosynthesis, the anaplerotic pathway linked to amino acid synthesis, homeostasis of cytosolic pH, electroneutrality and osmolarity. PEPc belongs to a small, nuclear-encoded, multigenic family, where the different isoforms are involved in specific metabolic contexts (Lepiniec et al., 1994). Since its discovery (Bandurski and Greiner, 1953), the wealth of accumulated data has led to the unraveling of the enzyme's functional and regulatory properties. At the transcriptional level, some PEPc genes respond to external and internal factors, e.g. light, hormones and metabolites, while at the protein level, the allosteric nature of the enzyme allows its activity to be fine-tuned in relation to a varying metabolic environment. The last decade has seen a renewed interest in PEPc, mainly due to the discovery that it undergoes posttranslational control by a phosphorylation process linked to a highly complex signal transduction cascade. Today, it is one of the best-described models of plant signaling. This chapter will focus on what is known about these processes in leaves of C_4 and CAM plants, the two systems that have been studied in detail so far (Chollet et al., 1996; Vidal and Chollet, 1997; Nimmo, 2000). The PEPc forms that have been the focus of these studies are the major C_4 or CAM forms of the enzyme, which we denote here as 'photosynthetic' PEPc. In addition, these plants contain a second form, which is shared with C_3 plants and which can be denoted the anaplerotic or C_3 PEPc. Based on the scattered information available, we will discuss whether information gathered on the regulation of the photosynthetic isoforms can be extended to the C_3 form, which may be considered heterotrophic in nature, as it functions notably in the anaplerotic pathway that generates C precursors for biosynthetic purposes.

Abbreviations: Asp – aspartate; BCECF-AM – 2′,7′-bis-(2-carboxyethyl)-5-(and-6)carboxyfluorescein,acetoxymethyl ester; BSC – bundle sheath cells; CAM – crassulacean acid metabolism; CDPK – calmodulin-like domain protein kinase; CHX – cycloheximide; DAG – 1,2-diacylglycerol; DCMU – 3-(3,4 dichlorophenyl)-1,1-dimethyl urea; G6P – glucose 6-phosphate; Gln – glutamine; Glu – glutamate; Gly – glycine; GOGAT – glutamate synthase; GS – glutamine synthetase; Ins(1,4,5)P_3 – inositol-1,4,5-trisphosphate; MC – mesophyll cells; ME – malic enzyme; NR – nitrate reductase; OAA – oxaloacetate; PEPc – phosphoenolpyruvate carboxylase; PEPcK – phosphoenolpyruvate carboxylase protein kinase; PGA – 3-phosphoglyceric acid; PI-PLC – phosphoinositide-specific phospholipase C; PKA – mammalian protein kinase type A; RPP – reductive pentose phosphate (RPP pathway = Calvin cycle); S(P) – phosphorylated serine 8 (in *Sorghum* PEPc); S8D – serine 8 replaced by aspartate; Ser – serine; TCA – tricarboxylic acid; TP – triose phosphate(s); U73122 – 1-(6-{[17β-3methoxyestra-1,3,5(10)-trien-17-xyl]amino}hexyl)-1H-pyrrole-2,5-dione (U-73122); U73343 – 1-(6-{[17β-3methoxyestra-1,3,5(10)-trien-17-xyl]amino}hexyl)-2,5-pyrrolidinedione; W-7 – N-[6-aminohexyl]-5-chloro-1-naphthalenesulfonamide

II. Properties of Phosphoenolpyruvate Carboxylase

PEPc is a homotetramer, each subunit having an approximate mass of 110 kDa (Chollet et al., 1996). Recently, X-ray crystallographic analysis has shed light on the three-dimensional structure of the *E. coli* enzyme. The four subunits of the bacterial PEPc are organized in a 'dimer-of-dimers' form resulting in an overall square arrangement (Kai et al., 1999). This work has led to the localization of the active site and the Asp regulatory domains in the *E. coli* PEPc subunit. In maize C_4 PEPc, the presence of most of the important structural determinants, including the sub-unit linking domains, supports the idea that this plant enzyme has a similar structure to that of its bacterial homolog (Kai et al., 1999). Since all plant PEPc isoforms are very similar in primary structure, this tetrameric organization could be the canonical structure of each plant enzyme. Although active dimeric PEPc species have been detected in plant protein extracts, it is not yet known whether the dimer/tetramer equilibrium has any physiological role in vivo (McNaughton et al., 1989; Willeford and Wedding, 1992).

It has long been known that plant PEPc is regulated by metabolites (Andreo et al., 1987). However, most of these studies were performed using a poorly defined enzyme in terms of integrity (PEPc is very susceptible to proteolysis in vitro), phosphorylation state and pH (see below). Due to recent technological advances, metabolite regulation of PEPc has been revisited in detail, especially with respect to the photosynthetic C_4 form (Echevarria et al., 1994; Duff et al., 1995). The use of intact, non-phosphorylated, recombinant *Sorghum* C_4 PEPc confirmed that this enzyme was subject to two opposing control mechanisms: feedback inhibition by the end-product, L-malate (K_i:0.17 mM), and allosteric activation by glucose 6-phosphate (G6P; K_a:1.3 mM), when assayed at 1 mM PEP, pH 7.3 (Duff et al., 1995). Furthermore, other sugar phosphates, such as triose-phosphates (TP), and amino acids like Gly are capable of activating the *Sorghum* (Bakrim et al., 1998) and maize (Doncaster and Leegood, 1987; Gao and Woo, 1996) C_4-enzymes. Since it is important to evaluate the importance of these kinetic parameters, determined in vitro, in the physiological context, attempts have been made to measure them in the conditions believed to prevail in plant cells. At conditions close to those believed to exist in vivo (pH 7.3, 0.4 mM Mg^{2+}, 0.1 mM HCO_3^-), the maize enzyme exhibited a high degree of cooperativity towards PEP, a much lower affinity for this substrate ($S_{0.5}$: 12.3 mM) and for activators (e.g., $A_{0.5}$ for G6P: 3.9 mM), and a greater affinity for L-malate (I_{50}: 0.27 mM) (Tovar-Méndez et al., 2000). Indeed, L-malate appears to act as a competitive inhibitor with respect to PEP (Duff et al., 1995), whereas G6P increases the apparent affinity of C_4 PEPc for PEP (Gao and Woo, 1996). Thus, the positive effectors enhance the ability of PEP to compete with L-malate. This situation appears to be true also for the CAM and C_3 PEPc forms (O'Leary, 1982; Andreo et al., 1987).

The 3D-structure of *E. coli* PEPc has helped explain the molecular mechanism of L-malate inhibition. As mentioned above, the primary structures of *E. coli* and plant PEPcs are similar, with the notable exception of the N-terminal phosphorylation domain, which is absent in the bacterial enzyme. Indeed, computerized modeling of plant PEPcs gives a structural conformation that is very close to that of the bacterial PEPc, except for some additional loops in the plant enzyme (Fig. 1). In the bacterial PEPc, Arg 587 is in a highly conserved Gly-rich loop shared by the Asp-binding site and the active site. Upon binding of the effector, Asp (equivalent to L-malate in the plant enzyme), the loop is displaced from the catalytic site, thus perturbing substrate binding and causing a loss of catalytic activity (Kai et al., 1999). This shows that L-malate is not a true competitive inhibitor with respect to PEP. Unfortunately, this study has not clarified the mechanism by which G6P binds to PEPc and how it affects the affinity of the enzyme for PEP and L-malate.

Finally, PEPc activity and its metabolic control are highly sensitive to pH (Andreo et al., 1987; Echevarria et al., 1994; Gao and Woo, 1996). An increase in pH within the physiological range (from 7.0 to 7.5) activates PEPc and partially desensitizes it against the effectors, notably L-malate. Thus, it appears that pH variations could also operate in rapid fine control of carboxylase activity in situ.

III. The Enzyme's Physiological Context

During plant evolution, the photosynthetic pathway has been adapted to various environmental conditions, thus giving rise to C_4 and CAM plants. C_4 plants exhibit specific anatomical and biochemical features. Their leaf architecture conforms, in most cases, to the classical 'Kranz' anatomy characterized by concentrically organized photosynthetic tissues, ie, outer mesophyll cells (MC) surrounding inner bundle sheath cells (BSC). In terms of metabolic adaptation, there exist diverse types of C_4 plants, but the general metabolic scheme of division of labor is conserved: two cycles, the C_4 CO_2-concentrating cycle and the RPP pathway, work in concert to assimilate CO_2. In 'L-malate formers', the primary fixation of CO_2 (in its hydrated form) is carried out by a specific PEPc isoform (C_4 PEPc) in the MC cytosol to form OAA, which is then reduced to L-malate in the MC chloroplast. Export of L-malate to the BSC and its subsequent decarboxylation by an NADP-dependent

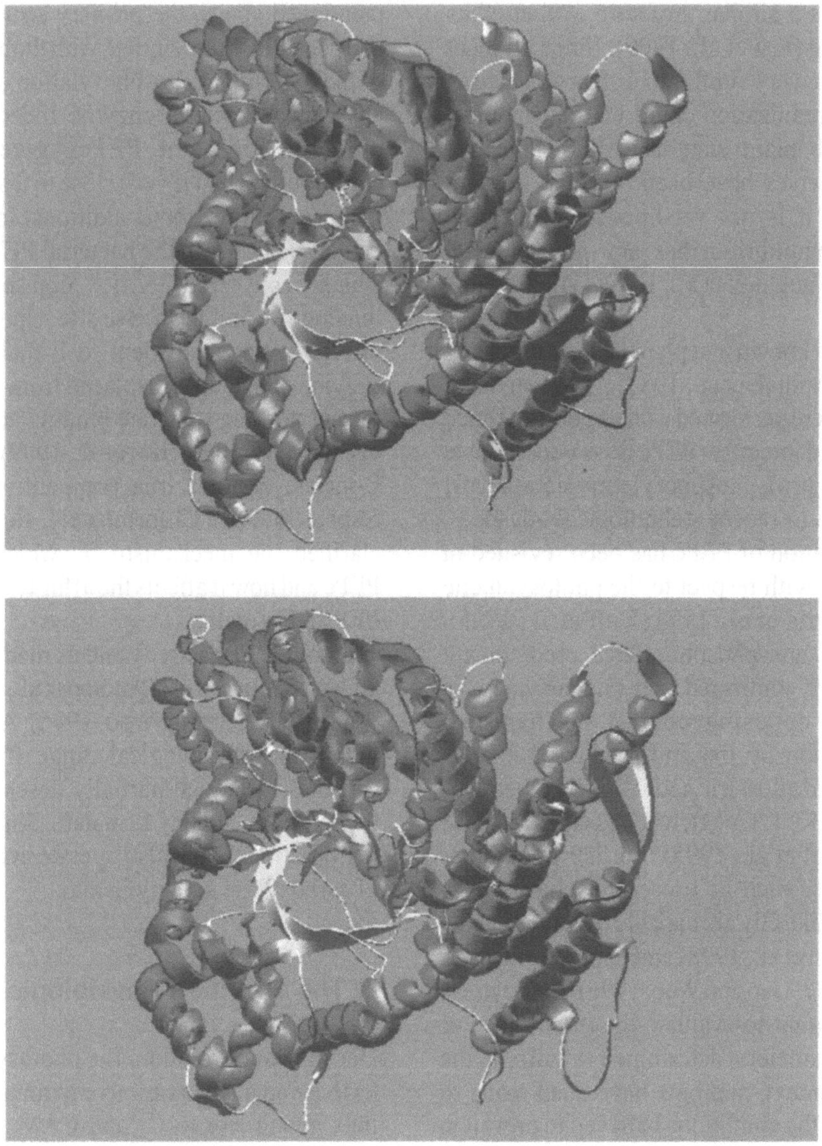

Fig. 1. Ribbon drawings showing the 3D-structure of the *E. coli* (upper panel, after Kai et al., 1999) and *Sorghum* (lower panel) PEPC monomers. The eight β-strands forming the β-barrel are shown in white and the α-helices are shown in gray. The *Sorghum* PEPC monomer was modelled using the Swiss Prot modelling service (Guex and Peitsch, 1997). See also Color Plate 3.

malic enzyme (NADP-ME) in the chloroplast stroma, generates reducing power (NADPH) and CO_2 to be used in the reductive pentose phosphate (RPP) pathway. Because in some C_4 plants, like *Sorghum* and sugar cane, BSC chloroplasts are deficient in photosystem II activity and energy production, the 3-phosphoglyceric acid (PGA) formed in this cell compartment moves to MC chloroplasts to be transformed to TP. The fate of TP is twofold: to supply C skeletons to sucrose synthesis in the MC and to return fixed C to the BSC where it re-enters the RPP pathway. This intense metabolite trafficking between photosynthetic cells is gradient-driven and depends on a network of plasmodesmata in the cell wall. This biochemical and anatomical adaptation largely prevents the wasteful production of CO_2 by photorespiration and, thus, loss of C from the leaf. In arid environments, this allows better water and N use efficiency and higher productivity relative to C_3 plants (Hatch, 1977). In C_4 plants, the photosynthetic PEPc is controlled by light (day)–dark (night) transitions so that CO_2 is efficiently fixed during the day.

In contrast to C_4 plants, CAM plants fix atmospheric CO_2 through a specific, photosynthetic PEPc during the night, when stomata are open. In this case, L-malate is accumulated and stored in the MC vacuole. This process is driven by an H^+-ATPase that pumps protons into the MC vacuole. During the following day, L-malate is released from the vacuole and the subsequent diurnal consumption of this metabolite is carried out by NAD/NADP-ME to meet RPP pathway CO_2 requirements. Flux through the CAM PEPc is controlled by a circadian oscillator rather than by light-dark transitions. This metabolic adaptation to very arid environments allows CAM plants to restrict water loss (Nimmo, 2000).

In both C_4 and CAM plants, PEPc participates in complex and highly integrated metabolic pathways. The spatial or temporal separation of two distinct CO_2 fixation steps requires a high degree of coordination and enzymatic control, the biochemical basis of which will now be discussed.

IV. Reversible Modulation in vivo by a Regulatory Phosphorylation Cycle

The existence of posttranslational mechanisms acting on the photosynthetic PEPc was indicated by observations that certain functional and regulatory properties of the enzyme were altered in protein extracts from leaves of CAM and C_4 plants during the day-night cycle. The phosphorylation/dephosphorylation-dependent regulation of PEPc was initially reported for the photosynthetic isoform of the CAM plant, *Bryophyllum* (Nimmo et al., 1984; Brulfert et al., 1986), and, shortly afterwards, of maize, a C_4 species (Budde and Chollet, 1986). Subsequently, a great deal of data on the enzyme's covalent control was gathered, radically advancing our understanding of the regulation of photosynthetic PEPc. More recently, much effort has been devoted to identifying the requisite PEPc protein kinase (PEPcK) and deciphering the cascade components that ultimately determine the phosphorylation status of PEPc.

A. Phosphoenolpyruvate Carboxylase as a Target for Phosphorylation

The first evidence that PEPc was phosphorylated on a Ser residue came from studies comparing the malate sensitivity and phosphorylation status of the day and night CAM-PEPc forms (Nimmo et al., 1984; Brulfert et al., 1986). However, the identification of the exact phosphorylated Ser residue site was determined by in planta radiolabeling of proteins and subsequent phosphopeptide analysis of the immuno-purified C_4 PEPc (*Sorghum*, maize). A comparison of the amino acid sequence of the purified peptide with sequences deduced from the known cDNAs and genes, revealed that the phospho-Ser was located close to the N-terminus. A consensus phosphorylation domain, E/D-R/K-X-X-S(P)-I-D-A-Q-L/M-R, was defined from a survey of all PEPc sequences available at the time. The phosphorylated Ser is at position 8 and 15 of the sequence of photosynthetic PEPc from *Sorghum* and maize, respectively. It is now clear that this domain is plant-invariant, whatever the physiological type (C_3, CAM, C_4), and that it is absent from bacterial and cyanobacterial PEPc that do not undergo phosphorylation (Chollet et al., 1996; Vidal and Chollet, 1997).

In vitro studies showed that phosphorylation of the specific Ser residue modulated the effects of metabolite regulation on C_4 PEPc activity. The extensive Ser phosphorylation (one per subunit) of recombinant *Sorghum* C_4 PEPc caused only a modest effect on the K_m for PEP but an approximately two-fold increase in V_{max}, a seven-fold increase in the K_i for L-malate, and a 4.5-fold decrease in the K_a for G6P (measured at suboptimal pH (pH 7.3) and PEP concentration (2.5 mM)) (Duff et al., 1995). These effects of phosphorylation have been observed in all PEPc forms investigated so far, whether C_3, CAM or other C_4 enzymes (see Chollet et al., 1996; Vidal and Chollet, 1997 and Nimmo, 2000, for reviews).

In vivo, light-induced phosphorylation of *Sorghum* and maize C_4 PEPc was complete within 1–2 h, as estimated by the decrease in the enzyme's sensitivity to L-malate or the increase in ^{32}P radiolabeling of the protein. The final ratio of the phosphorylated/nonphosphorylated enzyme was found to be dependent upon light intensity (Bakrim et al., 1992). Dephosphorylation, presumably by a type-2A protein phosphatase, as shown for the PEPc from the facultative CAM species *Kalanchoe fedtschenkoi* (Carter et al., 1990), followed a similar time course when the plants were returned to darkness.

The use of site-directed mutagenesis and recombinant protein technology clearly showed that the phosphorylation-induced changes in C_4 PEPc properties could be mimicked by the introduction of a negative charge (Ser8 to Asp-mutated PEPc) to the

N-terminal domain of the protein (Duff et al., 1995). Therefore, the additional negative charge on the N-terminal domain appeared to be involved in the regulatory process. Based on the 3-D structure of the bacterial enzyme, it has been proposed that the negatively charged N-terminus extension interacts with certain residues of the plant PEPc so as to block the access of L-malate to the inhibitor site, thereby decreasing the enzyme's sensitivity to the feedback inhibitor (Kai et al., 1999).

B. Identification of the Phosphoenolpyruvate Carboxylase Protein Kinase

Both Ca^{2+}-dependent and -independent protein kinases have been shown to phosphorylate the target C_4 PEPc in reconstitution assays in vitro (Jiao and Chollet 1989; Bakrim et al., 1992; Chollet et al., 1996; Nhiri et al., 1998; Ogawa et al., 1998). The identity of the PEPcK has long been a matter of debate and considerable efforts have been made to distinguish between physiologically relevant and other phosphorylation events (Chollet et al., 1996; Vidal and Chollet, 1997). A number of well established molecular and physiological characteristics of C_4 and CAM plant PEPc phosphorylation suggested that the authentic PEPcK must be a Ca^{2+}-independent, protein-Ser/Thr kinase phosphorylating the N-terminal Ser, thereby giving rise to the expected changes in the catalytic and regulatory properties of the enzyme (Chollet et al., 1996). Furthermore, studies of various plant extracts using renaturation on gels and subsequent activity staining suggested that the PEPcK had a molecular mass of 32 and/or 37–39 kDa and acted as a monomer (Li and Chollet, 1994). In addition, cycloheximide (CHX), an inhibitor of cytosolic protein synthesis, was a potent blocker of PEPcK upregulation in situ (Chollet et al., 1996; Nimmo, 2000). Such observations suggest that the protein must have a relatively high turnover rate and that regulatory protein factors must be involved. Recently, this issue has been resolved following the long-awaited cloning of a cDNA encoding the Ca^{2+}-independent PEPcK from the facultative CAM plants *K. fedtschenkoi* (in vitro transcription-translation screening approach; Hartwell et al., 1999) and *Mesembryanthemum crystallinum* (DDRT-PCR approach; Taybi et al., 2000). Such studies established that it was indeed the PEPcK that was regulated at the level of gene expression. Accumulation of CAM PEPcK transcripts was high during the night and matched the marked increase in PEPcK activity and the phosphorylation state of CAM-PEPc, all three parameters exhibiting a circadian rhythm under constant conditions (Hartwell et al., 1999). Therefore, in CAM plants, PEPcK expression is controlled both developmentally and by a circadian oscillator, whereas in C_4 plants light is the signal.

The PEPcK has a number of interesting features. 1) It is the smallest protein kinase known so far. The predicted molecular mass is around 31 kDa, made up of 274, 279 and 284 amino acids in the enzymes from *K. fedtschenkoi*, *M. crystallinum* and *Arabidopsis thaliana*, respectively. Indeed, it is made up of a kinase catalytic domain with minimal or no additions. 2) Although it belongs to the Ca^{2+}/calmodulin-regulated group of protein kinases, it lacks the regulatory auto-inhibitory region and extended finger (EF)-hands. 3) In reconstitution assays, it displays an alkaline pH optimum (pH 8, using the recombinant enzyme from *M. crystallinum*). It phosphorylates very specifically the N-terminal regulatory Ser of the target PEPc, and decreases the malate sensitivity of the enzyme (Taybi et al., 2000).

Another unique feature of this Ca^{2+}-independent PEPcK appears to be that its activity is not modulated directly by second messengers (such as Ca^{2+}/calmodulin or cyclic nucleotides) or by phosphorylation/dephosphorylation processes, but rather through rapid changes in its turnover rate (Bakrim et al., 1992; Chollet et al., 1996; Hartwell et al., 1996; Hartwell et al., 1999; Taybi et al., 2000). But what is the mechanism underlying this control? In *K. daigremontiana*, the observation that high cytosolic malate levels coincided with the decrease in transcript abundance led to the proposal that malate may affect PEPcK gene expression via feedback repression (Borland et al., 1999; Nimmo, 2000). However, this mechanism is not yet fully understood. It seems that this idea cannot be extended to the C_4 system as it is well established that high malate levels coincide with a high PEPcK content in the light. However, reconstitution assays using C_4 leaf extracts containing PEPcK activity or the mammalian type A protein kinase (PKA: previously shown to be able to phosphorylate C_4 PEPc in vitro; Terada et al., 1990), revealed that the phosphorylation of purified, recombinant C_4 PEPc was inhibited by L-malate (Wang and Chollet, 1993; Echevarria et al., 1994), and that G6P and PEP protected against this inhibition. The effect of these metabolites on C_4 PEPc phosphorylation was also suggested to occur in situ in MC protoplasts from *Digitaria sanguinalis* (see hereafter and Bakrim et al., 1998). CAM plant PEPcK has also been found to be inhibited by L-malate in vitro; however, G6P was not shown to antagonize this effect (Carter et al., 1991). In general, this indirect means of regulating protein phosphorylation might allow an individual, target-dependent control of a multi-substrate protein kinase. However, to date, all available evidence suggests that this highly regulated Ca^{2+}-independent PEPcK is specific for plant PEPc (Chollet et al., 1996).

The effect of metabolites on PEPc phosphorylation can be explained by a metabolite-induced modification in PEPc-PEPcK interactions via an alteration in PEPc conformation. In agreement with this idea, the recent investigation of the local structural requirements for phosphorylation of C_4 PEPc by the PEPcK suggests a secondary site of interaction with the target protein (Li et al., 1997), which appears to be located in the C-terminal end of the PEPc (C. Echevarria, personal communication). This anchoring site presumably ensures the precise positioning of the protein kinase for efficient phosphorylation and its orientation could be modified by metabolites binding to PEPc.

Based on in silico investigations, putative PEPcK genes have been identified in the genome of rice, rapeseed, soybean, alfalfa, tomato and banana. In *A. thaliana*, two genes, PEPcK1 and PEPcK2, are present. These share 66% identity and are located on chromosomes 1 and 3, respectively. Both genes are interrupted by a conserved single intron in the 3′ end. PEPcK1 is expressed more specifically in rosette leaves, transcript abundance being higher in the light than in the dark. In contrast, PEPcK2 transcripts were found mainly in flowers and roots (Fontaine et al., 2000). In *M. crystallinum*, Southern blot analysis revealed a second, less intense band hybridizing to the PEPcK probe (Taybi et al., 2000). This was suggested to be consistent with the existence of the two PEPcK isoforms (32 and 39 kDa) previously described in CAM-induced leaves of this plant (Li and Chollet, 1994). Furthermore, it appears from database information that two PEPcK genes are present in tomato. What is the physiological significance of multiple PEPcK genes? Are there specific PEPcK isoforms that interact with specific PEPc isoforms in specific plant tissues? Are all PEPcK genes regulated in the same manner by the same signal cascade? In the future, will further PEPcK genes be discovered? At the present time such questions have no answers. However, it is expected, with the screening of insertional mutant libraries and the availability of the complete *Arabidopsis* genome sequence, that these questions will be rapidly resolved.

C. The Transduction Cascade

Most data on the cascade controlling PEPcK activity and PEPc phosphorylation have been obtained for the C_4 photosynthetic enzyme. The dependence of PEPcK regulation on photosynthesis was demonstrated by inhibitor studies in planta. First, treatment with the electron transport inhibitor, DCMU, or the uncoupler, gramicidin, revealed that upregulation of PEPcK and phosphorylation of C_4 PEPc in the MC cytosol were dependent on functional electron transport and ATP synthesis (Bakrim et al., 1992; Chollet et al., 1996). Second, use of the triose phosphate isomerase inhibitor, DL-glyceraldehyde, demonstrated the requirement for a functional RPP pathway in the BSC chloroplasts. These results led to the working hypothesis that light transduction involved intercellular cross-talk, possibly mediated through changes in the level of a photosynthetic metabolite and/or energy charge. To address this question and to further identify the components of the light-transduction cascade, a cellular approach using isolated MC protoplasts from the C_4 grasses *Digitaria* and *Sorghum* was developed (Giglioli-Guivarc'h et al., 1996). Illuminated, isolated protoplasts showed a marked decrease in L-malate sensitivity and an increase in catalytic activity of endogenous C_4 PEPc when a weak base, such as NH_4Cl or methylamine, was added to the suspension medium. These changes were shown to be associated with a marked, light-induced stimulation of a Ca^{2+}-independent PEPcK (Bakrim et al., 1992) which was found to be sensitive to CHX in situ (Giglioli-Guivarc'h et al., 1996).

1. Alkalization of the Cytosol in C_4 Mesophyll Cells

The weak bases that trigger the in situ phosphorylation of C_4 PEPc permeate into protoplasts in their neutral form and subsequently tend to increase cytosolic pH following protonation. The weak base-induced alkalization of the cytosol was experimentally documented by loading MC protoplasts with the fluorescent pH-probe, BCECF-AM, and performing in-situ fluorescence imaging by confocal microscopy.

As an alternative approach, applying the 'null-point' method (Van der Veen et al., 1992) and monitoring induced changes in protoplast fluorescence by flow cytometry also provided estimations of the cytosolic pH. An increase in cytosolic pH from about 6.4 to 7.3 with the concomitant increase in the activity of PEPcK and the apparent phosphorylation state of C_4 PEPc were found to be well correlated with the concentration of the exogenous weak-base (Giglioli-Guivarc'h et al., 1996). Therefore, intracellular alkalization of MC protoplasts was implicated as an early signaling event in the C_4 PEPc phosphorylation circuitry. It could be argued that a protoplast system does not reflect the true physiological situation within the leaf. However, MC from an excised *Sorghum* leaf loaded with the pH probe, carboxyfluoroscein, emitted fluorescence soon after exposure to light. This effect was reversed upon return to darkness, as judged by confocal microscopy. Interestingly, such changes were blocked by DL-glyceraldehyde, which inhibits PEPcK accumulation and PEPc phosphorylation in the illuminated C_4 leaf (Giglioli-Guivarc'h et al., 1996). This observation established that an increase in leaf MC cytosolic pH depended on a functional RPP pathway, in good agreement with other data (Raghavendra et al., 1993; Giglioli-Guivarc'h et al., 1996), and provided an important clue as to the possible nature of the putative intercellular message. The most likely candidate was PGA, the RPP pathway intermediate that moves into MC chloroplasts for subsequent phosphorylation/reduction. As transport by the chloroplast phosphate translocator proceeds only via the partially protonated PGA^{2-}, pumping of protons from the cytosol into the stroma would ensue, causing a net alkalization of the cytosol. Indeed, when PGA was added to MC protoplasts, it produced similar effects to those elicited by methylamine or NH_4Cl, i.e., alkalization of cytosolic pH (as judged by confocal microscopy) and upregulation of the Ca^{2+}-independent PEPcK and phosphorylation state of C_4 PEPc (Giglioli-Guivarc'h et al., 1996). It must be noted that both light and alkalization of the MC cytosol were needed for the induction of C_4 PEPcK activity. Consistent with these findings are the observations that induction was blocked by the inhibitors gramicidin and DCMU, while the increase in cytosolic pH was unaffected (Giglioli-Guivarc'h et al., 1996). However, this model of the crosstalk between the RPP pathway and the PEPc phosphorylation cascade, derived from protoplast studies, is not consistent with observations in the maize *bsd2* mutant, which is deficient in Rubisco. Despite the absence of a functional RPP pathway, the mutant is able to induce PEPcK in the light and to phosphorylate C_4 PEPc (Smith et al., 1998). To account for these contradictory observations, one might suppose that the PGA entering the MC chloroplasts of the mutant is not of photosynthetic origin. Alternatively, the MC cytosolic pH might be increased by other unidentified processes such as the activation of tonoplast ATPase and pyrophosphatase, thereby allowing the activation cascade to function and the PEPcK to accumulate in the illuminated mutant leaf. These points need to be addressed further before conclusions can be drawn.

2. Phosphoinositide-Specific Phospholipase C and Inositol-1,4,5-Trisphosphate

Stimulus-response coupling in animal cells frequently involves the hydrolysis of phosphatidyl-inositol-4,5-bisphosphate generating the two second messengers inositol-1,4,5-trisphosphate ($Ins(1,4,5)P_3$) and 1,2-diacylglycerol. This reaction is catalysed by a phosphoinositide-specific phospholipase C (PI-PLC; EC 3.1.4.11). Most components in the PI-PLC signaling system have structural or functional equivalents in plants, and evidence is emerging that they are involved in signaling (for reviews see Drøbak, 1992; Coté and Crain, 1993; Munnik et al., 1998). It has been shown that phosphorylation of C_4 PEPc is inhibited by preincubation of illuminated, weak-base-treated MC protoplasts with the PI-PLC antagonists, neomycin and U-73122. In contrast, U-73343, an inactive analog of U-73122, has no inhibitory activity on phosphorylation (Coursol et al., 2000). Furthermore, phosphorylation of PEPc in MC protoplasts has been shown to be accompanied by a marked and transient increase in $Ins(1,4,5)P_3$ levels. This increase was dependent on both light and the presence of NH_4Cl, and specifically inhibited by U-73122. Such findings indicate that PI-PLC is potentially an upstream component of the C_4 PEPc phosphorylation cascade in MC protoplasts. But how might PI-PLC be activated and what would be the role of $Ins(1,4,5)P_3$ in the induced MC protoplasts? Little is known about the precise mode of action of plant PI-PLCs, except that the enzyme is totally dependent on Ca^{2+}, at physiological concentrations, when assayed in vitro (Munnik et al., 1998). Therefore, activation of PI-PLC by light and cytosolic alkalization may be mediated by Ca^{2+} in a process possibly involving

Ca^{2+} influx across the plasma membrane, or perhaps Ins(1,4,5)P_3-induced Ca^{2+} release from internal stores. Information concerning the localization and concentration of the Ins(1,4,5)P_3-mobilizable Ca^{2+} stores would be crucial for understanding how PI-PLC is activated in the C_4 PEPc phosphorylation cascade. Currently, the vacuole is considered to be the major Ins(1,4,5)P_3-sensitive Ca^{2+} store in higher plants (Schumaker and Sze, 1987; Ranjeva et al., 1988; Brosnan and Sanders, 1993). However, there is evidence for Ins(1,4,5)P_3-mediated Ca^{2+} release from stores other than the vacuole in plants (Muir and Sanders, 1997).

3. Calcium and Upstream Calcium-Dependent Protein Kinase(s)

It has been shown that pretreatment of C_4 MC protoplasts with the calcium ionophore A23187 (calcimycin) combined with EGTA inhibited phosphorylation of C_4 PEPc. Specific recovery, however, was achieved if excess Ca^{2+} was reintroduced to protoplasts in the presence of A23187 (Pierre et al., 1992). The origin of calcium and its mobilization into the cytosol of C_4 MC protoplasts have been investigated by testing various pharmacological reagents. TMB-8 is a tonoplast, Ins(1,4,5)P_3-gated Ca^{2+} channel blocker and when present in the protoplast suspension during induction, it severely inhibited the in situ up-regulation of PEPcK activity and C_4 PEPc phosphorylation. In contrast, nifedipine and diltiazem (considered to act as plasma membrane Ca^{2+} channel inhibitors) did not have any effect on PEPcK activity or C_4 PEPc phosphorylation (Giglioli-Guivarc'h et al., 1996). Such findings support the view that Ca^{2+} and Ca^{2+} channels are involved in the light-transduction pathway. Given that the PEPcK is a Ca^{2+}-independent enzyme, a multicyclic protein kinase cascade involving upstream Ca^{2+} elements, was suggested to be involved in transducing the light signal. In good agreement with this proposal, W-7 (an inhibitor of Ca^{2+}/calmodulin-regulated protein kinases or CDPK) was found to have a marked inhibitory effect on PEPcK upregulation and PEPc phosphorylation in situ (Giglioli-Guivarc'h et al., 1996). Thus, these results suggested that the transduction chain involved a Ca^{2+}-dependent protein-kinase(s) exerting an effect in the signaling pathway, possibly controlling the transcriptional activity of the PEPcK gene. A model for the spatio-temporal organization of the light-signal transduction chain controlling the activity of PEPcK and, thus the phosphorylation of C_4 PEPc, is illustrated in Fig. 2.

4. A Similar Cascade in Crassulacean Acid Metabolism Plants?

As mentioned above, in CAM plants the upregulation of PEPcK and PEPc phosphorylation during the night is governed by a circadian oscillator (Nimmo, 2000). Physiological-based investigations in which malate levels were manipulated (e.g., enclosure of the plant in an atmosphere of N_2 during the night, increase in temperature) led to the proposal that this metabolite exerts a negative feedback control on PEPcK gene expression and can override circadian control (Borland et al., 1999; Nimmo, 2000). In this model, the primary target of the oscillator is malate transport across the tonoplast. However, the possibility cannot be excluded that the circadian oscillator directly influences PEPcK gene expression, perhaps through a transcription factor similar to the CCA1 (Circadian Clock Associated) protein of *Arabidopsis* (Nimmo, 2000). On the other hand, we have recently obtained evidence that effectors of PEPcK induction in C_4 MC protoplasts (CHX, U73122, TMB-8, W7) are also powerful blockers of PEPcK accumulation and PEPc phosphorylation in darkened leaves of *M. crystallinum* (Bakrim et al., 2001). Based on this evidence, we hypothesize that the connection between the circadian oscillator and PEPcK gene expression is via the same cascade elements as those characterized in C_4 plants e.g., PI-PLC, Ins(1,4,5)P_3, calcium and a yet unknown Ca^{2+}-dependent protein kinase. A corollary of this hypothesis is that PEPcK gene expression in CAM plants would be the result of two opposing mechanisms involving the transduction cascade (positive) and L-malate (negative). It is of particular importance to understand how the CAM PEPc cascade is triggered in the dark. The possible involvement of an early change in MC cytosolic pH has been checked in illuminated *K. fedtschenkoi* leaf disks. No positive effect of a weak base (NH_4Cl) on the in situ phosphorylation of PEPc was detected (Paterson and Nimmo, 2000). However, it is possible that the circadian oscillator (which is not operative in the light) is needed for the pH response to be observed. In the darkened CAM leaf MC, cytosolic alkalization could be due to the tonoplast H^+/ATPase, which pumps protons into the vacuole, thus allowing malate influx. This would be at variance with observations

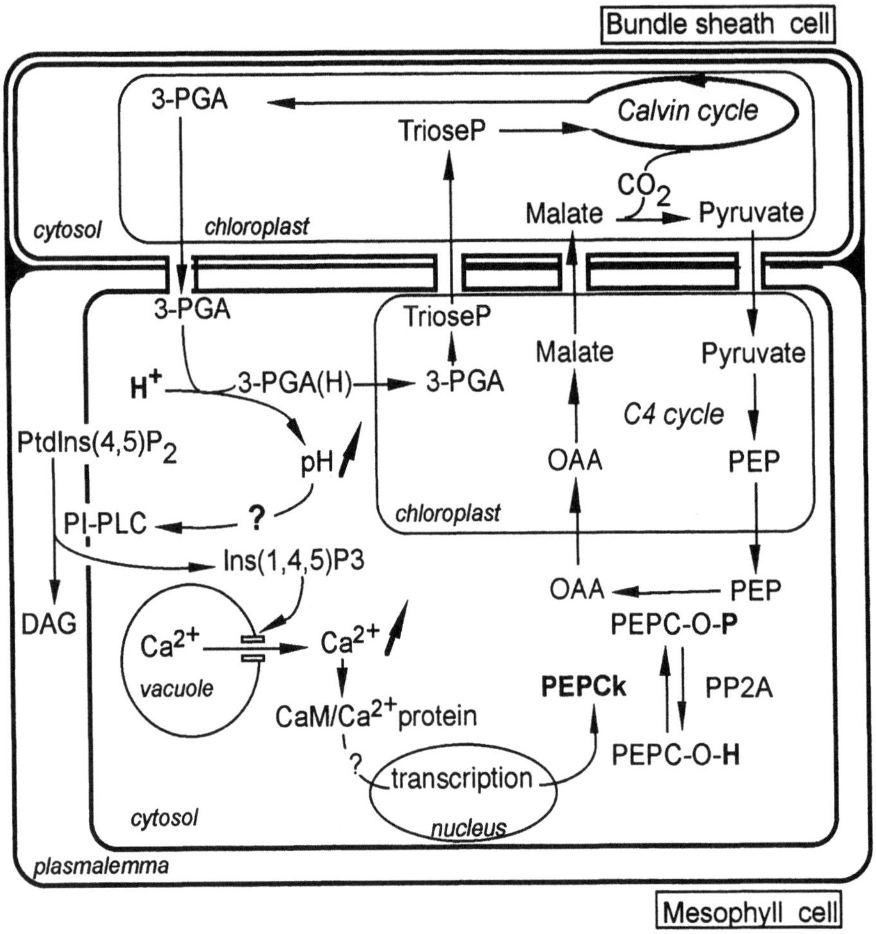

Fig. 2. Model of the light-transduction pathway leading to the phosphorylation of C_4 PEPC in an illuminated C_4 leaf. For details, see text. CaM: calmodulin; DAG: 1,2-diacylglycerol; Ins(1,4,5)P_3: inositol-1,4,5-trisphosphate; OAA: oxaloacetate; PEP: phosphoenolpyruvate; PEPC-O-H: nonphosphorylated phosphoenolpyruvate carboxylase; PEPC-O-P: phosphorylated phosphoenolpyruvate carboxylase; PEPCk: phosphoenolpyruvate carboxylase kinase; 3-PGA: 3-phosphoglyceric acid; 3-PGA(H): 3-phosphoglyceric acid(protonated); PI-PLC: phosphoinositide-specific phospholipase C; PP2A: type 2A protein phosphatase; PtdIns(4,5)P_2: phosphatidylinositol-4,5-bisphosphate.

in C_4 leaves, where PGA^{2-} entry into MC chloroplasts has been implicated. Therefore, the unifying characteristic between the two types of plants would be that the signaling cascade is triggered by cytosolic pH changes following activation of metabolite transport from the cytosol.

V. Significance of Regulatory Phosphorylation of the Photosynthetic Isoform

It has been shown that the uptake of CHX by an excised *Sorghum* or maize leaf performing steady state photosynthesis, caused a progressive decrease in C_4 PEPc phosphorylation state, that correlated with the reduction in the CO_2 assimilation rate (Bakrim et al., 1993). In a similar manner, treatment of detached CAM plant leaves with puromycin or CHX blocked the nocturnal rise in PEPcK activity, maintained PEPc in the dephosphorylated state and blocked periodic fixation of internal CO_2 by PEPc (Carter et al., 1991). Clearly, PEPc phosphorylation has a crucial regulatory role in the overall functioning of C_4 and CAM photosynthesis.

In an illuminated C_4 leaf at high irradiance, C_4 PEPc is faced with the millimolar concentrations of L-malate required for malate diffusion to the neighboring BSC. These levels of L-malate are sufficiently high (10 to 20 mM, as deduced from theoretical calculations) to severely impair the catalytic activity of the dephosphorylated enzyme

(K_i for L-malate being about 0.2 mM in C_4 PEPc). Reconstitution assays were performed in the presence of L-malate and positive effectors, at pH values around 7.3, in order to simulate the physiological conditions likely to prevail in the mesophyll cytosol in the light. It was observed that the phosphorylated form of C_4 PEPc had a markedly higher activity and a reduced sensitivity to L-malate when compared to the non-phosphorylated form. Interestingly, the sensitivity to L-malate was decreased further in the presence of positive effectors (Echevarria et al., 1994; Gao and Woo, 1996; Bakrim et al., 1998; Tovar-Mendez et al., 2000). Typical values for the *Sorghum* enzyme are depicted in Table 1. Neither the concentration of L-malate nor the MC cytosolic pH in CAM plants are known with precision. Since the CAM enzyme displays similar characteristics to the C_4 PEPc (feedback inhibition by L-malate, phosphorylation during the CO_2 fixation phase when L-malate is synthesized, antagonism of L-malate by positive effectors), we can probably assume a similar pattern of regulation for PEPc in the two types of plant. Therefore, the regulatory role of C_4 and CAM PEPc phosphorylation appears to be to attenuate the inhibitory effect of L-malate on the enzyme by modulating its affinity for the opposing metabolite effectors. This would enable the photosynthetic PEPc to continue to fix C when L-malate concentrations are high in the MC cytosol.

While phosphorylation of C_4 PEPc has an impact on the sensitivity of this enzyme to metabolite effectors, these metabolites in turn control the phosphorylation state of C_4 PEPc by modulating the catalytic activity of PEPcK (Wang and Chollet, 1993; Echevarria et al., 1994). Since the steady state phosphorylation of PEPc is dynamic, reflecting the balance between the activities of the PEPcK and the PEPc phosphatase, any imbalance in the ratio of positive/negative effectors would result in a corresponding change in PEPcK activity and, thus, C_4 PEPc phosphorylation state. Therefore, in contrast to the relatively slow upregulation (about one hour) of the PEPcK elicited by the light-dependent cascade, changing metabolite levels would be expected to rapidly modulate the phosphorylation state and catalytic properties of C_4 PEPc. Such a mechanism would allow the enzyme to respond to abrupt fluctuations in the light environment. In such a context, phosphorylation appears not only to protect C_4 PEPc against L-malate but also to help adjust PEPc catalytic activity according to the demand of the RPP pathway for a C_4 acid-derived supply of CO_2. This complex regulatory mechanism provides flexibility for adjusting C flow in C_4 plants to altered environmental conditions and ensures coordination of the two physically segregated metabolic cycles involved in C_4 photosynthesis. A similar reasoning might apply to CAM PEPcK, which has also been shown to be modulated by L-malate (Li and Chollet, 1994). Whether this is operative in vivo in the CAM plant requires further study.

VI. Regulatory Phosphorylation of the C_3 Form: Importance in Anaplerosis

In the C_3 plant leaf, PEPc is not directly involved in photosynthesis, but fulfils a variety of physiological roles. In the anaplerotic pathway, which must also occur in C_4 plants, it contributes to the replenishment of tricarboxylic acid (TCA) cycle intermediates when organic acids are directed towards other metabolic processes such as amino acid and protein synthesis (Huppe and Turpin, 1994). During amino acid synthesis, organic acids are used for NH_4^+ assimilation through glutamine synthetase (GS) and glutamate

Table 1. Kinetic analysis of purified non-phosphorylated (PEPc-OH) and phosphorylated (PEPc-OP) C_4 phosphoenolpyruvate carboxylase (PEPc). The purified, recombinant enzyme was fully phosphorylated in vitro by protein kinase A. The I_{50} for L-malate was determined at 2.5 mM phosphoenolpyruvate, 1 μg PEPc and pH 7.3. Glycine (Gly), glucose-6-phosphate (G6P) and triose phosphate (TP) were added in the assay medium to a concentration of 4, 5 and 15 mM, respectively.

Enzyme	$V(\times 10^{-3})$ (μmol/min)	I_{50} (L-malate, mM)				
		Control	Gly	G6P	TP	Σ
PEPC-OH	13.2	0.47	3.2	2.3	2	6
PEPC-OP	24	1.4	12.3	11	8.8	16

synthase (GOGAT) in the GS/GOGAT cycle (Gálvez et al., 1999). When both the nitrate assimilatory pathway and the production of photorespiratory ammonium are activated in the light, the C flux through PEPc is increased to provide OAA and/or malate to the mitochondria (Champigny and Foyer, 1992). In this respect, PEPc can be considered as a branch of the glycolytic pathway. As nitrate reduction consumes protons, C_3 PEPc activity also leads to an increase in organic acid content that reduces alkalization and thus contributes to cytosolic pH homeostasis. Furthermore, it has been proposed that C_3 PEPc may supply OAA to be used by the chloroplastic/mitochondrial OAA/malate shuttle to provide the cytosol with the reducing power required by nitrate reductase (NR) (Oaks, 1994). Therefore, PEPc displays an intimate relationship with nitrate reduction and ammonia assimilation. As N assimilation proceeds, primary metabolism is reset so that more C is diverted to respiratory metabolism by means of a complex coordinated regulation of many enzymes and transporters, including signaling networks and metabolites. In C_4 and CAM photosynthesis, PEPc phosphorylation has been demonstrated to profoundly influence the metabolic regulation of the enzyme and to be essential for the functioning of the pathway. This aspect of PEPc regulation will now be considered in the case of the C_3 plant PEPc and whether it plays a crucial role in the coordination of C/N metabolism will be discussed.

Intuitively, the concept that PEPc must be protected against malate, as proposed in C_4 and CAM plant photosynthesis, should apply to any system in which the production of this metabolite increases, as occurs in anaplerotic C flow. Indeed, regulatory phosphorylation of C_3 PEPc is supported by a number of observations. 1) The presence of the N-terminal phosphorylation domain in all plant PEPc sequences obtained so far, whatever the physiological type. 2) The presence of a Ca^{2+}-independent PEPc-kinase in leaves of C_3 plants (Vidal and Chollet, 1997) and the isolation of C_3-plant PEPcK cDNAs and genes. 3) The induction of a PEPcK activity in illuminated C_3 leaves and protoplasts that is blocked by CHX in a similar manner to that of the C_4 and CAM enzyme, suggesting that protein turnover is involved in the upregulation of this protein-Ser/Thr kinase in C_3 plants (Vidal and Chollet, 1997). Collectively, these data support the hypothesis that upregulation of a PEPcK controlling C_3-type PEPc occurs via a transduction cascade similar to that which operates for the C_4 and CAM photosynthetic enzymes. However, whether the upstream signaling elements identified in C_4 mesophyll cells are also key players remains poorly documented.

Recent experiments using barley leaf protoplasts have suggested that while PEPc phosphorylation occurs in situ in the light (Krömer et al., 1996a; Smith et al., 1996) and is modulated by protein synthesis and calcium (Smith et al., 1996), the mechanism leading to upregulation of the corresponding PEPcK might differ from that found in C_4 mesophyll protoplasts (Smith et al., 1996). In this respect, the following points merit further discussion. First, both an increase (weak base loading) and a decrease (weak acid loading) in cytosolic pH led to enhanced PEPc phosphorylation in situ (Lillo et al., 1996). Therefore, unlike C_4 protoplasts, it is not clear whether alkalization of the cytosol is a step of the C_3 cascade. However, light-dependent cytosolic alkalization in mesophyll cells from a variety of C_3 plants, including barley, has been reported by Yin et al. (1990). Because the vacuolar pH was concomitantly decreased, it was suggested that this reflected the activation of a tonoplast H^+-ATPase, at variance with the C_4 protoplast in which this mechanism has been attributed to PGA (Giglioli-Guivarc'h et al., 1996). Whatever the mechanism involved, a light-dependent increase in C_3 leaf cytosolic pH is well-established. Second, calcium was not found to play a role in the PEPc signaling system when investigated using a barley MC protoplast system (Smith et al., 1996). Indeed, when these protoplasts were depleted of calcium by means of the specific ionophore A23187 and EGTA, the sensitivity of PEPc to malate declined in the light thus indicating that PEPc phosphorylation was not abolished by the treatment. Third, the results indicated that two different PEPcK (a light-induced form and a constitutive form) could reside in barley MC protoplasts (Smith et al., 1996). The constitutive PEPcK could phosphorylate PEPc at another site and thus disguise the appearance of the inducible one. Additional experiments using a variety of C_3 plant systems are needed to confirm this observation and to provide a more precise description of the C_3 PEPc signaling circuit.

The transduction cascade involved in the reversible phosphorylation of C_3 PEPc must respond to various signals. There are experimental data to suggest that the rate of C flux through the anaplerotic PEPc is modulated by nitrate and/or amino acids via a change in PEPc phosphorylation status (Champigny and

Foyer, 1992). For instance, detailed studies have shown that C_3 PEPc undergoes a marked decrease in L-malate sensitivity (reflecting a higher phosphorylation status of the enzyme) in both N-sufficient wheat and tobacco leaves in the light, as well as in plants resupplied with N after deficiency (Duff and Chollet, 1995; Li et al., 1996). Furthermore, a Ca^{2+}-independent PEPcK has been shown to be present and reversibly light-activated in C_3 leaves. In wheat leaves, similar studies based on in vivo ^{32}P-labeling have shown that high nitrate nutrition increased the C_3 PEPc phosphorylation state and catalytic activity to a level above that induced by light alone (Van Quy and Champigny, 1992). These changes were inhibited by feeding mannose to the excised leaf, thereby decreasing PEPcK activity via a presumed reduction in ATP content (Van Quy and Champigny, 1992; Foyer et al., 1996). Furthermore, in reconstituted phosphorylation assays, the measurable PEPcK activity was found to be several-fold higher in the light than in the dark and this was further increased in N-sufficient plant extracts compared to N-deficient ones (Foyer et al., 1996). All these data, therefore, indicate that the C_3 leaf PEPcK content and activity increase in the light and that leaf N status can influence the regulatory phosphorylation of C_3 PEPc.

What are the N-linked metabolites that could control PEPcK activity and/or PEPc phosphorylation status in C_3 plants? One could be Gln, as in N-deficient maize re-supplied with N, where it has been shown to upregulate the C_4 PEPc transcript level (Suzuki et al., 1994). Indeed, wheat leaf PEPc and PEPcK activity have been found to be activated by Gln and inhibited by Glu in in vitro assays and reconstituted phosphorylation assays, respectively (Foyer et al., 1996). In contrast, Duff and Chollet (1995) could not detect any effect of either of these metabolites or nitrate on wheat PEPcK activity. However, upregulation of tobacco PEPcK in the light was markedly inhibited by the GS inhibitors, methionine sulfoximine and phosphinothricin, under both nonphotorespiratory and photorespiratory conditions, and this effect was specifically and significantly antagonized by feeding Gln to the excised leaf (Li et al., 1996). Since such compounds had no detectable effects on the light-activation of the maize PEPcK, the authors concluded that a disruption of leaf N metabolism did not have the same impact on the regulatory phosphorylation of PEPc in illuminated C_3 and C_4 leaves. Therefore, it was proposed that C_3 and C_4 PEPcKs might be regulated by similar but not identical light-signal transduction pathways. As in C_4 plants, DCMU was found to inhibit PEPcK upregulation while Gln was unable to overcome this effect. Furthermore, Gln could not replace light in promoting PEPcK activity in vivo. Gln appears not to be a C_3 cascade component but rather acts in C/N signaling by modulating the light effect on the expression of PEPcK. Indeed, Gln has been implicated previously as a positive and negative modulator in the control of gene expression of leaf PEPc and NR, respectively (Vincentz et al., 1993; Suzuki et al., 1994). Further research is needed to elucidate a precise role of Gln in the regulation of C_3 PEPcK.

Another signal regulating C_3 PEPc activity in its anaplerotic function could be nitrate (Stitt, 1999). This has been investigated using NR-deficient tobacco mutants that, as a consequence of their very low NR activity, accumulate large amounts of nitrate in their leaves when grown on 12 mM nitrate. Compared to wild-type tobacco, the NR-deficient mutant showed increased transcripts encoding NR, nitrite reductase, 'cytosolic NADP-dependent' isocitrate dehydrogenase, cytosolic pyruvate kinase and PEPc, and a dramatic accumulation of organic acids including L-malate. Interestingly, the sensitivity of PEPc to L-malate was found to be significantly reduced following nitrate addition. This emphasizes that N-mediated regulation of phosphorylation is an important aspect of C_3 PEPc control. Indeed, such changes could reflect a nitrate-dependent upregulation of PEPcK activity. However, it should be noted that the NR-deficient mutants were also depleted in Gln, so the exact importance of nitrate and Gln control in PEPcK/PEPc regulation remains unclear. Thus, although it appears that PEPcK synthesis can be altered by nitrate and/or N-metabolites, the underlying mechanism(s) controlling this effect remain(s) unknown.

It is tempting to speculate that the coordinated regulation of physiologically related genes involved in the C/N interaction (e.g., NR, PEPc, PEPcK) could be orchestrated by relatively few key metabolites. Regulatory systems that monitor cell metabolite status and control gene expression and enzyme activities are well characterized in bacteria and fungi. Two examples are the PII protein, which senses 2-oxoglutarate, and hexokinase, which senses sugars. However, in plants, concepts of control through such components are still emerging (Hsieh et al., 1998; Sheen et al., 1999; Stitt, 1999). For instance, to account for root-to-leaf N-signaling, a model involving cytokinins and a His-Asp phosphorelay similar to that found in bacteria has been proposed (Sakakibara et al., 2000). Whether this system is somehow connected to the modulation of PEPc activity via the PEPcK transduction cascade and/or involved in C/N signaling pathways that regulate C/N interactions remains to be elucidated.

VII. Conclusions and Perspectives

The intense research performed during the last decade has led to the view that PEPc phosphorylation is a common regulatory mechanism in all plant types. The best studied system is C_4 PEPc phosphorylation, a cardinal regulatory event in C_4 photosynthesis, in which the proposed transduction chain that links the light stimulus to upregulation of PEPcK involves several classical second messengers as found earlier in animal cells. These include cytosolic pH, PI-PLC, Ins(1,4,5)P_3 and calcium. However, several crucial questions remain unanswered. One of these concerns the molecular characterization of the mesophyll δ-type PI-PLC and the mechanism by which it undergoes a transient activation following the increase in cytosolic pH. Another poorly understood step in the C_4 cascade is the component directly involved in PEPcK gene upregulation in the MC nucleus. The recent cloning of CAM and C_3 plant PEPcK genes will allow us to investigate, in more detail, how these enzymes are transcriptionally controlled either by light and/or a variety of metabolic signals. Structural and functional studies of the PEPcK gene promoter will lead to the identification of *cis*-acting elements and interacting protein factors that are presumably modulated by phosphorylation. This analysis will facilitate evaluation of the role of key metabolites in the regulation of gene expression. Although much is known about the C_4 and CAM PEPc signaling cascades, the equivalent components of the C_3 PEPc system are still to be identified. This can now be undertaken in the model plant *A. thaliana*, where molecular genetics and cellular/pharmacological approaches can be developed to elucidate the cascade. Finally, unraveling the organization of the complex regulatory network involved in the posttranslational regulation of C/N enzymes (e.g., PEPc, NR, sucrose phosphate synthase) is a fascinating challenge. It remains to be discovered whether and how the different pathways are connected in the integration of various environmental and internal signals, including metabolites. The role of these signals in the network of controls that modulate the protein kinases and phosphatases that act on key enzymes to coordinate C/N metabolism awaits discovery.

References

Andreo CS, Gonzalez DH and Iglesias AA (1987) Higher plant phosphoenolpyruvate carboxylase: Structure and regulation. FEBS Lett 213: 1–8

Bakrim N, Echevarria C, Crétin C, Arrio-Dupont M, Pierre JN, Vidal J, Chollet R and Gadal P (1992) Regulatory phosphorylation of *Sorghum* leaf phosphoenolpyruvate carboxylase: Identification of the protein-serine kinase and some elements of the signal-transduction cascade. Eur J Biochem 204: 821–830

Bakrim N, Prioul JL, Deleens E, Rocher JP, Arrio-Dupont M, Vidal J, Gadal P and Chollet R (1993) Regulatory phosphorylation of C_4 phosphoenolpyruvate carboxylase: A cardinal event influencing the photosynthesis rate in *Sorghum* and maize. Plant Physiol 101: 891–897

Bakrim N, Nhiri M, Pierre JN and Vidal J (1998) Metabolite control of *Sorghum* C_4 phosphoenolpyruvate carboxylase catalytic activity and phosphorylation state. Photosynth Res 58: 153–162

Bakrim N, Brulfert J, Vidal J and Chollet R (2001) Phosphoenolpyruvate carboxylase kinase is controlled by a similar signaling cascade in CAM and C_4 plants. Biochem Biophys Res Commun 286: 1158–1162

Bandurski RS and Greiner CM (1953) The enzymatic synthesis of oxalacetate from phosphenolpyruvate and carbon dioxide. J Biol Chem 204: 781–786

Borland AM, Hartwell J, Jenkins GI, Wilkins MB and Nimmo HG (1999) Metabolite control overrides circadian regulation of phosphoenolpyruvate carboxylase kinase and CO_2 fixation in crassulacean acid metabolism. Plant Physiol 121: 889–896

Brosnan JM and Sanders D (1993) Identification and characterisation of high-affinity binding sites for inositol trisphosphate in red beet. Plant Cell 5: 931–940

Brulfert J, Vidal J, LeMaréchal P, Gadal P, Queiroz O, Kluge M and Kruger I (1986) Phosphorylation-dephosphorylation process as a probable mechanism for the diurnal regulatory changes of phosphoenolpyruvate carboxylase in CAM plants. Biochem Biophys Res Commun 136: 151–159

Budde RJA and Chollet R (1986) In vitro phosphorylation of maize leaf phosphoenolpyruvate carboxylase. Plant Physiol 82: 1107–1114

Carter PJ, Nimmo HG, Fewson CA and Wilkins MB (1990) *Bryophyllum fedtschenkoi* protein phosphatase type 2A can dephosphorylate phosphoenolpyruvate carboxylase. FEBS Lett 263: 233–236

Carter PJ, Nimmo HG, Fewson CA and Wilkins MB (1991) Circadian rhythm in the activity of a plant protein kinase. EMBO J 10: 2063–2068

Champigny M-L and Foyer CH (1992) Nitrate activation of cytosolic protein kinases diverts photosynthetic carbon from sucrose to amino acid biosynthesis. Basis for a new concept.

Plant Physiol 100: 7–12
Chollet R, Vidal J and O'Leary MH (1996) Phosphoenolpyruvate carboxylase: A ubiquitous, highly regulated enzyme in plants. Annu Rev Plant Physiol Plant Mol Biol 47: 273–298
Coté CG and Crain RC (1993) Biochemistry of phosphoinositides. Annu Rev Plant Physiol Plant Mol Biol 44: 333–356
Coursol S, Giglioli-Guivarc'h N, Vidal J and Pierre J-N (2000) An increase in phosphoinositide-specific phospholipase C activity precedes induction of C_4 phosphoenolpyruvate carboxylase phosphorylation in illuminated and NH_4Cl-treated protoplasts from *Digitaria sanguinalis*. Plant J 23: 497–506
Doncaster HD and Leegood RC (1987) Regulation of phosphoenolpyruvate carboxylase activity in maize leaves. Plant Physiol 84: 82–87
Drøbak BK (1992) The plant phosphoinositide system. Biochem J 288: 697–712
Duff SMG and Chollet R (1995) In vivo regulation of wheat-leaf phosphoenolpyruvate carboxylase by reversible phosphorylation. Plant Physiol 107: 775–782
Duff SMG, Andreo CS, Pacquit V, Lepiniec L, Sarath G, Condon SA, Vidal J, Gadal P and Chollet R (1995) Kinetic analysis of the non-phosphorylated, in vitro phosphorylated, and phosphorylation-site-mutant (Asp8) forms of intact recombinant C_4 phosphoenolpyruvate carboxylase from *Sorghum*. Eur J Biochem 228: 92–95
Echevarria C, Pacquit V, Bakrim N, Osuna L, Delgado B, Arrio-Dupont M and Vidal J (1994) The effect of pH on the covalent and metabolic control of C_4 phosphoenolpyruvate carboxylase from *Sorghum* leaf. Arch Biochem Biophys 315: 425–430
Fontaine V, Hartwell J, Jenkins GI and Nimmo HG (2000) Expression of two PEP carboxylase kinase genes in *Arabidopsis thaliana*. In: Current Topics in Plant Biochemistry, Physiology and Molecular Biology, Eighteenth Annual Symposium. Plant Protein Phosphorylation-Dephosphorylation, p 128. University of Missouri.
Foyer CH, Champigny M-L, Valadier M-H and Ferrario S (1996) Partitioning of photosynthetic carbon: The role of nitrate activation of protein kinases. In: Shewry PR, Halford NG and Hooley R (eds) Proceedings of the Phytochemical Society of Europe: Protein Phosphorylation in Plants, pp 35–51. Oxford Science Publications, Oxford
Gálvez S, Lancien M and Hodges M (1999) Are isocitrate dehydrogenases and 2-oxoglutarate involved in the regulation of glutamate synthesis? Trends Plant Sci 4: 465–507
Gao Y and Woo KC (1996) Regulation of phosphoenolpyruvate carboxylase in *Zea mays* by protein phosphorylation and metabolites and their roles in photosynthesis. Aust J Plant Physiol 23: 25–32
Giglioli-Guivarc'h N, Pierre JN, Brown S, Chollet R, Vidal J and Gadal P (1996) The light-dependent transduction pathway controlling the regulatory phosphorylation of C_4 phosphoenolpyruvate carboxylase in protoplasts from *Digitaria sanguinalis*. Plant Cell 8: 573–586
Guex N and Peitsch MC (1997) SWISS-MODEL and the Swiss-PdbViewer: An environment for comparative protein modelling. Electrophoresis 18: 2714–2723
Hartwell J, Smith L, Wilkins MB, Jenkins GI and Nimmo HG (1996) Higher plant phosphoenolpyruvate carboxylase kinase is regulated at the level of translatable mRNA in response to light or a circadian rhythm. Plant J 10: 1071–1078
Hartwell J, Gill A, Nimmo GA, Wilkins MB, Jenkins GI and Nimmo HG (1999) Phosphoenolpyruvate carboxylase kinase is a novel protein kinase regulated at the level of expression. Plant J 20: 333–342
Hatch MD (1977) C_4 pathway of photosynthesis: Mechanism and physiological function. Trends Biochem Sci 2: 199–201
Hsieh M-H, Lam H-M, Van de Loo FJ and Coruzzi G (1998) A PII-like protein in Arabidopsis: Putative role in nitrogen sensing. Proc Natl Acad Sci USA 95: 13965–13970
Huppe HC and Turpin DH (1994) Integration of carbon and nitrogen metabolism in plant and algal cells. Annu Rev Plant Physiol Plant Mol Biol 45: 577–607
Jiao JA and Chollet R (1989) Regulatory seryl-phosphorylation of C_4 phosphoenolpyruvate carboxylase by a soluble protein kinase from maize leaves. Arch Biochem Biophys 269: 526–535
Kai Y, Matsumura H, Inoue T, Terada K, Nagara Y, Yoshinaga T, Kihara A, Tsumura K and Izui K (1999) Three-dimensional structure of phosphoenolpyruvate carboxylase: A proposed mechanism for allosteric inhibition. Proc Natl Acad Sci USA 96: 823–828
Krömer S, Gardeström P and Samuelsson G (1996) Regulation of the supply of oxaloacetate for mitochondrial metabolism via phosphoenolpyruvate carboxylase in barley leaf protoplasts. I. The effect of covalent modification on PEP carboxylase activity, pH response, and kinetic properties. Biochim Biophys Acta 1289: 343–350
Lepiniec L, Vidal J, Chollet R, Gadal P and Crétin C (1994) Phosphoenolpyruvate carboxylase: Structure, regulation and evolution. Plant Sci 99: 111–124
Li B and Chollet R (1993) Resolution and identification of C_4 phosphoenolpyruvate-carboxylase protein-kinase polypeptides and their reversible light activation in maize leaves. Arch Biochem Biophys 307: 416–419
Li B and Chollet R (1994) Salt induction and the partial purification/characterization of phosphoenolpyruvate carboxylase protein-serine kinase from an inducible Crassulacean-acid-metabolism (CAM) plant, *Mesembryanthemum crystallinum* L. Arch Biochem Biophys 314: 247–254
Li B, Zhang X Q and Chollet R (1996) Phosphoenolpyruvate carboxylase kinase in tobacco leaves is activated by light in a similar but not identical way as in maize. Plant Physiol 111: 497–505
Li B, Pacquit V, Jiao J, Duff SMG, Maralihalli GB, Sarath G, Condon SA, Vidal J and Chollet R (1997) Structural requirements for phosphorylation of C_4-leaf phosphoenolpyruvate carboxylase by its highly regulated protein-serine kinase. A comparative study with synthetic-peptide substrates and native, mutant target proteins. Aust J Plant Physiol 24: 443–449
Lillo C, Smith LH, Nimmo HG and Wilkins MB (1996) Regulation of nitrate reductase and phosphoenolpyruvate carboxylase activities in barley leaf protoplasts. Planta 200: 181–185
McNaughton GAL, Fewson CA, Wilkins SMB and Nimmo HG (1989) Purification, oligomerisation state and malate sensitivity of maize leaf phosphoenolpyruvate carboxylase. Biochem J 261: 349–355
Muir SR and Sanders D (1997) Inositol 1,4,5-trisphosphate-sensitive Ca^{2+} release across non-vacuolar membranes in cauliflower. Plant Physiol 114: 1511–1521
Munnik T, Irvine RF and Musgrave A (1998) Phospholipid signalling in plants. Biochim Biophys Acta 1389: 222–272

Nhiri M, Bakrim N, Pacquit V, El Hachimi-Messouak Z, Osuna L and Vidal J (1998) Calcium-dependent and -independent phosphoenolpyruvate carboxylase kinases in *Sorghum* leaves: Further evidence for the involvement of the calcium-independent protein kinase in the in situ regulatory phosphorylation of C_4 phosphoenolpyruvate carboxylase. Plant Cell Physiol 39: 241–246

Nimmo GA, Nimmo HG, Fewson CA and Wilkins MB (1984) Diurnal changes in the properties of phosphoenolpyruvate carboxylase in *Bryophyllum* leaves: A possible covalent modification. FEBS Lett 178: 199–203

Nimmo HG (2000) The regulation of phosphoenolpyruvate carboxylase in CAM plants. Trends Plant Sci 5: 75–80

Oaks A (1994) Efficiency of nitrogen utilization in C_3 and C_4 cereals. Plant Physiol 106: 407–414

Ogawa N, Yabuta N, Ueno Y and Izui K (1998) Characterization of a maize Ca^{2+}-dependent protein kinase phosphorylating phosphoenolpyruvate carboxylase. Plant Cell Physiol 39: 1010–1019

O'Leary MH (1982) Phosphoenolpyruvate carboxylase: An enzymologist's view. Annu Rev Plant Physiol 33: 297–315

Paterson KM and Nimmo HG (2000) Effects of pH on the induction of phosphoenolpyruvate carboxylase kinase in *Kalanchoe fedtschenkoi*. Plant Sci 154 : 135–141

Pierre JN, Pacquit V, Vidal J and Gadal P (1992) Regulatory phosphorylation of phosphoenolpyruvate carboxylase in protoplasts from *Sorghum* mesophyll cell and the role of pH and Ca^{2+} as possible components of the light transduction pathway. Eur J Biochem 210: 531–537

Raghavendra AS, Yin Z and Heber U (1993) Light-dependent pH changes in leaves of C_4 plants. Planta 189: 278–287

Ranjeva R, Carrasco A and Boudet AM (1988) Inositol trisphosphate stimulates the release of calcium from intact vacuoles isolated from *Acer* cells. FEBS Lett 230: 137–141

Sakakibara H, Taniguchi M and Sugiyama T (2000) His-Asp phosphorelay signaling: A communication avenue between plants and their environment. Plant Mol Biol 42: 273–278

Schumaker K and Sze H (1987) Inositol 1,4,5-trisphosphate releases Ca^{2+} from vacuolar membrane vesicles of oat roots. J Biol Chem 262: 3944–3946

Sheen J, Zhou L and Jang J-C (1999) Sugars as signaling molecules. Curr Opin Plant Biol 2: 410–418

Smith LH, Lillo C, Nimmo HG and Wilkins MB (1996) Light regulation of phosphoenolpyruvate carboxylase in barley mesophyll protoplasts is modulated by protein synthesis and calcium, and not necessarily correlated with phosphoenolpyruvate carboxylase kinase activity. Planta 200: 174–180

Smith LH, Langdale JA and Chollet R (1998) A functional Calvin cycle is not indispensable for the light activation of C_4 phosphoenolpyruvate carboxylase kinase and its target enzyme in the maize mutant *bundle sheath defective 2-mutable 1*. Plant Physiol 118: 191–197

Stitt M (1999) Nitrate regulation of metabolism and growth. Curr Opin Plant Biol 2: 178–186

Suzuki I, Crétin C, Omata T and Sugiyama T (1994) Transcriptional and posttranscriptional regulation of nitrogen-responding expression of phosphoenolpyruvate carboxylase gene in maize. Plant Physiol 105: 1223–1229

Taybi T, Patil S, Chollet R and Cushman JC (2000) A minimal serine/threonine protein kinase circadianly regulates phosphoenolpyruvate carboxylase activity in crassulacean acid metabolism-induced leaves of the common ice plant. Plant Physiol 123: 1471–1481

Terada K, Kai T, Okuno S, Fujisawa H and Izui K (1990) Maize leaf phosphoenolpyruvate carboxylase: Phosphorylation of Ser15 with a mammalian cyclic AMP-dependent protein kinase diminishes sensitivity to inhibition by L-malate. FEBS Lett 259: 241–244

Tovar-Méndez A, Mujica-Jiménez C and Munoz-Clares R (2000) Physiological implications of the kinetics of maize leaf phosphoenolpyruvate carboxylase. Plant Physiol 123: 149–160

Van der Veen R, Heimovaara-Dijkstra S and Wang M (1992) Cytosolic alkalinization mediated by abscisic acid is necessary, but not sufficient, for abscisic acid-induced gene expression in barley aleurone protoplasts. Plant Physiol 100: 699–705

Van Quy L and Champigny M-L (1992) NO_3^- enhances the kinase activity for phosphorylation of phosphoenolpyruvate carboxylase and sucrose phosphate synthase proteins in wheat leaves. Plant Physiol 99: 344–347

Vidal J and Chollet R (1997) Regulatory phosphorylation of C_4 PEP carboxylase. Trends Plant Sci 2: 230–237

Vincentz M, Moureaux T, Leydecker M-T, Vaucheret H and Caboche M (1993) Regulation of nitrate and nitrite reductase expression in *Nicotiana plumbaginifolia* leaves by nitrogen and carbon metabolites. Plant J 3: 1027–1035

Wang YH and Chollet R (1993) Partial purification and characterization of phosphoenolpyruvate carboxylase protein-serine kinase from illuminated maize leaves. Arch Biochem Biophys 304: 496–502

Willeford KO and Wedding RT (1992) Oligomerization and regulation of higher plant phosphoenolpyruvate carboxylase. Plant Physiol 99: 755–758

Yin Z-H, Neimanis S, Wagner U and Heber U (1990) Light-dependent pH changes in leaves of C_3 plants. 1. Recording pH changes in various cellular compartments by fluorescence probes. Planta 182: 244–252

Chapter 10

Mitochondrial Functions in the Light and Significance to Carbon-Nitrogen Interactions

Per Gardeström*, Abir U. Igamberdiev
Umeå Plant Science Centre, Department of Plant Physiology, Umeå University, 901 87 Umeå, Sweden

A. S. Raghavendra
Department of Plant Sciences, School of Life Sciences, University of Hyderabad, Hyderabad 500 046, India

Summary 152
I. Introduction 152
II. Export of Photosynthate from the Chloroplast 153
A. Triose Phosphates 153
B. Reductants and ATP 153
C. Glycolate 154
III. Mitochondrial Products of Photorespiration 154
IV. Products of Glycolysis in the Light 155
V. Operation of the Tricarboxylic Acid Cycle 157
A. Entry of Glycolytic Substrates 157
B. Partial Tricarboxylic Acid Cycle Activity in the Light 158
C. Metabolic Shuttles between Mitochondria and other Compartments 158
VI. Electron Transport and Redox Levels in Plant Mitochondria 160
A. The Plant Mitochondrial Electron Transport Chain 160
B. Photosynthesis and Mitochondrial Electron Transport 160
C. Photorespiration and Mitochondrial Electron Transport 161
D. External NADH and NADPH 162
E. Mitochondrial Electron transport and Production of Active Oxygen Species 162
VII. Participation of Mitochondria in the Regulation of Metabolism during Transitions between Light and Darkness 163
A. The Role of Mitochondria in Photosynthetic Induction 163
B. The Role of Mitochondria during Light-Enhanced Dark Respiration 164
VIII. Mitochondrial Respiration and Photoinhibition 164
IX. The Role of Mitochondria in C_4 Photosynthesis 165
X. Glycolate Metabolism in Algal Mitochondria 165
XI. Concluding Remarks 166
Acknowledgments 166
References 167

*Author for correspondence, email: per.gardestrom@plantphys.umu.se

Christine H. Foyer and Graham Noctor (eds): Photosynthetic Nitrogen Assimilation and Associated Carbon and Respiratory Metabolism, pp. 151–172.

Summary

Nitrogen assimilation involves the cooperation of several subcellular compartments. The mitochondria play key roles in both primary nitrogen assimilation and photorespiratory ammonia recycling. Mitochondrial functions in the light depend on the export of substrates from the chloroplast. One of these substrates is glycolate, which is converted to glycine in the peroxisomes. Oxidation of glycine, which produces ammonia and generates NADH, is the main activity of leaf mitochondria of C_3 plants in the light. The products of photorespiratory glycine oxidation will have a pronounced influence on other mitochondrial activities. Chloroplasts also export triose phosphates, which in the cytosol are mainly utilized for sucrose synthesis. However, a portion of the triose phosphate is converted via glycolysis to substrates such as pyruvate, malate and oxaloacetate. It is argued that oxaloacetate may be the most important end product of glycolysis in the light. Regardless of the substrate entering mitochondria, citrate will be the first product in the tricarboxylic acid (TCA) cycle. Recent evidence indicates that the oxidation of substrates in the TCA cycle is not complete in the light. Limitations in isocitrate oxidation by increased mitochondrial NAD(P)H/NAD(P) ratios favor citrate export, and therefore deliver carbon skeletons to the rest of the cell for amino acid synthesis. Mitochondrial glycine oxidation can contribute to ATP formation for the cytosol, but other non-coupled pathways of electron transport also operate and may be more important in the light than in darkness. Photorespiratory and respiratory carbon fluxes in the light form a highly flexible system to balance the demands of energy (ATP) and reducing equivalents (NADH, NADPH) in different compartments. Thus, the function of leaf mitochondria in the light is not only to carry out oxidative phosphorylation, but also to redistribute metabolites, and to regulate the pH, redox and energy balances of the photosynthetic cell.

I. Introduction

During the last decade, several reports have established that the chloroplasts and mitochondria in green tissues interact very strongly with each other. Mitochondrial functions in the light depend on export of respiratory substrates from the chloroplast. Biosynthetic processes in the cytosol of photosynthetic tissues, including assimilation and metabolism of nitrogen (N), are highly demanding in terms of energy (ATP), reducing power and C skeletons. The requirements for ATP and NAD(P)H are met by the products exported from both chloroplasts and mitochondria. Therefore, the process of N assimilation is linked closely to chloroplast function as well as to mitochondrial oxidative metabolism (Champigny, 1995; Padmasree and Raghavendra, 1998). The detailed aspects of mitochondrial respiration in the light have been recently reviewed (Azcón-Bieto, 1992; Raghavendra et al., 1994; Gardeström and Lernmark, 1995; Krömer, 1995; Gardeström, 1996; Hoefnagel et al., 1998; Padmasree and Raghavendra, 1998, 2000; Atkin et al., 2000a). The present chapter focuses on the interdependence (and interaction) of photosynthetic metabolism and mitochondrial respiration in the light. We also briefly discuss transitions from darkness to light (photosynthetic induction) and from light to darkness (light-enhanced dark respiration, LEDR). Because of the impact of photorespiration on mitochondrial metabolism in the light, most of the discussion is connected with C_3 plants, but C_4 plants and algae are also briefly mentioned.

Abbreviations: 2-OG – 2-oxoglutarate; AOS – active oxygen species; AOX – alternative oxidase; Asp – aspartate; CoA – coenzyme A; CS – citrate synthase; F2,6BP – fructose-2,6-bisphosphate; F6P – fructose-6-phosphate; FBP – fructose-1,6-bisphosphate; FBPase – fructose-1,6-bisphosphatase; Fd – ferredoxin; G3P – glyceraldehyde-3-phosphate; G3PDH – glyceraldehyde-3-phosphate dehydrogenase; GDC – glycine decarboxylase; GDH – glutamate dehydrogenase; Glu – glutamate; Gly – glycine; GOGAT – glutamate synthase; GS – glutamine synthetase; ICDH – isocitrate dehydrogenase; LEDR – light-enhanced dark respiration; MDH – malate dehydrogenase; ME – malic enzyme; OAA – oxaloacetate; PDC – pyruvate dehydrogenase complex; PEP – phosphoenolpyruvate; PEPc – phosphoenolpyruvate carboxylase; PEPCK – phosphoenolpyruvate carboxykinase; PGA – 3-phosphoglyceric acid; Pi – phosphate; PK – pyruvate kinase; RPP – reductive pentose phosphate (RPP pathway = Calvin cycle); Rubisco – ribulose-1,5-bisphosphate carboxylase/oxygenase; RuBP – ribulose-1,5-bisphosphate; Ser – serine; SHAM – salicylhydroxamic acid; SHMT – serine hydroxymethyl transferase; SOD – superoxide dismutase; TCA – tricarboxylic acid; Td – thioredoxin; TP – triose phosphate

II. Export of Photosynthate from the Chloroplast

A. Triose Phosphates

In the light chloroplasts fix CO_2 in the carboxylation reaction catalyzed by ribulose-1,5-bisphosphate carboxylase/oxygenase (Rubisco) and C in excess of that required for RPP pathway operation is either stored in the chloroplast as starch or exported. Export of C occurs in the form of triose phosphates (TP) through the TP-Pi exchange translocator (Pi translocator; Flügge, 1999). In the cytosol, most of the TP are utilized for sucrose formation, but some will enter glycolysis (Fig. 1) and be converted to compounds that are potential substrates for the tricarboxylic acid (TCA) cycle in mitochondria. These compounds are principally pyruvate, oxaloacetate (OAA) or malate. An important point of regulation of carbon flow between sucrose synthesis and glycolysis is at the level of fructose-1,6-bisphosphate (FBP). Fructose-1,6-bisphosphatase (FBPase), which catalyzes the conversion of FBP to fructose-6-phosphate (F6P), is inhibited by fructose-2,6-bisphosphate (F2,6BP). Pyrophosphate-dependent phosphofructokinase, another enzyme that can catalyze the interconversion of FBP and F6P, is activated by F2,6BP. Previous considerations of the control of sucrose synthesis have incorporated detailed discussion of feed-forward and feed-back regulation by F2,6BP (Stitt, 1990).

B. Reductants and ATP

Chloroplasts possess translocators that can catalyze a highly active malate-OAA exchange (Hatch et al., 1984). The driving force for this exchange is photosynthetically formed NADPH in the chloroplast stroma which is associated with OAA reduction to malate by NADP-malate dehydrogenase (NADP-MDH) (Fridlyand et al., 1998). This enzyme is light-activated via the ferredoxin-thioredoxin (Fd-Td) system and thus its activation state depends on the redox state of the stroma (Miginiac-Maslow et al., 2000). Exchange of malate with OAA can operate between most cellular compartments, linking NADP-

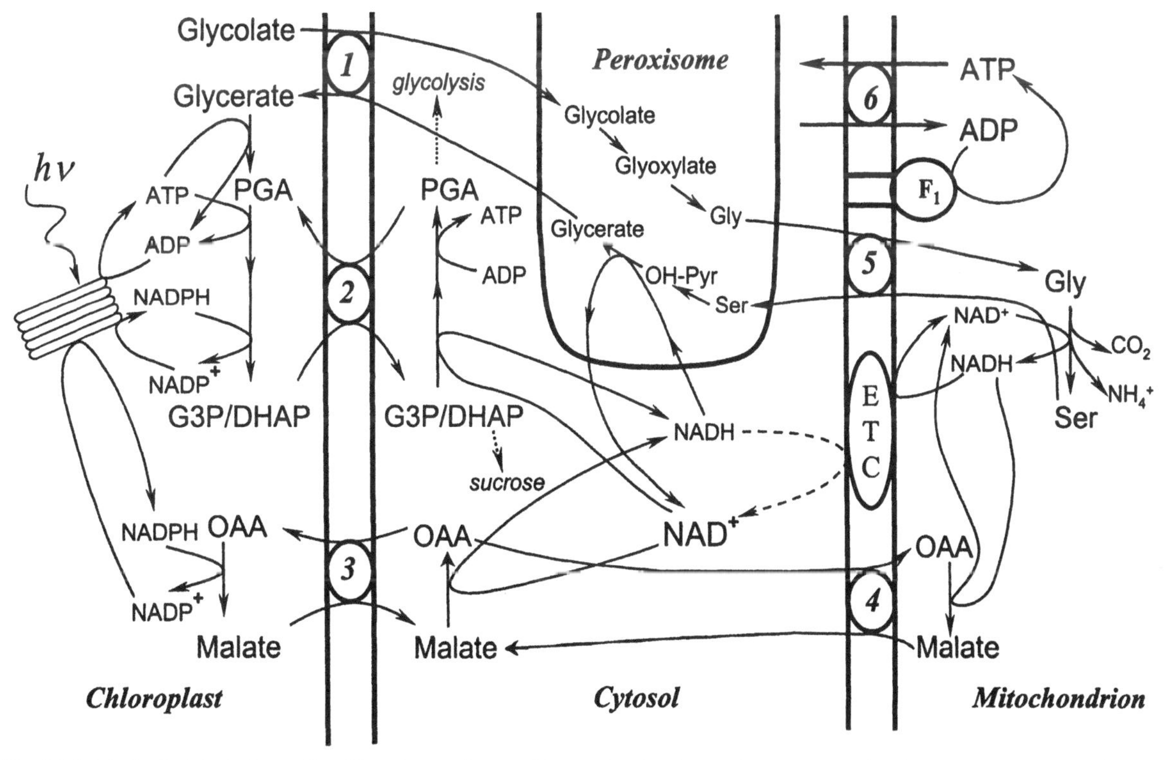

Fig. 1. Metabolic shuttles between the chloroplast, mitochondrion, cytosol and peroxisome. 1, glycolate/glycerate translocator; 2, triose-phosphate/Pi translocator; 3, chloroplast envelope dicarboxylate translocator; 4, mitochondrial OAA translocator; 5, Gly/Ser translocator; 6, mitochondrial adenylate translocator. DHAP, dihydroxyacetone phosphate; ETC, electron transport chain; G3P, glyceraldehyde 3-phosphate; Gly, glycine; OAA, oxaloacetate; OH-Pyr, hydroxypyruvate; PGA, 3-phosphoglyceric acid; Ser, serine.

dependent MDH in chloroplasts with NAD-dependent isoforms of MDH in the cytosol, mitochondria, chloroplasts and peroxisomes (Gietl, 1992). In the light this system, known as the 'malate valve', can function to transport excess reducing equivalents from chloroplasts to other parts of the cell (Krömer and Scheibe, 1996).

The Pi translocator can catalyze an exchange between TP and 3-phosphoglyceric acid (PGA) (Flügge, 1999), which allows the export of reducing power and potentially also ATP from the chloroplast to the cytosol (Fig. 1). Whether ATP is formed depends on whether TP is oxidized via the non-phosphorylating NADP-dependent glyceraldehyde-3-phosphate dehydrogenase (G3PDH), or the phosphorylating NAD-dependent G3PDH. The maximal activity of both enzymes is similar (Krömer, 1995), but their respective contribution is still uncertain. The NADP-dependent enzyme has a very high affinity for its substrates, NADP and G3P. However, it is inhibited by high concentrations of G3P and NADPH (Kelly and Gibbs, 1973; Scagliarini et al., 1990; Casati et al., 2000). This means that the enzyme only operates at low cytosolic G3P concentrations and when the cytosolic NADP(H) pool is not strongly reduced. The NAD-dependent enzyme has a low affinity for its substrates NAD and G3P (Duggleby and Dennis, 1974). Since the NAD(H) pool is present in the cytosol in millimolar concentration and is very oxidized (Wigge et al., 1993; Igamberdiev et al., 2001), the concentration of G3P will be limiting for conversion via the phosphorylating pathway. In photorespiratory conditions, TP concentrations in the cytosol are decreased and the NADP(H) pool is slightly more oxidized (Wigge et al., 1993) thus possibly suppressing operation of the phosphorylating pathway and activating the non-phosphorylating bypass.

Transport of ATP from the chloroplast to the cytosol is unlikely to proceed via the chloroplastic ATP/ADP translocator, which has a low activity and kinetic properties that favor import of ATP into chloroplasts (Noctor and Foyer, 1998). In isolated chloroplasts from C_3 plants, maximum rates of the chloroplast ATP translocator are about ten-fold lower than those of other transporters, such as P_i translocator (Flügge and Heldt, 1991). Chloroplasts can contribute to cytosolic ATP if TP is oxidized via phosphorylating G3PDH. This may have importance in stress conditions, for example, when CO_2 is depleted and thus RPP pathway intermediates are exhausted. This mechanism may also be important where cytosolic ATP cannot be generated from Gly oxidation, as in a barley mutant deficient in glycine decarboxylase (GDC) (Igamberdiev et al., 2001).

C. Glycolate

In C_3 plants the oxygenation reaction of Rubisco will lead to the formation of phosphoglycolate which, after conversion to glycolate, is transported out of the chloroplast (Fig. 1). The further metabolism of glycolate involves reactions in peroxisomes and mitochondria (and to some extent also in the cytosol) and is known as the photorespiratory carbon cycle (Keys and Leegood, this volume). At atmospheric CO_2 concentration, Rubisco will catalyze one oxygenation reaction for every 2–3 carboxylation reactions (Lorimer and Andrews, 1981) and so the flux through the pathway by far exceeds the flux through glycolysis. The photorespiratory carbon cycle ensures that 75% of the carbon in glycolate is recovered as PGA and returned to the RPP pathway. The remaining 25% is lost as CO_2 in the mitochondrial oxidation of Gly. Photorespiratory flux is determined by several factors such as irradiance, temperature and the relative concentrations of CO_2 and O_2 in the chloroplast stroma. Much of the C diverted into glycolate is returned to the chloroplast, with only minor use of intermediates of the photorespiratory cycle for other reactions. This steady flow of C through the mitochondria in the light will have a pronounced effect on other mitochondrial functions. We will therefore now consider the products of photorespiration in mitochondria and then discuss the consequences for other mitochondrial activities in the light.

III. Mitochondrial Products of Photorespiration

In the mitochondria Gly is converted to Ser by the combined action of GDC and Ser hydroxymethyltransferase (SHMT). Subunits of GDC and SHMT are the most abundant proteins in the mitochondrial matrix of C_3 plants. The products of these reactions, in addition to Ser, are CO_2, NH_3 and NADH (Fig. 1). The activity of GDC depends mainly on the availability of Gly, and is inhibited by NADH and Ser (Oliver, 1994). Thus, for active Gly oxidation in mitochondria, reoxidation of NADH and export of

Ser are necessary. The NH_3 will diffuse to the chloroplast where it is reassimilated by the glutamine synthetase/glutamate synthase (GS/GOGAT) system. Also some of the CO_2 will be refixed in the chloroplast while some will be lost to the atmosphere. Both GDC and the TCA cycle reactions reduce NAD to NADH and competition for NAD is a very important point of interaction between the two processes which otherwise do not share common substrates. In the glycolate pathway consumption of NADH in the peroxisomal hydroxypyruvate reductase reaction is stoichiometric with its production by GDC, and therefore the two reactions can be linked via a malate/OAA shuttle. Accordingly, in isolated mitochondria Gly decarboxylation is stimulated by OAA (Woo and Osmond, 1976) whereas oxygen consumption is eliminated (Lilley et al., 1987). Gly oxidation is also stimulated by malate (Woo and Osmond, 1976; Bergman and Ericson, 1983) which also may be consistent with the operation of such a shuttle.

In addition to export of photorespiratory NADH, some can also be reoxidized by the mitochondrial electron transport chain, either via the cytochrome pathway resulting in ATP production or via the alternative oxidase (AOX), which bypasses most or all of the coupling sites. To balance mitochondrial consumption of photorespiratory NADH, an equivalent amount of NADH must be supplied for hydroxypyruvate reduction from the chloroplast. In vivo the relative contribution by these pathways has important implications for the redox balance of the photosynthetic cell. Calculations based on experiments with isolated mitochondria and on estimated cytosolic NADH/NAD ratios, indicate that in steady state photosynthesis, 25–50% of the NADH formed in mitochondria can be exported to the cytosol via the malate/OAA shuttle (Krömer and Heldt, 1991a; Krömer, 1995).

In experiments with protoplasts incubated under photorespiratory conditions (limiting CO_2) as compared to non-photorespiratory conditions (high CO_2), a significant increase was observed in the mitochondrial ATP/ADP ratio (Gardeström and Wigge, 1988). An increase was also observed in the redox state of the mitochondrial NAD(H) pool (Wigge et al., 1993; Igamberdiev et al., 2001). Interestingly, an increased redox state was also observed in the mitochondrial NADP(H) pool (Igamberdiev et al., 2001). The increased photorespiration-dependent reduction of NADP(H) may be due to a transhydrogenation reaction (Bykova et al., 1999). Contrary to proton-translocating transhydrogenases of animal mitochondria, this reaction is not energy-linked but is associated with complex I and with another enzyme which may be similar to the soluble transhydrogenases of some bacteria. It was shown that during oxidation of Gly by isolated pea mitochondria, the internal NADPH dehydrogenase is operating, which could be explained by the transhydrogenation between NADH and NADP (Bykova and Møller, 2001). Photorespiration-linked increases in ATP, NADH and NADPH will affect other mitochondrial functions in the light, in particular the TCA cycle, as discussed below.

IV. Products of Glycolysis in the Light

In photosynthetic tissues, the glycolytic flux is maintained by export of TP from the chloroplasts. In the cytosol of all plant cells two different enzymes participating in glycolysis use phospho*enol*pyruvate (PEP) as substrate. These are pyruvate kinase (PK), which converts PEP to pyruvate, and PEP carboxylase (PEPc) which carboxylates PEP to OAA (Chapter 9, Vidal et al.). The OAA can be reduced to malate by cytosolic MDH (Fig. 2). In the mitochondrial inner membrane there are transporters for all three products (Laloi, 1999). The question is whether any of these can be identified as the main substrate taken up by mitochondria.

Several indirect pieces of evidence suggest that pyruvate is not the main substrate in the light. First, a monocarboxylate transporter (or pyruvate transport protein) in plant mitochondria (Vivekananda and Oliver, 1990) has lower activity compared to the dicarboxylate and OAA carriers in mitochondria from cucumber cotyledons, which is reflected in the lower rates of respiration with pyruvate (Hill et al., 1994). Second, PK is inhibited by high ATP/ADP ratios (Hu and Plaxton, 1996) which may be expected in the cytosol, especially under photorespiratory conditions. Third, transgenic plants with suppressed PK survived without any visible injuries in the light. They showed a similar net C gain and rate of photosynthesis to controls, over a range of light intensities. Furthermore, leaf growth was not suppressed (Knowles et al., 1998), although root growth was retarded (Grodzinski et al., 1999).

In contrast to PK, PEPc is more active in the light than in the dark, as a result of reversible phosphorylation (Van Quy et al., 1991; Krömer et al., 1996a,b;

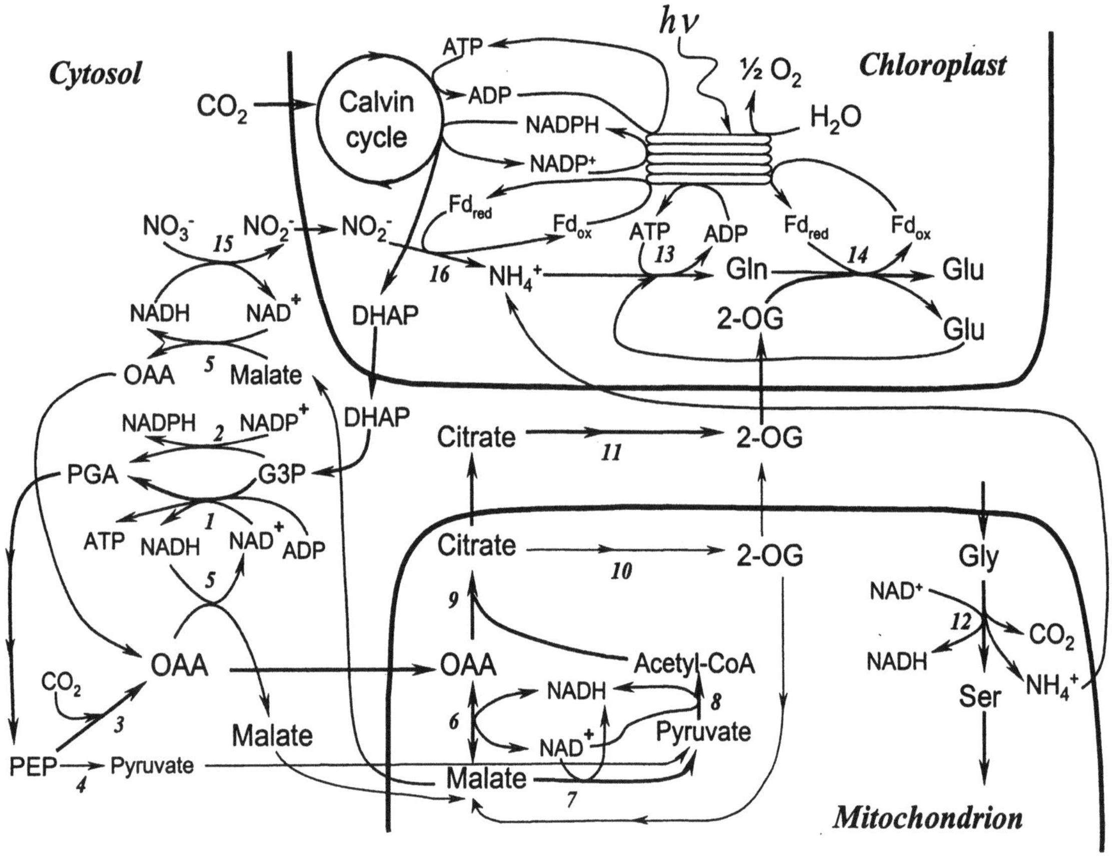

Fig. 2. Metabolic interactions between chloroplast, mitochondrion and cytosol. In the cytosol, dihydroxyacetone phosphate (DHAP) exported from the chloroplast is converted to glyceraldehyde 3-phosphate (G3P), which is oxidized to 3-phosphoglyceric acid (PGA) either by phosphorylating (1), or by non-phosphorylating G3P dehydrogenase (2). Phosphoenolpyruvate (PEP) formed in the glycolytic sequence from PGA is converted to oxaloacetate (OAA) by PEP carboxylase (3), or to pyruvate by pyruvate kinase (4). At increased cytosolic NADH/NAD ratio OAA may be converted to malate by cytosolic malate dehydrogenase (5). OAA, pyruvate or malate can enter the mitochondria where they form citrate by combined action of the mitochondrial malate dehydrogenase (6), NAD-malic enzyme (7), pyruvate decarboxylase complex (8) and citrate synthase (9). Citrate is converted to 2-oxoglutarate (2-OG) by aconitase and isocitrate dehydrogenase in the mitochondria (10) and cytosol (11). Photorespiratory Gly is converted to Ser in the mitochondria by glycine decarboxylase and serine hydroxymethyl transferase (12). 2-OG enters the chloroplast to be used for ammonium reassimilation via the glutamine synthetase/glutamate synthase system (13, 14). Nitrate is reduced by nitrate reductase (15) in the cytosol to nitrite, which is reduced to ammonium in the chloroplast by nitrite reductase (16). CoA, coenzyme A; Fd_{ox}, oxidized ferredoxin; Fd_{red}, reduced ferredoxin. Calvin cycle = RPP pathway.

Chapter 9 (Vidal et al.)). Furthermore, PEPc is activated by Gly, which increases its affinity for the activator glucose-6-phosphate and decreases sensitivity to the inhibitor, malate. This may be important, especially in photorespiratory conditions (Tovar-Méndez et al., 1998, 2000). The importance of PEPc in respiration was also shown in potato plants overexpressing the enzyme. PEPc over-expression led to enhanced respiration both in the dark and the light, and to accumulation of malate and increased sucrose biosynthesis (Häusler et al., 1999). Whereas Glu is inhibitory to both PK and PEPc, Asp inhibits PEPc and activates PK (Moraes and Plaxton, 2000). Photorespiratory ammonia may exert inhibition of PK by replacing K^+ necessary for its activity (Davies, 1979). Thus, the PEPc/PK branchpoint in glycolysis is strongly regulated by the cytosolic ATP/ADP ratio and by the products of N metabolism, particularly Gly, Glu, Asp and ammonia. PEPc and PK show opposite diurnal changes in activity and in transcript abundance (Scheible et al., 2000).

Cytosolic MDH will be a major determinant of whether malate or OAA is the main substrate available for the TCA cycle in the light. The equilibrium of this reaction is displaced towards malate but the NAD(H) pool of the cytosol has been shown to be very oxidized both by indirect (Heineke et al., 1991) and direct measurements (Igamberdiev et al., 2001). An oxidized NAD(H) pool makes a relatively high cytosolic OAA concentration possible. This was estimated to be around 0.1 mM in the light (slightly higher than the concentration estimated for darkness) (Heineke et al., 1991). The cytosolic malate concentration was estimated to about 1 mM (Heineke et al., 1991).

Plant mitochondria have an active OAA transporter with a capacity much higher than that of the dicarboxylate transporter transferring malate across the inner mitochondrial membrane (Ebbighausen et al., 1985). The kinetic properties favor OAA import in exchange for other acids, e.g. citrate or malate. In pea leaf mitochondria the OAA transporter had a very high affinity for OAA with a micromolar K_m and a high V_{max} (Ebbighausen et al., 1985). This shows a high capacity to import OAA from the cytosol to the mitochondria. Import of malate and OAA into mitochondria was sensitive to different inhibitors, indicating that they are transported on different carriers. Moreover, the uptake of OAA was not inhibited by a thousand-fold excess of malate (Douce and Neuburger, 1989 and references therein). OAA derived from PEPc activity may thus be the main respiratory substrate entering the mitochondria in the light. In swelling experiments with isolated pea leaf mitochondria, the uptake of malate was decreased significantly by the presence of physiological OAA concentrations (Zoglowek et al., 1988), indicating that, even in the presence of malate, OAA import to mitochondria may be favored (Krömer, 1995). Recent investigations on liposomes incorporating mitochondrial membrane proteins suggest that plant mitochondria contain an OAA translocator that differs from all other known mitochondrial translocators (Hanning et al., 1999).

V. Operation of the Tricarboxylic Acid Cycle

A. Entry of Glycolytic Substrates

Plant mitochondria have a unique enzyme, NAD-malic enzyme (NAD-ME), which allows conversion of malate to pyruvate in the mitochondrial matrix (Wedding, 1989). Because of this enzyme, the TCA cycle does not require the import of pyruvate as it can be formed from imported malate (Fig. 2). By the same mechanism OAA can be converted to pyruvate via malate. NAD-ME is activated by lower pH, coenzyme A (CoA) and its derivatives, fumarate and Mn^{2+} (Douce and Neuburger, 1989). Although the enzyme can use Mg^{2+}, this cation is less effective than Mn^{2+} (de Aragao et al., 1996). Fumarate activation may be important when the mitochondrial malate concentration increases and fumarate is formed in the fumarase reaction. Accumulation of fumarate is common in many plants (Chia et al., 2000), and it is possibly the only TCA cycle intermediate which has no transporter (Wiskich, 1977). NAD-ME has the lowest affinity for NAD of all TCA cycle dehydrogenases and is relatively insensitive to NADH. When fully activated it can operate at a high matrix NADH/NAD ratio and engage the rotenone-resistant internal NADH-dehydrogenase, whose affinity for NADH is lower than complex I (Pascal et al., 1990).

In photosynthetic tissues respiratory decarboxylation is usually inhibited in the light (Pärnik and Keerberg, 1995), and fine regulation of pyruvate dehydrogenase complex (PDC) becomes an important controlling step for TCA cycle activity. Mitochondrial PDC has a relatively low maximum catalytic activity in comparison to TCA cycle enzymes, with the exception of isocitrate dehydrogenases (ICDH) (Wiskich and Dry, 1985). It was proposed that in non-photosynthetic tissues carbon entry to TCA is limited by the maximal activity of PDC (Millar et al., 1998). In darkness, its maximum activity is close to that required to catalyze the TCA flux. This may explain why it has not been possible to recover transgenic plants with less than 80% wild type PDC activity (Rocha-Sosa et al., 1989; Grof et al., 1995).

Pyruvate entry into the TCA cycle is strongly regulated at PDC by substrate availability. Product inhibition, as well as reversible inhibition through phosphorylation of the mitochondrial PDC, may also be important. The activity and activation of PDC are also modified by different C and N metabolites. The complex has been reported to be inactivated in the light, particularly in photorespiratory conditions (Budde and Randall, 1987, 1990). This may be due to photorespiratory NH_3, which stimulates the protein kinase that phosphorylates PDC and also to a rise in intramitochondrial ATP/ADP and NADH/NAD ratios, which inhibit the enzyme (Moore et al., 1993). Pyruvate has a stimulatory effect on PDC activity, leading to abolition of the effects of ammonium and other inhibitors (Schuller and Randall, 1989). This may explain the observation that no inactivation of PDC occurs in barley protoplasts under photorespiratory conditions (Krömer et al., 1994). Similar to other 2-oxoacid dehydrogenases, redox balance could also exert control over PDC. Td-mediated activation occurs at the high NADPH/NADP ratios found in mitochondria when sufficient pyruvate is present (Bunik et al., 1997).

B. Partial Tricarboxylic Acid Cycle Activity in the Light

Regardless of which substrate is imported into mitochondria, OAA will be condensed with acetyl-CoA in the citrate synthase (CS) reaction to produce citrate, which is then converted to isocitrate by aconitase. The next step of the TCA cycle is conversion of isocitrate to 2-oxoglutarate (2-OG). This step has a lower maximal capacity than other TCA cycle reactions and so might well be limiting for the overall rate of the cycle. It may therefore be an important step for regulation of C flow in the TCA cycle. Mitochondria contain two isocitrate dehydrogenases (ICDH), one NAD-dependent and the other NADP-dependent. In photosynthetic tissues, the activity of the latter is equal to or even higher than that of the NAD-dependent form, whereas in non-photosynthetic tissues the NAD-dependent form predominates (Mackenzie and McIntosh, 1999).

Plant NAD-ICDH is not regulated by ADP, AMP or calcium, as are the enzymes from animals and microorganisms. It has a narrow pH optimum (around pH 7.5) and is allosterically activated by its substrate, with a $K_{0.5}$ of 0.1-0.3 mM, and non-competitively inhibited by NADPH (K_i 0.3 mM) (Rasmusson and Møller, 1990; McIntosh and Oliver, 1992a). The NADP-ICDH has a broad pH optimum and is saturated at very low concentrations of NADP and isocitrate (K_m in the micromolar range) (Rasmusson and Møller, 1990). Contrary to NAD-ICDH, the NADP enzyme can catalyze the reverse reaction at appreciable rates (Dalziel and Londesborough, 1968; Des Rosiers et al., 1994). The two forms of ICDH constitute a sensitive system responding to the mitochondrial NAD(P)H/NAD(P) ratios. At high ratios, NADP-ICDH can catalyze the reverse reaction while the NAD-ICDH is inhibited by NAD(P)H. Thus, in these conditions, isocitrate oxidation is suppressed, and isocitrate or citrate may be exported from the mitochondria (Fig. 3). In model experiments by Hanning and Heldt (1993), the main product released by mitochondria incubated with OAA was citrate, but a significant amount of 2-OG was also produced. This may be important especially in non-photorespiratory conditions, when NAD(P)H/NAD(P) ratios in mitochondria are relatively low. In these conditions, the limiting step will be oxidation of 2-OG or the succinyl-CoA synthetase reaction, which is dependent on ADP and inhibited by ATP. It was shown that maize root mitochondria contain a 2-OG transporter that exchanges this compound for malate, malonate or OAA (Genchi et al., 1991).

A tricarboxylate (or citrate) carrier was purified from mitochondrial membranes and characterized by two different groups (McIntosh and Oliver, 1992b; Genchi et al., 1999). It can exchange citrate with different TCA intermediates, including 2-OG, malate and OAA. The citrate carrier from pea was shown to be inactive with isocitrate (McIntosh and Oliver, 1992b), while the maize citrate carrier exhibited high capacity for isocitrate transport. In any case, citrate export from mitochondria may be more important than isocitrate export, since the equilibrium of the reaction catalysed by aconitase is displaced strongly towards citrate formation (Day and Wiskich, 1977). This is also supported by ^{13}C nuclear magnetic resonance studies using intact leaves, confirming that citrate is a major mitochondrial product in the light (Gout et al., 1993).

In the cytosol citrate can be converted to 2-OG by cytosolic aconitase and ICDH and used for Glu synthesis. Alternatively, it can return to mitochondria and be metabolized through the TCA cycle, which in this case operates between mitochondria and cytosol. Glu formed from 2-OG can also re-enter the TCA cycle, either via mitochondrial glutamate dehydrogenase (GDH) or by Glu decarboxylase in the cytosol forming γ-aminobutyric acid which is readily oxidized in mitochondria. Formation and oxidation of γ-aminobutyric acid is regulated by Glu availability, which is increased when Gly oxidation in mitochondria is suppressed, e.g., by its specific inhibitor aminoacetonitrile, and by high cytosolic Ca^{2+} concentration (Scott-Taggart et al., 1999).

C. Metabolic Shuttles between Mitochondria and other Compartments

In the light mitochondria import OAA (or malate) and export citrate. This may be important for maintaining the cytosolic NADPH/NADP ratio at values appropriate to biosynthetic purposes. Since NADP-ICDH isozymes are also present in peroxisomes, isocitrate can also be used to generate NADPH in this compartment. The chloroplastic isoform of

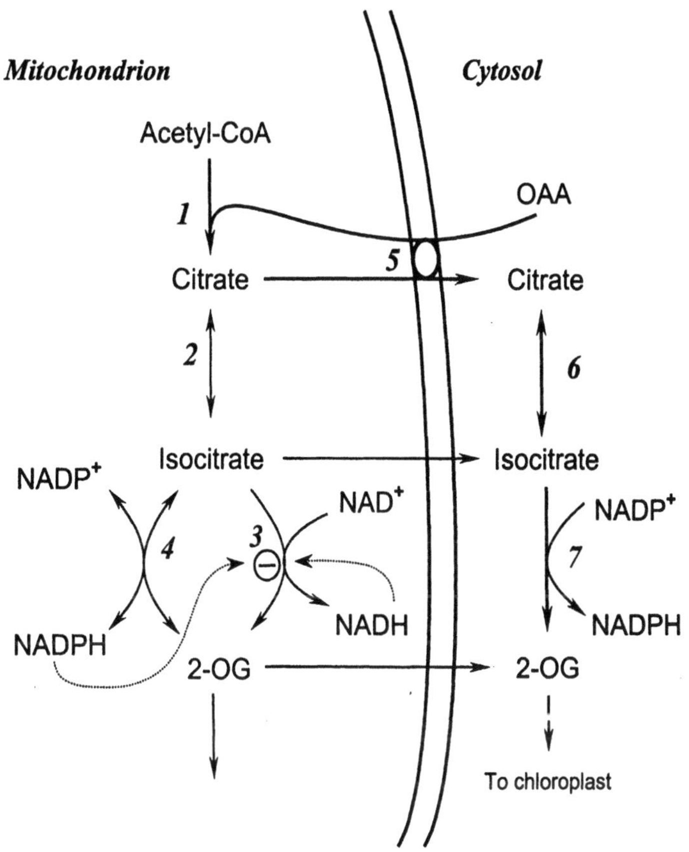

Fig. 3. Isocitrate dehydrogenase (ICDH) system and citrate efflux from mitochondria. Increase in the intramitochondrial NAD(P)H/NAD(P) ratio inhibits NAD-ICDH and stimulates the reverse reaction of NADP-ICDH. Since the aconitase reaction equilibrium favors citrate formation, this will lead to citrate efflux from mitochondria in exchange for OAA or other organic acid. 1, citrate synthase; 2, mitochondrial aconitase; 3, NAD-dependent ICDH; 4, mitochondrial NADP-dependent ICDH; 5, tricarboxylate transporter; 6, cytosolic aconitase; 7, cytosolic ICDH. Inhibition of NAD-ICDH by NADH and NADPH is shown by dotted arrows. CoA, coenzyme A; OAA, oxaloacetate; 2-OG, 2-oxoglutarate.

ICDH may be important for maintaining NADPH in the chloroplast in darkness, which is required for some of the biosynthetic functions of the chloroplast. Two valves may operate, one driven by chloroplasts, and the other driven by mitochondria. The malate valve, driven by photosynthetic electron transport, increases NADH/NAD ratios in different cell compartments, whereas the (iso)citrate valve, driven by the increased reduction level in mitochondria, tends to reduce the NADP(H) pools in the cytosol and peroxisomes. Crucially for N assimilation, the (iso)citrate valve supplies 2-OG for amino acid synthesis (Fig. 2). In addition to the mitochondrial contribution via the (iso)citrate valve, cytosolic NADPH can also be derived from the chloroplast via non-phosphorylating G3PDH.

Major functions of the OAA translocator are the export of reducing equivalents from the mitochondria via the malate-OAA shuttle and the export of citrate via the citrate-OAA shuttle (Hanning et al., 1999). Operation of a malate/OAA shuttle between mitochondria and cytosol/peroxisomes is important for reduction of hydroxypyruvate formed in the photorespiratory cycle (Krömer and Heldt, 1991a). High MDH activity in peroxisomes, which is of the same order as in mitochondria, is sufficient to sustain the photorespiratory flow (Heupel et al., 1991).

The Glu/Asp transporter in mitochondria (Vivekananda and Oliver, 1989) can provide interchange of these amino acids between mitochondria and other cell compartments. Gly/Ser counterexchange may be facilitated by a specific transporter. However, at concentrations higher than 0.5 mM, these amino acids rapidly diffuse through the mitochondrial membrane (Yu et al., 1983). Mitochondria also produce acetate via acetyl CoA hydrolase (Zeiher and Randall, 1990), the acetate formed can diffuse without any transporter to the chloroplasts and be used for biosynthetic purposes via the mevalonate pathway or in fatty acid synthesis.

VI. Electron Transport and Redox Levels in Plant Mitochondria

A. The Plant Mitochondrial Electron Transport Chain

The oxidation reactions in the TCA cycle yield NADH in the mitochondrial matrix. This NADH can be reoxidized by complex I. Besides complex I, where electron transport is linked to proton pumping, plant mitochondria possess four non-coupled NAD(P)H dehydrogenases, two on the external side, and two on the internal side (Melo et al., 1996; Agius et al., 1998; Møller, 2001). The internal NADH dehydrogenase has a much higher K_m for NADH than complex I and can operate only at elevated NADH/NAD ratios (Møller and Lin, 1986). Both the external NADH and NADPH dehydrogenases and the internal NADPH dehydrogenase are stimulated by Ca^{2+}. The molecular structure of two of these dehydrogenases was reported for potato mitochondria: they were found to be homologous to the rotenone-insensitive NADH dehydrogenases of *E. coli* and yeast (Rasmusson et al., 1999). One of them contains a sequence resembling calcium-binding motifs. It was proposed that these two dehydrogenases correspond to internal and external rotenone-insensitive NADH dehydrogenases of plant mitochondria. An NADH dehydrogenase of 43 kDa was shown to be located inside the inner membrane and to contain FAD (Menz and Day, 1996). An NAD(P)H dehydrogenase of 26 kDa has also been purified (Rasmusson et al., 1993). More investigations are needed to characterize mitochondrial NAD(P)H dehydrogenases and their physiological functions.

From the different dehydrogenases electrons are transferred to ubiquinone. The path of electron transport from ubiquinol to oxygen can be coupled or non-coupled to proton pumping, proceeding either via the cyanide-sensitive cytochrome pathway or the cyanide-insensitive AOX (Lambers, 1985; Vanlerberghe and McIntosh, 1997; Mackenzie and McIntosh, 1999). The AOX has been purified, characterized, and its genes isolated (Siedow and Umbach, 2000 and references therein). In spite of the recent progress in our knowledge of the molecular structure and regulation of AOX both at molecular and biochemical levels (McIntosh, 1994; Siedow and Umbach, 1995, 2000; Vanlerberghe and McIntosh, 1997), information on the physiological significance and detailed metabolic function of AOX is still limited. However, AOX is believed to function as an overflow mechanism (Lambers, 1985). The AOX participates in thermogenesis in *Araceae* (Wagner et al., 1998), in maintaining respiration in conditions where ATP synthesis is restricted such as Pi deficiency (Vanlerberghe and McIntosh, 1997; Parsons et al., 1999), in ameliorating chilling injury (Purvis and Shewfelt, 1993), and in preventing formation of active oxygen species (AOS) (Purvis, 1997; Maxwell et al., 1999). For further discussion of the physiological roles of AOX, see Chapter 11 (Vanlerberghe and Ordog).

B. Photosynthesis and Mitochondrial Electron Transport

The enzyme composition of leaf mitochondria differs significantly depending on the developmental stage of leaf. The rate of photorespiration gradually increases during leaf development (Tobin et al., 1989; Lennon et al., 1995; Vauclare et al., 1996). The amount of mitochondrial GDC increased at least five-fold during the development of wheat leaves (Tobin et al., 1988). It was proposed that GDC and AOX are coordinated during development, whereas cytochrome oxidase is more closely coordinated with the TCA cycle enzymes (Lennon et al., 1995; Finnegan et al., 1997). In a GDC-deficient mutant of barley, the AOX protein was present in very low amounts and a compensatory increase of respiratory capacity of the cytochrome pathway was observed (Igamberdiev et al., 2001).

The relative proportion of cytochrome and alternative pathways is flexible and varies with environmental conditions and developmental stage (temperature, age of the tissue and injury/wounding). The alternative pathway is thought to be particularly active in photosynthetic tissues. Several pieces of data indicate that AOX activity is important in the light, e.g., the level of AOX was observed to increase during greening of etiolated leaves (Atkin et al., 1993). Direct evidence for the involvement of the AOX in respiration in the light was obtained using oxygen isotope fractionation techniques. The increase in alternative pathway electron flux accounted for all of the increased respiration in the light phase in plants with crassulacean acid metabolism (Robinson et al., 1992). In light-grown soybean cotyledon mitochondria, increased partitioning to the alternative pathway in state 4 was observed. This was further increased by the addition of either pyruvate or dithiothreitol. In etiolated cotyledon mitochondria, the alternative pathway showed little ability to compete for electrons with the cytochrome pathway under any circumstances (Ribas-Carbo et al., 1997). Similarly, there was no engagement of the AOX in darkness in green cotyledons of soybean. In the light, however, 60% of the respiratory flux occurred through the alternative pathway. When green cotyledons were transferred back to darkness, the engagement of the AOX decreased (Ribas-Carbo et al., 2000).

Mitochondrial respiration is essential for optimal photosynthesis. Low concentrations of oligomycin, which strongly inhibit mitochondrial oxidative phosphorylation but do not affect chloroplast photophosphorylation, caused an inhibition of photosynthesis by 30–40% in barley leaf protoplasts, but not in isolated chloroplasts (Krömer et al., 1988). Oligomycin caused a decrease in the ATP/ADP ratio and an increase in the content of glucose-6-phosphate and F6P. Subcellular analysis of protoplasts revealed that oligomycin caused a larger decrease in the cytosolic ATP/ADP ratio than in the stromal ratio. Moreover, the increase in hexose monophosphates was restricted to the cytosol, whereas the stromal hexose monophosphates decreased upon the addition of oligomycin (Krömer et al., 1993). Oligomycin caused an increase in the TP/ PGA ratio (Krömer and Heldt, 1991b). Thus, during photosynthesis, mitochondrial oxidative phosphorylation contributes to the ATP supply of the cell and prevents overreduction of the chloroplast redox carriers by oxidizing reducing equivalents generated by photosynthetic electron transport (Krömer and Heldt, 1991a,b). Sucrose phosphate synthase activity was also reduced by oligomycin treatment. Under high irradiances, the inhibition of sucrose synthesis by oligomycin apparently caused a feedback inhibition of the RPP pathway and, thus, photosynthetic activity. At saturating light, mitochondrial oxidation of excess photosynthetic redox equivalents is required to sustain high rates of photosynthesis (Krömer et al., 1993).

The relative contribution of cytochrome and alternative pathways during photosynthesis was studied in mesophyll protoplasts of pea and barley, using low concentrations of the inhibitors of mitochondrial electron transport antimycin A (an inhibitor of the cytochrome pathway) and salicylhydroxamic acid (SHAM, an inhibitor of the alternative pathway). Both these compounds decreased the rate of photosynthetic O_2 evolution in mesophyll protoplasts, but did not affect photosynthetic rate in isolated chloroplasts (Padmasree and Raghavendra 1999a,b,c). These results demonstrate that both the cytochrome pathway and AOX are essential for equilibration of the redox balance in photosynthetic cells.

C. Photorespiration and Mitochondrial Electron Transport

Photorespiration increases the reduction of NAD and NADP in leaf mitochondria (Igamberdiev et al., 2001). Active oxidation of Gly induces non-coupled pathways of electron transport, i.e., rotenone-insensitive NAD(P)H oxidation in mitochondria and cyanide-insensitive electron transport (Igamberdiev et al., 1997a; Bykova and Møller, 2001). This may be important in order to allow photorespiratory flux to proceed at maximal rates without control by the ATP/ADP and NAD(P)H/NAD(P) ratios. The increase in NADH switches on the rotenone-insensitive NADH dehydrogenase (whose K_m for NADH is much higher than the K_m of complex I). Similarly, the increase of NADPH can switch on the rotenone-insensitive NADPH dehydrogenase, if the Ca^{2+} concentration is sufficient for its operation. Increased activity of AOX is observed in these conditions. This effect is facilitated by NADPH, possibly via Td and pyruvate (Vanlerberghe and McIntosh, 1997). Pyruvate can be formed in the ME reaction, which is not strongly inhibited by NADH (Pascal et al., 1990). This allows electron transport to proceed independently of the high ATP/ADP ratio observed in mitochondria oxidizing Gly (Gardeström and Wigge, 1988).

Glycine decarboxylase (GDC) is very strongly inhibited by NADH with a K_i of 15 μM and a K_m for NAD of 75 μM, i.e., it has a five-fold higher affinity for NADH than for NAD (Oliver, 1994). Transgenic plants with defective complex I exhibit severe limitations in the oxidation of glycine because of an increased NADH/NAD ratio which inhibits GDC (Sabar et al., 2000). However, the NADH/NAD ratio is increased in mitochondria under photorespiratory conditions, even in normal plants (Wigge et al., 1993; Igamberdiev et al., 2001). This increase will restrict GDC operation and require rapid removal of NADH. This can occur via the malate/OAA shuttle

(Krömer and Heldt, 1991a,b) or via active oxidation through the mitochondrial electron transport chain, including alternative dehydrogenases and AOX (Igamberdiev et al., 1997a).

It is possible that some NADH is reoxidized via transhydrogenation of NADP, thus increasing the NADPH/NADP ratio (Bykova and Møller, 2001). High NADPH/NADP ratio in photorespiratory conditions may be important for fatty acid biosynthesis. All the enzymes necessary for fatty acid biosynthesis, which requires large quantities of NADPH, reside in plant mitochondria. The H protein of GDC accounts for a considerable proportion of leaf lipoic acid, which is synthesized from octanoic acid, one of the major intermediates in the mitochondrial synthesis of fatty acids (Wada et al., 1997; Gueguen et al., 2000). Plant mitochondria are also a major site of NADPH-dependent folate biosynthesis, which is also required in the photorespiratory conversion of Gly to Ser (Neuburger et al., 1996).

It has been shown that glutathione biosynthesis can use photorespiratory Gly (Noctor et al., 1999), and the increase in NADPH/NADP ratio in photorespiratory conditions may be important to maintain reduction of the mitochondrial glutathione pool through glutathione reductase. This enzyme is an NADPH-dependent flavoprotein present in plant mitochondria and other compartments (Rasmusson and Møller, 1990; Creissen et al., 1995). It participates in the ascorbate-glutathione cycle and is a key component in the detoxification of AOS. Reduced glutathione may be an important antioxidant during photorespiration, when the increase in mitochondrial reduction state may favor formation of AOS (Møller and Rasmusson, 1998).

A high NADPH/NADP ratio is also important for reduction of mitochondrial Td, which is a low molecular weight protein with possible antioxidative functions. Two forms of Td, as well as an NADPH-Td reductase, have been identified in plant mitochondria (Konrad et al., 1996; Banze and Follman, 2000). Reduced Td may protect mitochondria against oxidative stress and cause reductive activation of CS and AOX (Schürmann and Jacquot, 2000). It could also activate 2-oxoacid dehydrogenases, i.e., PDC and 2-OG dehydrogenase. This could mitigate PDC inhibition under photorespiratory conditions, and favor oxidation of 2-OG that is imported into mitochondria or that is formed as a product of GDH activity.

D. External NADH and NADPH

The presence of significant cytosolic OAA, produced by PEPc, shows that the cytosolic NAD(H) pool must be highly oxidized, since the reaction catalyzed by MDH strongly favors malate formation (Gietl, 1992). Thus, the cytosolic NADH/NAD ratio was estimated to be extremely low, indirect measurements giving a value of about 10^{-3} in photosynthetic tissues (Heineke et al., 1991). Although plant mitochondria possess external NADH and NADPH dehydrogenases on the inner membrane, and NADH dehydrogenase on the outer membrane, oxidation of cytosolic NADH and NADPH depends on the presence in the outer membranes of pores that permit passage of these molecules. The permeability of these pores may be regulated in vivo (Vander Heiden et al., 2000). Since external NADH and NADPH dehydrogenases are activated by Ca^{2+}, the concentration of which increases in stress conditions, it seems that reoxidation of cytosolic NADH could be important primarily under stress. Cytosolic NADPH is perhaps more likely to be oxidized than cytosolic NADH, since the cytosolic NADPH/NADP ratio is about 1 (Wigge et al., 1993; Igamberdiev et al., 2001), and so [NADPH] is relatively high. However, NADPH is oxidized only when sufficient Ca^{2+} is available. Thus, external NADH and NADPH oxidations, like many other Ca^{2+}-dependent processes, are probably of importance in extreme situations, when their metabolic utilization is suppressed. The cytosolic Ca^{2+} concentration may be low in the light and increase in darkness (Millar and Sanders, 1987).

E. Mitochondrial Electron transport and Production of Active Oxygen Species

Mitochondria are the major sites of AOS production in animal cells. In plants, the electron transport chains of both chloroplasts and mitochondria are responsible for AOS formation. Even under optimal conditions, AOS formation in plants occurs at rates that are orders of magnitude higher than in mammalian cells (Puntarulo et al., 1988) and, in mitochondria, at least 1% total O_2 consumption leads to their production. An increased reduction state of electron transport chains increases the probability of AOS formation (Rich and Bonner, 1978; Purvis et al., 1995). The major sites of AOS production in mitochondria are complexes I and III, but the internal NADPH dehydrogenase may also contribute to this process (Møller, 2001).

AOS are important mediators in signal transduction pathways, and lead to increased expression of several genes involved in antioxidant defense, including the AOX genes (Wagner, 1995). In addition, plant mitochondria contain different systems to eliminate AOS, such as Mn-SOD which scavenges the superoxide radical (Zhu and Scandalios, 1993). H_2O_2 can be scavenged by catalase or various peroxidases, although the presence of these enzymes in plant mitochondria has not been established with certainty (Foyer and Noctor, 2000). Glutathione reductase can operate in connection with the ascorbate-glutathione cycle, where ascorbate peroxidase scavenges H_2O_2. This cycle was shown to be present in plant mitochondria, although its activity is lower than in chloroplasts (Jiménez et al., 1997). The ascorbate concentration in mitochondria was determined to be about 24 mM, and glutathione about 6 mM if we consider the protein concentration in mitochondrial matrix to be 1 mg μl^{-1} (Jiménez et al., 1997; Møller, 2001).

At high reduction levels, AOX becomes important for avoiding increased AOS formation (Purvis, 1997; Maxwell et al., 1999). It becomes engaged at elevated ubiquinone reduction states (Hoefnagel et al., 1995). At increased pyruvate concentrations inside the mitochondria (which occur in vivo in photorespiratory conditions when PDC may be suppressed), it successfully competes with the cytochrome pathway for electrons (Hoefnagel et al., 1995). Glyoxylate activates AOX as effectively as pyruvate, but it remains to be established whether high rates of photorespiration lead to increased glyoxylate concentrations inside mitochondria.

VII. Participation of Mitochondria in the Regulation of Metabolism during Transitions between Light and Darkness

A. The Role of Mitochondria in Photosynthetic Induction

Following a period of darkness, photosynthesis does not begin immediately but takes minutes to hours to attain the rate set by the prevailing conditions. This effect is known as induction, and is associated with the activation of enzymes and readjustment of metabolite pools involved in the RPP pathway and sucrose synthesis (Gardeström, 1993). During this period, the stromal NADP(H) pool becomes very reduced and therefore allows the malate valve to operate at maximal capacity. Two recent reports suggest that the restriction of mitochondrial metabolism leads to the prolongation of photosynthetic induction in barley mesophyll protoplasts (Igamberdiev et al., 1998) and pea (Padmasree and Raghavendra, 1999b). This effect could be due to suppression of malate oxidation in the mitochondria, and restriction of flux through the malate valve by inadequate resupply of OAA to the chloroplast. A decrease in PEPc can also prolong the photosynthetic induction period (Gehlen et al., 1996), implicating this enzyme in the supply of OAA for malate valve operation between the chloroplast and other compartments.

A related observation is that in the presence of inhibitors of the mitochondrial cytochrome pathway (antimycin A) and of oxidative phosphorylation (oligomycin), there is a marked decrease in the levels of ribulose-1,5-bisphosphate (RuBP), the primary substrate for C assimilation (Padmasree and Raghavendra, 1999b). These results imply that mitochondrial electron transport is important in maintaining RuBP concentrations in the chloroplast, by allowing efficient activation of chloroplastic enzymes. It is unclear how this effect is mediated since inhibition of the malate valve should increase the reduction state of the chloroplast stroma and thereby favor enzyme activation through the Td system. A possible explanation for the delay of activation of NADP-MDH and the RPP pathway enzymes may be a slower alkalization of the chloroplast stroma, when oxidation of malate is suppressed (Igamberdiev et al., 1998). Alkalization, together with a high reduction state, are important in activation of NADP-MDH (Kagawa and Hatch, 1977) and other chloroplastic enzymes. Mitochondrial inhibitors such as antimycin and oligomycin appear to lead to a more reduced photosynthetic electron transport chain and to slower acidification of the thylakoid lumen during photosynthetic induction period, as evidenced by chlorophyll fluorescence measurements (Igamberdiev at el., 1998). One possibility is that overreduction of photosystem I leads to an increased Mehler reaction and so higher chloroplastic H_2O_2 production. The Mehler reaction is facilitated in the absence of OAA (Hoefnagel et al., 1998), although enzyme inactivation by H_2O_2 would require Td and enzyme thiol groups to be able to compete with the highly active chloroplastic ascorbate peroxidase for H_2O_2 reduction.

The above results suggest that mitochondrial oxidation of malate formed in chloroplasts is important for coordination of chloroplast and mitochondrial function. Moreover, rotenone, which is an inhibitor of mitochondrial complex I, also prolongs the photosynthetic induction period and affects the activation state of the stromal NADP-MDH, suggesting that the reoxidation of chloroplastic malate proceeds in the matrix and involves complex I.

B. The Role of Mitochondria during Light-Enhanced Dark Respiration

Following the transition from light to darkness, NADP-MDH remains partly active during the first few minutes, providing transport of assimilatory power from the chloroplast (Nakamoto and Edwards, 1983). Oxidation of Gly continues for a limited period after illumination (2–5 min), and the associated CO_2 evolution is known as the post-illumination burst. Following this very rapid rate of CO_2 release, respiration continues at rates that are still higher than those seen after a prolonged period of darkness. This effect is defined as light-enhanced dark respiration (LEDR), and may continue for up to 30–60 min (Hoefnagel et al., 1998; Atkin et al., 2000a). Since malate is not used for reduction of hydroxypyruvate in the peroxisomes, and nitrate reduction is rapidly suppressed (Riens and Heldt, 1992), the cytosolic and peroxisomal utilization of NADH is decreased. In these conditions, NADH will support OAA conversion to malate. The situation is similar to that observed in photosynthetic induction, when the main substrate oxidized in mitochondria is not OAA, but malate. The external oxidation of NADH and NADPH by mitochondria may also be possible in this period.

Malic enzyme and PDC are possibly more active during LEDR than in the light, providing huge amounts of malate to be oxidized in the mitochondria (Hill and Bryce, 1992; Atkin et al., 2000a). As a result, the concentration of malate decreases in darkness (Hampp et al., 1984; Heineke et al., 1991; Hill and Bryce, 1992). Citrate produced by mitochondria may be utilized in the TCA cycle since the reduction level in mitochondria drops, allowing ICDH to produce 2-OG at high rates. A significant part of LEDR appears to be connected to the alternative pathway. The importance of AOX during the interaction between respiration and photosynthesis is evidenced by the sensitivity of LEDR to SHAM in barley mesophyll protoplasts (Igamberdiev et al., 1997b). A similar response to SHAM was shown for the algae *Selenastrum minutum*, *Chlamydomonas reinhardtii* and *Euglena gracilis* (Lynnes and Weger, 1996; Xue et al., 1996; Ekelund, 2000).

VIII. Mitochondrial Respiration and Photoinhibition

Mitochondrial respiration optimizes chloroplast function, particularly in the prevention of overenergization and overreduction of chloroplasts. This may occur in four ways: (i) integration and maintenance of metabolite movement, facilitating the export of excess energy and/or reductant from the chloroplasts, (ii) Promotion of sucrose biosynthesis (a carbon sink) and feed-forward enhancement of photosynthetic rate, (iii) Minimization of the photosynthetic induction period, and (iv) Maintenance of enzyme activation in the chloroplast. These effects can be due to either direct intervention or through feedback or feed-forward regulation. Such interactions involve extensive metabolite traffic between subcellular compartments, including peroxisomes as well as chloroplasts, cytosol, and mitochondria.

Photoinhibition of photosynthesis occurs under conditions which either overload photochemical capacity (excess light) or limit carbon fixation (e.g. low temperature or deficiency in RPP pathway enzymes). Mitochondrial respiration is one of the defense mechanisms that protect plant cells against photoinhibition, by providing an outlet for dissipation and recycling of reducing equivalents generated by the chloroplast (Raghavendra et al., 1994; Padmasree and Raghavendra, 1998). At limiting CO_2, protoplasts of the barley mutant deficient in GDC were shown to exhibit increased ATP/ADP and NADPH/NADP ratios in the chloroplasts (Igamberdiev et al., 2001). This indicates that photorespiration, and particularly Gly oxidation, is important for preventing overreduction and overenergization of the chloroplast. The GDC mutant showed an increased malate valve capacity, as well as enhanced capacity for scavenging chloroplastic reducing equivalents, as shown by a higher activation state of chloroplast NADP-MDH and an increased activity of mitochondrial and cytosolic isozymes of NAD-MDH. Even a marginal interference by respiratory inhibitors makes the protoplasts highly susceptible to photoinhibition (Saradadevi and Raghavendra, 1992). The restriction of mitochondrial respiration by inhibitors leads to the accumulation of reducing equivalents in the form of triose-P and/or malate (Igamberdiev et al., 1998; Padmasree and Raghavendra, 1999c).

An increase in mitochondrial respiratory capacity has been shown to be important in protecting photosynthesis against over-reduction of chloroplast electron carriers during cold-hardening of winter rye plants (Hurry et al., 1995). An increase in respiratory capacity is very common in plants exposed to cold temperatures (Körner and Larcher, 1988) and could be an important mechanism to cope with the susceptibility of photosynthetic apparatus to excess light. In a related study, Atkin et al. (2000b) observed that the cold-acclimation of dark respiration in snow gum leaves is characterized by changes in both the temperature sensitivity and apparent 'capacity' of the respiratory apparatus, and that such changes will have an important impact on the C economy of snow gum plants.

Mitochondrial respiration can be an important source of ATP and thus can help in the recovery of photosynthesis after photoinhibition analogous to what has been shown in the cyanobacterium, *Anacystis nidulans* (Shyam et al., 1993; Singh et al., 1996). It is conceivable that the mitochondria supply ATP for the chloroplast, but to date there is little experimental evidence to support this view. However, mitochondrial respiration has been shown to be the source of ATP in a mutant of *C. reinhardtii* deficient in chloroplastic ATP synthase (Raghavendra et al., 1994).

IX. The Role of Mitochondria in C_4 Photosynthesis

In C_3-C_4 intermediate and C_4 plants oxidation of Gly is confined to the bundle sheath mitochondria (Leegood and von Caemmerer, 1994; Devi et al., 1995; Dai et al., 1996). This allows more effective refixation of photorespiratory CO_2. C_3-C_4 intermediate plants with higher PEPc activity are more similar to C_4 plants than those with less active PEPc (Byrd et al., 1992). C_4 plants differ in their site of decarboxylation in bundle sheath cells, which can occur in mitochondria (NAD-ME type plants), the cytosol (PEP carboxykinase (PEPCK) type plants) or chloroplasts (NADP-ME type plants) (Hatch and Carnal, 1992). The direct role of bundle sheath cell mitochondria in malate decarboxylation in NAD-ME type plants is reflected by an approximate 50-fold increase in NAD-ME and 20-fold increase in Asp aminotransferase compared to C_3 plants, while CS and cytochrome oxidase activities are more or less the same (Hatch and Carnal, 1992).

In PEPCK-type plants also, the bundle sheath mitochondria contain about six times more ME than mitochondria in C_3 plants. This is explained by an increased requirement for cytosolic ATP for PEPCK activity, which is produced through the oxidation of malate in the mitochondria. In these plants NAD-ME is activated by ATP, which is not observed in plants whose primary decarboxylating enzyme is NAD-ME (Furbank et al., 1991). In addition, increased transport of metabolites, particularly malate, across the mitochondrial membranes is necessary in these two types of C_4 plant. Only in NADP-ME type plants do mitochondria have no direct role in C_4 photosynthesis.

Photorespiratory flux still occurs in the bundle sheath cells of C_4 plants, liberating CO_2 in the mitochondria (Leegood and von Caemmerer, 1994). When C_4 photosynthesis is limited by the supply of atmospheric CO_2, photorespiration in bundle sheath cells serves as a pump to concentrate CO_2 inside the leaf (Laisk and Edwards, 1997). The rates of C_3 and C_4 cycles are coordinated through the pool sizes of the C_4 cycle, which are in equilibrium with the PGA pool. At low CO_2, the C_4 pools decrease and are slowly regenerated by CO_2 from Gly oxidation in bundle sheath cells.

X. Glycolate Metabolism in Algal Mitochondria

Many algae contain mitochondrial glycolate dehydrogenase instead of peroxisomal glycolate oxidase. Thus, the metabolism of photorespiratory glycolate to Ser occurs in algal mitochondria (Stabenau, 1992). Hydroxypyruvate reduction in some algae is located in peroxisomes, but in other species, like *Dunaliella*, mitochondria contain all the enzymes of the photorespiratory cycle (Stabenau et al., 1993). Only the most advanced algae, the *Charophyceae*, which are most likely to be the direct ancestors of higher plants (Graham and Kaneko, 1991), contain higher plant-type peroxisomes and photorespiration in which the only photorespiratory reactions occurring in the mitochondria are those involved in the conversion of Gly to Ser. Glycolate oxidase may be present in peroxisomes of other groups of algae, e.g. in

Heterocontophyta; however, photorespiratory glyoxylate in these algae is condensed with acetyl-CoA in the malate synthase reaction, and only a small proportion is aminated to Gly (Stabenau, 1992).

Glycolate dehydrogenase activity was shown to be confined to the outer mitochondrial membrane (Beezley et al., 1976). Its operation is linked to electron transport with O_2 uptake and generation of a proton gradient for ATP synthesis (Paul et al., 1975). Operation of glycolate dehydrogenase will increase the reduction state, which will in turn limit its activity. This is in contrast to peroxisomal glycolate oxidase, which is not regulated by redox state. Thus, at high photorespiration rates, oxidation of glycolate is suppressed and it is excreted into the surrounding medium (Stabenau et al., 1993). Low CO_2 conditions, however, induce CO_2 concentrating mechanisms based on carbonic anhydrase. These limit loss of C (Badger and Price, 1994). Oxidation of Gly in most algal mitochondria proceeds only at a limited rate. This rate is determined by glycolate dehydrogenase activity and by the CO_2 concentrating mechanism.

XI. Concluding Remarks

It is logical that different organelles within the plant cell interact in a way that optimizes cellular functions (Figs. 1 and 2). The dependence of chloroplast photosynthesis on mitochondrial metabolism is therefore not surprising. However, the relative importance of different pathways of mitochondrial electron transport, and the flexibility of switching between coupled and non-coupled pathways of electron transport, remain to be clearly established. Preliminary evidence from the use of metabolic inhibitors suggested that the balance between coupled and non-coupled pathways is important for the functioning of chloroplast metabolism. The use and specificity of metabolic inhibitors are debatable due to the possible unspecificity and limited permeability of inhibitors. Further experiments are needed on the interaction between various components, particularly between chloroplasts and the alternative pathway of mitochondrial electron transport. Transgenic plants and mutants deficient in specific proteins/enzymes would be useful tools with which to test key concepts.

There must be a network of signals between organelles that triggers and coordinates changes in their respective metabolic status. Metabolite concentration is one such possible type of signal. It has already been shown that the relative ratios of TP/PGA and malate/OAA could be important in mediating the interaction of mitochondria and chloroplasts (Padmasree and Raghavendra, 1999c). There could, however, be additional signals such as cytosolic pH, N status, phosphate level, superoxide radicals or even secondary messengers such as calcium.

Nitrogen itself is an important signal for modulating C metabolism and subsequently the functioning of cellular organelles (Champigny, 1995; Stitt, 1999; Lewis et al., 2000). The effects of nitrate or ammonia on leaf tissue are phenomenonal, particularly in the modulation of gene expression and the diversion of C skeletons from carbohydrate into amino acid metabolism. Supply of nitrate or ammonia to N-starved leaves upregulates the biosynthesis not only of nitrate reductase, but also PEPc and carbonic anhydrase. At the same time nitrate down-regulates the activity of sucrose phosphate synthase. Reciprocal changes in the activity of PEPc and sucrose phosphate synthase are linked to the increase in the phosphorylation status of these two enzymes (Champigny, 1995; Toroser and Huber, 2000).

Plant cells have developed a strategy to meet the demands for energy (ATP) and reducing equivalents (NADH, NADPH) of different compartments. Supply and demand patterns are dynamic depending on the microenvironment of the cell. For example, upon illumination the chloroplasts can generate ATP as well as NADPH in excess of their own need and can export to other compartments. Under limiting light and high CO_2, the chloroplast may have to supplement its needs by either import or by restricting export. Mitochondria are geared to export ATP and citrate, leading to reduction of NADP in the cytosol. The import of reducing equivalents by peroxisomes from both chloroplasts and mitochondria demonstrates the flexibility of interorganellar dependence within the photosynthetic cell. It is very important to examine the C/N interaction involving multiple organelles, in transgenic plants and mutants deficient in specific reactions in chloroplasts, mitochondria or peroxisomes.

Acknowledgments

This work was supported by grants from the Swedish Natural Science Research Council and the European Union Biotechnology Framework IV (P.G.), from the Swedish Royal Academy (P.G. and A.U.I.), and from the Department of Science and Technology (SP/SO/A-12/98), New Delhi (A.S.R.).

References

Agius SC, Bykova NV, Igamberdiev AU and Møller IM (1998) The internal rotenone-insensitive NADPH dehydrogenase contributes to malate oxidation by potato tuber and pea leaf mitochondria. Physiol Plant 104: 329–336

Atkin OK, Cummins WR and Collier DE (1993) Light induction of alternative pathway capacity in leaf slices of *Belgium endive*. Plant Cell Environ 16: 231–235

Atkin OK, Millar AH, Gardeström P and Day DA (2000a) Photosynthesis, carbohydrate metabolism and respiration in leaves of higher plants. In: Leegood RC, Sharkey TD and von Caemmerer S (eds) Photosynthesis: Physiology and Metabolism, pp 153–175. Kluwer, Dordrecht

Atkin OK, Evans JR, Ball MC, Lambers H and Pons TL (2000b) Leaf respiration of snow gum in the light and dark. Interactions between temperature and irradiance. Plant Physiol 122: 915-923

Azcón-Bieto J (1992) Relationship between photosynthesis and respiration in the dark in plants. In: Barber J, Guerrero MG and Medrano H (eds) Trends in Photosynthesis Research, pp 241–253. Intercept, Andover

Badger MR and Price GD (1994) The role of carbonic anhydrase in photosynthesis. Annu Rev Plant Physiol Plant Mol Biol 45: 369–392

Banze M and Follman H (2000) Organelle-specific NADPH thioredoxin reductase in plant mitochondria. J Plant Physiol 156: 126–129

Beezley BB, Gruber PJ and Frederick SE (1976) Cytochemical localization of glycolate dehydrogenase in mitochondria of *Chlamydomonas*. Plant Physiol 58: 315–319

Bergman A and Ericson J (1983) Effect of pH, NADH, succinate and malate on the oxidation of glycine in spinach leaf mitochondria. Physiol Plant 59: 421–427

Budde RJA and Randall DD (1987) Regulation of pea mitochondrial pyruvate dehydrogenase complex activity: Inhibition of ATP-dependent inactivation. Arch Biochem Biophys 258: 600–606

Budde RJA and Randall DD (1990) Pea leaf mitochondrial pyruvate dehydrogenase complex is inactivated in vivo in a light-dependent manner. Proc Natl Acad Sci USA 87: 673–676

Bunik V, Follmann H and Bisswanger H (1997) Activation of mitochondrial 2-oxoacid dehydrogenases by thioredoxin. Biol Chem 378: 1125–1130

Bykova NV and Møller IM (2001) Involvement of matrix NADP turnover in the oxidation of NAD-linked substrates by pea leaf mitochondria. Physiol Plant 111: 448-456

Bykova NV, Rasmusson AG, Igamberdiev AU, Gardeström P and Møller IM (1999) Two separate transhydrogenase activities are present in plant mitochondria. Biochem Biophys Res Commun 265: 106–111

Byrd GT, Brown RH, Bouton JH, Bassett CL and Black CC (1992) Degree of C_4 photosynthesis in C_4 and C_3-C_4 *Flaveria* species and their hybrids. I. CO_2 assimilation and metabolism and activities of phospho*enol*pyruvate carboxylase and NADP-malic enzyme. Plant Physiol 100: 939–946

Casati DFG, Sesma JI and Iglesias AA (2000) Structural and kinetic characterization of NADP-dependent, non-phosphorylating glyceraldehyde-3-phosphate dehydrogenase from celery leaves. Plant Sci 154: 107–115

Champigny M-L (1995) Integration of photosynthetic carbon and nitrogen metabolism in higher plants. Photosynth Res 46: 117–127

Chia DW, Yoder TJ, Reiter W-D and Gibson SI (2000) Fumaric acid: An overlooked form of fixed carbon in *Arabidopsis* and other plant species. Planta 211: 743–751

Creissen G, Reynolds H, Xue YB and Mullineaux P (1995) Simultaneous targeting of pea glutathione reductase and of a bacterial fusion protein to chloroplasts and mitochondria in transgenic tobacco. Plant J 8: 167–175

Dai ZY, Ku MSB and Edwards GE (1996) Oxygen sensitivity of photosynthesis and photorespiration in different photosynthetic types in the genus *Flaveria*. Planta 198: 563–571

Dalziel K and Londesborough JC (1968) The mechanisms of reductive carboxylation reactions. Carbon dioxide or bicarbonate as substrate of nicotinamide-adenine dinucleotide phosphate-linked isocitrate dehydrogenase and 'malic' enzyme. Biochem J 110: 223–230

Davies DD (1979) The central role of phospho*enol*pyruvate in plant metabolism. Annu Rev Plant Physiol 30: 131–158

Day DA and Wiskich JT (1977) Factors limiting respiration by isolated cauliflower mitochondria. Phytochemistry 16: 1499–1502

de Aragao MEF, Jolivet Y, de Melo DF, Lima MDGS and Dizengremel P (1996) Purification and partial characterization of NAD-dependent malic enzyme isolated from *Vigna unguiculata* hypocotyls. Plant Physiol Biochem 34: 363–368

Des Rosiers C, Fernandez CA, David F and Brunengraber H (1994) Reversibility of the mitochondrial isocitrate dehydrogenase reaction in the perfused rat liver. Evidence from isotopomer analysis of citric acid cycle intermediates. J Biol Chem 269: 27179–27182

Devi MT, Rajagopalan AV and Raghavendra AS (1995) Predominant localization of mitochondria enriched with glycine-decarboxylating enzymes in bundle-sheath cells of *Alternanthera tenella*, a C_3-C_4 intermediate species. Plant Cell Environ 18: 589–594

Douce R and Neuburger M (1989) The uniqueness of plant mitochondria. Annu Rev Plant Physiol Plant Mol Biol 40: 371–414

Duggleby RG and Dennis DT (1974) Nicotinamide adenine dinucleotide-specific glyceraldehyde 3-phosphate dehydrogenase from *Pisum sativum*. Assay and steady state kinetics. J Biol Chem 249: 167–174

Ebbighausen H, Chen J and Heldt HW (1985) Oxaloacetate translocator in plant mitochondria. Biochim Biophys Acta 810: 184–199

Ekelund NGA (2000) Interactions between photosynthesis and 'light-enhanced dark respiration' (LEDR) in the flagellate *Euglena gracilis* after irradiation with ultraviolet radiation. J Photochem Photobiol 55: 63–69

Finnegan PM, Whelan J, Millar AH, Zhang QS, Smith MK, Wiskich JT and Day DA (1997) Differential expression of the multigene family encoding the soybean mitochondrial

alternative oxidase. Plant Physiol 114: 455–466
Flügge U-I (1999) Phosphate translocators in plastids. Annu Rev Plant Physiol Plant Mol Biol 50: 27–45
Flügge U-I and Heldt HW (1991) Metabolite translocators of the chloroplast envelope. Annu Rev Plant Physiol Plant Mol Biol 42: 129–144
Foyer CH and Noctor G (2000) Oxygen processing in photosynthesis: Regulation and signalling. New Phytol 146: 359–388
Fridlyand LE, Backhausen JE and Scheibe R (1998) Flux control of the malate valve in leaf cells. Arch Biochem Biophys 349: 290–298
Furbank RT, Agostiño A and Hatch MD (1991) Regulation of C_4 photosynthesis: Modulation of mitochondrial NAD-malic enzyme by adenylates. Arch Biochem Biophys 289: 376–381
Gardeström P (1993) Metabolite levels in the chloroplast and extrachloroplast compartments of barley leaf protoplasts during the initial phase of photosynthetic induction. Biochim Biophys Acta 1183: 327–332
Gardeström P (1996) Interactions between mitochondria and chloroplasts. Biochim Biophys Acta 1275: 38–40
Gardeström P and Lernmark U (1995) The contribution of mitochondria to energetic metabolism in photosynthetic cells. J Bioenerg Biomembr 27: 415–421
Gardeström P and Wigge B (1988) Influence of photorespiration on ATP/ADP ratios in the chloroplasts, mitochondria, and cytosol, studied by rapid fractionation of barley (*Hordeum vulgare*) protoplasts. Plant Physiol 88: 69–76
Gehlen J, Panstruga R, Smets H, Merkelbach S, Kleines M, Porsch P, Fladung M, Becker I, Rademacher T, Häusler RE and Hirsch HJ (1996) Effects of altered phospho*enol*pyruvate carboxylase activities on transgenic C_3 plant *Solanum tuberosum*. Plant Mol Biol 32: 831–848
Genchi G, de Santis A, Ponzone C and Palmieri F (1991) Partial purification and reconstitution of the 2-oxoglutarate carrier from corn (*Zea mays* L.) mitochondria. Plant Physiol 96: 1003–1007
Genchi G, Spagnoletta A, De Santis A, Stefanizzi L and Palmieri F (1999) Purification and characterization of the reconstitutively active citrate carrier from maize mitochondria. Plant Physiol 120: 841–847
Gietl C (1992) Malate dehydrogenase isoenzymes—cellular locations and role in the flow of metabolites between the cytoplasm and cell organelles. Biochim Biophys Acta 1100: 217–234
Gout E, Bligny R, Pascal N and Douce R (1993) C_{13} nuclear magnetic resonance studies of malate and citrate synthesis and compartmentation in higher plant cells. J Biol Chem 268: 3986–3992
Graham LE and Kaneko Y (1991) Subcellular structures of relevance to the origin of land plants (embryophytes) from green algae. Crit Rev Plant Sci 10: 323–342
Grodzinski B, Jiao J, Knowles VL and Plaxton WC (1999) Photosynthesis and carbon partitioning in transgenic tobacco plants deficient in leaf cytosolic pyruvate kinase. Plant Physiol 120: 887–895
Grof CPL, Winning BM, Scaysbrook TP, Hill SA and Leaver CJ (1995) Mitochondrial pyruvate dehydrogenase: Molecular cloning of the E1 subunit and expression analysis. Plant Physiol 108: 1623–1629
Gueguen V, Macherel D, Jaquinod M, Douce R and Bourguignon J (2000) Fatty acid and lipoic acid biosynthesis in higher plant mitochondria. J Biol Chem 275: 5016–5025
Hampp R, Goller M and Füllgraf H (1984) Determination of compartmented metabolite pools by a combination of rapid fractionation of oat mesophyll protoplasts and enzymic cycling. Plant Physiol 75: 1017–1021
Hanning I and Heldt HW (1993) On the function of mitochondrial metabolism during photosynthesis in spinach leaves (*Spinacia oleracea* L.). Partitioning between respiration and export of redox equivalents and precursors for nitrate assimilation products. Plant Physiol 103: 1147–1154
Hanning I, Baumgarten K, Schott K and Heldt HW (1999) Oxaloacetate transport into plant mitochondria. Plant Physiol 119: 1025–1031
Hatch MD and Carnal NW (1992) The role of mitochondria in C_4 photosynthesis. In: Lambers H and van der Plas LHW (eds) Molecular, Biochemical and Physiological Aspects of Plant Respiration, pp. 135–148. SPB Academic Publishing, The Hague
Hatch MD, Dröscher L, Flügge U-I and Heldt HW (1984) A specific translocator for oxaloacetate transport in chloroplasts. FEBS Lett 178: 15–19
Häusler RE, Kleines M, Uhrig H, Hirsch HJ and Smets H (1999) Overexpression of phospho*enol*pyruvate carboxylase from *Corynebacterium glutamicum* lowers the CO_2 compensation point (Γ^*) and enhances dark and light respiration in transgenic potato. J Exp Bot 50: 1231–1242
Heineke D, Riens B, Grosse H, Hoferichter P, Peter U, Flugge UI and Heldt HW (1991) Redox transfer across the inner chloroplast envelope membrane. Plant Physiol 95: 1131–1137
Heupel R, Markgraf T, Robinson DG and Heldt HW (1991) Compartmentation studies on spinach leaf peroxisomes. Evidence for channeling of photorespiratory metabolites in peroxisomes devoid of intact boundary membrane. Plant Physiol 96: 971–979
Hill SA and Bryce JH (1992) Malate metabolism and light-enhanced dark respiration in barley mesophyll protoplasts. In: Lambers H and van der Plas LHW (eds) Molecular, Biochemical and Physiological Aspects of Plant Respiration, pp. 221–230. SPB Academic Publishing, The Hague
Hill SA, Bryce JH and Leaver CJ (1994) Pyruvate metabolism in mitochondria from cucumber cotyledons during early seedling development. J Exp Bot 45: 1489–1491
Hoefnagel MHN, Millar AH, Wiskich JT and Day DA (1995) Cytochrome and alternative respiratory pathways compete for electrons in the presence of pyruvate in soybean mitochondria. Arch Biochem Biophys 318: 394–400
Hoefnagel MHN, Atkin OK and Wiskich JT (1998) Interdependence between chloroplasts and mitochondria in the light and the dark. Biochim Biophys Acta 1366: 235–255
Hu ZY and Plaxton WC (1996) Purification and characterization of cytosolic pyruvate kinase from leaves of the castor oil plant. Arch Biochem Biophys 333: 298–307
Hurry V, Tobiæson M, Krömer S, Gardeström P and Öquist G (1995) Mitochondria contribute to increased photosynthetic capacity of leaves of winter rye (*Secale cereale* L.) following cold-hardening. Plant Cell Environ 18: 69–76
Igamberdiev AU, Bykova NV and Gardeström P (1997a) Involvement of cyanide-resistant and rotenone-insensitive pathways of mitochondrial electron transport during oxidation of glycine in higher plants. FEBS Lett 412: 265–269

Igamberdiev AU, Zhou GQ, Malmberg G and Gardeström P (1997b) Respiration of barley protoplasts before and after illumination. Physiol Plant 99: 15–22

Igamberdiev AU, Hurry V, Krömer S and Gardeström P (1998) The role of mitochondrial electron transport during photosynthetic induction. A study with barley (*Hordeum vulgare*) protoplasts incubated with rotenone and oligomycin. Physiol Plant 104: 431–439

Igamberdiev AU, Bykova NV, Lea PJ and Gardeström P (2001) The role of photorespiration in redox and energy balance of photosynthetic plant cells: A study with a barley mutant deficient in glycine decarboxylase. Physiol Plant 111: 427–438

Jiménez A, Hernández JA, del Río LA and Sevilla F (1997) Evidence for the presence of the ascorbate-glutathione cycle in mitochondria and peroxisomes of pea leaves. Plant Physiol 114: 275–284

Kagawa T and Hatch MD (1977) Regulation of C_4 photosynthesis: Characterization of a protein factor mediating the activation and inactivation of NADP-malate dehydrogenase. Arch Biochem Biophys 184: 290–297

Kelly GJ and Gibbs M (1973) Nonreversible D-glyceraldehyde 3-phosphate dehydrogenase of plant tissues. Plant Physiol 52: 111–118

Knowles VL, McHugh SG, Hu ZY, Dennis DT, Miki BL and Plaxton WC (1998) Altered growth of transgenic tobacco lacking leaf cytosolic pyruvate kinase. Plant Physiol 116: 45–51

Konrad A, Banze M and Follman H (1996) Mitochondria of plant leaves contain two thioredoxins. Completion of the thioredoxin profile of higher plants. J Plant Physiol 149: 317–321

Körner C and Larcher W (1988) Plant life in cold environments. In: Long SF, Woodward FI (eds) Symposium of the Society for Experimental Biology, Vol 42. Plants and Temperature, pp 25–57. University Press, Cambridge

Krömer S (1995) Respiration during photosynthesis. Annu Rev Plant Physiol Plant Mol Biol 46: 45–70

Krömer S and Heldt HW (1991a) Respiration of pea leaf mitochondria and redox transfer between the mitochondrial and extramitochondrial compartment. Biochim Biophys Acta 1057: 42–50

Krömer S and Heldt HW (1991b) On the role of mitochondrial oxidative phosphorylation in photosynthesis metabolism as studied by the effect of oligomycin on photosynthesis in protoplasts and leaves of barley (*Hordeum vulgare*). Plant Physiol 95: 1270–1276

Krömer S and Scheibe R (1996) Function of the chloroplastic malate valve for respiration during photosynthesis. Biochem Soc Trans 24: 761–766

Krömer S, Stitt M and Heldt HW (1988) Mitochondrial oxidative phosphorylation participating in photosynthetic metabolism of a leaf cell. FEBS Lett 226: 352–356

Krömer S, Malmberg G and Gardeström P (1993) Mitochondrial contribution to photosynthetic metabolism. A study with barley (*Hordeum vulgare* L.) leaf protoplasts at different light intensities and CO_2 concentrations. Plant Physiol 102: 947–955

Krömer S, Lernmark U and Gardeström P (1994) In vivo mitochondrial pyruvate dehydrogenase activity, studied by rapid fractionation of barley leaf protoplasts. J Plant Physiol 144: 485–490

Krömer S, Gardeström P and Samuelsson G (1996a) Regulation of the supply of cytosolic oxaloacetate for mitochondrial metabolism via phospho*enol*pyruvate carboxylase in barley leaf protoplasts. 1. The effect of covalent modification on PEPC activity, pH response, and kinetic properties. Biochim Biophys Acta 1289: 343–350

Krömer S, Gardeström P and Samuelsson G (1996b) Regulation of the supply of oxaloacetate for mitochondrial metabolism via phospho*enol*pyruvate carboxylase in barley leaf protoplasts. 2. Effects of metabolites on PEPC activity at different activation states of the protein. Biochim Biophys Acta 1289: 351–361

Laisk A and Edwards GE (1997) CO_2 and temperature-dependent induction in C_4 photosynthesis: An approach to the hierarchy of rate-limiting processes. Aust J Plant Physiol 24: 505–516

Laloi M (1999) Plant mitochondrial carriers: An overview. Cell Mol Life Sci 56: 918–944

Lambers H (1985) Respiration in intact plants and tissues: Its regulation and dependence on environmental factors, metabolism and invaded organisms. In: Douce R and Day DA (eds) Encyclopedia of Plant Physiology, New Series, Vol 18. Higher Plant Cell Respiration, pp 418–473. Springer Verlag, Berlin

Leegood RC and von Caemmerer S (1994) Regulation of photosynthetic carbon assimilation in leaves of C_3-C_4 intermediate species of *Moricandia* and *Flaveria*. Planta 192: 232–238

Lennon AM, Pratt J, Leach G and Moore A (1995) Developmental regulation of respiratory activity in pea leaves. Plant Physiol 107: 925–932

Lewis CE, Noctor G, Causton D and Foyer CH (2000) Regulation of assimilate partitioning in leaves. Aust J Plant Physiol 27: 507–519

Lilley RM, Ebbighausen H and Heldt HW (1987) The simultaneous determination of carbon-dioxide release and oxygen-uptake in suspensions of plant leaf mitochondria oxidizing glycine. Plant Physiol 83: 349–353

Lorimer GH and Andrews TJ (1981) The C_2 chemo- and photorespiratory carbon oxidation cycle. In: Stumpf PK and Conn EE (eds) The Biochemistry of Plants, Vol 8, pp 329–374. Academic Press, New York

Lynnes JA and Weger HG (1996) Azide-stimulated oxygen consumption by the green alga *Selenastrum minutum*. Physiol Plant 97: 132–138

Mackenzie S and McIntosh L (1999) Higher plant mitochondria. Plant Cell 11: 571–585

Maxwell DP, Wang Y and McIntosh L (1999) The alternative oxidase lowers mitochondrial reactive oxygen production in plant cells. Proc Natl Acad Sci USA 96: 8271–8276

McIntosh CA and Oliver DJ (1992a) NAD-linked isocitrate dehydrogenase: Isolation, purification, and characterization of the protein from pea mitochondria. Plant Physiol 100: 69–75

McIntosh CA and Oliver DJ (1992b) Isolation and characterization of the tricarboxylate transporter from pea mitochondria. Plant Physiol 100: 2030–2034

McIntosh L (1994) Molecular biology of the alternative oxidase. Plant Physiol 105: 781–786

Melo AMP, Roberts TH and Møller IM (1996) Evidence for the presence of two rotenone insensitive NAD(P)H dehydrogenases on the inner surface of the inner membrane of potato tuber mitochondria. Biochim Biophys Acta 1276: 133–139

Menz RI and Day DA (1996) Purification and characterization of a 43-kDa rotenone-insensitive NADH dehydrogenase from

plant mitochondria. J Biol Chem 271: 23117–23120
Miginiac-Maslow M, Johansson K, Ruelland E, Issakidis-Bourguet E, Schepens I, Goyer A, Lemaire-Chamley M, Jacquot J-P, le Maréchal P and Decottignies P (2000) Light-activation of NADP-malate dehydrogenase: A highly controlled process for an optimized function. Physiol Plant 111: 322–329
Miller AH, Saeed S, Jenner HL, Knorpp C, Leaver CJ and Hill SA (1998) Control and regulation of the tricarboxylic acid cycle in potato tubers. In: Møller IM, Gardeström P, Glimelius K and Glaser E (eds) Plant Mitochondria: From Gene to Function, pp. 551-557. Backhuys, Leiden
Millar AJ and Sanders D (1987) Depletion of cytosolic free calcium induced by photosynthesis. Nature 326: 397-400
Møller IM (2001) Plant mitochondria and oxidative stress: Electron transport, NADPH turnover and metabolism of reactive oxygen species. Annu Rev Plant Physiol Plant Mol Biol 52: 561–591
Møller IM and Lin W (1986) Membrane-bound NAD(P)H dehydrogenases in higher plant cells. Annu Rev Plant Physiol 37: 309–334
Møller IM and Rasmusson AG (1998) The role of NADP in the mitochondrial matrix. Trends Plant Sci 3: 21–27
Moore AL, Gemel J and Randall DD (1993) The regulation of pyruvate dehydrogenase activity in pea leaf mitochondria. The effect of respiration and oxidative phosphorylation. Plant Physiol 103: 1431–1435
Moraes TF and Plaxton WC (2000) Purification and characterization of phospho*enol*pyruvate carboxylase from *Brassica napus* (rapeseed) suspension cell cultures. Implications for phospho*enol*pyruvate carboxylase regulation during phosphate starvation, and the integration of glycolysis with nitrogen metabolism. Eur J Biochem 267: 4465–4476
Nakamoto H and Edwards GE (1983) Influence of oxygen and temperature on the dark inactivation of pyruvate, orthophosphate dikinase and NADP-malate dehydrogenase in maize. Plant Physiol 71: 568–573
Neuburger M, Rebeille F, Jourdain A, Nakamura S and Douce R (1996) Mitochondria are a major site for folate and thymidylate synthesis in plants. J Biol Chem 271: 9466–9472
Noctor G and Foyer CH (1998) A re-evaluation of the ATP: NADPH budget during C_3 photosynthesis: A contribution from nitrate assimilation and its associated respiratory activity? J Exp Bot 49: 1895–1908
Noctor G, Arisi ACM, Jouanin L and Foyer CH (1999) Photorespiratory glycine enhances glutathione accumulation in both the chloroplastic and cytosolic compartments. J Exp Bot 50: 1157–1167
Oliver DJ (1994) The glycine decarboxylase complex from plant mitochondria. Annu Rev Plant Physiol Plant Mol Biol 45: 323–337
Padmasree K and Raghavendra AS (1998) Interaction with respiration and nitrogen metabolism. In: Raghavendra AS (ed) Photosynthesis: A Comprehensive Treatise, pp 197–211. University Press, Cambridge
Padmasree K and Raghavendra AS (1999a) Importance of oxidative electron transport over oxidative phosphorylation in optimizing photosynthesis in mesophyll protoplasts of pea (*Pisum sativum* L.). Physiol Plant 105: 546–553
Padmasree K and Raghavendra AS (1999b) Prolongation of photosynthetic induction as a consequence of interference with mitochondrial oxidative metabolism in mesophyll protoplasts of the pea (*Pisum sativum* L.). Plant Sci 142: 29–36
Padmasree K and Raghavendra AS (1999c) Response of photosynthetic carbon assimilation in mesophyll protoplasts to restriction on mitochondrial oxidative metabolism: Metabolites related to the redox status and sucrose biosynthesis. Photosynth Res 62: 231–239
Padmasree K and Raghavendra AS (2000) Photorespiration and interaction between chloroplasts, mitochondria and peroxisomes. In: Yunus M, Pathre U and Mohanty P (eds) Probing Photosynthesis: Mechanism, Regulation and Adaptation, pp 245–261. Taylor and Francis, UK
Pärnik T and Keerberg O (1995) Decarboxylation of primary and end-products of photosynthesis at different oxygen concentrations. J Exp Bot 46: 1439–1447
Parsons HL, Yip JYH and Vanlerberge GC (1999) Increased respiratory restriction during phosphate-limited growth in transgenic tobacco cells lacking alternative oxidase. Plant Physiol 121: 1309–1320
Pascal N, Dumas R and Douce R (1990) Comparison of the kinetic behavior toward pyridine nucleotides of NAD-linked dehydrogenases from plant mitochondria. Plant Physiol 94: 189–193
Paul JS, Sullivan CW and Vulcani BE (1975) Mitochondrial glycolate dehydrogenase in *Cylindritheca fusiformis* and *Nitzschia alba*. Arch Biochem Biophys 169: 152–159
Puntarulo S, Sánchez RA and Boveris A (1988) Hydrogen peroxide metabolism in soybean embryonic axes at the onset of germination. Plant Physiol 86: 626-630
Purvis AC (1997) Role of the alternative oxidase in limiting superoxide production by plant mitochondria. Physiol Plant 100: 165–170
Purvis AC and Shewfelt RL (1993) Does the alternative pathway ameliorate chilling injury in sensitive plant tissues? Physiol Plant 88: 712–718
Purvis AC, Shewfelt RL and Gegogeine JW (1995) Superoxide production by mitochondria isolated from green bell pepper fruit. Physiol Plant 94: 743–749
Raghavendra AS, Padmasree K and Saradadevi K (1994) Interdependence of photosynthesis and respiration in plant cells: Interactions between chloroplasts and mitochondria. Plant Sci 97: 1–14
Rasmusson AG and Møller IM (1990) NADP-utilizing enzymes in the matrix of plant mitochondria. Plant Physiol 94: 1012–1018
Rasmusson AG, Fredlund KM and Møller IM (1993) Purification of a rotenone-insensitive NAD(P)H dehydrogenase from the inner surface of the inner membrane of red beetroot mitochondria. Biochim Biophys Acta 1141: 107–110
Rasmusson AG, Svensson ÅS, Knoop V, Grohmann L and Brennicke A (1999) Homologues of yeast and bacterial rotenone-insensitive NADH dehydrogenases in higher eucaryotes: two enzymes are present in potato mitochondria. Plant J 20: 79–87
Ribas-Carbo M, Lennon AM, Robinson SA, Giles L, Berry JA and Siedow JN (1997) The regulation of electron partitioning between the cytochrome and alternative pathways in soybean cotyledon and root mitochondria. Plant Physiol 113: 903–911
Ribas-Carbo M, Robinson SA, Gonzàlez-Meler MA, Lennon AM, Giles L, Siedow JN and Berry JA (2000) Effects of light on respiration and oxygen isotope fractionation in soybean cotyledons. Plant Cell Environ 23: 983–990

Rich PR and Bonner WD Jr (1978) The sites of superoxide anion generation in higher plant mitochondria. Arch Biochem Biophys 188: 206–213

Riens B and Heldt HW (1992) Decrease of nitrate reductase activity in spinach leaves during a light-dark transition. Plant Physiol 98: 573–577

Robinson SA, Yakir D, Ribas-Carbo M, Giles L, Osmond CB, Siedow JN and Berry JA (1992) Measurements of the engagement of cyanide-resistant respiration in the Crassulacean acid metabolism plant *Kalanchoë daigremontiana* with the use of online oxygen isotope discrimination. Plant Physiol 100: 1087–1091

Rocha-Sosa M, Sonnewald U, Frommer W, Stratmann M, Schell J and Willmitzer L (1989) Both developmental and metabolic signals activate the promoter of a class I patatin gene. EMBO J 8: 23–29

Sabar M, de Paepe R and de Kouchkovsky Y (2000) Complex I impairment, respiratory compensations, and photosynthetic decrease in nuclear and mitochondrial male sterile mutants of *Nicotiana sylvestris*. Plant Physiol 124: 1239–1250

Saradadevi K and Raghavendra AS (1992) Dark respiration protects photosynthesis against photoinhibition in mesophyll protoplasts of pea (*Pisum sativum*). Plant Physiol 99: 1232–1237

Scagliarini S, Trost P and Pupillo P (1990) Glyceraldehyde 3-phosphate:NADP reductase of spinach leaves. Steady state kinetics and effect of inhibitors. Plant Physiol 94: 1337–1344

Scheible W-R, Krapp A and Stitt M (2000) Reciprocal diurnal changes of phospho*enol*pyruvate carboxylase expression and cytosolic pyruvate kinase, citrate synthase and NADP-isocitrate dehydrogenase expression regulate organic acid metabolism during nitrate assimilation in tobacco leaves. Plant Cell Environ 23: 1155–1168

Schuller KA and Randall DD (1989) Regulation of pea mitochondrial pyruvate dehydrogenase complex. Does photorespiratory ammonium influence mitochondrial carbon metabolism? Plant Physiol 89: 1207–1212

Schürmann P and Jacquot JP (2000) Plant thioredoxin systems revisited. Annu Rev Plant Physiol Plant Mol Biol 51: 371–400

Scott-Taggart CP, van Cauwenberghe OR, McLean MD and Shelp BJ (1999) Regulation of γ-aminobutyric acid synthesis in situ by glutamate availability. Physiol Plant 106: 363–369

Shyam R, Raghavendra AS and Sane PV (1993) Role of dark respiration in photoinhibition of photosynthesis and its reactivation in the cyanobacterium *Anacystis nidulans*. Physiol Plant 88: 446–452

Siedow JN and Umbach AL (1995) Plant mitochondrial electron-transfer and molecular-biology. Plant Cell 7: 821–831

Siedow JN and Umbach AL (2000) The mitochondrial cyanide-resistant oxidase: Structural conservation amid regulatory diversity. Biochim Biophys Acta 1459: 432–439

Singh KK, Shyam R and Sane PV (1996) Reactivation of photosynthesis in the photoinhibited green alga *Chlamydomonas reinhardtii*: Role of dark respiration and of light. Photosynth Res 49: 11–20

Stabenau H (1992) Evolutionary changes of enzymes in peroxisomes and mitochondria of green algae. In: Stabenau (ed) Phylogenetic Changes in Peroxisomes in Algae. Phylogeny of Plant Peroxisomes, pp. 63–79. University Publishers, Oldenburg

Stabenau H, Winkler U and Saftel W (1993) Localization of glycolate dehydrogenase in two species of *Dunaliella*. Planta 191: 362–364

Stitt M (1990) Fructose-2,6-bisphosphate as a regulatory molecule in plants. Annu Rev Plant Physiol Plant Mol Biol 41: 153–185

Stitt M (1999) Nitrate regulation of metabolism and growth. Curr Opin Plant Biol 2: 178–186

Tobin AK, Sumar N, Patel M, Moore AL and Stewart GR (1988) Development of photorespiration during chloroplast biogenesis in wheat leaves. J Exp Bot 39: 833–843

Tobin AK, Thorpe JR, Hylton CM and Rawsthorne S (1989) Spatial and temporal influences on the cell-specific distribution of glycine decarboxylase in leaves of wheat (*Triticum aestivum* L.) and pea (*Pisum sativum* L.). Plant Physiol 91: 1219–1225

Toroser D and Huber SC (2000) Carbon and nitrogen metabolism and reversible protein phosphorylation. Adv Bot Res 32: 435–458

Tovar-Méndez A, Rodriguez-Sotres R, Lopez-Valentin DM and Muñoz-Clares RA (1998) Re-examination of the roles of PEP and Mg^{2+} in the reaction catalysed by the phosphorylated and non-phosphorylated forms of phospho*enol*pyruvate carboxylase from leaves of *Zea mays*. Effects of the activators glucose 6-phosphate and glycine. Biochem J 332: 633–642

Tovar-Méndez A, Mújica-Jiménez C and Muñoz-Clares RA (2000) Physiological implications of the kinetics of maize leaf phospho*enol*pyruvate carboxylase. Plant Physiol 123: 149–160

Vander Heiden MG, Chandel NS, Li XX, Schumacker PT, Colombini M and Thompson CB (2000) Outer mitochondrial membrane permeability can regulate coupled respiration and cell survival. Proc Natl Acad Sci USA 97: 4666–4671

Van Quy L, Foyer C and Champigny ML (1991) Effect of light and NO_3^- on wheat leaf phospho*enol*pyruvate carboxylase activity. Evidence for covalent modulation of the C_3 enzyme. Plant Physiol 97: 1476–1482

Vanlerberghe GC and McIntosh L (1997) Alternative oxidase: From gene to function. Annu Rev Plant Physiol Plant Mol Biol 48: 703–734

Vauclare P, Diallo N, Bourguignon J, Macherel D and Douce R (1996) Regulation of the expression of the glycine decarboxylase complex during pea leaf development. Plant Physiol 112: 1523–1530

Vivekananda J and Oliver DJ (1989) Isolation and partial characterization of the glutamate/aspartate transporter from pea leaf mitochondria using a specific monoclonal antibody. Plant Physiol 91: 272–277

Vivekananda J and Oliver DJ (1990) Detection of the monocarboxylate transporter from pea mitochondria by means of a specific monoclonal antibody. FEBS Lett 260: 217–219

Wada H, Shintani D and Ohlrogge J (1997) Why do mitochondria synthesize fatty acids? Evidence for involvement in lipoic acid production. Proc Natl Acad Sci USA 94: 1591–1596

Wagner AM (1995) A role for active oxygen species as second messengers in the induction of alternative oxidase gene expression in *Petunia hybrida* cells. FEBS Lett 368: 339–342

Wagner AM, Wagner MJ and Moore AL (1998) In vivo ubiquinone reduction levels during thermogenesis in *Araceae*. Plant Physiol 117: 1501–1506

Wedding RT (1989) Malic enzymes of higher plants. Characteristics, regulation, and physiological function. Plant Physiol 90: 367–371

Wigge B, Krömer S and Gardeström P (1993) The redox levels

and subcellular distribution of pyridine nucleotides in illuminated barley leaf protoplasts studied by rapid fractionation. Physiol Plant 88: 10–18

Wiskich JT (1977) Mitochondrial metabolite transport. Annu Rev Plant Physiol 28: 45–69

Wiskich JT and Dry IB (1985) The tricarboxylic acid cycle in plant mitochondria: its operation and regulation. In: Douce R and Day DA (eds) Encyclopedia of Plant Physiology, Vol. 18. Higher Plant Cell Respiration, pp 281–313. Springer-Verlag, Berlin

Woo KC and Osmond CB (1976) Glycine decarboxylation in mitochondria isolated from spinach leaves. Aust J Plant Physiol 3: 771–785

Xue XP, Gauthier DA, Turpin DH and Weger HG (1996) Interactions between photosynthesis and respiration in the green alga *Chlamydomonas reinhardtii*. Characterization of light-enhanced dark respiration. Plant Physiol 112: 1005–1014

Yu C, Claybrook DL and Huang AHC (1983) Transport of glycine, serine and proline into spinach leaf mitochondria. Arch Biochem Biophys 227: 180–187

Zeiher CA and Randall DD (1990) Identification and characterization of mitochondrial acetyl-coenzyme A hydrolase from *Pisum sativum* L. seedlings. Plant Physiol 94: 20–27

Zhu DH and Scandalios JG (1993) Maize mitochondrial manganese superoxide dismutases are encoded by a differentially expressed multigene family. Proc Natl Acad Sci USA 90: 9310–9314

Zoglowek C, Krömer S and Heldt HW (1988) Oxaloacetate and malate transport by plant mitochondria. Plant Physiol 87: 109–115

Chapter 11

Alternative Oxidase: Integrating Carbon Metabolism and Electron Transport in Plant Respiration

Greg C. Vanlerberghe* and Sandi H. Ordog
Division of Life Science and Department of Botany, University of Toronto at Scarborough, 1265 Military Trail, Scarborough, ON, Canada M1C1A4

Summary 173
I. Integration in Plant Respiration 174
II. The Alternative Oxidase in Plant Mitochondrial Electron Transport 174
III. Regulation of Alternative Oxidase 176
A. Biochemical Regulation of Alternative Oxidase Activity by the Carbon and Redox Status of the Mitochondrion 176
B. Regulation of Alternative Oxidase Gene Expression—Links to the Carbon and Redox Status of the Mitochondrion? 180
IV. Physiological Function of Alternative Oxidase 181
A. A General Role to Integrate Carbon Metabolism with Mitochondrial Electron Transport and to Prevent the Excessive Mitochondrial Generation of Active Oxygen Species 181
B. Roles in Specific Cellular and Developmental Processes 185
1. Thermogenesis 186
2. Root Development 186
3. Reproductive Development 186
4. Plant-Pathogen Interactions and Cell Death 186
Acknowledgments 188
References 188

Summary

The plant mitochondrial electron transport chain (ETC) is branched. Electrons pass along the phosphorylating cytochrome pathway or the non-phosphorylating alternative oxidase (AOX) pathway, which represents the CN-resistant component of respiration. The production of monoclonal antibodies, isolation of cDNA and genomic clones, and generation of transgenic plants have dramatically increased our understanding of AOX. The partitioning of electrons to AOX is regulated in a dynamic manner which is dependent upon both the carbon and redox status of the mitochondrion and it is likely that the contribution of AOX to total plant respiration has often been dramatically underestimated. Both matrix pyruvate level and the redox state of matrix NAD(P) alter the kinetic properties of AOX, modulating its ability to compete with the cytochrome pathway for electrons. Site-directed mutagenesis studies are revealing the mechanisms of this biochemical regulation. AOX is encoded, in some species at least, by a multi-gene family and, while the genes are differentially expressed, the functional significance of the different gene products is not yet understood. AOX gene expression may be dependent upon

*Author for correspondence, email: gregv@utsc.utoronto.ca

Christine H. Foyer and Graham Noctor (eds): Photosynthetic Nitrogen Assimilation and Associated Carbon and Respiratory Metabolism, pp. 173–191.

signals which reflect the carbon and redox status of the mitochondrion. Both citrate and active oxygen species (AOS) are potentially important in the signal transduction from mitochondrion to nucleus that controls AOX expression. Metabolic conditions that lead to accumulation of reducing equivalents and pyruvate in the mitochondrial matrix will favor partitioning of electrons toward AOX. Such conditions arise when there is an imbalance between upstream carbon metabolism and downstream electron transport, for example during shifts in metabolism, developmental change, nutrient availability, abiotic or biotic stress. Hence, the general function of AOX may be to integrate the coupled processes of carbon metabolism and electron transport, so as to correct for such imbalances. Experiments with transgenic cells lacking AOX have shown that such integration is critical in preventing both excessive mitochondrial AOS generation and redirections in carbon metabolism. This role for AOX may be particularly important under conditions such as phosphorus-limited growth. Recent data also suggest that AOX plays a role in resistance responses to pathogen attack and in cell death processes.

I. Integration in Plant Respiration

Carbon oxidation in respiratory pathways (glycolysis, oxidative pentose phosphate pathway, tricarboxylic acid (TCA) cycle) is coupled to reduction of pyridine nucleotides. An important route by which the reducing equivalents are subsequently oxidized is through the mitochondrial electron transport chain (ETC). Here, electron transport to O_2 is coupled (through the generation of proton motive force) to the synthesis of ATP from ADP and inorganic phosphate (Pi) by the process of oxidative phosphorylation (Siedow and Day, 2000). Because carbon metabolism and electron transport are coupled processes, there must be mechanisms to integrate them such as to accommodate for changes in the supply of, or demand for, carbon, reducing power and ATP by cell metabolism. The need for integration may be particularly important in plants, where another organelle (the chloroplast) is intimately involved in energy metabolism and where respiration plays a major role in both catabolic and anabolic processes. Recent reviews describing the catabolic and anabolic roles of respiration and the integration of respiration into the whole of cell metabolism include Huppe and Turpin (1994), Plaxton (1996), Noctor and Foyer (1998), Hoefnagel et al. (1998) and Siedow and Day (2000). This chapter will review our current understanding of a unique component of the plant mitochondrial ETC, the alternative oxidase (AOX). The enzyme may play an important general role in the integration of carbon metabolism and electron transport, as well as having a role in specific cellular and developmental processes. It is a component of primary metabolism for which a wealth of new information has appeared in recent years.

Abbreviations: AA – antimycin A; AOS – active oxygen species; AOX – alternative oxidase; Cyt – cytochrome; cytOX cytochrome oxidase; ETC – electron transport chain; HR – hypersensitive response; PCD – programed cell death; PEP – phospho*enol*pyruvate; Pi – inorganic phosphate; PK – pyruvate kinase; Q – ubiquinone; Q_r – ubiquinol; SA – salicylic acid; SHAM – salicylhydroxamic acid; TCA – tricarboxylic acid; TMV – tobacco mosaic virus; wt – wild-type

II. The Alternative Oxidase in Plant Mitochondrial Electron Transport

AOX is a mitochondrial inner membrane protein which functions as a component of the plant ETC (see Vanlerberghe and McIntosh, 1997; Simons and Lambers, 1999 for recent reviews). It catalyzes the O_2-dependent oxidation of reduced ubiquinone (Q_r, ubiquinol), producing ubiquinone (Q) and water (Fig. 1). Plant-like AOX's are found in some algae (Weger et al., 1990), fungi (Yukioka et al., 1998), yeast (Minagawa et al., 1992) and protists (Clarkson et al., 1989) but it is amongst higher plants that this ETC component appears to be ubiquitous. Importantly, electron flow from Q_r to AOX is not coupled to the generation of proton motive force and hence is a non-phosphorylating branch of the ETC, bypassing the last two sites of energy conservation associated with the Cyt pathway (Fig. 1). Central questions regarding the branched nature of electron transport in plant metabolism are: 1) What factors determine the partitioning of electrons in the Q pool between the energy-coupled Cyt pathway and the non-coupled AOX pathway? 2) What is the function of the AOX pathway in metabolism and/or other cell processes?

Until the mid-1980s, AOX was best described as the CN-resistant component of plant respiration. Subsequently, Elthon et al. (1989a) developed a

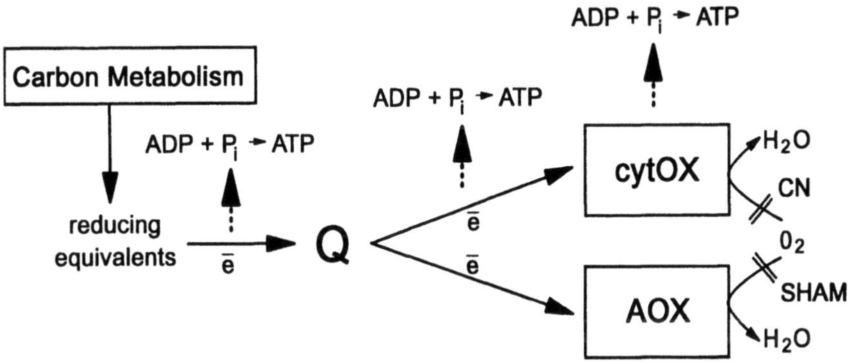

Fig. 1. The branched nature of electron transport in plant mitochondria. Reducing equivalents generated by carbon metabolism supply electrons to reduce the Q pool. Subsequent electron transfer to oxygen can occur via the Cyt pathway (which includes Complex III, Cyt *c* and cytOX) or electrons may be transferred directly to the alternative oxidase (AOX). Dashed arrows represent three sites of energy conservation, where electron transport is coupled to the generation of proton motive force used to synthesize ATP. Note that the AOX pathway bypasses two of these three sites. Chemical inhibitors of electron transfer components include antimycin A (inhibiting Complex III), CN (inhibiting cytOX) and salicylhydroxamic acid (SHAM), which inhibits AOX.

monoclonal antibody (AOA) raised against a *Sauromatum guttatum* AOX protein. This antibody has since been used to facilitate the identification and quantification of AOX in a wide range of species, its usefulness arising from the fact that it recognizes a highly conserved sequence among plant AOX proteins (Finnegan et al., 1999). The AOA antibody was used to identify the nuclear gene encoding AOX when cDNA and genomic clones were isolated from *S. guttatum* (Rhoads and McIntosh, 1991, 1993). Many other AOX genes have since been isolated and multi-gene families have been identified in some plant species (Whelan et al., 1996; Saisho et al., 1997). Sense and antisense constructs of AOX genes have been used to generate transgenic plants with increased and decreased levels of AOX protein (Vanlerberghe et al., 1994). Such plants are proving useful in the study of both the biochemical regulation and physiological function of AOX (Vanlerberghe et al., 1995, 1997, 1998, 1999; Kitashiba et al., 1999; Maxwell et al., 1999; Parsons et al., 1999). Sequence data have been used to generate models of AOX structure and it is thought that the active site contains a binuclear iron center (Siedow et al., 1995; Andersson and Nordlund, 1999). The sequence motifs required for import of AOX into the mitochondrion have also been extensively examined (Tanudji et al., 1999). Recently, a thylakoid membrane protein has been identified which shows significant homology to the mitochondrial AOX and may represent the terminal oxidase in chlororespiration (Carol et al., 1999; Wu et al., 1999; Cournac et al., 2000).

Regarding AOX activity, particularly in vivo, it is important to distinguish between AOX *capacity* and AOX *engagement*. The AOX capacity of a plant cell is generally measured by the addition of a Cyt pathway inhibitor (such as CN) followed by the addition of an AOX inhibitor, such as salicylhydroxamic acid (SHAM) or n-propyl gallate. Then, capacity is generally defined as the O_2 uptake resistant to the Cyt pathway inhibitor and sensitive to the AOX inhibitor. AOX capacity is thus a measure of the *maximum possible flux* of electrons to AOX, a measure which is probably most often dependent upon AOX protein level but which could be dependent upon other limiting components in respiration, particularly when AOX protein levels are high. AOX capacity does not give any indication of the actual flux of electrons to AOX in the cell (prior to the introduction of inhibitor), but is useful to give an indication of the level of AOX expression in a cell. Alternatively, AOX engagement is a measure of the *actual flux* of electrons to AOX within a cell. This measure is much more difficult to determine and it is probably fair to say that we still have a paucity of such data. One approach to measuring engagement is to examine the ability of an AOX inhibitor (in the absence of a Cyt pathway inhibitor) to decrease O_2 uptake. However, it must be realized that this approach may underestimate the engagement of AOX or even indicate a lack of AOX engagement under conditions in which AOX was indeed engaged (see Day et al., 1996 for a critical discussion of these points). At best then, this approach can only give an indication that some level of engagement was taking place. The most reliable way to measure engagement would appear to be an oxygen isotope discrimination technique originally developed by Guy et al. (1989). This noninvasive method is based on the observation that AOX and cytochrome oxidase (cytOX) discriminate to different extents against heavy labeled O_2 ($^{18}O^{16}O$). Systems for both gas-phase and aqueous phase measurement of plant respiration using this technique have been further developed (Robinson et al., 1995) but more data of this type are clearly needed if we are to understand the physiological and developmental conditions in which AOX is being utilized.

III. Regulation of Alternative Oxidase

A. Biochemical Regulation of Alternative Oxidase Activity by the Carbon and Redox Status of the Mitochondrion

Q_r is the common substrate of the energy-coupled Cyt pathway and the non-coupled AOX pathway. Earlier experiments showed that while the activity of the Cyt pathway varied linearly with the redox poise of the Q pool, AOX was not active until the level of Q_r reached a threshold level (Moore et al., 1988). Such studies strengthened the view that AOX acted as an 'energy overflow' pathway, only becoming engaged in respiration when Cyt pathway activity was saturated (Lambers, 1982). Such metabolic conditions might arise when respiratory substrate is plentiful, leading to a flood of reducing equivalents into the ETC, and/or when cell adenylate energy charge is high, such that oxidative phosphorylation is restricted by the availability of ADP. Further with this hypothesis, it was envisioned that continued carbon flux through the TCA cycle and supported by AOX might be critical to provide carbon intermediates during periods of extensive biosynthesis (Lambers, 1982).

New insight into the biochemical regulation of AOX activity have now provided a more refined view of what factors determine the partitioning of electrons between AOX and the Cyt pathway. Importantly, this view suggests that the AOX pathway is not simply an *overflow* of the Cyt pathway but rather that partitioning is regulated in a more dynamic manner which is dependent upon both the carbon and redox status of the mitochondrion. Specifically, both matrix pyruvate level and the redox state of the matrix pyridine nucleotide pool act to alter the kinetic properties of AOX, hence modulating its ability to actually *compete* with the Cyt pathway for electrons

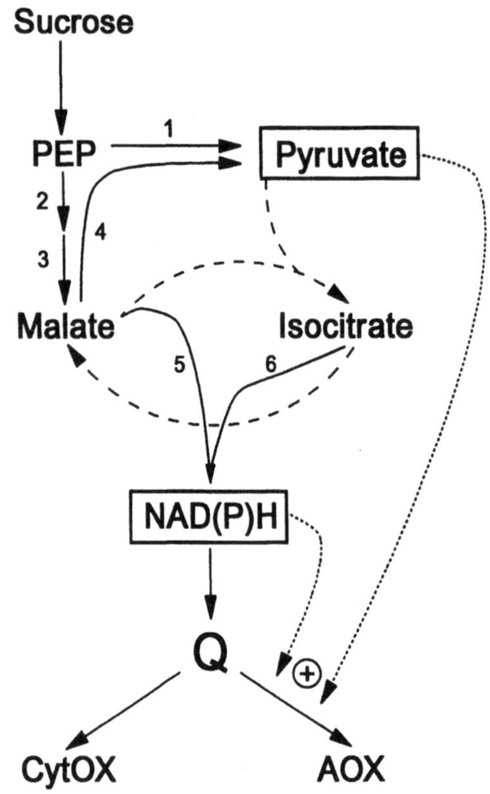

Fig. 2. Feedforward activation of alternative oxidase (AOX) by carbon metabolism. The kinetic properties of AOX can be modulated by the carbon (pyruvate) and redox (pyridine nucleotide pool) status of the mitochondrion, thereby altering the partitioning of electrons between cytOX and AOX. See text for further details. This regulation functions to balance upstream carbon metabolism and downstream electron transport when these coupled processes become imbalanced by changes in the supply of, or demand for, carbon, reducing power, and ATP. Numbers refer to enzymatic reactions discussed in the text: (1) pyruvate kinase; (2) phosphoenolpyruvate (PEP) carboxylase; (3) malate dehydrogenase (producing malate); (4) malic enzyme; (5) malate dehydrogenase (producing oxaloacetate); (6) isocitrate dehydrogenase.

(Fig. 2). Below we will describe our current understanding of the biochemical mechanisms by which this is achieved and then describe how such mechanisms act to integrate carbon metabolism in glycolysis and the TCA cycle with the ETC.

AOX can exist in the inner mitochondrial membrane as either a non-covalently linked or covalently linked dimer, which is thought to consist of similar or identical subunits (Umbach and Siedow, 1993). The dimer, when covalently linked by a disulfide bond *between* the two subunits, is a less active form of AOX (as determined by in organello assays), while reduction of the disulfide bond to its component sulfhydryls produces a more active form. In other words, there is a redox modulation of AOX activity by sulfhydryl/disulfide interconversion. The two forms can be interconverted artificially by treatment of mitochondria with the reductant dithiothreitol and the oxidant diamide. The forms can then be visualized by non-reducing SDS-PAGE and immunoblot analysis (Umbach and Siedow, 1993).

The in organello mechanism of reduction of AOX to its more active form is mediated by the oxidation of specific TCA cycle substrates, notably isocitrate and malate (Vanlerberghe et al., 1995)(Fig. 2). Assays with tobacco leaf mitochondria showed that AOX reduction in response to isocitrate or malate oxidation occurred rapidly, indicating that the sulfhydryl/disulfide system was capable of providing short-term fine regulation of AOX activity. Other substrates (succinate, glycine, 2-oxoglutarate, pyruvate, external NAD(P)H), while effectively oxidized, were ineffective at reducing AOX. A plausible explanation is that intramitochondrial reducing equivalents generated by the activity of isocitrate dehydrogenase (when mitochondria are given citrate or isocitrate) or malate dehydrogenase (when mitochondria are given malate) supports AOX reduction. The substrate specificity suggests that specifically NADPH is required for AOX reduction since, among the substrates tested, only isocitrate and malate oxidation are potentially coupled to reduction of NADP in plant mitochondria. This is because both the mitochondrial NAD-malate dehydrogenase and NAD-malic enzyme can utilize NADP effectively and because plant mitochondria have an NADP-specific isocitrate dehydrogenase in addition to the NAD-specific enzyme (Møller and Rasmusson, 1998). Recently, the first plant cDNA encoding a mitochondrial NADP-specific isocitrate dehydrogenase was cloned (Gálvez et al., 1998). It will be of interest to establish whether it plays a critical role in AOX reduction.

The above findings also suggest that AOX reduction is mediated by a mitochondrial thioredoxin or glutathione system, both of which require specifically NADPH. Components of each of these systems are identified in plant mitochondria, but their specific roles in such mitochondria are poorly understood and their role in AOX reduction is not confirmed (Møller and Rasmusson, 1998). As well, it has recently been reported that plant mitochondria contain appreciable non-energy linked transhydrogenase activity, an activity which could couple the oxidation of strictly NAD-linked substrates with NADPH production, thus bypassing a strict requirement for NADP-linked substrate oxidation in the TCA cycle (Bykova et al., 1999).

In addition to the sulfhydryl/disulfide regulatory system, AOX activity is strongly dependent upon the presence of particular α-keto acids, most notably pyruvate, but also including glyoxylate, hydroxypyruvate, and 2-oxoglutarate (Millar et al., 1993). Pyruvate activation takes place from within the mitochondrial matrix, is fully reversible, and is not dependent upon pyruvate metabolism (Millar et al., 1993, 1996). Further, only the more active reduced form of AOX is subject to pyruvate activation (Umbach et al., 1994). Hence, significant AOX activity in tobacco mitochondria was dependent upon *both* reduction of the regulatory disulfide bond and the presence of pyruvate (Vanlerberghe et al., 1995). Pyruvate acts to increase the V_{max} of AOX (without any significant effect on its affinity for Q_r), possibly by preventing inhibition of the enzyme by Q (Hoefnagel and Wiskich, 1998). Early studies also suggested that pyruvate action was due to its interaction with a Cys sulfhydryl to form a thiohemiacetal since activation was mimicked by iodoacetate (Umbach and Siedow, 1996) and evidence has since emerged to further support this hypothesis (see below).

Given that particular Cys residues might be involved in both the sulfhydryl/disulfide regulatory system and the mechanism of pyruvate activation, several studies have utilized site-directed mutagenesis of cloned AOX genes to further investigate these regulatory mechanisms. Based on a cDNA sequence, a tobacco AOX protein was shown to include two Cys, at positions 126 and 176 in the N-terminal hydrophilic domain (Vanlerberghe and McIntosh, 1994). These were candidates for involvement in the redox modulation and/or pyruvate activation of AOX because they were predicted to reside in the matrix and were the only two Cys completely conserved among the known plant sequences. Hence, site-directed mutagenesis was employed and transgenic tobacco plants expressing high levels of different mutated AOX proteins were generated (Vanlerberghe et al., 1998). The regulatory properties of these AOX proteins were then studied in mitochondria isolated from the plants. Mutation of Cys-126 to Ala produced an AOX that could no longer be converted to the disulfide-linked less active form, thus identifying the more N-terminal Cys as being responsible for redox modulation of AOX. This mutation also resulted in complete loss of pyruvate activation, providing circumstantial evidence that pyruvate activation was dependent upon the Cys-126 sulfhydryl (such as for the formation of a thiohemiacetal). Mutation of Cys-176 indicated that it did not play any apparent role in either redox modulation or pyruvate activation (Vanlerberghe et al., 1998).

Rhoads et al. (1998), expressing mutated *Arabidopsis* AOX proteins in *Escherichia coli*, provided more direct evidence that pyruvate interacts with the more N-terminal Cys residue to form a thiohemiacetal. They substituted this Cys residue with the acidic residue Glu, a residue that might substitute for the thiohemiacetal if the carboxyl group on the thiohemiacetal is the activating moiety. Indeed, the resultant AOX enzyme displayed significant activity in the absence of pyruvate. It has also been shown that pyruvate can protect the tobacco Cys-126 sulfhydryl against oxidation during mitochondrial isolation, additional evidence that pyruvate interacts directly with this Cys sulfhydryl (Vanlerberghe et al., 1999).

It has also been confirmed for soybean AOX that the more N-terminal Cys residue is responsible for both redox modulation and pyruvate activation (Djajanegara et al., 1999). This study also found that substitution of the N-terminal Cys of either the soybean or *Arabidopsis* AOX with Ser generated an enzyme which could be specifically activated by succinate, whereas the native enzymes did not readily respond to succinate. It has also been confirmed for tobacco AOX that substitution of Cys-126 by Ser generates an AOX protein subject to succinate activation (G. C .Vanlerberghe, unpublished). Interestingly, there has now been reported a rice AOX in which Ser occurs naturally in place of the N-terminal Cys (Ito et al., 1997). Presumably, this protein will not be subject to the same redox modulation and pyruvate activation described for other AOX proteins, but rather may be regulated by matrix succinate concentration. As well, the AOX in fungi is monomeric and not activated by pyruvate but rather strongly activated by GMP (Umbach and Siedow, 2000). Since several plant species appear to contain a family of AOX genes (see below), it is possible that different isoforms will be both differentially expressed and differentially regulated.

The sulfhydryl/disulfide regulatory system and its effect on AOX activity have been extensively studied in organello but the in vivo significance of this regulation is more difficult to evaluate. An approach taken with pea leaf tissue was to infer the redox state of the protein in vivo by examining its redox state following mitochondrial isolation (Lennon et al., 1995). However, Umbach and Siedow (1997) found that the regulatory sulfhydryl can undergo oxidation during mitochondrial isolation and that the inclusion of sulfhydryl reagents in the mitochondrial isolation media, while preventing this oxidation, also led to a reduction of the oxidized form. Another approach has been to analyze the protein from a total cellular protein extract, thus bypassing the mitochondrial isolation step (Millar et al., 1998; Millenaar et al., 1998). For example, Millar et al. (1998) found a good correlation between AOX protein form and in vivo AOX activity (measured by oxygen isotope discrimination) during root development.

Several approaches were taken to determine the in vivo redox status of tobacco AOX and to evaluate the physiological significance of the sulfhydryl/disulfide system for short-term regulation of AOX activity (Vanlerberghe et al., 1999). Results obtained after mitochondrial isolations were compared with those obtained by a rapid, whole-cell protein extraction procedure. Also, pyruvate was included in mitochondrial and whole-cell protein extraction buffers as this metabolite was shown to protect against oxidation of AOX, presumably due to its interaction with the Cys-126 sulfhydryl (see above). As a whole, the results indicated that the sulfhydryl/disulfide system is predominantly in the reduced form in vivo under a range of respiratory conditions. Nonetheless, it was shown that increases in AOX activity in suspension cells (such as after inhibition of the Cyt pathway with antimycin A) correlated with a slight further reduction of AOX (Vanlerberghe et al., 1999).

The use of whole-cell protein extracts to examine the redox state of the *leaf* protein has not been reported, probably due to difficulties in visualizing the protein on immunoblots from such extracts (G. C. Vanlerberghe, unpublished). Nonetheless, when leaf mitochondria are isolated in the presence of pyruvate to protect against oxidation of the regulatory sulfhydryl, AOX is again present predominantly in the reduced form (Vanlerberghe et al., 1999). Clearly, more work is required to establish the degree to which the AOX sulfhydryl/disulfide system regulates in a dynamic way the partitioning of electrons to AOX. For example, it is possible that the short-term fine regulation of AOX activity is predominantly dependent upon matrix levels of pyruvate and that the sulfhydryl/disulfide system is utilized for more long-term coarse regulation of AOX, such as during tissue developmental changes (Millar et al., 1998).

Given our current understanding of the biochemical regulation of AOX, metabolic conditions that lead to accumulation of mitochondrial NAD(P)H and/or pyruvate have the potential to favor the partitioning of electrons in the Q pool toward AOX (Fig. 2). Indeed, it has now been demonstrated with isolated mitochondria that when AOX is fully activated, it does not behave as an 'overflow' pathway but rather *competes* with the Cyt pathway for electrons (Hoefnagel et al., 1995; Ribas-Carbo et al., 1995). In other words, the partitioning of electrons to AOX is not a static switch (overflow) but rather a dynamic system responding to the availability of carbon and reducing power in the mitochondrion. Given these insights, it is likely that the contribution of AOX to plant respiration has been widely underestimated in the past (see Day et al., 1996 for a critical discussion of this point).

Conversion of AOX to its active form in response to reduction of the mitochondrial pyridine nucleotide pool provides a mechanism to integrate electron transport with carbon metabolism in the TCA cycle. For example, a limitation of TCA cycle turnover by the ETC will result in a more reduced pyridine nucleotide pool and favor conversion of AOX to its more active reduced form. This will effectively increase the capacity of electron transport, favoring oxidation of the pyridine nucleotide pool and allowing increased turnover of the TCA cycle. Alternatively, activation of AOX by matrix pyruvate level provides a mechanism to integrate electron transport with carbon metabolism in glycolysis. First, pyruvate synthesis from phospho*enol*pyruvate (PEP) by the enzyme pyruvate kinase (PK) is a key regulatory step in plant glycolysis (Plaxton, 1996). The reaction is far removed from equilibrium, such that increases in glycolytic flux result in decreases in PEP and increases in pyruvate. Hence, a high rate of glycolytic flux could effectively increase the capacity of electron transport by activating AOX. Second, PK requires ADP as substrate, and the synthesis of pyruvate may depend on the degree to which glycolytic flux is restricted by ADP availability (adenylate control). Hence, a strict limitation of glycolytic flux by ADP limitation of PK may lower the pyruvate level, leading to inactivation of AOX under conditions when substrate supply to the mitochondrion is limiting. Note that this contrasts with adenylate control of oxidative phosphorylation in the mitochondrion, in which case AOX activity could be favored by an accumulation of pyruvate and/or increased reduction of the pyridine nucleotide pool. It must also be kept in mind that, in plants, the combined action of PEP carboxylase, malate dehydrogenase and mitochondrial NAD-malic enzyme provides a potential route to generate pyruvate while bypassing PK. The interaction of AOX with these steps in carbon metabolism will be discussed below.

AOX catalyzes a non-phosphorylating pathway of electron transport and, as such, its unregulated activity could have a negative impact on carbon balance. It is not surprising then that AOX appears to be subject to complex and tight biochemical regulation. The importance of this is underscored in studies in which transgenic organisms expressing AOX have been generated. When a plant AOX was functionally expressed in *Schizosaccharomyces pombe* (a yeast which normally lacks AOX), AOX was highly engaged in respiration, competing effectively with the Cyt pathway for electrons (Affourtit et al., 1999). The authors suggested that mechanisms which control AOX engagement in plants under physiological conditions were non-operative in the yeast. As a result of the unregulated AOX activity, growth rate and growth yield of the yeast were both lowered significantly. Alternatively, when tobacco AOX was constitutively expressed at high levels in tobacco, it had no obvious impact on growth, at least under normal growth conditions (Vanlerberghe et al., 1994) and it did not significantly increase the partitioning of electrons to AOX under different conditions (R. D. Guy and G. C. Vanlerberghe, unpublished). Hence, while the level of AOX protein in a tissue will likely determine the maximum possible partitioning of electrons to the AOX pathway, it is the biochemical regulatory mechanisms which ultimately determine the level of AOX engagement. It has also been observed that, at least in some tissues, the absolute concentration of Q in the mitochondrial membrane (in addition to the redox poise of Q) may be an important factor regulating AOX engagement (Ribas-Carbo et al., 1997).

Finally, it should be noted that while substitution of the N-terminal regulatory Cys in tobacco AOX led to a dramatic loss of in organello AOX activity (due to a lack of pyruvate activation), the mutant enzyme nonetheless showed high activity in vivo (Vanlerberghe et al., 1998). Hence, it is likely that there are still unknown mechanisms capable of promoting AOX activity in vivo and possibly substituting for α-keto acid activation.

B. Regulation of Alternative Oxidase Gene Expression—Links to the Carbon and Redox Status of the Mitochondrion?

The expression of AOX has been examined in cells and tissues by measures of CN-resistant respiration, AOX protein level and AOX mRNA level. As a whole, such studies indicate that most plant tissues express the pathway but that the level of expression is highly variable, tissue specific, responsive to both biotic and abiotic stress, and dependent upon developmental processes and growth conditions (Vanlerberghe and McIntosh, 1997).

In both soybean and *Arabidopsis*, a family of differentially expressed AOX genes has been identified (Whelan et al., 1996; Saisho et al., 1997). For example, in soybean, three AOX genes are identified (AOX1, AOX2, AOX3) but only AOX1 expression is rapidly induced in response to inhibition of the Cyt pathway at Complex III by antimycin A (AA)(Whelan et al., 1996). Similarly, only one of four *Arabidopsis* AOX genes responds to AA (Saisho et al., 1997). In soybean, the different AOX genes are expressed in a tissue-dependent manner (Finnegan et al., 1997) and are differentially expressed in the cotyledons during postgerminative development (McCabe et al., 1998). At present, the functional significance of the different AOX gene products is unclear. There is some evidence that the soybean AOX2 and AOX3 gene products display different sensitivity to pyruvate activation (Finnegan et al., 1997). It is also possible that AOX (which functions as a dimer) could consist of a mixture of homodimeric and heterodimeric proteins, which may have different properties (Finnegan et al., 1997).

Rapid induction of the AOX pathway at the gene expression level by chemical inhibition of the Cyt pathway is a phenomenon which occurs in both plants (Vanlerberghe and McIntosh, 1992, 1994; Wagner and Wagner, 1997) and other organisms (Bertrand et al., 1983; Sakajo et al., 1991). Clearly, a mechanism exists whereby AOX expression responds to changes in Cyt pathway activity. How this status of electron transport is perceived and then transmitted to the nucleus is unknown but both physiological signals (Vanlerberghe and McIntosh, 1996) and the products of other genes (Bertrand et al., 1983) are likely involved. We summarize below our current understanding of what signal(s) may be involved in regulating AOX expression.

AOX gene expression may respond to a particular metabolite, the level of which reflects some key parameter of respiratory status. To examine this possibility, different TCA cycle and related metabolites were supplied exogenously to tobacco suspension cells and their effect on AOX expression was determined (Vanlerberghe and McIntosh, 1996). Within two hours of the addition of 10 mM citrate, AOX mRNA had increased almost four-fold and this was followed by a large increase in AOX capacity and protein. Similarly, when cellular citrate levels were elevated by inhibiting aconitase with mono-fluoroacetate, AOX was induced. These results are interesting in several respects. Citrate is the first organic acid of the TCA cycle and its accumulation (e.g. because of slowed carbon flow through the TCA cycle) could represent an important physiological signal to integrate TCA cycle metabolism with AOX expression. Also, the citrate treatments shown to rapidly induce AOX do so without inhibiting growth or the capacity of the Cyt pathway (G. C. Vanlerberghe, unpublished) and without inhibiting the respiration rate of the cells (Vanlerberghe and McIntosh, 1996), indicating that induction of AOX can occur independently of such changes. Finally, it is interesting that AOX expression might be linked, via the level of citrate, to aconitase activity. In a wide range of organisms including plants, aconitase has been implicated as a particularly sensitive mitochondrial target to inactivation by AOS (Verniquet et al., 1991; Melov et al., 1999). Citrate accumulation as a result of oxidative inactivation of aconitase could act as an important signal to induce the synthesis of additional AOX protein. Additional AOX protein might then act to alleviate the intramitochondrial generation of AOS (see below for a description of this function of AOX) and hence alleviate the inhibition of aconitase. Supporting this model, we showed that treatment of tobacco suspension cells with H_2O_2 resulted in citrate accumulation in the cell (presumably as a result of aconitase inactivation) and that this was accompanied by increased AOX expression (Vanlerberghe and McIntosh, 1996). Also, treatment of *Petunia hybrida* cells with H_2O_2 results in elevated levels of AOX protein (Wagner, 1995) and nuclear run-on assays showed that H_2O_2 stimulates transcription of a fungal AOX gene (Yukioka et al., 1998).

Studies in fungi have found that AA induction of AOX is suppressed by low oxygen conditions or by the addition of scavengers of AOS such as plant flavonoids (Minagawa et al., 1992 and references therein). Flavonoids can also block AA-induced AOX expression in tobacco cells (G. C. Vanlerberghe, unpublished). These observations suggest that AOS (presumably generated as a result of the over-reduction of ETC components following AA addition) is an important intermediate in the induction. However, inhibition of the Cyt pathway by AA, while strongly inducing AOX, does not result in citrate accumulation (Vanlerberghe and McIntosh, 1996). Rather, carbon accumulation occurs further upstream at pyruvate (Vanlerberghe et al., 1997).

The above findings suggest that AOS generation in the mitochondrion can also act independently of the inhibition of aconitase and accumulation of citrate to induce AOX. Studies in a wide range of organisms suggest that AOS are important signaling molecules, taking part in the regulation of diverse cellular functions. However, little is known of the specific AOS involved in particular signaling pathways or the specific mechanism by which such oxidants act. It is likely that such oxidants have direct protein targets (redox sensors?), the function of which can be reversibly altered by exposure to the AOS. Examples in the literature of how such alteration of function could occur include the oxidation of specific reactive cysteine residues in proteins (Zheng et al., 1998), modification of particular protein-protein interactions (Saitoh et al., 1998) and protein S-glutathiolation (Klatt and Lamas, 2000). Whether a particular redox sensor is an important component in the signal transduction pathway from mitochondrion to nucleus which regulates AOX gene expression is not known.

Recently, Tsuji et al. (2000) reported that submergence (hypoxic treatment) of rice increased the level of gene transcript for alcohol dehydrogenase while decreasing the transcript level of an AOX gene. Both effects could be blocked by ruthenium red which is believed to inhibit Ca^{2+} fluxes from organelles, including the mitochondrion. The effect of ruthenium red could be overcome by supplementing the medium with Ca^{2+}. This result suggests that calcium flux from mitochondrion to cytosol may be an important signal regulating AOX expression.

When AOX was being primarily viewed as an 'overflow' of the Cyt pathway, it was hypothesized that it may act to oxidize 'excess' carbohydrate (Lambers, 1982). Given such a role, one might expect AOX gene expression to correlate positively with the carbohydrate status of plant tissue. However, despite the wealth of new information on AOX expression, there are no indications that AOX expression correlates positively with carbohydrate status or the pool size of a particular carbohydrate. Hence, while particular carbohydrates are recognized as important signal molecules regulating the expression of diverse genes (Smeekens, 2000), it is unclear whether they represent a primary signal controlling AOX expression.

IV. Physiological Function of Alternative Oxidase

A. A General Role to Integrate Carbon Metabolism with Mitochondrial Electron Transport and to Prevent the Excessive Mitochondrial Generation of Active Oxygen Species

Given our current understanding of the biochemical regulation of the AOX enzyme, metabolic conditions that lead to accumulation of Q_r and/or matrix NAD(P)H and/or matrix pyruvate will favor electron flow toward AOX (Fig. 2). In general, such conditions will arise when there is an imbalance between upstream carbon metabolism and downstream electron transport. Such an imbalance could arise as a result of changes in energy and carbon metabolism, changes in Cyt pathway activity or some combination of both. The resulting changes in Q_r, NAD(P)H or pyruvate levels could then act to increase or decrease AOX activity in order to correct the imbalance. Such imbalances may also generate mitochondrial signals (citrate, AOS etc.) which provide for coarse regulation of AOX via changes in gene expression. Finally, a wide range of developmental, metabolic and environmental factors could trigger such imbalances. These concepts are summarized in Fig. 3.

Transgenic tobacco are being utilized to critically assess the role of AOX in balancing carbon metabolism and electron transport under different physiological conditions such as during P-limitation (Parsons et al., 1999). P is a macronutrient which commonly limits the growth of plants (Raghothama, 1999) and P-limitation had been shown to increase AOX capacity in plant and algal cells (Rychter and Mikulska, 1990 and other references in Parsons et al., 1999) suggesting a role for the pathway under

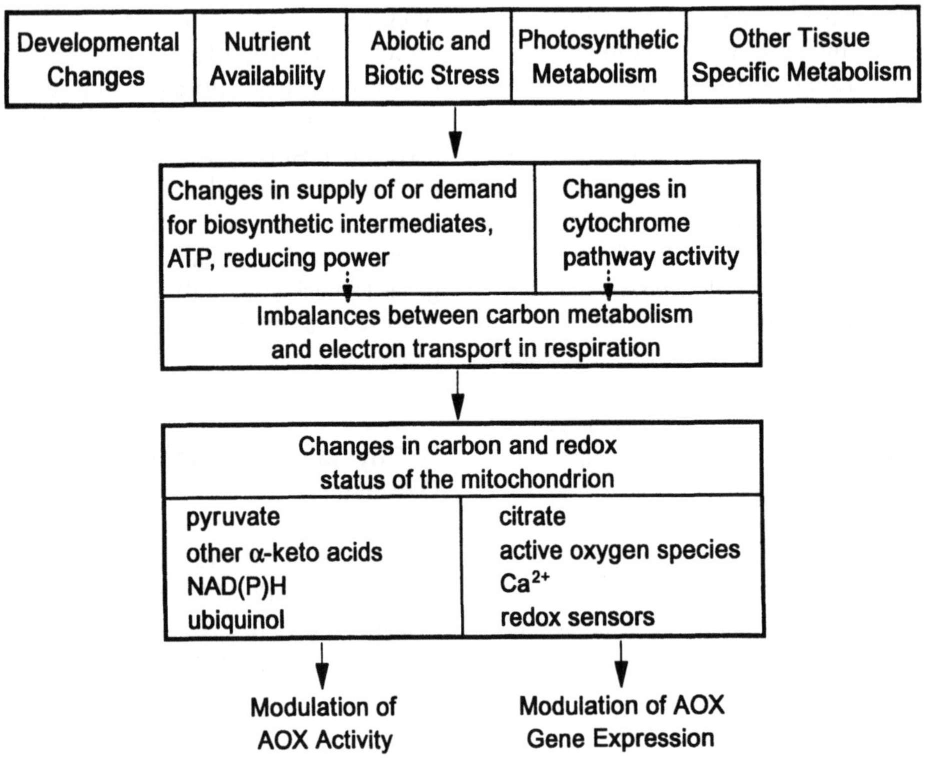

Fig. 3. The modulation of alternative oxidase (AOX) activity (fine regulation) and gene expression (coarse regulation), in response to changes in the carbon and redox status of the mitochondrion. A wide range of developmental, metabolic and environmental factors may impact AOX in such a manner. See text for further details.

these conditions. A common metabolic consequence of P limitation is a significant reduction in the cellular levels of adenylates and Pi (see references in Parsons et al., 1999). Since both key glycolytic reactions and oxidative phosphorylation require ADP and/or Pi as substrate, the absolute concentration of these compounds in the cytosol and mitochondrion is a critical factor controlling flux through respiratory pathways (adenylate control). Nonetheless, extensive studies have shown that plant glycolysis responds in an adaptive manner to P limitation by the induction of alternate pathways that effectively bypass each of the adenylate and/or Pi-dependent steps (see Theodorou and Plaxton, 1995 for a comprehensive review). For example, conversion of PEP to pyruvate (usually associated with the ADP-dependent reaction catalyzed by PK) is functionally replaced by two alternate routes. One route is via a PEP phosphatase, while the second involves the combined action of PEP carboxylase, malate dehydrogenase, and NAD-malic enzyme (Theodorou and Plaxton, 1995). These glycolytic adaptations will allow carbon flow in glycolysis to continue without being subject to severe adenylate control. However, carbon flow beyond glycolysis in the TCA cycle will still be dependent upon continued ETC activity (for turnover of the pyridine nucleotide pool) which itself could be subject to tight adenylate control due to the ADP and Pi requirements of oxidative phosphorylation. Such a limitation, however, could be overcome by induction of the non-phosphorylating AOX pathway.

Hence, P-limited growth represents a physiological condition in which an imbalance between carbon metabolism and electron transport could manifest itself in the absence of AOX. This possibility was investigated by a comparison of wild-type (wt) tobacco suspension cells with transgenic cells which lack AOX expression as a result of the constitutive expression of an AOX antisense transgene (Vanlerberghe et al., 1994). It was found that AOX protein and capacity were indeed dramatically induced in wt cells in response to growth under P limitation, while induction was completely suppressed in the antisense (AS8) cells (Parsons et al., 1999). The lack of AOX in AS8 during P-limited growth resulted in a restriction of respiration, the consequences of which were examined at the level of both carbon metabolism and electron transport. Some results from these studies

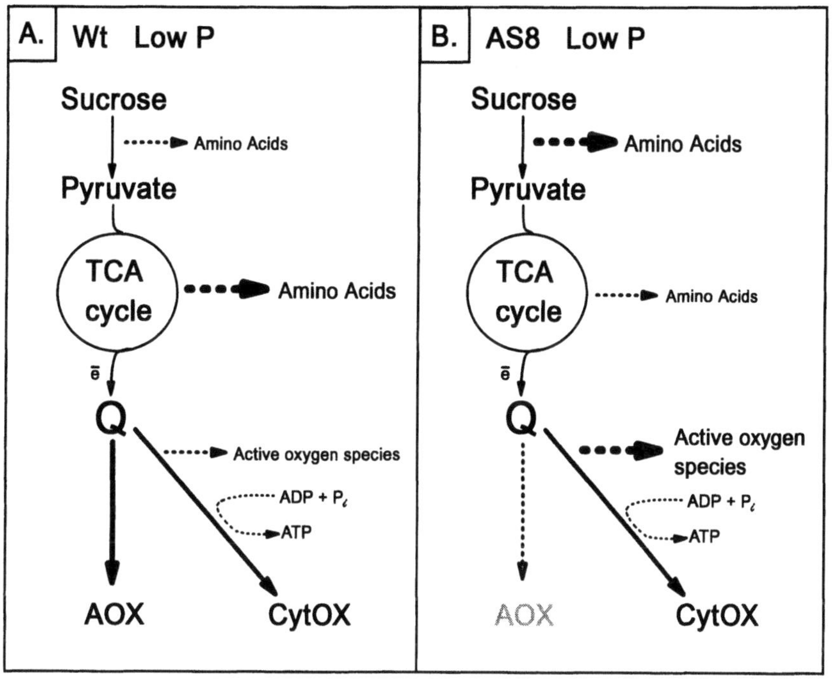

Fig. 4. A model of alternative oxidase (AOX) function during phosphorus (P)-limitation. Under P-limitation, wt tobacco cells (panel A) induce high levels of AOX protein while antisense (AS8) cells (panel B) lack AOX protein. In wt cells, carbon flow and electron transport can proceed with low levels of ADP and Pi that would otherwise be restrictive, due to the activity of AOX. In AS8 cells, the lack of AOX results in a restriction of electron transport and carbon flow through the TCA cycle. As a result, the pool size of amino acids derived from glycolytic intermediates increases while those derived from TCA cycle intermediates decreases (in comparison to wt cells). Also, over-reduction of electron transport chain components in AS8 increases the mitochondrial generation of AOS (in comparison to wt cells). See text for further details.

are summarized in Fig. 4.

At the carbon metabolism level, the lack of AOX resulted in altered patterns of carbon flow, as evidenced by the pool sizes of amino acids whose carbon skeletons are derived from respiratory intermediates. Compared with low-P-grown wt cells, low-P-grown AS8 cells maintained much larger pools of Ser and Tyr and a significantly smaller pool of Gln. For example, while Ser and Tyr accounted for less than 2% of the total amino acid pool of low-P-grown wt cells, they accounted for 27% of the amino acid pool of low-P-grown AS8 cells. The data indicate that with a lack of AOX, there was a shift from the accumulation of amino acids derived from down-stream carbon intermediates to those derived from upstream carbon intermediates (Fig. 4). This is an indication that, in the absence of AOX, respiratory carbon flow beyond glycolysis was restricted under P-limitation.

Interestingly, under normal growth conditions, there were no significant differences in the amino acid pools of the two cell types except that AS8 cells maintained a dramatically smaller pool of Ala, an amino acid derived from pyruvate. While Ala represented 44% of the total amino acid pool of wt cells, it was only 3% of the pool in AS8 cells. This is an indication that pyruvate availability under these conditions was limited and suggests that the low level of AOX present in wt cells under normal growth conditions is critical to relieving the adenylate control of PK (Parsons et al., 1999). During P-limitation, relief of the adenylate control of PK by AOX may not be necessary, if alternate routes of PEP to pyruvate conversion are induced (Theodorou and Plaxton, 1995). Consistent with this, no difference in Ala level was found between the two cell types when grown under P-limitation. While induction of AOX may not be critical to the relief of adenylate control in glycolysis during P-limitation, it may still be critical to the relief of adenylate control at the level of oxidative phosphorylation and account for the observed restrictions in downstream carbon metabolism in AS8 cells under P-limitation. Below, we consider further the approach used to examine whether low-P-grown cells lacking AOX display a restriction at the level of electron transport.

Both in organello and in vivo studies indicate that the mitochondrial ETC is a major source of generation of AOS in eukaryotic cells, including plant cells. The major sites of generation of these AOS are at Complex I and Complex III but the relative contribution of these two sites to AOS generation in plant mitochondria is not completely understood and may depend upon prevailing physiological conditions (Braidot et el., 1999; Casolo et al., 2000). At Complex III, the Q cycle includes a transient quinone radical (ubisemiquinone) which can participate in a one electron transfer to O_2 to generate the superoxide anion (Boveris et al., 1976). The superoxide anion may then be rapidly converted to H_2O_2 by mitochondrial superoxide dismutase (Bowler et al., 1989). The rate of superoxide generation by Complex III is highly dependent upon the proton motive force across the inner mitochondrial membrane since increasing the proton motive force increases the half-life of ubisemiquinone. Hence, chemical inhibition of downstream ETC components or an ADP or Pi limitation of oxidative phosphorylation strongly promotes AOS generation, while the addition of ADP/Pi or protonophorous uncouplers strongly inhibits such AOS generation (Budd et al., 1997; Korshunov et al., 1997).

Parsons et al. (1999) hypothesized that restricted activity of the Cyt pathway as a result of severe adenylate control of oxidative phosphorylation during P-limited growth could promote an over-reduction of ETC components and the associated generation of AOS. However, induction of AOX respiration under low P might function to prevent such over-reduction, hence dampening the generation of AOS. To examine the in vivo generation of AOS by tobacco cells over time with high sensitivity, the cell-permeable probe 2′,7′-dichlorodihydrofluorescin diacetate was used. Indeed, it was found that low-P-grown AS8 cells lacking AOX had dramatically higher rates of generation of AOS than low-P-grown wt cells (Parsons et al., 1999). Further, the high rate of AOS generation in AS8 could be reduced to wt rates by the addition of the uncoupler FCCP. These results are consistent with the idea that AOX, by balancing carbon metabolism and electron transport during P-limited growth, plays an important role in preventing the excessive generation of AOS (Fig. 4).

Interestingly, even under normal growth conditions, AS8 cells appeared to have slightly greater rates of generation of AOS (Parsons et al., 1999). Such an effect was also noted by Maxwell et al. (1999), who showed that the major intracellular site of generation of these AOS was indeed the mitochondrion. Further, they showed that the expression of a catalase gene was dramatically elevated in AS8 cells. Catalase is one of a number of important antioxidant enzymes induced in cells to combat oxidative stress. Taken together, the above studies provide strong in vivo evidence that AOX is a necessary component in the plant mitochondrial ETC to prevent the excessive generation of AOS, a concept originally outlined by Purvis and Shewfelt (1993).

Two other dramatic differences were seen between the low-P-grown wt and AS8 cells (Parsons et al., 1999). First, the cell dimensions (length and width) of AS8 cells were dramatically altered during P-limited growth while the wt cell dimensions showed little response. At present, the significance of this is not known but it is tempting to speculate that it relates to the increased level of oxidative stress being experienced by AS8 cells. Second, the AS8 culture accumulated significantly more dry weight during P-limited growth than did the wt (Parsons et al., 1999). This suggests that during periods of P limitation, the induction of the non-phosphorylating AOX pathway in wt cells acts to compromise overall growth (by decreasing the efficiency of carbon conversion into biomass) while this does not occur in AS8. At first glance, such a compromising of growth in wt cells would appear to be a negative consequence of AOX. However, perhaps such modulation of growth is critically important to ensure an appropriate match between biomass accumulation and P availability, such that the *tissue* concentration of P does not fall below some critical threshold (S. Sieger and G. C. Vanlerberghe, unpublished).

In organello studies also support a role for AOX to lessen the generation of AOS. Such studies have monitored the rate of generation of H_2O_2 by mitochondria during substrate oxidation. It is found that when AOX is chemically inhibited (such as by SHAM) or converted to its inactive form by diamide, the rate of H_2O_2 production increases (Popov et al., 1997; Purvis, 1997; Casolo et al., 2000). Alternatively, when AOX is converted to its active form by dithiothreitol or activated by pyruvate, the rate of H_2O_2 production decreases (Purvis, 1997; Braidot et al., 1999; Casolo et al., 2000).

Another observation is consistent with the idea that AOX acts to prevent the generation of AOS by preventing over-reduction of ETC components. Rapid tissue extractions have been used to directly quantify the in vivo reduction state of the Q pool under different respiratory conditions. When Cyt pathway activity declines, either artificially by the use of chemical inhibitors (Wagner and Wagner, 1997; Millenaar et al., 1998) or as a normal consequence of development (Millar et al., 1998), the reduction state of the Q pool is maintained at a remarkably constant level, correlating with increased AOX activity. An exception to this may be found in specialized thermogenic floral organs which heat up to temperatures well above ambient due to extremely high rates of respiration, occurring primarily via AOX (see below). For example, in *Arum maculatum*, this period of thermogenesis is accompanied by levels of ubiquinone reduction much higher than is typically seen (Wagner et al., 1998).

As outlined in Fig. 3, a wide range of factors could cause general imbalances between carbon metabolism and electron transport, leading to induction and activation of AOX. For example, AOX respiration may have an important role(s) during photosynthesis since photosynthetic metabolism impacts the energy and redox balance of the cell. The potential role(s) of AOX during photosynthesis is discussed in Chapter 10 (Gardeström et al.) and so will not be further discussed here.

Abiotic stresses such as temperature fluctuations may also impact carbon metabolism and electron transport. In this regard, several studies have shown that AOX protein levels increase when plants are grown at low temperature or shifted from high to low temperature (Gonzalez-Meler et al., 1999 and references therein). One hypothesis was that AOX could play a general thermoregulatory role (in tissues other than the specialized floral organs of Arum species, see below) to protect plants from exposure to cold (Moynihan et al., 1995 and references therein). However, based upon models of heat production by metabolic pathways and models of heat dissipation, a general thermoregulatory role for AOX can be excluded (Breidenbach et al., 1997).

The role of AOX in relation to temperature remains unresolved but two recent reports, both utilizing oxygen isotope discrimination to measure AOX engagement, have provided much needed new information (Gonzalez-Meler et al., 1999; Ribas-Carbo et al., 2000). Gonzalez-Meler et al. (1999) examined the respiratory characteristics of soybean and mung bean plants grown at high or low temperature and measured over a temperature range. The respiratory responses were both tissue and species dependent but, in general, this study did not find that the contribution of AOX to total respiration increased at lower measurement temperatures or in plants grown at low temperature. However, the low temperature induced increase in AOX protein seen in some plant tissues may nonetheless be important to maintain the steady-state levels of electron flux to AOX as temperature declines (Gonzalez-Meler et al., 1999). In another study, respiratory characteristics were defined in chilling-sensitive and chilling-tolerant cultivars of maize during the recovery period after a chill treatment (Ribas-Carbo et al., 2000). This study found that there was a large increase in AOX engagement only in the chilling-sensitive cultivar. Also, only the chilling-sensitive cultivar displayed decreased Cyt pathway activity and a lack of growth during the recovery period. This study indicates that AOX engagement may prevail during periods when the Cyt pathway has suffered stress-induced damage or when plant growth has been negatively impacted by stress.

There are other examples in which AOX respiration would appear to be associated with stress conditions in which Cyt pathway activity has declined and/or growth has been curtailed. Such conditions include high salt (Jolivet et al., 1990; Miyasaka et al., 2000), herbicide treatment (Aubert et al., 1997), excess copper (Padua et al., 1999), high CO_2 (Palet et al., 1991), and nitric oxide treatment (Millar and Day, 1996). As a whole, such studies indicate that AOX may play an important general role in the plant response to stress (Simons and Lambers, 1999). Again, such a general role may be to balance carbon metabolism and electron transport when these coupled processes are differentially impacted by the stress condition or when the stress condition alters the demands on metabolism for carbon, reducing power and ATP. More studies utilizing oxygen isotope discrimination or utilizing transgenic plants lacking AOX should provide further insight into the general importance of the AOX pathway in different stress conditions.

B. Roles in Specific Cellular and Developmental Processes

This section will review roles of AOX in particular cellular or developmental processes. In these cases, AOX respiration may again play a general role to integrate the ETC with carbon metabolism or may have a more specific and/or still ill-defined role.

1. Thermogenesis

A well defined example of AOX involvement in development is its role in the thermogenic inflorescense of Arum lilies such as *S. guttatum* (Meeuse, 1975). In the specialized floral organs of such species, extremely high rates of respiration during anthesis generates heat to volatilize insect-attracting chemicals for pollination. This respiration occurs predominantly via AOX due to a developmental increase in AOX capacity and concomitant decrease in Cyt pathway capacity (Elthon et al., 1989b). An important signal molecule involved in these changes in ETC components is salicylic acid (SA) but its mechanism of action is unknown (Raskin et al., 1989). It is interesting, however, that SA can bind and inactivate aconitase (Ruffer et al., 1995) and that aconitase inactivation (resulting in citrate accumulation) is implicated in the induction of AOX (see above). Most studies of thermogenesis have concentrated on the upregulation of AOX, but perhaps investigation of the mechanisms involved in the observed decrease in Cyt pathway capacity could shed important new information on what appears to be a coordinate regulation of ETC components.

2. Root Development

The contribution of the Cyt pathway and AOX to the respiration of developing soybean roots was investigated using the oxygen isotope discrimination technique (Millar et al., 1998). This study found that young root systems (4 day old seedlings) respired almost exclusively via the Cyt pathway but that in older root systems (17 day), more than 50% of total respiration occurred via AOX. This dramatic change in partitioning of electrons occurred concomitant with a 60% decrease in overall respiration rate as well as a seven-fold decrease in growth rate. The respiratory changes did not appear to be due to a dramatic induction of AOX protein but rather to a decrease in maximum cytOX activity and possibly a shift in AOX toward its more active reduced form (Millar et al., 1998). This study is consistent with the idea that AOX may prevail during periods of relatively slow growth when demand for ATP is curtailed and cytOX is downregulated. Under such conditions, the contribution of AOX to total root respiration appears to be very high, at least in soybean.

3. Reproductive Development

Immunohistochemical work in petunia (Conley and Hanson, 1994) and bean (Johns et al., 1993) have shown that AOX protein is abundant in the tapetum and meiocytes during microsporogenesis. Further, Saisho et al. (1997) found that the *Arabidopsis* AOX1b gene (one of four AOX genes identified in *Arabidopsis*) was expressed exclusively in floral tissues. To examine a potential role of AOX in floral development, an *Arabidopsis* AOX gene was expressed in tobacco in antisense orientation and under the control of a tapetum-specific promoter (Kitashiba et al., 1999). At least one plant had a reduced level of AOX in the anthers and this plant displayed dramatically reduced pollen viability. This suggests that AOX plays some critical, but as yet undefined role in pollen development. It also suggests the possibility of using antisense AOX genes to produce male sterility in economically important plants.

In many fruits, the onset of ripening is accompanied by a marked rise in respiration rate known as the climacteric, the function of which is unclear. The climacteric is triggered by the endogenous production of ethylene. In mango (Cruz-Hernández and Gómez-Lim, 1995) and in apple (Duque and Arrabaca, 1999) ripening is associated with increased AOX protein but this is not the case in tomato (Almeida et al., 1999). Also, the level of engagement of AOX during the climacteric is not known for any fruit. Hence, the involvement of AOX in this burst of respiration is unclear. Interestingly, an ethylene-response mutant of *Arabidopsis* did not show a normal induction of AOX (in response to pathogen infection, see below) indicating that ethylene may be an important signal for AOX expression (Simons et al., 1999).

4. Plant-Pathogen Interactions and Cell Death

Recently, several studies examined the potential role of AOX respiration in plant responses to pathogen attack. For example, both pharmacological and correlative evidence suggests that AOX has an active role in the resistance response of tobacco to tobacco mosaic virus (TMV)(Murphy et al., 1999, and references therein). This evidence includes: 1) SA treatment of susceptible (nn-genotype) tobacco induces the expression of AOX and increases resistance to TMV. 2) The N-gene mediated resistance response of tobacco to TMV is associated with increased expression of AOX. 3) Both SA-induced resistance and N-gene mediated resistance are attenuated by the AOX inhibitor SHAM. 4) Inhibitors of the Cyt pathway (CN and AA) induce AOX expression and increase resistance to TMV.

Lennon et al. (1997) used oxygen isotope discrimination to examine AOX engagement following TMV infection of tobacco plants. While TMV infection increased the level of AOX protein in both infected and systemic leaves, there was no evidence that AOX engagement differed in infected versus uninfected plants. However, it is likely that a very localized change in AOX engagement (for example, in or around sites of infection) would have gone undetected by the technique used. Also, further study is required to establish whether the AOX induced in systemic leaves might play an important role if these leaves were subsequently challenged with virus. Clearly, more definitive evidence is required to establish whether AOX plays any active role in resistance of tobacco to TMV (Murphy et al., 1999).

Infection of a plant by a virulent pathogen results in disease while infection by an avirulent pathogen results in plant resistance responses. An important resistance response is the hypersensitive response (HR), in which plant cells in the area of infection undergo a rapid form of programmmed cell death (PCD). The mitochondrion is known to play a critical role in a common form of PCD in animals known as apoptosis (Green and Reed, 1998) but its role in plant PCD (such as during the HR) is poorly understood (Jones, 2000). An important early event in animal apoptosis is the release of Cyt *c* from the inner mitochondrial membrane (where it is involved in electron transport from Complex III to cytOX) to the cytosol (Green and Reed, 1998). This event has several consequences, each of which may play a role in at least some cell death programs. Cyt *c* in the cytosol triggers a cascade of events which culminates in the activation of specific cysteine proteases (caspases) which are required both to further execute the death program and to take part in the ordered disassembly of the cell (Green and Reed, 1998). Cyt *c* loss from the mitochondrion also results in a decline in ATP production and an increase in the generation of AOS due to over-reduction of ETC components upstream of Cyt *c*.

There is some evidence that release of Cyt *c* and activation of caspase-like proteases may also occur in plant PCD (Balk et al., 1999; Stein and Hansen, 1999; Tian et al., 2000). However, given that plants contain an AOX which can be strongly induced when Cyt pathway respiration is blocked (Vanlerberghe et al., 1992, 1994), it is possible that the role of the mitochondrion in plant PCD is different than in animal apoptosis. For example, induction of AOX could maintain some ATP production and alleviate AOS generation. In this case, AOX may act to attenuate cell death programs involving Cyt *c* release.

Studies indicate that AOX expression may in fact be induced in or around tissues undergoing the HR. A differential screening strategy used to identify *Arabidopsis* genes induced early in the HR to a bacterial pathogen identified both AOX and a mitochondrial anion channel gene (Lacomme and Roby, 1999). The early induction of these genes closely paralleled one another, was transient in nature and was specifc to an avirulent interaction. Interestingly, mitochondrial anion channels are involved in the mechanism of Cyt *c* release during animal apoptosis (Green and Reed, 1998). In another study, induction of AOX occurred in the interaction of *Arabidopsis* with either virulent or avirulent bacteria although the induction with the virulent bacteria was significantly delayed (Simons et al., 1999). In this case, AOX induction appeared to correlate with the rapid PCD response associated with the avirulent bacteria and with the delayed and disease associated cell death caused by the virulent bacteria.

Interestingly, an immunohistochemical study in differentiating soybean root showed that AOX protein strongly localized to developing xylem tissue (Hilal et al., 1997). Also, when primary xylem differentiation was delayed by an NaCl treatment of the roots, there was a corresponding delay in the temporal pattern of AOX protein level (Hilal et al., 1998). These studies suggest a possible link between AOX expression and xylem differentiation, a developmental process which culminates in PCD (Groover and Jones, 1999). Recently, Amor et al. (2000) found that an anoxia pretreatment of soybean cells both increased AOX expression and protected cells against death during a subsequent insult with H_2O_2. This protective effect was reversed by AOX inhibitors suggesting that AOX activity during the H_2O_2 treatment was critical to cell survival. While not examined, one possibility is that the oxidative stress had crippled the Cyt pathway and that the induced AOX pathway was then capable of maintaining cell viability, just as it can following chemical inhibition of Cyt pathway activity in tobacco cells (Vanlerberghe et al., 1997).

The above studies are an indication that AOX may play some role in the HR or other PCD responses in plants. Interestingly, nitric oxide has been recently implicated as an important signal in plant defense responses such as HR (Klessig et al., 2000 and references therein) and it is known that nitric oxide is a potent inhibitor of the plant cytOX but not AOX (Millar and Day, 1996). The role of nitric oxide in the regulation of mitochondrial electron transport is reviewed in Chapter 12 (Millar et al.).

Acknowledgments

Our research program on AOX is funded by grants from the Natural Sciences and Engineering Research Council of Canada and we gratefully acknowledge that support.

References

Affourtit C, Albury MS, Krab K and Moore AL (1999) Functional expression of the plant alternative oxidase affects growth of the yeast *Schizosaccharomyces pombe*. J Biol Chem 274: 6212–6218

Almeida AM, Jarmuszkiewicz W, Khomsi H, Arruda P, Vercesi AE and Sluse FE (1999) Cyanide-resistant, ATP-synthesis-sustained, and uncoupling-protein-sustained respiration during postharvest ripening of tomato fruit. Plant Physiol 119: 1323–1329

Amor Y, Chevion M and Levine A (2000) Anoxia pretreatment protects soybean cells against H_2O_2-induced cell death: Possible involvement of peroxidases and of alternative oxidase. FEBS Lett 477: 175–180

Andersson ME and Nordlund P (1999) A revised model of the active site of alternative oxidase. FEBS Lett 449: 17–22

Aubert S, Bligny R, Day DA, Whelan J and Douce R (1997) Induction of alternative oxidase synthesis by herbicides inhibiting branched-chain amino acid synthesis. Plant J 11: 649–657

Balk J, Leaver CJ and McCabe PF (1999) Translocation of cytochrome *c* from the mitochondria to the cytosol occurs during heat-induced programmed cell death in cucumber plants. FEBS Lett 463: 151–154

Bertrand H, Argan CA and Szakacs NA (1983) Genetic control of the biogenesis of cyanide insensitive respiration in *Neurospora crassa*. In: Schweyen RJ, Wolf K and Kaudewitz F (eds) Mitochondria 1983. pp 495–507. Walter de Gruyter, Berlin

Boveris A, Cadenas E and Stoppani AOM (1976) Role of ubiquinone in the mitochondrial generation of hydrogen peroxide. Biochem J 156: 435–444

Bowler C, Alliotte T, De Loose M, Van Montagu M and Inze D (1989) The induction of manganese superoxide dismutase in response to stress in *Nicotiana plumbaginifolia*. EMBO J 8: 31–38

Braidot E, Petrussa E, Vianello A and Macri F (1999) Hydrogen peroxide generation by higher plant mitochondria oxidizing complex I or complex II substrates. FEBS Lett 451: 347–350

Breidenbach RW, Saxton, MJ, Hansen LD and Criddle RS (1997) Heat generation and dissipation in plants: Can the alternative oxidative phosphorylation pathway serve a thermoregulatory role in plant tissues other than specialized organs? Plant Physiol 114: 1137–1140

Budd SL, Castilho RF and Nicholls DG (1997) Mitochondrial membrane potential and hydroethidine-monitored superoxide generation in cultured cerebellar granule cells. FEBS Lett 415: 21–24

Bykova NV, Rasmusson AG, Igamberdiev AU, Gardeström P and Møller IM (1999) Two separate transhydrogenase activities are present in plant mitochondria. Biochem Biophys Res Commun 265: 106–111

Carol P, Stevenson D, Bisanz C, Breitenbach J, Sandmann G, Mache R, Coupland G and Kuntz M (1999) Mutations in the Arabidopsis gene *IMMUTANS* cause a variegated phenotype by inactivating a chloroplast terminal oxidase associated with phytoene desaturation. Plant Cell 11: 57–68

Casolo V, Braidot E, Chiandussi E, Macri F and Vianello A (2000) The role of mild uncoupling and non-coupled respiration in the regulation of hydrogen peroxide generation by plant mitochondria. FEBS Lett 474: 53–57

Clarkson AB, Bienen EJ, Pollakis G and Grady RW (1989) Respiration of bloodstream forms of the parasite *Trypanosoma brucei brucei* is dependent on a plant-like alternative oxidase. J Biol Chem 264: 17770–17776

Conley CA and Hanson MR (1994) Tissue-specific protein expression in plant mitochondria. Plant Cell 6: 85–91

Cournac L, Redding K, Ravenel J, Rumeau D, Josse EM, Kuntz M and Peltier G (2000) Electron flow between photosystem II and oxygen in chloroplasts of photosystem I-deficient algae is mediated by a quinol oxidase involved in chlororespiration. J Biol Chem 275: 17256–17262

Cruz-Hernández A and Gómez-Lim MA (1995) Alternative oxidase from mango (*Mangifera indica*, L.) is differentially regulated during fruit ripening. Planta 197: 569–576

Day DA, Krab K, Lambers H, Moore AL, Siedow JN, Wagner AM and Wiskich JT (1996) The cyanide-resistant oxidase: To inhibit or not to inhibit, that is the question. Plant Physiol 110: 1–2

Djajanegara I, Holtzapffel R, Finnegan PM, Hoefnagel MHN, Berthold DA, Wiskich JT and Day DA (1999) A single amino acid change in the plant alternative oxidase alters the specificity of organic acid activation. FEBS Lett 454: 220–224

Duque P and Arrabaca JD (1999) Respiratory metabolism during cold storage of apple fruit. II. Alternative oxidase is induced at the climacteric. Physiol Plant 107: 24–31

Elthon TE, Nickels RL and McIntosh L (1989a) Monoclonal antibodies to the alternative oxidase of higher plant mitochondria. Plant Physiol 89: 1311–1317

Elthon TE, Nickels RL and McIntosh L (1989b) Mitochondrial events during development of thermogenesis in *Sauromatum guttatum* (Schott). Planta 180: 82–89

Finnegan PM, Whelan J, Millar AH, Zhang Q, Smith MK, Wiskich JT and Day DA (1997) Differential expression of the multigene family encoding the soybean mitochondrial alternative oxidase. Plant Physiol 114: 455–466

Finnegan PM, Wooding AR and Day DA (1999) An alternative oxidase monoclonal antibody recognizes a highly conserved sequence among alternative oxidase subunits. FEBS Lett 447: 21–24

Gálvez S, Roche O, Bismuth E, Brown S, Gadal P and Hodges M (1998) Mitochondrial localization of a NADP-dependent isocitrate dehydrogenase isoenzyme by using the green fluorescent protein as a marker. Proc Natl Acad Sci USA 95: 7813–7818

Gonzalez-Meler MA, Ribas-Carbo M, Giles L and Siedow JN (1999) The effect of growth and measurement temperature on the activity of the alternative respiratory pathway. Plant Physiol 120: 765–772

Green DR and Reed JC (1998) Mitochondria and apoptosis. Science 281: 1309–1312

Groover A and Jones AM (1999) Tracheary element differentiation uses a novel mechanism coordinating programmed cell death and secondary cell wall synthesis. Plant Physiol 119: 375–384

Guy RD, Berry JA, Fogel ML and Hoering TC (1989) Differential fractionation of oxygen isotopes by cyanide-resistant and cyanide-sensitive respiration in plants. Planta 177: 483–491

Hilal M, Castagnaro A, Moreno H and Massa EM (1997) Specific localization of the respiratory alternative oxidase in meristematic and xylematic tissues from developing soybean roots and hypocotyls. Plant Physiol 115: 1499–1503

Hilal M, Zenoff AM, Ponessa G, Moreno H and Massa EM (1998) Saline stress alters the temporal patterns of xylem differentiation and alternative oxidase expression in developing soybean roots. Plant Physiol 117: 695–701

Hoefnagel MHN and Wiskich JT (1998) Activation of the plant alternative oxidase by high reduction levels of the Q-pool and pyruvate. Arch Biochem Biophys 355: 262–270

Hoefnagel MHN, Millar AH, Wiskich JT and Day DA (1995) Cytochrome and alternative respiratory pathways compete for electrons in the presence of pyruvate in soybean mitochondria. Arch Biochem Biophys 318: 394–400

Hoefnagel MHN, Atkin OK and Wiskich JT (1998) Interdependence between chloroplasts and mitochondria in the light and the dark. Biochim Biophys Acta 1366: 235–255

Huppe HC and Turpin DH (1994) Integration of carbon and nitrogen metabolism in plant and algal cells. Annu Rev Plant Physiol Plant Mol Biol 45: 577–607

Ito Y, Saisho D, Nakazono M, Tsutsumi N and Hirai A (1997) Transcript levels of tandem-arranged alternative oxidase genes in rice are increased by low temperature. Gene 203: 121–129

Johns C, Nickels R, McIntosh L and Mackenzie S (1993) The expression of alternative oxidase and alternative respiratory capacity in cytoplasmic male sterile common bean. Sex Plant Reprod 6: 257–265

Jolivet Y, Pireaux JC and Dizengremel P (1990) Changes in properties of barley leaf mitochondria isolated from NaCl-treated plants. Plant Physiol 94: 641–646

Jones A (2000) Does the plant mitochondrion integrate cellular stress and regulate programmed cell death? Trends Plant Sci 5: 225–230

Kitashiba H, Kitazawa E, Kishitani S and Toriyama K (1999) Partial male sterility in transgenic tobacco carrying an antisense gene for alternative oxidase under the control of a tapetum-specific promoter. Mol Breed 5: 209–218

Klatt P and Lamas S (2000) Regulation of protein function by S-glutathiolation in response to oxidative and nitrosative stress. Eur J Biochem 267: 4928–4944

Klessig DF, Durner J, Noad R, Navarre DA, Wendehenne D, Kumar D, Zhou JM, Shah J, Zhang S, Kachroo P, Trifa Y, Pontier D, Lam E and Silva H (2000) Nitric oxide and salicylic acid signaling in plant defense. Proc Natl Acad Sci USA 97: 8849–8855

Korshunov SS, Skulachev VP and Starkov AA (1997) High protonic potential actuates a mechanism of production of reactive oxygen species in mitochondria. FEBS Lett 416: 15–18

Lacomme C and Roby D (1999) Identification of new early markers of the hypersensitive response in *Arabidopsis thaliana*. FEBS Lett 459: 149–153

Lambers H (1982) Cyanide-resistant respiration: a non-phosphorylating electron transport pathway acting as an energy overflow. Physiol Plant 55: 478–485

Lennon AM, Pratt J, Leach G and Moore AL (1995) Developmental regulation of respiratory activity in pea leaves. Plant Physiol 107: 925–932

Lennon AM, Neuenschwander UH, Ribas-Carbo M, Giles L, Ryals JA and Siedow JN (1997) The effects of salicylic acid and tobacco mosaic virus infection on the alternative oxidase of tobacco. Plant Physiol 115: 783–791

Maxwell DP, Wang Y and McIntosh L (1999) The alternative oxidase lowers mitochondrial reactive oxygen production in plant cells. Proc Natl Acad Sci USA 96: 8271–8276

McCabe TC, Finnegan PM, Millar H, Day DA and Whelan J (1998) Differential expression of alternative oxidase genes in soybean cotyledons during postgerminative development. Plant Physiol 118: 675–682

Meeuse BJD (1975) Thermogenic respiration in aroids. Annu Rev Plant Physiol 26: 117–126

Melov S, Coskun P, Patel M, Tuinstra R, Cottrell B, Jun AS, Zastawny TH, Dizdaroglu M, Goodman SI, Huang TT, Miziorko H, Epstein CJ and Wallace DC (1999) Mitochondrial disease in superoxide dismutase 2 mutant mice. Proc Natl Acad Sci USA 96: 846–851

Millar AH and Day DA (1996) Nitric oxide inhibits the cytochrome oxidase but not the alternative oxidase of plant mitochondria. FEBS Lett 398: 155–158

Millar AH, Wiskich JT, Whelan J and Day DA (1993) Organic acid activation of the alternative oxidase of plant mitochondria. FEBS Lett 329: 259–262

Millar AH, Hoefnagel MHN, Day DA and Wiskich JT (1996) Specificity of the organic acid activation of alternative oxidase in plant mitochondria. Plant Physiol 111: 613–618

Millar AH, Atkin OK, Menz I, Henry B, Farquhar G and Day DA (1998) Analysis of respiratory chain regulation in roots of soybean seedlings. Plant Physiol 117: 1083–1093

Millenaar FF, Benschop JJ, Wagner AM and Lambers H (1998) The role of the alternative oxidase in stabilizing the in vivo reduction state of the ubiquinone pool and the activation state of the alternative oxidase. Plant Physiol 118: 599–607

Minagawa N, Koga S, Nakano M, Sakajo S and Yoshimoto A (1992) Possible involvement of superoxide anion in the induction of cyanide-resistant respiration in *Hansenula anomala*. FEBS Lett 302: 217–219

Miyasaka H, Kanaboshi H and Ikeda K (2000) Isolation of several anti-stress genes from the halotolerant green alga *Chlamydomonas* by simple functional expression screening with *Escherichia coli*. World J Microbiol Biotechnol 16: 23–29

Møller IM and Rasmusson AG (1998) The role of NADP in the mitochondrial matrix. Trends Plant Sci 3: 21–27

Moore AL, Dry IB and Wiskich JT (1988) Measurement of the redox state of the ubiquinone pool in plant mitochondria. FEBS Lett 235: 76–80

Moynihan MR, Ordentlich A and Raskin I (1995) Chilling-induced heat evolution in plants. Plant Physiol 108: 995–999

Murphy AM, Chivasa S, Singh DP and Carr JP (1999) Salicylic acid-induced resistance to viruses and other pathogens: A parting of the ways? Trends Plant Sci 4: 155–160

Noctor G and Foyer CH (1998) A re-evaluation of the ATP:NADPH budget during C_3 photosynthesis: A contribution from nitrate assimilation and its associated respiratory activity? J Exp Bot 49: 1895–1908

Padua M, Aubert S, Casimiro A, Bligny R, Millar AH and Day DA (1999) Induction of alternative oxidase by excess copper in sycamore cell suspensions. Plant Physiol Biochem 37: 131–137

Palet A, Ribas-Carbo M, Argiles JM and Azcón-Bieto J (1991) Short-term effects of carbon dioxide on carnation callus cell respiration. Plant Physiol 96: 467–472

Parsons HL, Yip JYH and Vanlerberghe GC (1999) Increased respiratory restriction during phosphate-limited growth in transgenic tobacco cells lacking alternative oxidase. Plant Physiol 121: 1309–1320

Plaxton WC (1996) The organization and regulation of plant glycolysis. Annu Rev Plant Physiol Plant Mol Biol 47: 185–214

Popov VN, Simonian RA, Skulachev VP and Starkov AA (1997) Inhibition of the alternative oxidase stimulates H_2O_2 production in plant mitochondria. FEBS Lett 415: 87–90

Purvis AC (1997) Role of the alternative oxidase in limiting superoxide production by plant mitochondria. Physiol Plant 100: 165–170

Purvis AC and Shewfelt RL (1993) Does the alternative pathway ameliorate chilling injury in sensitive plant tissues? Physiol Plant 88: 712–718

Raghothama KG (1999) Phosphate acquisition. Annu Rev Plant Physiol Plant Mol Biol 50: 665–693

Raskin I, Turner IM and Melander WR (1989) Regulation of heat production in the inflorescences of an *Arum* lily by endogenous salicylic acid. Proc Natl Acad Sci USA 86: 2214–2218

Rhoads DM and McIntosh L (1991) Isolation and characterization of a cDNA clone encoding an alternative oxidase protein of *Sauromatum guttatum* (Schott). Proc Natl Acad Sci USA 88: 2122–2126

Rhoads DM and McIntosh L (1993) The salicylic acid-inducible alternative oxidase gene *aox1* and genes encoding pathogenesis-related proteins share regions of sequence similarity in their promoters. Plant Mol Biol 21: 615–624

Rhoads DM, Umbach AL, Sweet CR, Lennon AM, Rauch GS and Siedow JN (1998) Regulation of the cyanide-resistant alternative oxidase of plant mitochondria: Identification of the cysteine residue involved in α-keto acid stimulation and intersubunit disulfide bond formation. J Biol Chem 273: 30750–30756

Ribas-Carbo M, Berry JA, Yakir D, Giles L, Robinson SA, Lennon AM and Siedow JN (1995) Electron partitioning between the cytochrome and alternative pathways in plant mitochondria. Plant Physiol 109: 829–837

Ribas-Carbo M, Lennon, AM, Robinson SA, Giles L, Berry JA and Siedow JN (1997) The regulation of electron partitioning between the cytochrome and alternative pathways in soybean cotyledon and root mitochondria. Plant Physiol 113: 903–911

Ribas-Carbo M, Aroca R, Gonzalez-Meler MA, Irigoyen JJ and Sanchez-Diaz M (2000) The electron partitioning between the cytochrome and alternative respiratory pathways during chilling recovery in two cultivars of maize differing in chilling sensitivity. Plant Physiol 122: 199–204

Robinson SA, Ribas-Carbo M, Yakir D, Giles L, Reuveni Y and Berry JA (1995) Beyond SHAM and cyanide: Opportunities for studying the alternative oxidase in plant respiration using oxygen isotope discrimination. Aust J Plant Physiol 22: 487–496

Ruffer M, Steipe B and Zenk MH (1995) Evidence against specific binding of salicylic acid to plant catalase. FEBS Lett 377: 175–180

Rychter AM and Mikulska M (1990) The relationship between phosphate status and cyanide-resistant respiration in bean roots. Physiol Plant 79: 663–667

Saisho D, Nambara E, Naito S, Tsutsumi N, Hirai A and Nakazono M (1997) Characterization of the gene family for alternative oxidase from *Arabidopsis thaliana*. Plant Mol Biol 35: 585–596

Saitoh M, Nishitoh H, Fujii M, Takeda K, Tobiume K, Sawada Y, Kawabata M, Miyazono K and Ichijo H (1998) Mammalian thioredoxin is a direct inhibitor of apoptosis signal-regulating kinase (ASK) 1. EMBO J 17: 2596–2606

Sakajo S, Minagawa N, Komiyama T and Yoshimoto A (1991) Molecular cloning of cDNA for antimycin A-inducible mRNA and its role in cyanide-resistant respiration in *Hansenula anomala*. Biochim Biophys Acta 547: 138–148

Siedow JN and Day DA (2000) Respiration and photorespiration. In: Buchanan B, Gruissem W and Jones R (eds) Biochemistry and Molecular Biology of Plants, pp 676–728. American Society of Plant Physiologists, Rockville, Maryland

Siedow JN, Umbach AL and Moore AL (1995) The active site of the cyanide-resistant oxidase from plant mitochondria contains a binuclear iron center. FEBS Lett 362: 10–14

Simons BH and Lambers H (1999) The alternative oxidase: Is it a respiratory pathway allowing a plant to cope with stress? In: Lerner HL (ed) Plant Responses to Environmental Stresses: From Phytohormones to Genome Reorganization, pp 265–286. Marcel Dekker Inc, New York

Simons BH, Millenaar FF, Mulder L, Van Loon LC and Lambers H (1999) Enhanced expression and activation of the alternative oxidase during infection of *Arabidopsis* with *Pseudomonas syringae* pv tomato. Plant Physiol 120: 529–538

Smeekens S (2000) Sugar-induced signal transduction in plants. Annu Rev Plant Physiol Plant Mol Biol 51: 49–81

Stein J and Hansen G (1999) Mannose induces an endonuclease responsible for DNA laddering in plant cells. Plant Physiol 121: 71–79

Tanudji M, Sjoling S, Glaser E and Whelan J (1999) Signals required for the import and procesing of the alternative oxidase into mitochondria. J Biol Chem 274: 1286–1293

Theodorou ME and Plaxton WC (1995) Adaptations of plant

respiratory metabolism to nutritional phosphate deprivation. In: Smirnoff N (ed) Environment and Plant Metabolism: Flexibility and Acclimation, pp 79–109. Bios Scientific Publishers, Oxford

Tian RH, Zhang GY, Yan CH, Dai YR (2000) Involvement of poly(ADP-ribose) polymerase and activation of caspase-3-like protease in heat shock-induced apoptosis in tobacco suspension cells. FEBS Lett 474: 11–15

Tsuji H, Nakazono M, Saisho D, Tsutsumi N and Hirai A (2000) Transcript levels of the nuclear-encoded respiratory genes in rice decrease by oxygen deprivation: Evidence for involvement of calcium in expression of the alternative oxidase 1a gene. FEBS Lett 471: 201–204

Umbach AL and Siedow JN (1993) Covalent and noncovalent dimers of the cyanide-resistant alternative oxidase protein in higher plant mitochondria and their relationship to enzyme activity. Plant Physiol 103: 845–854

Umbach AL and Siedow JN (1996) The reaction of the soybean cotyledon mitochondrial cyanide-resistant oxidase with sulfhydryl reagents suggests that α-keto acid activation involves the formation of a thiohemiacetal. J Biol Chem 271: 25019–25026

Umbach AL and Siedow JN (1997) Changes in the redox state of the alternative oxidase regulatory sulfhydryl/disulfide system during mitochondrial isolation: Implications for inferences of activity in vivo. Plant Sci 123: 19–28

Umbach AL and Siedow JN (2000) The cyanide-resistant alternative oxidases from the fungi *Pichia stipitis* and *Neurospora crassa* are monomeric and lack regulatory features of the plant enzyme. Arch Biochem Biophys 378: 234–245

Umbach AL, Wiskich JT and Siedow JN (1994) Regulation of alternative oxidase kinetics by pyruvate and intermolecular disulfide bond redox status in soybean seedling mitochondria. FEBS Lett 348: 181–184

Vanlerberghe GC and McIntosh L (1992) Coordinate regulation of cytochrome and alternative pathway respiration in tobacco. Plant Physiol 100: 1846–1851

Vanlerberghe GC and McIntosh L (1994) Mitochondrial electron transport regulation of nuclear gene expression: Studies with the alternative oxidase gene of tobacco. Plant Physiol 105: 867–874

Vanlerberghe GC and McIntosh L (1996) Signals regulating the expression of the nuclear gene encoding alternative oxidase of plant mitochondria. Plant Physiol 111: 589–595

Vanlerberghe GC and McIntosh L (1997) Alternative oxidase: from gene to function. Annu Rev Plant Physiol Plant Mol Biol 48: 703–734

Vanlerberghe GC, Vanlerberghe AE and McIntosh L (1994) Molecular genetic alteration of plant respiration: Silencing and overexpression of alternative oxidase in transgenic tobacco. Plant Physiol 106: 1503–1510

Vanlerberghe GC, Day DA, Wiskich JT, Vanlerberghe AE and McIntosh L (1995) Alternative oxidase activity in tobacco leaf mitochondria: Dependence on tricarboxylic acid cycle-mediated redox regulation and pyruvate activation. Plant Physiol 109: 353–361

Vanlerberghe GC, Vanlerberghe AE and McIntosh L (1997) Molecular genetic evidence of the ability of alternative oxidase to support respiratory carbon metabolism. Plant Physiol 113: 657–661

Vanlerberghe GC, McIntosh L and Yip JYH (1998) Molecular localization of a redox-modulated process regulating plant mitochondrial electron transport. Plant Cell 10: 1551–1560

Vanlerberghe GC, Yip JYH and Parsons HL (1999) In organello and in vivo evidence of the importance of the regulatory sulfhydryl/disulfide system and pyruvate for alternative oxidase activity in tobacco. Plant Physiol 121: 793–803

Verniquet F, Gaillard J, Neuburger M and Douce R (1991) Rapid inactivation of plant aconitase by hydrogen peroxide. Biochem J 276: 643–648

Wagner AM (1995) A role for active oxygen species as second messengers in the induction of alternative oxidase gene expression in *Petunia hybrida* cells. FEBS Lett 368: 339–342

Wagner AM and Wagner MJ (1997) Changes in mitochondrial respiratory chain components of Petunia cells during culture in the presence of antimycin A. Plant Physiol 115: 617–622

Wagner AM, Wagner MJ and Moore AL (1998) In vivo ubiquinone reduction levels during thermogenesis in Araceae. Plant Physiol 117: 1501–1506

Weger HG, Chadderton AR, Lin M, Guy RD and Turpin DH (1990) Cytochrome and alternative pathway respiration during transient ammonium assimilation by N-limited *Chlamydomonas reinhardtii*. Plant Physiol 94: 1131–1136

Whelan J, Millar AH and Day DA (1996) The alternative oxidase is encoded in a multigene family in soybean. Planta 198: 197–201

Wu D, Wright DA, Wetzel C, Voytas DF and Rodermel S (1999) The *IMMUTANS* variegation locus of *Arabidopsis* defines a mitochondrial alternative oxidase homolog that functions during early chloroplast biogenesis. Plant Cell 11: 43–55

Yukioka H, Inagaki S, Tanaka R, Katoh K, Miki N, Mizutani A and Masuko M (1998) Transcriptional activation of the alternative oxidase gene of the fungus *Magnaporthe grisea* by a respiratory-inhibiting fungicide and hydrogen peroxide. Biochim Biophys Acta 1442: 161–169

Zheng M, Aslund F and Storz G (1998) Activation of the OxyR transcription factor by reversible disulfide bond formation. Science 279: 1718–1721

Chapter 12

Nitric Oxide Synthesis by Plants and its Potential Impact on Nitrogen and Respiratory Metabolism

A. Harvey Millar*[1,2], David A. Day[1] and Christel Mathieu[1]

[1]*Department of Biochemistry and* [2]*Plant Sciences Group, The University of Western Australia, Nedlands 6009, Western Australia, Australia*

Summary 193
I. Nitric Oxide as a Biological Messenger Molecule 194
II. Evidence of Nitric Oxide Synthesis and Accumulation in Plants 194
A. Nitric Oxide Production from Nitrite in Plant Cells 194
B. Nitric Oxide Synthase Homologs in Plants 195
III. Evidence of Nitric Oxide Modulation of Plant Signaling, Metabolism and Development 196
A. cGMP Dependent Pathways of Nitric Oxide Action 196
B. cGMP Independent Pathways of Nitric Oxide Action 198
1. Stimulation of Cell Elongation 198
2. Roles as a Protectant/Antioxidant 199
3. Binding to Hemoglobins 199
4. Modification of Aconitase 199
5. Inhibition of Respiration 200
IV. So What is the Role of Nitric Oxide in Plants? 201
Acknowledgments 202
References 202

Summary

Nitric oxide (NO) undertakes important roles as a signaling molecule, as a cytotoxic agent and also as an antioxidant in animals. It is now clear that this gaseous molecule plays similar roles in plants. Evidence for plant NO synthesis both by L-arginine-dependent nitric oxide synthase enzymes and also via nitrite-dependent nitrate reductase enzymes is rapidly accumulating. Several plant defense strategies involve NO in cGMP dependent signaling pathways, and developmental processes such as cell elongation and senescence have been shown to be modulated by NO. The potential of NO as both a pro-oxidant and an anti-oxidant in plants has been highlighted and a number of important metabolic enzymes are reportedly inhibited by NO in plants. These studies highlight the potential impact of NO on the development, defense and metabolism of plants and call for a concerted effort to unravel the importance of this nitrogen radical in the biochemistry of plant function.

*Author for correspondence, email: hmillar@cyllene.uwa.edu.au

Christine H. Foyer and Graham Noctor (eds): Photosynthetic Nitrogen Assimilation and Associated Carbon and Respiratory Metabolism, pp. 193–204.

I. Nitric Oxide as a Biological Messenger Molecule

The gaseous nitrogen radical, nitric oxide (NO), is an important mediator of physiological and pathophysiological processes in animals and has intrigued researchers over the past decade. This interest has resulted in the birth of a new field of research, the award of the 1998 Nobel Prize in Medicine, and in the appearance of a number of international journals dedicated to the what, where, how and why of NO biochemistry. Prominent roles for NO have been established in the regulation of blood vessel relaxation, the control of synaptic transmission, and the response of macrophages to infectious agents and to tumor cells (Stamler et al., 1992; Wink and Mitchell, 1998). Central to these studies has been the identification and characterization of a class of enzymes called nitric oxide synthases (NOSs) responsible for the release of NO during their five-electron oxidation of the guanidino nitrogen of L-arginine to yield L-citrulline (Knowles and Moncada, 1994). This reaction has mechanistic parallels to cytochrome P450 catalysis and it requires a number of cofactors including NADPH, flavin and tetrahydrobiopterin. Two constitutive NOS classes have been identified. The first are referred to as endothelial NOS (eNOS) and were identified as enzymes responsible for producing the endothelial relaxing factor (now known to be NO), which is essential in the regulation of vasodilatation (Palmer et al., 1987; Knowles and Moncada, 1994). The second class are neural NOS (nNOS) which were initially identified in brain tissue but have since been observed in other mammalian tissues (Frandsen et al., 1996). Most of the work on the role of NO in pathogenic situations has focused on an inducible form of the enzyme, iNOS (or macNOS), which is found in activated macrophages of the immune system (Clark and Rockett, 1996). A variety of targets for NOS-synthesized NO have been identified in animals. Most notably, the NO-activation of soluble guanylate cyclase leads to elevated intercellular cGMP concentrations and links NO with a diverse set of signal transduction pathways (Schmidt and Walter, 1994).

Abbreviations: AOS – active oxygen species; $cADP_R$ – cyclic ADP ribose; cGMP – cyclic GMP; CytOX – cytochrome *c* oxidase; eNOS – endothelial nitric oxide synthase; Hb – hemoglobin; iNOS – inducible form of nitric oxide synthase; IRE – iron-response element; IRP – iron-regulatory protein; Lb – leghemoglobin; L-NMMA – L-N^G-monomethyl-L-arginine monoacetate; LPS – lipopolysaccharides; NiR – nitrite reductase; nNOS – neural nitric oxide synthase; NO – nitric oxide; NOS – nitric oxide synthase; NR – nitrate reductase; PAL – phenylalanine ammonia lyase; PR-1 – pathogenesis-related protein 1; SIPK – salicylate induced protein kinase; TMV – tobacco mosaic virus

II. Evidence of Nitric Oxide Synthesis and Accumulation in Plants

The concept of small gaseous regulators of signaling and metabolism should be no surprise to plant researchers after many years of investigation of ethylene as a plant hormone. However, research over the past few decades into the effects of nitrogen oxides on plants has focused on their abundance as atmospheric pollutants rather than their potential as plant growth regulators (Wellburn, 1990). Since the discovery of the role of NO in mammals, the search for NO as a biologically significant regulator in higher plants has begun. Investigation of NO in plants is complicated by the difficulty of measuring short-lived nitrogen radicals in a biological system where large amounts of nitrogen are converted between oxidation states. Further, the stable breakdown products of NO, namely nitrite and nitrate, which are often used to monitor NO synthesis in animals, are present in large abundance in plants. Despite these impediments, the last five years have yielded evidence that NO is produced and is biologically active in higher plants, and that it is formed both as a consequence of nitrogen metabolism and also via L-arginine specific enzymatic processes analogous to those in animals (Delledonne et al., 1998; Durner et al., 1998; Hausladen and Stamler, 1998; Durner and Klessig, 1999; Wojtaszek, 2000).

A. Nitric Oxide Production from Nitrite in Plant Cells

Measurements of gaseous emissions from the leaves of legume species first demonstrated that NO could be synthesized at high levels by plant cells and released to the atmosphere (Klepper, 1979, 1987; Dean and Harper, 1986; Leshem, 1996). Rates of NO formation were greatly enhanced by conditions that maintained nitrate reductase (NR) activity in the absence of nitrite reductase (NiR) activity, such as anaerobic conditions and the application of herbicides, which lead to accumulation of nitrite in

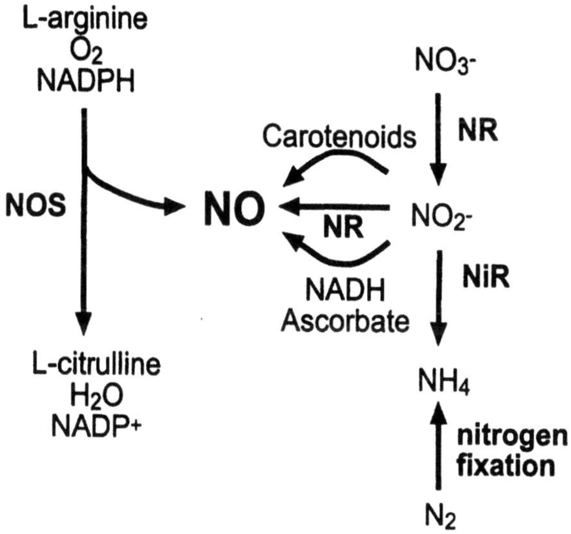

Fig. 1. Mechanisms of nitric oxide (NO) formation in plants. NO detected in plant can be formed via nitrogen metabolism through enzymatic or non-enzymatic pathways or can be synthesized from L-arginine by a nitric oxide synthase (NOS). NR: nitrate reductase; NiR: nitrite reductase.

plant tissues. This NO was produced enzymatically from nitrite in leaves of legumes by the constitutive NAD(P)H nitrate reductase (cNR) that is found exclusively in these species (Dean and Harper, 1986) (Fig. 1). Day et al. (1998) measured a transient burst of NO production by nitrate fertilized soybean leaves immediately post-illumination. This phenomenon correlated with the transient accumulation of nitrite under these conditions observed by Reins and Heldt (1992) and implicates either cNR or direct ascorbate reduction of nitrite in the chloroplast as the source of NO (Day et al., 1998).

Soybean mutants and other plant species that lack cNR do not appear capable of these high levels of enzymatic conversion of nitrite to NO, but do produce lower levels of NO (Churchill and Klepper, 1979; Klepper, 1990). Recently, working with sunflower, maize, rape, spruce, sugar cane, tobacco and spinach, Wildt et al. (1997) demonstrated a light-dependent NO synthesis by plants which was correlated with the rate of CO_2 fixation in the light. This study drew a link between active photosynthesis, and thus the potential for nitrogen assimilation, and the accumulation of NO. NO formation in these plant tissues could be due to NR activities, direct chemical reduction of nitrite via NADH or ascorbate (Evans and McAuliffe, 1956), or through catalysis stimulated by carotenoids (Cooney et al., 1994) (Fig. 1). In more detailed studies of the common inducible NR enzyme in higher plants it has recently been shown that both NO and also the toxic peroxynitrite molecule can be produced by the nitrite reducing activity of this enzyme (Yamasaki et al., 1999; Yamasaki and Sakihama, 2000). These authors called for a re-assessment of our interpretation of the mechanisms governing regulation of NR in plants to consider whether some of these mechanisms may act primarily to regulate NO formation rather than to control nitrate reduction in vivo (Yamasaki and Sakihama, 2000).

B. Nitric Oxide Synthase Homologs in Plants

Several recent studies have focused on the detection of NOS-like activities in plants, analogous to the enzyme activities identified in mammals. NOS activity was first detected in green husks in the leguminous plant *Mucuna hassjoo* (Ninnemann and Maier, 1996). The same year, the presence of NOS in roots and nodules of *Lupinus albus* was reported (Cueto et al., 1996). The accumulation of NO and the detection of NOS activity in response to an avirulent pathogen in soybean suspension cells and in leaves of *Arabidopsis thaliana* has also been highlighted (Delledonne et al., 1998). NOS activity was also detected in leaves of resistant, but not sensitive, tobacco plants infected by tobacco mosaic virus (TMV) (Durner et al., 1998). The presence of NOS in roots tips and young leaves of maize seedlings was also reported (Ribeiro et al., 1999), and NOS activity has been detected in pea plants where it was localized to peroxisomes (Barroso et al., 1999).

Different methods have been used to assay these NOS activities and to detect NOS proteins in plants. All of the reports cited above have measured the conversion of radioactive (^{14}C or ^{3}H) L-arginine to radioactive L-citrulline in the plant extracts. The formation of NO has been monitored also by the reduction of ferrous hemoglobin to methemoglobin (Cueto et al., 1996; Ninnemann and Maier, 1996; Delledonne et al., 1998; Durner et al., 1998; Barroso et al., 1999; Ribeiro et al., 1999). These NOS activities were inhibited by classical inhibitors of animal NOS, such as L-N^G-monomethyl-L-arginine monoacetate (L-NMMA) (Cueto et al., 1996; Durner et al., 1998; Barroso et al., 1999), PBITU and L-NNA (Delledonne et al., 1998) or L-NAME and D-aminoguanidine (Barroso et al., 1999; Ribeiro et al., 1999).

The calcium dependence of plant NOS varies.

Some of the enzymes detected were calcium dependent, such as those from soybean cell suspension (Delledonne et al., 1998), roots and leaves of maize seedlings (Ribeiro et al., 1999), and pea peroxisomes (Barroso et al., 1999). Interestingly, in *Lupinus albus*, the NOS activity detected in the roots is calcium dependent but the one present in the nodules is not significantly dependent on calcium (Cueto et al., 1996). In animals, the constitutive forms of NOS (nNOS, eNOS) are calcium dependent while the inducible form of NOS (iNOS/macNOS) is calcium independent. Thus the data from *Lupinus albus* may indicate that a constitutive enzyme is present in the roots while the major isoform in the nodules could be inducible (Cueto et al., 1996). In mammalian cells, macNOS is induced after activation of macrophages by cytokines and lipopolysaccharides (LPS) (Radomski et al., 1990; Hortelano et al., 1993). It is interesting to speculate whether the calcium independent form present in the nodules is induced by *Rhizobium* LPS. These molecules are produced during infection and are essential for the initial interaction between the two symbiotic partners and also later in nodule development. A role for LPS-dependent NO in the establishment and maintenance of the symbiotic relationship needs further consideration (Cueto et al., 1996).

NOS activity in plants has also been assayed as NADPH-diaphorase activity by histochemical detection, a commonly employed marker for NOS in mammals. This assay allows in vivo localization of NOS to be determined. While NADPH-diaphorase activity is not specific to NOS, its activity in nodules was significantly decreased by the antagonist of L-arginine, L-NMMA, suggesting that at least part of the activity observed was due to a nitric oxide synthase (Cueto et al., 1996). The use of antibodies raised against animal NOS proteins (typically 130 kDa to 160 kDa in size) has also yielded valuable information in plants. A 166 kDa protein was identified in homogenates of young leaves and root tips of maize using antibodies raised against mouse macrophage and rabbit brain NOS (Ribeiro et al., 1999). A 130 kDa NOS has also been identified in pea peroxisomes using two different antibodies, one raised against the NADPH-binding region of murine iNOS and one raised against the 14 residues of the C-terminal end of murine iNOS which is specific for iNOS (Barroso et al., 1999). The same report also presented the immunolocalization of NOS to peroxisomes and chloroplasts of pea using these antibodies (Barroso et al., 1999).

This literature provides strong biochemical and immunological evidence that NOS enzyme(s) exist in plants, and are likely to produce NO (Fig. 1). The enzyme has not yet been purified from plants, but the purification of mammalian NOS was also very difficult (Bredt and Snyder, 1990). No gene encoding NOS in plants has been identified to date, but an EST from *Arabidopsis* shows homology to an nNOS inhibitor protein (PIN) (Jaffrey and Snyder, 1996). The fact that inhibitors of mammalian NOS also inhibit plant NOS activity, together with the cross-reactivity in plants to antibodies raised against mammalian NOS, suggest that the animal and plant NOS may share similarities which could be used to identify genes encoding the plant enzymes. However, mammalian NOSs share very high sequence homology with cytochrome P450 reductases and lack regions of homology specific to NOS which can make it difficult to discriminate between NOS and cytochrome P450 reductase sequences (Bredt et al., 1991).

III. Evidence of Nitric Oxide Modulation of Plant Signaling, Metabolism and Development

A. cGMP Dependent Pathways of Nitric Oxide Action

Much of the ability of NO to modulate signaling pathways in mammalian systems results from its activation of guanylate cyclase and the consequent stimulation of cGMP formation. This results in the so-called cGMP-dependent NO signaling pathways (Schmidt and Walter, 1994). NO activates soluble guanylate cyclase in mammals by binding to its heme and/or by S-nitrosylating critical cysteine residues of the enzyme (McDonald and Murad, 1996). In plants, NO treatment of spruce (*Picea abies*) needles led to a 10,000-fold elevation of in vivo cGMP levels (Pfeiffer et al., 1994) strongly suggesting that NO also activates guanylate cyclase in plants. Further evidence has been provided by Durner et al. (1998) who showed that cGMP concentration was greatly elevated by NO treatment of tobacco leaves and also that induction of gene expression by NO could be blocked by guanylate cyclase inhibitors.

The involvement of NO in plant signaling cascades has been implicated by studies of the hypersensitive response of plants to infectious agents. During the infection of a resistant plant by an avirulent pathogen, the pathogen expresses an *Avr* gene whose product is recognized by the product of a resistance (R) gene in the plant. This recognition leads to the massive production of active oxygen species (AOS), a response known as the oxidative burst. The AOS so produced cross-link the cell-wall components and induce several defense genes. The host cell's death may also be induced by the hypersensitive reaction in order to restrict the pathogen to the infection site, leading to its destruction and the resistance of the plant to the infection. Two separate reports have shown that NO is involved in the signaling pathway leading to the establishment of the hypersensitive response in plants (Fig. 2). Delledonne et al. (1998) reported that the inoculation of soybean suspension cells with *Pseudomonas syringae* pv *glycinae* provoked significant H_2O_2 accumulation in culture. They further showed that exogenous H_2O_2 provoked only a small proportion of cells to die, suggesting that the oxidative burst by itself was not sufficient to support a strong disease resistance response; other signals are obviously necessary. Treatment of the cells with an NO donor, on the other hand, did not provoke cell death either, but the combination of both NO and H_2O_2 did, mimicking the response to the pathogen infection. The pathogen-induced signaling pathway was also shown to involve a protein kinase, as cell death was abolished by treatment with cantharidin, an inhibitor of type-2a protein phosphatases. Moreover, the treatment of *Pseudomonas syringae*-infected *Arabidopsis* leaves with NOS inhibitors abolished the hypersensitive response (Delledonne et al., 1998). This inhibition led to the growth and spread of the bacteria and disease of the host plant (Delledonne et al., 1998).

A second independent report showed that infection of resistant, but not sensitive, tobacco plants with TMV, led to an increase of NOS activity (Durner et al., 1998). Moreover, treatment with NO donors induces the expression of defense related genes encoding a pathogenesis-related protein (PR-1) and phenylalanine ammonia lyase (PAL). PR-1 and PAL were also induced in tobacco by cGMP and $cADP_R$ which are well known second messengers for NO signaling in mammals. Treatment of these tobacco plants with NO provoked a transient and dramatic increase in endogenous cGMP concentration, suggesting that cGMP could also act as a second messenger for NO signaling in plants (Durner et al., 1998). Very recently, NO was found to activate a salicylic acid-induced MAP kinase (SIPK) in tobacco, in response to pathogen attack and other stresses (Kumar and Klessig, 2000). *Arabidopsis* plants expressing a recombinant aequorin have been used successfully to monitor cytosolic free Ca^{2+} concen-

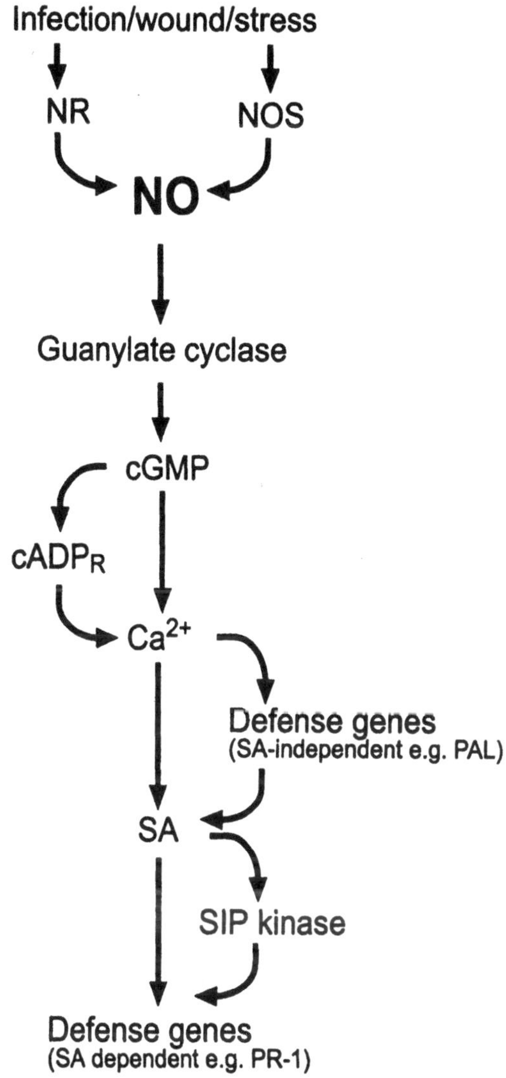

Fig. 2. cGMP dependent pathways of nitric oxide (NO) action in plants. NO accumulates in response to stress such as infection by a pathogen, which activates a guanylate cyclase and leads to increased level of cGMP. The accumulation of cGMP leads to higher levels of cytosolic calcium, via $cADP_R$ dependent or independent pathways. Ca^{2+} induces salicylic acid biosynthesis possibly via activation of phenylalanine ammonia lyase (PAL) gene expression. SA accumulation induces the expression of defense genes such as PR-1. This pathway could involve SIP kinase. SA: salicylic acid; SIP kinase: Salicylic acid Induced Protein kinase; PR-1: pathogenesis-related protein-1; PAL: phenylalanine ammonia lyase; $cADP_R$: cyclic ADP ribose; NR: nitrate reductase; NOS: nitric oxide synthase.

tration in vivo by a non-destructive light emission reaction (Knight et al., 1991). In preliminary experiments using these plants, a significant but transient increase in cytosolic free Ca^{2+} concentration was observed in intact seedlings within 60 sec of treatment with fresh preparations of the NO releasing substance, NOC-18. No increase in Ca^{2+} concentration was recorded with seedlings treated with an NO-depleted NOC-18 solution (A. H. Millar and M. Knight, unpublished). This study provides the first direct evidence that Ca^{2+} operates in a signal transduction pathway downstream of NO in plants, analogous to the pathways identified in mammals and consistent with the implications of the results of Delledonne et al. (1998) and Durner et al. (1998) (Fig. 2).

B. cGMP Independent Pathways of Nitric Oxide Action

A variety of cGMP-independent pathways of NO action have also been identified in mammals. These pathways involve the modification of target proteins. This modification can be via S-nitrosylation of thiol-containing amino acid residues and nitration of tyrosine, the direct interaction of NO with metal-centers, or the action of NO as a structural analog of O_2 and inhibitor of O_2 binding and O_2 consuming enzymes (Kroncke et al., 1997; Ischiropoulos, 1998). Similar targets have been identified in plants and several physiological effects of NO addition have been highlighted in the literature, suggesting a multi-faceted role for NO in plants (Fig. 3).

1. Stimulation of Cell Elongation

A variety of effects of NO on plant cell growth have been investigated by applying NO-donor molecules, or by spraying NO gas directly onto plant tissues. A potential role for NO in cell growth during seed germination was first proposed after both NOS and calmodulin proteins were detected in pea and wheat embryo tissues, using antibodies raised against the rabbit brain-NOS and sheep calmodulin respectively (Sen and Cheema, 1995). Subsequently, convincing evidence for the role of NO in both lettuce and *Arabidopsis* as a regulator of light-mediated mechanisms leading to seed germination, de-etiolation and inhibition of hypocotyl elongation has been reported (Beligni and Lamattina, 2000).

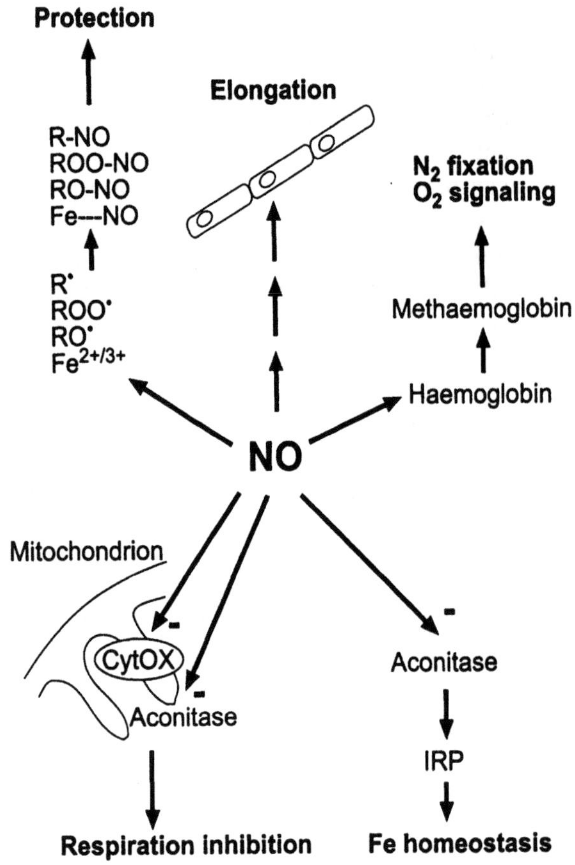

Fig. 3. cGMP independent pathways of NO action in plants. NO can act directly as an antioxidant by reacting with active oxygen species, can regulate cell elongation and could be involved in the regulation of N_2 fixation and signaling of O_2 concentration by binding to plant hemoglobins. NO also inhibits aconitase resulting in the perturbation of Fe homeostasis and inhibits CytOX leading to perturbation of respiration. CytOX: cytochrome *c* oxidase, IRP : iron regulatory protein.

As has been described in animals, it seems that NO can have contrasting effects on plant physiology depending on the amount of NO provided. Low concentrations of NO applied to pea leaf discs were able to promote cell growth and decrease ethylene production (Leshem, 1996). These authors suggest that NO, which is a highly diffusible molecule, is probably transported mainly to the apoplast, where it may come in contact with and subsequently weaken intermicellar cell wall links. This weakening would result in a loosening of the cell wall, allowing cell turgor to cause cell expansion. In contrast, high concentrations of NO inhibited cell growth apparently by the same mechanism (Leshem, 1996).

2. Roles as a Protectant/Antioxidant

While NO is often considered a cytotoxic compound, it can also function as an antioxidant in cells by its reaction with other radical molecules, thereby breaking the chain of free radical propagation (Wink et al., 1993; Halliwell et al., 1999). Several reports of NO as a protectant against senescence and oxidative stress in plants have appeared. In a variety of both climacteric and non-climacteric fruits and flowers, of vegetables and legume species, NO emission decreases with maturation and during senescence (Leshem and Haramaty, 1996; Leshem et al., 1998). Moreover, exogenous application of NO markedly delays senescence and maturation of these tissues (Leshem et al., 1998). NO is thought to delay senescence both by down-regulating ethylene emission and by acting as an antioxidant. NO can, therefore, be regarded as a naturally occurring plant growth effector. More recently, the same authors (Leshem et al., 1998) have shown that NO fumigation can be advantageously replaced, in the case of cut flowers, by the NO donor, sildenafil citrate (marketed under the trade name Viagra). Viagra application increases the vase life of cut flowers by as much as a week (Siegel-Itzkovich, 1999).

Other studies have demonstrated more directly that NO is able to counteract the toxicity of oxidative stress in plants. For example, NO-treated potato leaves were resistant to chlorosis, ion leakage, DNA fragmentation and apoptotic-like cell death produced by treatment with the AOS-generating herbicide, diquat, and by invasion with the pathogen *Phytophtora infestans* (Laxalt et al., 1997; Beligni and Lamattina, 1999a,b).

3. Binding to Hemoglobins

In mammals, NO binds to hemoglobin and causes its reduction to methemoglobin. Plants also contain hemoglobins that can be separated into two groups.

The symbiotic-type hemoglobins, or leghemoglobins (Lb), are found in infected cells of nitrogen-fixing nodules of both legume and non-legumes (Bergersen, 1980), where they facilitate diffusion of oxygen to the vigorously respiring nitrogen-fixing *Rhizobium*, while maintaining very low and stable oxygen tension inside the nodule (Appleby, 1992). Lb, therefore, plays an essential role in the functioning of the nitrogen-fixing nodule. The formation of Lb-NO results in inactivation of the Lb which consequently cannot fulfill its role as an oxygen carrier. The accumulation of Lb-NO has been correlated with a decrease of nitrogen-fixation activity in nodules of plants supplied with nitrate (Kanayama and Yamamoto, 1990). A recent study also detected Lb-NO complexes by electron paramagnetic resonance (EPR) spectrophotometry in intact frozen soybean nodules (Mathieu et al., 1998). The level of this complex was found to vary with nodule age, being highest in young nodules, lower in mature nodules and completely absent in old or senescent nodules (Mathieu et al., 1998). Thus the formation of Lb-NO could be involved in the regulation of the nitrogen fixation activity of the nodule.

The second group of plant hemoglobins (Hb) appears to be ancestral to the symbiotic forms and are present in non-symbiotic organs of legumes and in non-legumes (Arredondo-Peter et al., 1998). They also have a high affinity for oxygen and their role could be as oxygen sensor—for example, in signaling low oxygen concentration in plant cells (Anderson et al., 1996). Plant Hbs are able to bind NO and these Hb-NO complexes are remarkably stable due to their very slow dissociation rate. Binding of NO to the Hb molecule may, therefore, extend the half-life and distance over which NO can act in cells, as is thought to occur in mammals (Jia et al., 1996).

Some of the non-symbiotic plant Hbs have a much higher affinity for oxygen than their mammalian and symbiotic counterparts (Trevaskis et al., 1997). This is also true of the Hb from the parasitic nematode *Ascaris lumbricoides* which has recently been shown to catalyze oxygenase activity upon binding NO (Minning, 1999). It is possible that NO-binding to plant Hb also allows it to act as an oxygenase.

4. Modification of Aconitase

The iron-sulfur center of aconitase has been identified as a target for NO in mammals (Hentze and Kuhn, 1996). This enzyme in found in the cytosol and in the mitochondrial matrix, and both isozymes catalyze the reversible isomerization of citrate to isocitrate. In addition, the cytosolic enzyme can operate as an iron-regulatory protein (IRP) that binds to mRNAs containing an iron-response element (IRE) consensus sequence. IRP-binding inhibits the translation of IRE mRNAs when the IRE motif resides in the 5′ region and improves the stability of IRE mRNAs when the motif is in a 3′ region (Hentze and Kuhn, 1996). In this manner, IRP regulates the iron homeostasis of cells and leads to an accumulation of free iron. NO promotes the loss of the Fe-S cluster of aconitase, causing it to lose its enzyme activity but gain IRP activity. Plants also contain both mitochondrial and cytosolic aconitases, and NO has been shown recently to rapidly inhibit citrate to isocitrate conversion by this enzyme (Navarre et al., 2000). Analysis of a tobacco cDNA reveals the presence of mRNA binding motifs in the plant aconitase which are analogous to those identified in the mammalian IRP (Navarre et al., 2000). An increase in free iron upon NO production in response to pathogen attack may play a role in the hypersensitive response by enhancing AOS formation.

5. Inhibition of Respiration

Studies of the role of NO in synaptic transmission in mammals have shown that NO is a potent, reversible inhibitor of mitochondrial electron transport. This effect has been localized to the inhibition of cytochrome *c* oxidase (CytOX) through competition at the site of O_2 binding (Brown and Cooper, 1994; Borutaité and Brown, 1996). An eNOS homologue has also been localized to mammalian mitochondria (Bates et al., 1995) and addition of L-arginine to isolated mitochondria has been shown to inhibit respiration (Kobzik et al., 1995). This work has suggested that NO functions as an endogenous regulator of mitochondrial electron transport and oxidative phosphorylation in mammalian cells (Bates et al., 1996).

The plant mitochondrial electron transport chain is branched and contains two terminal oxidases: CytOX and the alternative oxidase (AOX). Unlike the cytochrome pathway, which is coupled to oxidative phosphorylation via proton translocation, electron transport from ubiquinol to AOX is non-phosphorylating and releases energy as heat (Day et al., 1995). The two terminal oxidases compete for electrons in plant mitochondria, with inhibition of one pathway redirecting flux to the other (Hoefnagel et al., 1995). The two oxidases can be differentiated by inhibitors such as cyanide and carbon monoxide (acting on CytOX) and n-propyl gallate or salicylhydroxamic acid (acting on AOX). AOX is thought to act as a bypass of the proton-translocating cytochrome pathway under conditions when the latter is overwhelmed or disrupted, thereby avoiding over-reduction of respiratory chain components and the

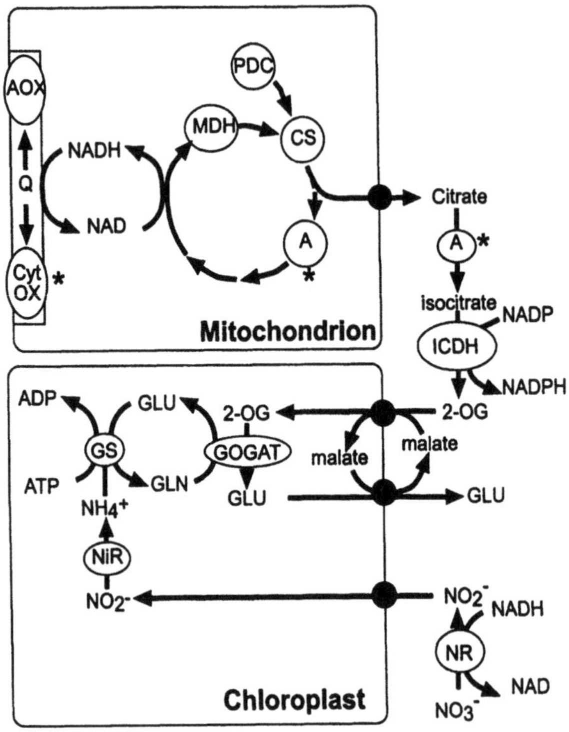

Fig. 4. Impact of nitric oxide (NO) on mitochondrial function during nitrogen assimilation. Perturbed NiR function results in NO formation which leads to CytOX inhibition and redirection of electron flux to AOX in the mitochondrion. Inhibition of aconitase increases citrate concentrations leading to AOX synthesis and decreases the availability of 2-oxoglutarate for assimilation of ammonium in the chloroplast. ICDH : isocitrate dehydrogenase, 2-OG: 2-oxoglutarate, GOGAT: glutamate synthase, GS: glutamine synthetase, MDH: malate dehydrogenase, PDC: pyruvate decarboxylase, CS: citrate synthase, A: aconitase, Q: ubiquinone, CytOX: cytochrome *c* oxidase, AOX: alternative oxidase, NiR: nitrite reductase, NR: nitrate reductase.

concomitant production of AOS (Wagner and Krab, 1995; Chapter 11, Vanlerberghe and Ordog).

The discovery that NO is a potent inhibitor of CytOX but not AOX in plant mitochondria (Millar and Day, 1996) raises the possibility that NO is an endogenously synthesized inhibitor of CytOX that has selected for the maintenance of AOX in higher plant species. The $K_{0.5}$ for cytochrome pathway inhibition by NO is approximately 0.1 μM compared to a $K_{0.5}$ of approximately 5 μM for alternative pathway inhibition. Caro and Puntarulo (1999) have also reported much greater inhibition of CytOX ($K_{0.5}$ = 0.3 μM) than AOX ($K_{0.5}$=3.6 μM) by NO in soybean embryonic axes mitochondria. These authors suggest than the endogenous burst of NO synthesis (up to 0.2

μM) in this developing tissue could differentially inhibit CytOX and increase operation of AOX. The decrease in AOS production in mitochondria when AOX is active (Popov et al., 1997; Maxwell et al., 1999) limits the reaction of NO with superoxide preventing the formation of highly destructive peroxynitrite and hydroxyl radicals. This may indicate a novel physiological role for AOX in preventing deleterious oxidative damage as a result of cytochrome pathway limitation upon NO production in plants (Millar and Day, 1997). Interestingly, by inhibiting aconitase NO will also lead to accumulation of citrate which will in turn stimulate AOX synthesis (Vanlerberghe and McIntosh, 1996).

IV. So What is the Role of Nitric Oxide in Plants?

The production of NO in plants during nitrogen metabolism, and especially during perturbation of nitrogen assimilation, adds a novel dimension to the biological function of NO in plants that is not evident in mammalian systems. Several situations can be envisaged which highlight the potential of NO production to affect nitrogen assimilation and associated carbon metabolism in plants, in addition to its obvious role in plant defense and cell signaling discussed above.

1) An increase in nitrite concentration and consequent NO production in the cytosol has the potential to act as a signal that nitrogen metabolism is inhibited. NO accumulated under these circumstances could feedback to inhibit mitochondrial metabolism that provides not only the carbon skeletons for nitrogen assimilation but also the ATP needed for further metabolism of amino acids and/or the export of recently assimilated nitrogen to other plant cells. NO, by inhibiting CytOX, would redirect electron flux to AOX, alter the ADP/ATP ratio in the cytosol and increase AOS formation by the respiratory electron transport chain. Aconitase in the mitochondrial matrix would also be inhibited, perturbing the tricarboxylic acid cycle and producing citrate. As citrate accumulation is known to induce AOX synthesis (Vanlerberghe and McIntosh, 1996), this provides a feedback loop for induction of additional oxidase if insufficient AOX protein was available. Inhibition of both mitochondrial and cytosolic aconitase activity by NO would greatly decrease the availability of 2-oxoglutarate for assimilation of ammonium (Fig. 4).

2) Nitrate reductase is intricately regulated in plants, not only by a kinase and phosphatase, but also by the differential association of inhibitor proteins with the phosphorylated and dephosphorylated forms (Huber et al., 1996; Mackintosh, 1998). These attributes indicate that NR could be a component in a protein kinase signaling cascade resulting in NO formation and subsequent initiation of gene expression through pathways dependent on Ca^{2+}, cGMP and salicylic acid.

3) NO produced during nitrogen assimilation could act as an antioxidant to defend the cell against increased radical production when superoxide and singlet oxygen are produced in the chloroplast. Production of these radicals can be greatly increased by, for example, the application of herbicides acting at the chloroplast photosystems. NO reaction with oxygen and lipid radicals will break the chain of free radical propagation.

4) Application of nitrates and nitrites inhibit and ultimately destroy nitrogen fixing symbioses between legume plants and strains of rhizobia. NO formed from nitrite in the nodule has the potential to inhibit the oxygen carrying function of Lb and also to inhibit the terminal oxidases of both the host mitochondria and the differentiated bacteria. Such NO targets might explain the inhibition of nitrogen fixation and the triggering of senescence of the nodule tissue when nitrate is applied.

Clearly NO, by virtue of its chemical reactivity as a radical, has the potential to affect plant cells in a variety of ways to the betterment or detriment of the plant as a whole. As well as the above possible roles of NO in respiratory carbon metabolism and nitrogen assimilation, this species may have many other effects in plants. For instance, NO has been shown to inhibit the water-splitting activity of Photosystem II in a manner that is reversible by bicarbonate (Ioannidis et al., 1998). The present review of an emerging field of research has not attempted to provide an exhaustive or definitive account of the processes in which NO is involved in plants, and much work remains to be done. We have attempted to present some fundamentals, outlined the current skeleton of our knowledge and provided a range of hypotheses we consider worthy of future investigation to further elucidate the role of this gaseous messenger in the interplay of signaling and carbon/nitrogen metabolism in plants.

Acknowledgments

The Australian Research Council is thanked for the Australian Post-Doctoral Fellowship to AHM and for research grants to DAD.

References

Anderson CR, Jensen EO, Lewellyn DJ, Dennis ES and Peacock WJ (1996) A new hemoglobin gene from soybean: A role for hemoglobin in all plants. Proc Natl Acad Sci USA 93: 5682–5687

Appleby CA (1992) The origin and function of hemoglobins in plants. Sci Prog 76: 365–398

Arredondo-Peter R, Hargrove MS, Moran JF, Sarath G and Klucas RV (1998) Plant hemoglobins. Plant Physiol 118: 1121–1125

Barroso JB, Corpas FJ, Carreras A, Sandalio LM, Valderrama R, Palma JM, Lupianez JA and del Rio LA (1999) Localization of nitric-oxide synthase in plant peroxisomes. J Biol Chem 274: 36729–36733

Bates TE, Loesch A, Burnstock G and Clark JB (1995) Immunocytochemical evidence for a mitochondrially located nitric oxide synthase in brain and liver. Biochem Biophys Res Commun 213: 896–900

Bates TE, Loesch A, Burnstock G and Clark JB (1996) Mitochondrial nitric oxide synthase: A ubiquitous regulator of oxidative phosphorylation. Biochem Biophys Res Commun 218: 40–44

Beligni MV and Lamattina L (1999a) Nitric oxide counteracts cytotoxic processes mediated by reactive oxygen species in plant tissues. Planta 208: 337–344

Beligni MV and Lamattina L (1999b) Nitric oxide protects against cellular damage produced by methylviologen herbicides in potato plants. Nitric Oxide Biol Ch 3: 199–208

Beligni MV and Lamattina L (2000) Nitric oxide stimulates seed germination and de-etiolation, and inhibits hypocotyl elongation, three light-inducible responses in plants. Planta 210: 215–221

Bergersen FJ (1980) Leghaemoglobin, oxygen supply and nitrogen fixation: Studies with soybean nodules. In: Stewart WDP, Gallon JR (eds) Nitrogen fixation, pp 139–160. Academic Press, London

Borutaité V and Brown GC (1996) Rapid reduction of nitric oxide by mitochondria, and reversible inhibition of mitochondrial respiration by nitric oxide. Biochem J 315: 295–299

Bredt DS and Snyder SH (1990) Isolation of nitric oxide synthetase, a calmodulin-requiring enzyme. Proc Natl Acad Sci USA 87: 682–685

Bredt DS, Hwang PM, Glatt CE, Lowenstein C, Reed RR and Snyder SH (1991) Cloned and expressed nitric oxide synthase structurally resembles cytochrome P-450 reductase. Nature 351: 714–718

Brown GC and Cooper CE (1994) Nanomolar concentrations of nitric oxide reversibly inhibit synaptosomal respiration by competing with oxygen at the cytochrome oxidase. FEBS Lett 356: 295–298

Caro A and Puntarulo S (1999) Nitric oxide generation by soybean embryonic axes. Possible effect on mitochondrial function. Free Rad Res 31: 5205–5212

Churchill KA and Klepper LA (1979) Effects of ametryn on nitrate reductase activity and nitrite content of wheat. Pestic Biochem Physiol 12: 156–162

Clark IA and Rockett KA (1996) Nitric oxide and parasitic disease. Adv Parisitol 37: 1–56

Cooney RV, Harwood PJ, Custer LJ and Franke AA (1994) Light mediated conversion of nitrogen dioxide to nitric oxide by carotenoids. Environ Health Perspect 102: 460–462

Cueto M, Hernández-Perera O, Martin R, Bentura ML, Rodrigo J, Lamas S and Golvano MP (1996) Presence of nitric oxide synthase activity in roots and nodules of *Lupinus albus*. FEBS Lett 398: 159–164

Day DA, Whelan J, Millar AH, Siedow JN and Wiskich JT (1995) Regulation of the alternative oxidase in plants and fungi. Aust J Plant Physiol 22: 497–509

Day DA, Whelan J, McCabe TC, Millar AH, Atkin OK, Djajanergara I, Schuller LJ and Finnegan PM (1998) Developmental and environmental regulation of alternative oxidase expression in soybean. In: Møller IM, Gardestrom P, Glimelius K, Glaser E (eds) Plant Mitochondria: Gene to Function pp 429–434. Backhuys Publishers, Leiden

Dean JV and Harper JE (1986) Nitric oxide and nitrous oxide production by soybean and winged bean during the in vivo nitrate reductase assay. Plant Physiol 82: 718–723

Delledonne M, Xia Y, Dixon RA and Lamb C (1998) Nitric oxide functions as a signal in plant disease resistance. Nature 394: 585–588

Durner J and Klessig DF (1999) Nitric oxide as a signal in plants. Curr Opin Plant Biol 2: 369–374

Durner J, Wendhenne D and Klessig DF (1998) Defense gene induction in tobacco by nitric oxide, cyclic GMP, and cyclic ADP-ribose. Proc Natl Acad Sci USA. 95: 10328–10333

Evans H and McAuliffe C (1956) Identification of NO, N_2O and N_2 as products of the non-enzymatic reduction of nitrite by ascorbate or reduced diphosphopyridine nucleotide. In McElory WD, Glass B (eds) Organic Nitrogen Metabolism, pp 189–197. John Hopkins Press, Baltimore

Frandsen U, Lopez-Figueroa M and Hellsten Y (1996) Localization of nitric oxide synthase in human skeletal muscle. Biochem Biophys Res Commun 227: 88–93

Halliwell B, Zhao K and Whiteman M (1999) Nitric oxide and peroxynitrite. The ugly, the uglier and the not so good—A personal view of recent controversies. Free Rad Res 31: 651–669

Hausladen A and Stamler JS (1998) Nitric oxide in plant immunity. Proc Natl Acad Sci USA 95: 10345–10347

Hentze MW and Kuhn LC (1996) Molecular control of vertebrate iron metabolism—mRNA based regulatory circuits operated by iron, nitric oxide, and oxidative stress. Proc Natl Acad Sci USA 93: 8175–8182

Hoefnagel MHN, Millar AH, Wiskich JT and Day DA (1995) Cytochrome and alternative respiratory pathways compete for electrons in the presence of pyruvate in soybean mitochondria. Arch Biochem Biophys 318: 394–400

Hortelano S, Genaro AM and Bosca L (1993) Phorbol esters induce nitric oxide synthase and increase arginine influx in cultured peritoneal macrophages. FEBS Lett 320: 135–139

Huber SC, Bachmann M and Huber JL (1996) Post-translational regulation of nitrate reductase activity: A role for Ca^{2+} and 14-3-3 proteins. Trends Plant Sci 1: 432–438

Ioannidis N, Sarrou J, Schansker G and Petrouleas V (1998) NO reversibly reduces the water-oxidizing complex of Photosystem II through S_0 and S_{-1} to the state characterized by the Mn(II)-Mn(III) multiline EPR signal. Biochemistry 37: 16445-16451

Ischiropoulos H (1998) Biological tyrosine nitration: A pathophysiological function of nitric oxide and reactive oxygen species. Arch Biochem Biophys 356: 1–11

Jaffrey SR and Snyder SH (1996) PIN—an associated protein inhibitor of neuronal nitric oxide synthase. Science 274: 774–777

Jia L, Bonaventura C, Bonaventura J and Stamler JS (1996) S-nitrosohaemoglobin: A dynamic activity of blood involved in vascular control. Nature 380: 221–226

Kanayama Y and Yamamoto Y (1990) Inhibition of nitrogen fixation in soybean plants supplied with nitrate. II. Accumulation and properties of nitrosylhemoglobin in nodules. Plant Cell Physiol 31: 207–214

Klepper LA (1979) Nitric oxide (NO) and nitrogen dioxide (NO_2) emissions from herbicide-treated soybean plants. Atmos Environ 13: 537–542

Klepper LA (1987) Nitric oxide emissions from soybean leaves during in vivo nitrate reductase assays. Plant Physiol 85: 96–99

Klepper LA (1990) Comparison between NO_x evolution mechanisms of wild-type and nr_1 mutant soybean leaves. Plant Physiol 93: 26–32

Knight MR, Campbell AK, Smith SM and Trewavas AJ (1991) Transgenic plant aequorin reports the effects of touch, cold-shock and elicitors on cytoplasmic calcium. Nature 352: 524–526

Knowles RG and Moncada S (1994) Nitric oxide synthases in mammals. Biochem J 298: 249–258

Kobzik L, Stringer B, Balligand J-C, Reid MB and Stamler JS (1995) Endothelial type nitric oxide synthase in skeletal muscle fibers: Mitochondrial relationships. Biochem Biophys Res Commun 211: 375–381

Kroncke KD, Fehsel K and Kolb-Bachofen V (1997) Nitric oxide: Cytotoxicity verses cytoprotection: How, why, when, and where? Nitric Oxide Biol Ch 1: 107–120

Kumar D and Klessig DF (2000) Differential induction of tobacco MAP kinases by the defense signals nitric oxide, salicylic acid, ethylene, and jasmonic acid. Mol Plant Microb Interact 13: 347–351

Laxalt AM, Beligni MV and Lamattina L (1997) Nitric oxide preserves the level of chlorophyll in potato leaves infected by *Phytophthora infestans*. Eur J Plant Pathol 103: 643–651

Leshem YY (1996) Nitric oxide in biological systems. Plant Growth Regul 18: 155–159

Leshem YY and Haramaty E (1996) The characterization and contrasting effects of the nitric oxide free radical in vegetative stress and senescence of *Pisum sativum* foliage. J Plant Physiol 148: 258–263

Leshem YY, Wills RBH and Ku VVV (1998) Evidence for the function of the free radical gas—nitric oxide (NO·)—as an endogenous maturation and senescence regulating factor in higher plants. Plant Physiol Biochem 36: 825–833

Mackintosh C (1998) Regulation of plant nitrate assimilation—from ecophysiology to brain proteins. New Phytol 139: 153–159

Mathieu C, Moreau S, Frendo P, Puppo A and Davies MJ (1998) Direct detection of radicals in intact soybean nodules: presence of nitric oxide-leghemoglobin complexes. Free Rad Biol Med 24: 1242–1249

Maxwell D, Wang Y and McIntosh L (1999) The alternative oxidase lowers mitochondrial reactive oxygen production in plant cells. Proc Natl Acad Sci USA 96: 8271–8276

McDonald LJ and Murad F (1996) Nitric oxide and cyclic GMP signaling. Proc Soc Exp Biol Med 211: 1–6

Millar AH and Day DA (1996) Nitric oxide inhibits the cytochrome oxidase but not the alternative oxidase of plant mitochondria. FEBS Lett 398: 155–158

Millar AH and Day DA (1997) Alternative solutions to radical problems. Trends Plant Sci 2: 289–290

Minning DM (1999) Ascaris haemoglobin is a nitric oxide activated 'deoxygenase'. Nature 401: 497–502

Navarre DA, Wendehenne D, Durner J, Noad R and Klessig DF (2000) Nitric oxide modulates the activity of tobacco aconitase. Plant Physiol 122: 573–582

Ninnemann H and Maier J (1996) Indications for the occurrence of nitric oxide synthases in fungi and plants and the involvement in photoconidiation of *Neurospora crassa*. Photochem Photobiol 64: 393–398

Palmer RMJ, Ferrige, AG and Moncada S (1987) Nitric oxide release accounts for the biological activity of the endothelium-derived relaxing factor. Nature 329: 524–526

Pfeiffer S, Janistyn B, Jessner G, Pichorner H and Ebermann R (1994) Gaseous nitric oxide stimulates guanosine-3′,5′-cyclic monophosphate (cGMP) formation in spruce needles. Phytochemistry 36: 259–262

Popov VN, Simonian RA, Skulachev VP and Starkov AA (1997) Inhibition of the alternative oxidase stimulates H_2O_2 production in plant mitochondria. FEBS Lett 415: 87–90

Radomski MW, Palmer RM and Moncada S (1990) An L-arginine/nitric oxide pathway present in human platelets regulates aggregation. Proc Natl Acad Sci USA 87: 5193–5197

Ribeiro EA, Cunha FQ, Tamashiro W and Martins IS (1999) Growth phase-dependent subcellular localization of nitric oxide synthase in maize cells. FEBS Lett 445: 283–286

Riens B and Heldt HW (1992) Decrease of nitrate reductase activity in spinach leaves during a light-dark transition. Plant Physiol 98: 573–577

Schmidt HHW and Walter U (1994) NO at work. Cell 78: 919–925

Sen S and Cheema IR (1995) Nitric oxide synthase and calmodulin immunoreactivity in plant embryonic tissue. Biochem Arch 11: 221–227

Siegel-Itzkovich J (1999) Viagra makes flowers stand up straight [news]. West J Med 171: 380

Stamler JS, Singel DJ and Loscalzo J (1992) Biochemistry of nitric oxide and its redox-activated forms. Science 258: 1898–1902

Trevaskis B, Watts RA, Andersson CR, Llewellyn DJ, Hargrove MS, Olson JS, Dennis ES and Peacock WJ (1997) Two hemoglobin genes in *Arabidopsis thaliana*—the evolutionary origins of leghemoglobins. Proc Natl Acad Sci USA 94: 12230–12234

Vanlerberghe GC and McIntosh L (1996) Signals regulating the expression of the nuclear gene encoding alternative oxidase of plant mitochondria. Plant Physiol 111: 589–595

Wagner AM and Krab K (1995) The alternative respiration pathway in plants: Role and regulation. Physiol Plant 95: 318–325

Wellburn AR (1990) Why are atmospheric oxides of nitrogen phytotoxic and not alternative fertilisers? New Phytol 115: 395–429

Wildt J, Kley D, Rockel A, Rockel P and Segschneider HJ (1997) Emission of NO from several higher plant species. J Geophys Res Atmos 102: 5919–5927

Wink DA and Mitchell JB (1998) Chemical biology of nitric oxide: Insights into regulatory, cytotoxic, and cytoprotective mechanisms of nitric oxide. Free Radical Biol Med 25: 434–456

Wink DA, Hanbauer I, Krishna MC, DeGraff W, Gamson J and Mitchell JB (1993) Nitric oxide protects against cellular damage and cytotoxicity from reactive oxygen species. Proc Natl Acad Sci USA 90: 9813–9817

Wojtaszek P (2000) Nitric oxide in plants—To NO or not to NO. Phytochemistry 54: 1–4

Yamasaki H and Sakihama Y (2000) Simultaneous production of nitric oxide and peroxynitrite by plant nitrate reductase: In vitro evidence for the NR-dependent formation of active nitrogen species. FEBS Lett 468: 89–92

Yamasaki H, Sakihama Y and Takahashi S (1999) An alternative pathway for nitric oxide production in plants: New features of an old enzyme. Trends Plant Sci 4: 128–129

Chapter 13

Nitrogen and Signaling

Anne Krapp* and Sylvie Ferrario-Méry
Laboratoire de la Nutrition Azotée des Plantes, Route de St. Cyr, F-78026 Versailles, France

Bruno Touraine
Laboratoire Biochimie et Physiologie Moléculaire des Plantes, UMR 5004
INRA/CNRS/Agro-M/UM 2, Place Viala, F-34060 Montpellier cedex 1, France

Summary 206
I. Introduction 206
II. Processes Regulated by Nitrate and Reduced Nitrogen-Compounds 206
A. Nitrate As a Signal 206
1. Morphology and Development 207
2. Nitrate and Ammonium Uptake 208
3. Conversion of Nitrate to Glutamine 210
4. Carbon Metabolism 212
B. Glutamine and Other Reduced Nitrogen-Compounds As Signals 213
1. Morphology and Development 213
2. Nitrate and Ammonium Uptake 213
3. Conversion of Nitrate to Glutamine 215
4. Carbon Metabolism 215
III. Molecular Mechanisms of Nitrogen Signal Perception and Transduction 216
A. Transcriptional Mechanisms 216
1. *Cis*-Acting Elements 216
2. *Trans*-Acting Elements 216
B. Post-Transcriptional Mechanisms 217
1. Ser-Protein Kinases/Phosphatases 217
2. 14-3-3 and PII-Like Proteins 218
3. Two-Component Regulatory Systems 219
C. Mechanisms of Nitrogen Sensing 220
IV. Concluding Remarks 220
Acknowledgments 220
References 220

*Author for correspondence, email: akrapp@versailles.inra.fr

Christine H. Foyer and Graham Noctor (eds): Photosynthetic Nitrogen Assimilation and Associated Carbon and Respiratory Metabolism, pp. 205–225.

Summary

In addition to their role as nutrients, nitrogen (N)-containing compounds are considered to be signaling molecules in plants. Plant development is modified by N-metabolites. Root architecture and root-to-shoot allocation are particularly sensitive to soil nitrate and these processes respond to nitrate via several mechanisms. Metabolic pathways are also influenced by N-compounds at several levels. The molecular mechanisms that exert this control are not yet understood but recent evidence suggests that N-effectors act by regulating gene expression as well as by exerting post-transcriptional and post-translational effects. Like the processes of nitrate and ammonium uptake and assimilation, organic acid synthesis and starch biosynthesis are modified by nitrate, glutamine and other products of N assimilation. In this chapter, we discuss the evidence for the role of nitrate and nitrogen metabolites, such as glutamine, as signals regulating plant morphology and metabolism.

I. Introduction

Application of nitrate fertilizer leads to a stimulation of all the steps in the pathway of nitrogen (N) assimilation. This results in increases in nitrate, ammonium, amino acids, proteins and other N-containing constituents in the plant (Marschner, 1995; Scheible et al., 1997a,b). Nitrate assimilation is closely integrated with other branches of plant metabolism. When N is supplied organic acid and carbohydrate metabolism is also affected. As a direct effect or consequence of these interactions plant growth is increased and there are marked changes in the allocation of resources and in whole plant morphology (Brouwer, 1962; Bloom et al., 1985; Lambers et al., 1990).

The pronounced modifications in metabolism and development that result from quantitative and qualitative changes in the availability of N are either a consequence of N assimilation or are due to signaling by either nitrate or by metabolites that are downstream of nitrate assimilation. It has long been suspected that nitrate is not only a resource but that it also acts, directly or indirectly, to trigger signals that modulate gene expression, metabolism and development (Redinbaugh and Campbell, 1991; Hoff et al., 1994; Crawford, 1995; Stitt, 1999). By analogy to mechanisms observed in microorganisms, we can predict that N metabolites such as ammonium, glutamine (Gln) and asparagine (Asn) act in this manner. In the following discussion we will present evidence that metabolic and developmental processes are regulated by N signals and review recent results concerning the molecular mechanisms underlying N signaling.

II. Processes Regulated by Nitrate and Reduced Nitrogen-Compounds

A. Nitrate As a Signal

Nitrate is the major substrate for the N assimilatory pathway in plants. This makes it difficult to distinguish between the effects of nitrate per se and those of other N-compounds. Experimentally changing the amount of nitrate supplied to plants of necessity results in the modification of the tissue concentrations of all subsequent compounds derived from nitrate assimilation (reduced N-compounds including ammonium and amino acids). However, genotypes with an altered capacity for nitrate assimilation by virtue of changed nitrate reductase (NR) activity have proved to be excellent experimental tools in this regard. Such plants have greatly aided the elucidation of nitrate signaling in plants and also provided insights into more general N signaling events. When the amount of N supply to these plants is modified tissue nitrate contents can be varied independently of the rate of nitrate assimilation which is determined by NR activity. Consequently effects of tissue nitrate alone on amino acid and protein contents, as well as on the overall rate of plant growth, can be determined (Scheible et al., 1997a,b).

Abbreviations: 2 OG – 2-oxoglutarate; AGPase – ADP-glucose-pyrophosphorylase; AMT – ammonium transporter; Asn – asparagine; C – carbon; c(i)HATS – constitutive (inducible) high affinity transport system; Fd – ferredoxin; Gln – glutamine; GluR – glutamate receptor; GOGAT – glutamine-oxoglutarate-aminotransferase; GS – glutamine synthetase; HATS – high affinity nitrate transport systems; LATS – low affinity transport system; N – nitrogen; NADP-ICDH – NADP-dependent isocitrate dehydrogenase; NR – nitrate reductase; NiR – nitrite reductase; NRT – nitrate transporter; OPPP – oxidative pentose phosphate pathway; PEP – phosphoenolpyruvate; PEPc – phosphoenolpyruvate carboxylase; SPS – sucrose phosphate synthase

1. Morphology and Development

N fertilization not only leads to overall increased growth and biomass production, but also results in alterations in the allocation of resources and in plant morphology. Recent experiments using transgenic plants with low NR activity have revealed that nitrate-mediated signaling triggers at least some of these changes in resource allocation and development.

Plants growing on low nitrate supply typically display a higher root-to-shoot ratio than plants adequately fed with this nutrient (Brouwer, 1962; Van de Werf and Nagel, 1996). This functional adjustment in the allocation of dry matter tends to reduce the demand for nitrate at the root surface when the external concentration of this anion is low, helping the plant to adapt to the decrease in N availability. Nitrate availability affects both root and shoot morphogenesis, shifting biomass allocation by concomitant decreases in shoot growth and increases in root growth. Leaf area development is poor when the nitrate supply is low while the root system becomes more finely branched (Grime et al., 1991; Fichtner and Schulze, 1992). Globally, these morphological changes are not restricted to nitrate but they resemble those observed by limitations in other nutrients, e.g. phosphate or sulfate. This suggests that they are general mechanisms of adaptation to low nutrient availability rather than specific responses to nitrate or nitrate-derived signals (Clarkson and Touraine, 1994). On the other hand, changing the availability of another inorganic N source, ammonium, does not trigger similar phenotypic responses. On the contrary, the growth of various plant species is inhibited when ammonium is supplied instead of nitrate as an exclusive N source (Chaillou et al., 1986; Raab and Terry, 1994). Therefore, while plants supplied with low nitrate are able to adapt their morphology in such a way as to enable better management of low N resources, they cannot achieve this display of phenotypic flexibility in response to limiting ammonium. In a recent study, Walch-Liu et al. (2000) showed that supplying tobacco with ammonium resulted in decreased rates of cell division and cell elongation in comparison to nitrate-fed plants. These authors concluded that the effects triggered by ammonium were not due to the ammonium ion per se (that is they ruled out the 'ammonium toxicity' hypothesis), but rather to lack of nitrate. Nitrate is hence required to maintain a sufficient flux of root-to-shoot cytokinin transport, as cytokinin mediates leaf morphogenesis. Lowered concentrations of cytokinins have been observed in the xylem sap of N-deprived potato plants compared to those supplied with nitrate (Sattelmacher and Marschner, 1978). Furthermore, abscisic acid increases in the xylem sap of nitrate-deficient plants, suggesting that changes in the hormonal balance in the xylem sap control leaf morphogenesis in response to low nitrate (Clarkson and Touraine, 1994).

The hypothesis that nitrate ions, rather than a metabolite more downstream in the N assimilatory pathway, are involved in phenotypic adaptations to changes in external nitrate concentration is consistent with the results obtained using plants affected in NR activity. The higher accumulation of nitrate observed in low-NR tobacco transformants than in N-replete wild-type plants, was accompanied by higher shoot-to-root ratios (Scheible et al., 1997a), even though the plants with low NR activity were severely N-limited with respect to organic N. More precisely, split-root experiments have shown that the inhibition of root growth was triggered by the accumulation of nitrate in the shoot, but not in the root. This is indicative of systemic regulation involving inter-organ signaling. However, considering that nitrate is quasi-excluded from the sieve sap, this ion is unlikely to be the signal translocated in the phloem from shoot to root. The nature of the nitrate-related signal that is translocated to roots and the mechanisms involved in the transduction of such an inter-organ signal, remain to be elucidated. The inhibition of root growth triggered by nitrate accumulation correlates with decreased allocation of carbon to the root (Scheible et al., 1997a) and with a decrease in the number of lateral roots (Stitt and Scheible, 1998; Stitt and Feil, 2000). Interestingly, Lexa and Cheeseman (1997) did not find any difference in the shoot-to-root ratio in a NR-deficient pea mutant. This result may, however, be connected with the ability of pea roots to form nodules, even though nodule formation is inhibited in the presence of nitrate.

The responses of root growth to nitrate availability are complex. In addition to the feedback inhibition of root growth at high nitrate, low nitrate has a positive effect on root development. Indeed, localized application of low nitrate leads to a localized stimulation of lateral root proliferation (Drew and Saker, 1976; Granato and Raper, 1989; Robinson, 1994). Localized application of low nitrate leads to the proliferation of lateral roots in tobacco (Scheible et al., 1997a) and *Arabidopsis thaliana* (Zhang and Forde, 1998), even in genotypes with very low NR activity. This response is not accompanied by increases in the local concentrations of either amino acids or proteins (Scheible et al., 1997a). This suggests that the effect involves nitrate-mediated signaling rather than a mechanism driven by the nutrient-mediated growth stimulation alone. Recently, a nitrate-inducible MADS-box transcription factor gene (*ANR1*) has been identified as a component of the signaling pathway. This is involved in the stimulation of lateral root growth by localized nitrate supply in *A. thaliana* (Zhang and Forde, 1998). The role of ANR1 in eliciting the developmental response to nitrate has been demonstrated using reverse genetics. *ANR1*-repressed lines (antisense or co-suppressed sense) lack the capacity to respond via lateral root proliferation to localized nitrate supply. In the *A. thaliana* wild type, nitrate -stimulated lateral root development is due to increased root elongation. This was attributed to an increase in the rate of cell production in the lateral root meristem (Zhang et al., 1999). The stimulatory effect of nitrate was blocked in the *axr4* auxin-resistant mutant, indicating that nitrate and auxin share common signaling pathways or components (Zhang et al., 1999). The sensitivity of lateral root development to inhibition by high nitrate concentrations was also higher in the *ANR1* antisense lines. Nitrate sensitivity increased with the degree of *ANR1* repression in the transgenic lines (Zhang and Forde, 1998). This result is consistent with the existence of dual mechanisms of nitrate regulation of root branching. The inhibitory effect of nitrate is pronounced in *ANR1*-deficient plants because the ANR1-dependent localized stimulatory effect is blocked. The development of lateral roots in the *nia1nia2* NR-deficient mutants was more sensitive to inhibition by high nitrate than it was in the wild-type (Zhang et al., 1999). These observations support the hypothesis that tissue nitrate plays a role in the production of an inhibitory signal (Scheible et al., 1997a). An overall model for the regulation of root branching by two opposite signals (external nitrate and internal plant N status) has been proposed by Zhang et al. (1999; Fig. 1).

2. Nitrate and Ammonium Uptake

Plants that have been grown without nitrate for several days exhibit low rates of nitrate uptake. This low flux rate continues over the first hour period when nitrate is re-supplied. Within hours to days of the onset of nitrate re-supply, however, depending on the species, the rate of nitrate uptake subsequently increases to a peak that is several-fold greater than the initial rate (Lee and Drew, 1986; Siddiqi et al., 1990; Aslam et al., 1992; Kronzucker et al., 1995a). This induction of increased nitrate transport in the presence of external nitrate is observed at relatively low external nitrate concentrations, and falls within the range of the high affinity nitrate transport systems (HATS). Careful kinetic analyses in roots of non-induced or induced plants have shown that the V_{max} values for transport and also the K_m, increase after several hours of exposure to nitrate (Lee and Drew, 1986; Hole et al., 1990; Siddiqi et al., 1990; Aslam et al., 1992; Kronzucker et al., 1995b). Based on these observations and the results of experiments where induction by transcription and translation was blocked by inhibitors (Tompkins et al., 1978; Lainé et al., 1995), two different transport systems were distinguished. These are the low capacity constitutive high-affinity transport system (cHATS) and the high capacity inducible high-affinity transport system (iHATS). In contrast, nitrate flux measurements (Siddiqi et al., 1990) and electrophysiology studies (Glass et al., 1992) have shown that the low-affinity transport system (LATS), which becomes apparent at external nitrate concentration higher than 1 mM, is not induced by nitrate.

Physiological studies have shown that nitrite, which is not found in significant amounts, is also able to induce nitrate uptake (Aslam et al., 1993). In contrast, ammonium cannot induce nitrate uptake (Aslam et al., 1993; King et al., 1993). Therefore, the down-stream products of ammonium assimilation are not potent inducers of nitrate uptake.

To date, two families of genes that encode nitrate transporters have been cloned from plants. These are denoted as *NRT1* and *NRT2*. Most of the data available to date concern the genes referred to as *NRT1.1* and *NRT2.1*. When the roots of N-deficient barley plants were exposed to nitrate, the steady-state level of *NRT2* transcripts rapidly increased (Trueman et al., 1996; Vidmar et al., 2000a). Similar observations were made in *Nicotiana plumbaginifolia* (Krapp et al., 1998), soybean (Amarasinghe et al., 1998) and *A. thaliana* (Filleur and Daniel-Vedele, 1999; Zhuo et al., 1999). Since NRT2.1 has been characterized as a high affinity transporter, it is considered to be an iHATS. The *A. thaliana AtNRT1.1* gene, which has LATS characteristics (Huang et al., 1996; Touraine

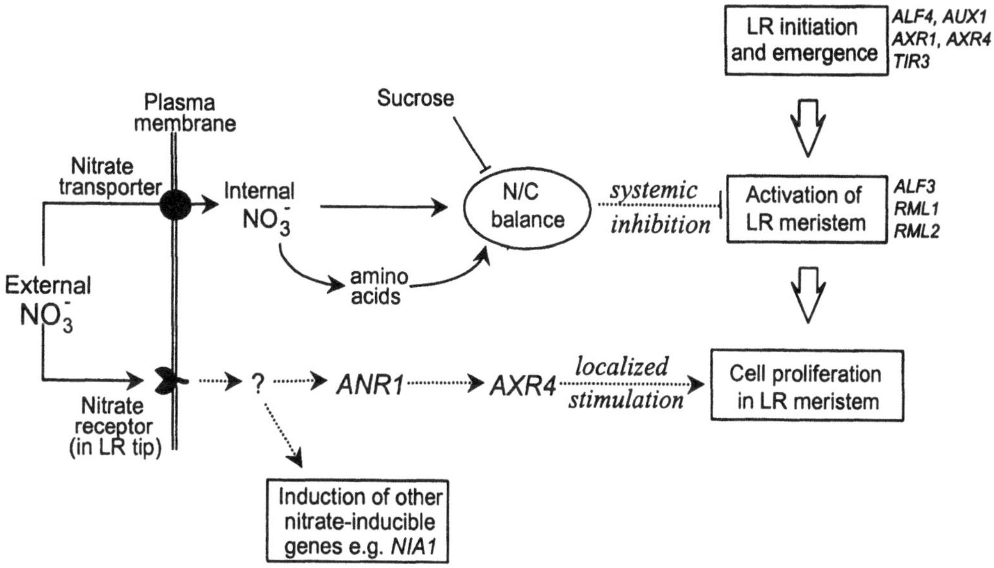

Fig. 1. Overall model for the regulation of root branching by nitrate (proposed by Zhang et al., 1999). LR, lateral root.

and Glass, 1997), is also inducible by nitrate (Tsay et al., 1993; Filleur and Daniel-Vedele, 1999). This observation is not consistent with the data derived from kinetic studies. On the other hand, studies using *chl1* mutants where the *NRT1.1* gene is altered, and expression studies in *Xenopus* oocytes, have indicated that AtNRT1.1 might actually be a dual-affinity nitrate transporter (Wang et al., 1998; Liu et al., 1999). However, these observations do not allow resolution of the puzzling properties of *NRT1.1* because the *chl1* mutants should be defective in the cHATS component, not in the iHATS (Wang et al., 1998). The available functional evidence cannot explain the observed up-regulation of *NRT1.1* transcripts upon exposure to nitrate.

When nitrate is continuously supplied to roots for periods of one to several days nitrate transport systems are induced and nitrate influx within the high-affinity range subsequently decreases (Zhuo et al., 1999; Vidmar et al., 2000a). Consistent with these observations, nitrate influx is known to be up regulated in N-deficient plants (e.g. Siddiqi et al., 1989; Lee, 1993). Most of the published physiological evidence suggests that feedback-regulation of iHATS in N-replete plants is exerted by a product of ammonium assimilation (as discussed below). However, a negative correlation between nitrate influx and total root nitrate content was found in barley (Siddiqi et al., 1989). This prompted King et al. (1993) to use a *nar1a/nar7w* NR-deficient mutant to distinguish between the effects due to nitrate per se and those of the products of nitrate assimilation. Since nitrate influx in these mutants was strongly inhibited within five days of exposure to nitrate, King et al. (1993) concluded that tissue nitrate exerts influence over nitrate influx by feedback regulation. These authors mentioned that this did not appear to be an exclusive mechanism of feedback regulation and could involve other N-compounds in addition to nitrate. By contrast, other studies using NR-deficient plants led to the conclusion that tissue nitrate is probably not a repressor of its own uptake. For example, large amounts of nitrate accumulate in NR-deficient mutants and transformants of barley (Warner and Huffaker, 1989), tobacco (Scheible et al., 1997a) and *A. thaliana* (Meyer zu Hörste, 1998). Gojon et al. (1998) reported unaltered (or only slightly lower) nitrate uptake rates in *N. plumbaginifolia* and tobacco genotypes with small decreases in NR activity. Such results indicate that high tissue nitrate contents do not lead to strong feedback inhibition of nitrate uptake. However, as discussed by Gojon et al. (1998), such genotypes exhibit lower rates of nitrate assimilation. This would be predicted to stimulate nitrate uptake (due to the release of feedback regulation by amino acids, as discussed below). This effect could mask any weak feedback regulation on nitrate uptake by nitrate.

Expression studies have shown that *NRT2.1* transcripts increased in the first hours after exposure to nitrate. They then decreased once more back to a level close to that found in roots of non-induced plants. The overall pattern in *NRT2.1* transcript levels paralleled the induction/repression pattern of nitrate influx (Zhuo et al., 1999; Vidmar et al., 2000a). Filleur and Daniel-Vedele (1999) reported variations of *NRT1.1* mRNA in *A. thaliana* roots similar to those of *NRT2.1* when nitrate was supplied to plants previously fed with Gln as the sole N source. The induction of *NRT1.1* was less marked than observed with *NRT2.1*. These expression profiles for *NRT1.1* showed marked differences with time compared to either *NRT2.1* transcript abundance or nitrate influx in other reports. In this case the expression patterns for *NRT1.1* and *NRT2.1* were repressed within 8 h of the onset of nitrate supply, compared to several days in other reports. Furthermore, the amount of *NRT1.1* transcripts in roots decreased when plants grown on nitrate were transferred to a N-free medium (Filleur and Daniel-Vedele, 1999; Lejay et al., 1999). This indicates that *NRT1.1* induction is essentially reversed when nitrate is withdrawn and that this occurs when plants become N-deficient. This pattern contrasts with that of *NRT2.1* expression, which is up regulated in plants grown on N-free medium. In the experiments of Filleur and Daniel-Vedele (1999), the observed decline in *NRT1.1* and *NRT2.1* mRNA abundance after the induction step, is probably linked to differences other than those resulting from feedback regulation. The different responses of *NRT1.1* and *NRT2.1* to N starvation, are explained by different sensitivities of these two systems to feedback regulation by amino-N compounds (Lejay et al., 1999), as discussed below. In a thorough investigation of the signals that could be involved in *HvNRT2.1* down-regulation, Vidmar et al. (2000b) ruled out the hypothesis that nitrate could be responsible for the feedback regulation. Indeed, addition of tungsten, which blocks the synthesis of the molybdenum-pterin cofactor and inhibits NR activity, resulted in a slight increase in the *NRT2.1* transcript level in roots and a decrease in nitrate influx. This indicates that nitrate may exert post-transcriptional control on nitrate transporters, but that it is not responsible for the feedback regulation of *NRT2.1* transcript abundance.

Net nitrate import is the result of two opposite fluxes, the influx that draws the nitrate ions into the root and the efflux processes by which these ions are released outside the root cells. Most of the studies on nitrate uptake specifically concern the influx component. However, a high proportion of the nitrate taken up by the root can be lost by efflux (Lee, 1993; Devienne et al., 1994; Muller et al., 1995). Thus, theoretically these inverse transport processes provide possibilities of nitrate uptake control via regulation of both efflux and influx transport systems. None of the nitrate efflux transporters, whether carriers or channels, have been identified to date, although recent electrophysiological studies using plasma membrane vesicles have given new insights into these processes (Pouliquin et al., 1999, 2000). At the physiological level, the regulation of nitrate efflux is still poorly understood. There are some indications that efflux is inducible by nitrate (Aslam et al., 1996). Furthermore, electrophysiological data obtained in planta using specific nitrate microelectrodes show that nitrate efflux in barley is dependent upon both external and internal nitrate concentrations (Van der Leij et al., 1998). Nitrate efflux may result from a simple concentration-activity relationship that must occur against the electrochemical gradient, and this need not imply that nitrate ions would act as a signal.

In *A. thaliana*, ammonium influx increased when plants were transferred from a medium containing ammonium to a medium containing nitrate as the sole N-source. However, there was no corresponding change in the mRNA abundance of any of the three *AMT1* genes (Gazzarini et al., 1999). On the other hand, when plants were deprived of any N-source, both the *AtAMT1.1* transcript level and the ammonium influx declined. It is not known whether the differences between nitrate nutrition on one hand and either ammonium nutrition or N-deficiency on the other, involve specific effects of nitrate. Contrary to the observations in *A. thaliana*, *LeAMT1* transcripts in tomato roots were not markedly influenced by the source of N (nitrate or ammonium) or even the absence of N from the medium (Lauter et al., 1996).

3. Conversion of Nitrate to Glutamine

Both NR and nitrite reductase (NiR) respond to nitrate at the level of gene expression. Transcription of NR coding sequences (*NIA*) is induced very rapidly by nitrate (Pouteau et al., 1989; Cheng et al., 1991; Lin et al., 1994). It is also subject to diurnal regulation, the *NIA* expression pattern correlating with tissue nitrate contents. A decrease in the *NIA* transcripts occurs during the photoperiod and is accompanied by a decrease in tissue nitrate (Scheible et al., 1997c). This is followed by a gradual recovery during the night that is correlated with a gradual increase in tissue nitrate. Addition of nitrate does not affect the

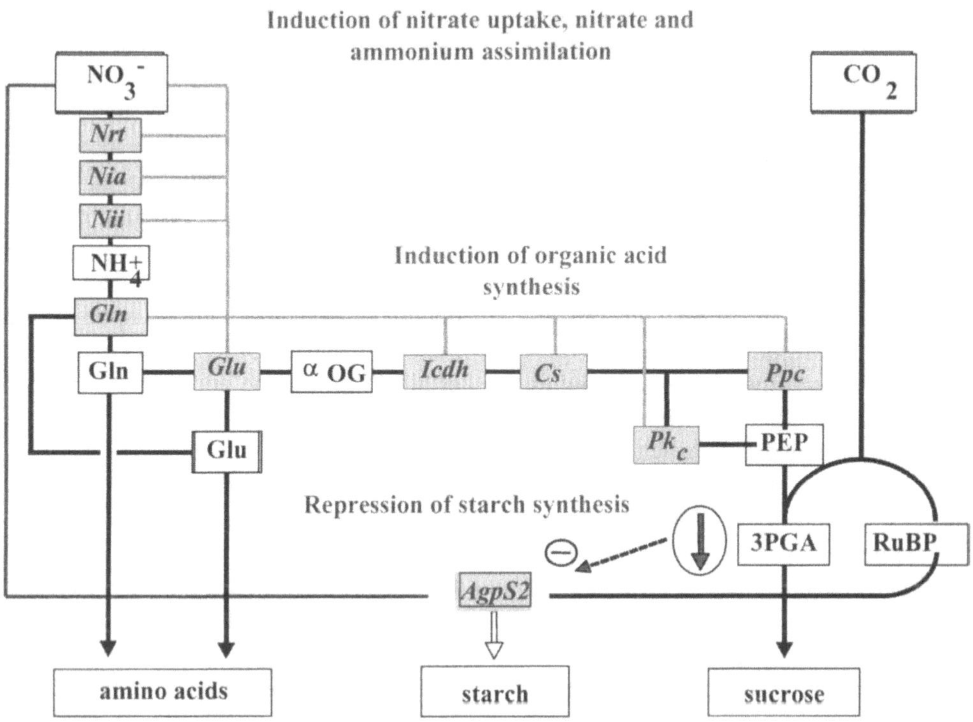

Fig. 2. Coordinated regulation of carbon and nitrogen metabolism by nitrate (Scheible et al., 1997b). AgpS2, ADP-glucose pyrophosphorylase; Cs, citrate synthase; Gln, glutamine; Glu, glutamate; Nia, nitrate reductase; Nii, nitrite reductase; Nrt, nitrate transporter; αOG, 2-oxoglutarate; PEP, phosphoenolpyruvate; 3PGA, 3-phosphoglycerate; PKc, pyruvate kinase; Ppc, PEP carboxylase; RuBP, ribulose-1,5-bisphosphate.

activation state of NR (Ferrario et al., 1996). Not only are the genes encoding the NR protein induced by nitrate, but the enzymes responsible for the synthesis of NR cofactors are also induced. This was shown by microarray analysis of *A. thaliana* cells grown in liquid culture (W. Scheible, personal communication). Nitrite reductase is co-regulated with NR (Faure et al., 1991). The induction factor for NiR is even higher than that for many other nitrate-induced genes (Wang et al., 2000). *UPM1* encodes a protein catalyzing a branch point in the biosynthesis of siroheme, an essential cofactor for NiR. In maize and *A. thaliana, UPM1* is very strongly induced by the addition of nitrate (Sakakibara et al., 1996; Wang et al., 2000). Given the high toxicity of nitrite, it is perhaps not surprising that the proteins necessary for nitrite reduction are very efficiently regulated.

In plants ammonium arises as a result of nitrite reduction, as a result of ammonium uptake or from photorespiration. Ammonium is assimilated into the organic N compounds Gln and glutamate, in the glutamine synthetase (GS)/glutamate synthase (GOGAT) catalyzed reaction sequence. Glutamine and glutamate are the N donors for almost all biosynthetic reactions involving N. Nitrate enhances abundance of transcripts encoding GS and GOGAT. Examples are: *GLN1* (encoding the cytosolic glutamine synthetase; GS1), *GLN2* (encoding plastid glutamine synthetase; GS2) and *GLU* (ferredoxin-dependent glutamate synthase; Fd-GOGAT). This was observed in maize roots (Redinbaugh and Campbell, 1993), tobacco roots and leaves (Scheible et al., 1997b) and *A. thaliana* (Wang et al., 2000). Nitrate-induced enhancement of *GLU* transcript has also been described in illuminated barley and *A. thaliana* leaves (Pajuelo et al., 1997; Wang et al., 2000). The nitrate-induced changes in cytosolic GS transcripts were much more marked than those of transcripts encoding the plastidic GS isoform. In addition, nitrate induced the appearance of a second type of *GLN1* transcript (Scheible et al., 1997b). Nitrate-induced increases in GS transcripts were accompanied by an increase in shoot and root GS activity (Scheible et al., 1997b). Nitrate induced the appearance of a second plastidic GS form in tomato leaves (Migge et al., 1997), which was absent when tomato plants were grown on ammonium as the sole N source. In maize roots, nitrate led to an increase in transcripts encoding one of the GS isoforms and Fd-GOGAT (Sivasankar and Oaks, 1995). In anaerobic rice seedlings, the addition of nitrate led to increased Fd-GOGAT protein but did not cause increases in the cytosolic or plastidic GS proteins (Mattana et al., 1996).

Evidence that changes in expression of the GS/GOGAT pathway are due to direct signaling by nitrate, rather than to effects of the products of N assimilation, has been obtained in several studies. The transcripts encoding GS2, Fd-GOGAT and to a lesser extent, GS1 also increase when nitrate is supplied to NR-deficient genotypes or genotypes with very low expression of NR (Vaucheret et al., 1990; Kronenberger et al., 1993; Scheible et al., 1997b). Addition of tungsten had no marked effect on either GS2 and Fd-GOGAT transcript or protein abundance (Migge et al., 1997).

During rapid nitrate assimilation the demand of reducing equivalents is high. When nitrate is assimilated in leaves in the light, the reducing equivalents are delivered by photosynthetic electron transport. In both algae and higher plants, there is evidence that addition of nitrate leads to increases in the rate of non-cyclic electron transport (Turpin and Bruce, 1990; Foyer et al., 1994a,b). In the alga *Scenedesmus minutum*, the addition of nitrate led to a severe oxidation of the photosynthetic electron transport chain. Moreover, the thioredoxin-mediated activation of reductive pentose phosphate pathway enzymes was prevented, leading to a marked inhibition of photosynthesis (Huppe and Turpin, 1994; Turpin et al., 1997). In the longer term, nitrate also exerts other effects that modify the electron transport processes. For example, nitrate accumulation in transformed plants with low NR activity was accompanied by a decrease in the chlorophyll *a/b* ratio (Lauerer, 1996). This indicates that nitrate-mediated signals modify the light-harvesting processes associated with Photosystem (PS) II. Enhanced PS II activity would tend to favor enhanced non-cyclic electron flow with greater production of reducing power.

Nitrate assimilation also occurs in the dark both in leaves and roots. The oxidative pentose phosphate pathway (OPPP) provides reducing equivalents in this situation. The reduced Fd required by NiR is produced from NADPH via ferredoxin-NADP-oxidoreductase. Nitrate addition leads to rapid induction of Fd (Matsumara et al., 1997) and ferredoxin-NADP-oxidoreductase (Ritchie et al., 1994) in maize roots. The importance of this cross-talk between nitrate metabolism and the redox state of the cell is underlined by the finding that NR expression and activation state increase in response to anaerobiosis. The addition of nitrate, but not ammonium, to *S. minutum* cultures leads to rapid activation of glucose-6-phosphate dehydrogenase. Glucose-6-phosphate contents decrease and 6-phosphogluconate levels increase upon nitrate addition, indicating that the OPPP has been stimulated (Huppe et al., 1992; Huppe and Turpin, 1994; Botrel and Kaiser, 1997; Turpin et al., 1997).

4. Carbon Metabolism

The presence of nitrate not only triggers changes in the nitrate assimilation pathway but it also reprograms several pathways of carbon metabolism. In the presence of nitrate carbohydrate synthesis is decreased and more carbon is converted via glycolysis to phosphoenolpyruvate (PEP) and enters organic acid metabolism. The synthesis of organic acids plays a major role in nitrate assimilation, because organic acids provide the C-skeletons for amino acid biosynthesis. In addition organic acids are involved in the maintenance of cellular pH which is important because hydroxide ion is produced during nitrate assimilation. When nitrate is added to *S. minutum* cultures, pyruvate kinase and PEP carboxylase (PEPc) are activated and the flux of carbon into organic acids is stimulated (Huppe and Turpin, 1994; Turpin et al., 1997). There is also evidence in higher plants that signals from N metabolism allow coordinate transcriptional regulation of a group of key enzymes of organic acid synthesis in higher plants. For example, the addition of nitrate led to increase level of transcripts encoding PEPc, the cytosolic pyruvate kinase, citrate synthase and the NADP-dependent isocitrate dehydrogenase (NADP-ICDH) in tobacco (Scheible et al., 1997b). The abundance of these transcripts also increases after the addition of nitrate to transformed plants with low NR activity. This suggests that the increase in transcripts is triggered by nitrate per se rather than by other compounds that are formed as a result of nitrate assimilation (Scheible et al., 1997b). Nitrate-induced changes in transcript levels occur within 2–4 h of nitrate addition, suggesting that the effect of nitrate is rapid and that high internal concentrations are not required (Scheible et al., 1997b, 2000). Significantly, nitrate addition does not lead to a significant increase in fumarase transcripts. Fumarase is an enzyme catalyzing a reaction in the section of the tricarboxylic acid cycle that is not required for net synthesis of malate or 2-oxoglutarate (2-OG) (Scheible et al., 2000). Observed changes in transcripts are frequently accompanied by increases in enzyme activity. Examples are: PEPc (Foyer et al., 1994b; Scheible et al., 1997b), pyruvate kinase, citrate synthase and NADP-ICDH (Scheible et al., 2000). Moreover, these effects lead to an accumulation of 2-OG and other organic acids (Scheible et al., 1997b). Other evidence suggests that many enzymes catalyzing steps in the organic acid biosynthesis pathway are controlled by post-translational regulation. For example, PEPc is regulated by protein phosphorylation. The degree of phosphorylation of the PEPc protein is largely determined by PEPc-kinase activity. This is controlled via de novo synthesis of the protein (Hartwell et al., 1996). Phosphorylation of the PEPc protein results in decreased sensitivity to inhibition by malate (Chollet et al., 1996). In transgenic tobacco plants with low NR activity, the sensitivity of PEPc to malate is also decreased by the addition of nitrate (Scheible et al., 1997b). This indicates that nitrate-mediated signals contribute to the post-translational regulation of PEPc.

Nitrate acts as a signal regulating starch biosynthesis. Starch biosynthesis is repressed by nitrate. This ensures that the carbon flux to amino acid synthesis is increased in the presence of nitrate. *AGPS2* transcripts, encoding large subunits of ADP-glucose pyrophosphorylase (AGPase), a key enzyme in starch synthesis, decrease within 2–4 h of the addition of nitrate to N-deficient tobacco. Removal or use of the added nitrate allows *AGPS2* transcripts to increase again (Scheible et al., 1997b). Similar changes are seen after adding nitrate to tobacco transformants with low NR activity. This suggests that the signal is related to nitrate rather than the metabolism of nitrate (Scheible et al., 1997b). In these transformants, nitrate addition leads to a rapid decrease in starch even though the rate of growth is not significantly altered. This result indicates that nitrate may also affect the rate of starch degradation.

Microarray analysis of nitrate-induced *A. thaliana* seedlings showed that two key genes of the OPPP (glucose-6-phosphate dehydrogenase and 6-phosphogluconate dehydrogenase) are strongly induced by nitrate (Wang et al., 2000). The induction of two other genes involved in this pathway: transaldolase and transketolase, was less pronounced. The role of nitrate in these interactive metabolic pathways is illustrated in Fig. 2.

B. Glutamine and Other Reduced Nitrogen-Compounds As Signals

Ammonium, Gln and other amino acids are the end products of nitrate assimilation. In the past it has been tempting to discuss the regulatory role of Gln as a signal molecule for the control of metabolic and morphologic processes in plants, by analogy to these roles in fungi. To date, no roles for Gln in the control of plant morphology have been shown.

1. Morphology and Development

As discussed in Section II.A.1, nitrate availability affects root architecture by regulating lateral root growth and development. It does so by stimulating lateral root expansion in regions exposed to low nitrate and by the systemic repression of lateral root formation in the presence of high nitrate (Fig. 1). Root nitrate is considered to be the signal responsible for the localized stimulation, whereas shoot nitrate is likely to be the signal that triggers systemic regulation. This is reinforced by the observation that in *A. thaliana* grown on solid medium, neither Gln nor Asn inhibit lateral root growth (T. J. Tranbarger and B. Touraine, unpublished). The nitrate-dependent systemic signal that is translocated from shoot to root in the phloem is suggested to be abscisic acid (Chapter 1, Foyer and Noctor). The possibility that N metabolites contribute to other steps of the inter-organ signaling pathway remains to be evaluated.

2. Nitrate and Ammonium Uptake

Short-term experiments have clearly shown that nitrate uptake is dependent on the presence of external nitrate. However, laboratory experiments and field studies have consistently shown that nitrate uptake is often independent of nitrate availability and that nitrate uptake is adjusted to growth rate when other factors become limiting (Touraine et al., 1994). Furthermore, both nitrate and ammonium uptake are stimulated by imposing periods of N-deficiency (Lee and Drew, 1986; Morgan and Jackson, 1988; Hole et al., 1990; Lee, 1993). This effect is considered to be due to the relief of negative feedback exerted by whole plant N status. The nutritional status-dependent repression/derepression phenomenon is a common feature for ion uptake by plant roots. For example, potassium, phosphorus and sulfur deficiency lead to enhanced uptake rates of the respective ions (Cogliatti and Clarkson, 1983; Drew et al., 1984; Lee, 1993). However, experiments to determine the effect of withdrawing single components from the nutrient solution on the uptake of other ions have shown that regulation is specific to single elements or ions (Lee and Rudge, 1986; Lappartient and Touraine, 1996).

Split root experiments where parts of root systems are exposed to local supplies of inorganic N (Touraine et al., 1994) have revealed that nitrate uptake is regulated by the general demand for N by the plant and not simply by the N status of the root tissues. This implies that the shoot N status is sensed and that this information is transmitted to the roots. Nitrate is very unlikely to be involved in this inter-organ signaling process because it is known to be quasi immobile in the phloem. It is universally agreed that products whose concentrations are positively linked to N monitor plant N status. Moreover, these products exert negative feedback on nitrate uptake systems. The favored model considers that these products are certain amino acids. Imsande and Touraine (1994) have reviewed physiological evidence in support of this hypothesis. The consensus, in brief, is: (*i*) phloem sap contains high concentrations of amino acids. ^{15}N-labeling experiments showed the existence of a pool of N which cycles rapidly from roots to shoots and back to roots (Cooper and Clarkson, 1989). This agrees with the requirements of the inter-organ signaling model. (*ii*) Certain amino acids are capable of inhibiting nitrate uptake when supplied directly to plant roots (addition to the external medium; Muller and Touraine, 1992). (*iii*) Changing the amino acid composition of the phloem sap experimentally inhibits nitrate transport in roots in a similar manner (Muller et al., 1995). The Achilles heel of this model lies in the relationships between shoot N status and the amino acid composition of the phloem sap. Information on the subject is generally lacking because of the experimental problems encountered when measuring actual amino acid concentrations in the phloem sap and translocation rates (for further discussion, see Chapter 15, Lohaus and Fischer). Split root experiments in castor bean, a species where phloem sap can be collected via stem incision, failed to show a decrease in phloem amino acid levels in response to feeding part of the root system with a N-free medium (Tillard et al., 1998). However, even in this species, the true translocation rate of amino acids (which depends on the concentration of extracted sap but also on the velocity of the sap) cannot be measured with reliability. Such limitations prevent the testing of the key questions concerning the relationships between amino acid export by leaves and N status.

Evidence for a feedback regulation of nitrate uptake by downstream products formed during nitrate assimilation has been provided by Gojon et al. (1998). *Nicotiana* spp. transformants with increased NR expression (and consequently increased levels of Gln) have lower nitrate uptake rates than wild-type plants (Gojon et al., 1998). Inhibitors of GS and GOGAT prevented the inhibitory effect of ammonium on nitrate uptake by dwarf bean roots (Breteler and Siegerist, 1984), indicating that amino acids rather than ammonium are responsible for feedback regulation.

The feedback regulation of nitrate uptake by amino acids is due to an inhibition of the influx, not a stimulation of the efflux component (Muller et al., 1995). This effect is likely to be due, at least in part, to decreased synthesis of transporter proteins as a consequence of transcriptional regulation of transporter gene(s). Addition of ammonium or Gln to nutrient medium resulted in a rapid decline (more dramatic with Gln than with ammonium) of *NRT2.1* mRNA abundance in roots of *N. plumbaginifolia* and soybean (Krapp et al., 1998; Amarasinghe et al., 1998). Using azaserine to inhibit GOGAT activity in barley, Vidmar et al. (2000b) showed that a concomitant increase in Gln and decrease in glutamate result in a decline in both *HvNRT2* transcripts and nitrate influx in the HATS range. The authors concluded that Gln is likely to be responsible for down-regulation of *HvNRT2* expression. Although several amino acids are potent repressors of *NRT2.1* and nitrate influx in *A. thaliana*, these results are consistent with observations that the abundance of the transcript negatively correlates with the average concentration of Gln in roots. The latter changes as a consequence of amino acids inter-conversion (K. Mouline, J. J. Vidmar and B. Touraine, unpublished). As already mentioned, *NRT1.1* is unaffected by plant N-status (Lejay et al., 1999) and is not subject to inhibition by reduced N compounds. *NRT2.1* expression studies in barley showed that in addition to nitrate (inducer) and Gln (repressor) ammonium is a probable signal in the regulation of this nitrate transporter. Ammonium may down-regulate *NRT2* both at transcriptional and post-transcriptional levels (Vidmar et al., 2000b). Evidence for the occurrence of post-transcriptional regulation by reduced N compounds (ammonium or a product of its assimilation) has also been obtained using transgenic *N. plumbaginifolia* plants with constitutive expression of *NpNRT2.1* (Fraisier et al., 2000). Addition of ammonium to the nutrient medium supplied to these plants resulted in decreased nitrate influx while transgene expression remained unchanged.

High-affinity ammonium uptake is subject to feedback regulation (Wang et al., 1993). This type of regulation was not observed for the ammonium LATS. Both ammonium and amino acids decreased ammonium uptake rate in wheat (Causin and Barneix, 1993). This would suggest that amino acids, rather than ammonium, are the repressors. However, the feedback regulation of the ammonium HATS has received less attention than the corresponding nitrate transporter. It is therefore not possible to draw conclusions concerning the roles of ammonium or amino acids as signal(s) responsible for the inhibition of the ammonium transport system in root cells. Consistent with the physiological evidence, *AMT1.1*, which encodes the AMT1 transporter with the highest affinity for ammonium, is strongly induced (de-repressed) in *A. thaliana* by N starvation (Gazzarini et al., 1999). Conversely, the re-supply of ammonium nitrate to N-deficient plants led to lower *AMT1.1* mRNA abundance and ammonium influx into roots (Rawat et al., 1999). The observations that methionine sulfoximine, a GS inhibitor, reverses the inhibitory effect of ammonium and that ammonium influx is negatively correlated with root Gln concentrations suggest that Gln could be the signal responsible for feedback regulation of *AMT1.1* expression and ammonium transport. *LeAMT1.1* expression is up-regulated by low N availability in tomato, while *LeAMT1.2* is induced by the supply of ammonium or nitrate (von Wirén et al., 2000). This suggests that *LeAMT1.1* and *LeAMT1.2* do not respond to the same signal(s).

3. Conversion of Nitrate to Glutamine

There are many indications that *NIA* transcription is repressed by Gln (Hoff et al., 1994). Feeding Gln or inhibiting its utilization led to an inhibition of *NIA* expression. As described previously the decrease of *NIA* transcripts during the photoperiod correlates with a decrease in nitrate and an accumulation of Gln (Scheible et al., 1997c). This may not hold true in all conditions, however, as Migge et al. (1999) have shown that an increase in tissue Gln is followed by inhibition of *NIA* expression. Furthermore ammonium has also been considered to be a negative signal for *NIA* expression. The effects of Gln have therefore to be considered with care, because plants that are grown on NH_4NO_3 have a high expression of NR (P. Matt, A. Krapp and M. Stitt, unpublished). Plants probably have to establish an appropriate balance between alkalizing nitrate reduction and acidifying ammonium assimilation in order to maintain cellular pH. Also the post-translational regulation of NR is affected by reduced N-components. Scheible et al. (1997c) showed that dark inactivation of NR is partially or completely reversed in nitrate-limited wild-type plants and in mutants and transformants with decreased expression of NR. Abolition of dark inactivation was correlated with decreased Gln and ammonium, both of which are products of nitrate assimilation. NR activation was also decreased by feeding Gln to detached leaves (Scheible et al., 1997c; Morcuende et al., 1998). In addition to its major role in primary N assimilation, the GS/GOGAT cycle also plays a crucial role in re-assimilating ammonium released during photorespiration. No direct effects of Gln and Asn on GS and Fd-GOGAT activities in maize roots were observed, even though these amino acids lead to a marked decrease of NR activity (Sivanskar and Oaks, 1995).

4. Carbon Metabolism

As discussed previously, PEPc transcripts and PEPc activity are rapidly enhanced by the addition of nitrate or ammonium to plants such as maize (Sugiharto et al., 1992; Sugiharto and Sugiyama, 1992; Suzuki et al., 1994). A comparison of the kinetics of N-induced changes in various metabolite pools with those of PEPc transcripts indicates that Gln plays a key role in the induction of PEPc. The nitrate and ammonium-induced increases in PEPc activity were prevented when phosphinothricin was added to inhibit GS activity in barley (Diaz et al., 1996). This again indicates that Gln induces PEPc. A correlation between foliar Gln content and PEPc activity has been observed (Murchie et al., 2000). Mitochondrial NAD-dependent ICDH is likely to be the major source of 2-OG (Lancien et al., 1999).

III. Molecular Mechanisms of Nitrogen Signal Perception and Transduction

A. Transcriptional Mechanisms

It is now well established that nitrate, ammonium and/or Gln act as signals regulating plant development and metabolism. Some progress in the elucidation of mechanisms involved in N-dependent control of gene expression and plant development has already been achieved, particularly at the transcriptional and post-transcriptional levels, as described below. However, by analogy with known sugar signal transduction pathways in other systems it is clear that N-signaling in plants probably involves several levels of regulation with many factors still undiscovered.

1. Cis-*Acting Elements*

Motifs important in N-dependent regulation of gene expression have not yet been completely characterized. To determine the nature of N-dependent regulation more precisely, promoter analyses have been developed using reporter gene expression in transgenic plants. These have involved deletions in the *NIA* promoter region and analysis of the minimal promoter sequence necessary for N-dependent regulation. In tobacco and *A. thaliana*, 1.4 and 2.7 kb of the *NIA* promoter respectively were sufficient for the nitrate inducibility (Cheng et al., 1992; Vincentz et al., 1993). While, experiments designed to identify relevant *cis*-acting sequences for the bean *NIA* promoter were unsuccessful in transgenic tobacco (Jensen et al., 1996), nitrate-responsive sequences were identified in the –238 and –188 bp 5′ flanking regions of the *NIA*1 and *NIA*2 genes in *A. thaliana* (Lin et al., 1994). Furthermore, a consensus region of 12 bp was discovered in the *A. thaliana NIA*1 and *NIA*2 gene promoters, and this sequence is conserved in the 5′ flanking regions of other nitrate-inducible genes (Hwang et al., 1997). Using gel mobility shift experiments a protein binding activity was observed for this conserved region. However, the protein binding activity is not directly affected by nitrate treatments (Hwang et al., 1997), suggesting that several factors and/or post-transcriptional regulation are involved. Similar to the promoter of the *NII* gene (encoding NiR) in spinach (Neininger et al., 1994) nitrate-response-elements have been reported in the tobacco *NII1* promoter. Analysis of 5′ deletions in the tobacco *NII1* promoter, fused to a reporter gene, revealed a 200 bp sequence upstream of the transcription start codon containing nitrate-response-elements (Dorbe et al., 1998). In addition, 30 bp localized between –230 and –200 bp seemed crucial (necessary but not sufficient) for nitrate-regulated expression of *NII* (Rastogi et al., 1997). Similarly, ammonium stimulation of the soybean GS (GS15) promoter appears not to be controlled by isolated regions of the promoter alone but rather may result from interactions between regulatory elements located all along the promoter (Tercé-Laforgue et al., 1999).

A fusion between the tobacco *NII*1 gene and a reporter gene was used to screen for mutants affected in *NII* gene expression in *A. thaliana* (Leydecker et al., 2000). However, mutants that overexpressed both the reporter gene and *NII*1mRNA in the absence of nitrate turned out to be impaired in molybdenum cofactor biosynthesis. Similarly, screens involving insensitivity to chlorate have produced mutants impaired in the structural genes encoding NR and the nitrate transporter (Hoff et al., 1994). A regulatory mutant (cr88) defective in photomorphogenesis and in the pathway of transduction of the light signal (Cao et al., 2000) leading to the regulation of *NIA2, CAB* and *RbcS* genes was produced in this way (Lin and Cheng, 1997).

2. Trans-*Acting Elements*

Although some regulatory proteins have been identified in fungi, very little is known about the *trans*-acting factors that are required for the N-regulation of genes in higher plants. In *Aspergillus nidulans* and in *Neurospora crassa*, nitrate inducibility of the *NRT*, *NIA* and *NII* genes is controlled by regulatory proteins, NIRA (*A. nidulans*) and NIT4 (*N. crassa*) (Unkles et al., 1991; Marzluf et al., 1997). Ammonium-dependent repression of genes involves another regulatory protein. This is called AREA in *A. nidulans* (Unkles et al., 1991) and NIT2 (AREA homolog) in *N. crassa* (Marzluf et al., 1997). In *N. crassa* a negative regulator of NIT2 (NMR) has been identified. This binds to the NIT2 protein and interferes with the DNA binding of NIT2 during N-repression (Xiao et al., 1995). Moreover, evidence of synergy between NIT2 and NIT4 in the induction of *NIA* gene (*NIT3*) has been recently been obtained. NIT2 and NIT4 bind to specific regions of the *NIT3* promoter. In addition, direct protein-protein interaction between NIT2 and NIT4 is required for optimal expression of *NIT3* (Feng and Marzluf, 1998). A NIT2 homolog has been identified in *Chlamydomonas reinhardtii* (Quesada et al., 1993). *AREA* and *NIT2* genes encode regulatory proteins which are members of the GATA family of transcription factors. They contain highly conserved DNA binding domains which consist of single zinc finger motifs and adjacent basic regions (Marzluf, 1997). Searches for sequences identical or homologous to the motifs for the binding site of NIT2 in plants have not yet been successful and the role of the NIT-2 motif remains controversial. An NIT-2 binding motif has been found in genes involved in N metabolism in higher plants. It is found in the 5′ flanking regions of genes that are induced by nitrate such as *NIA* (Lin et al., 1994) and *NII* (Tanaka et al., 1994) in *A. thaliana*, and FNIA in rice (Aoki et al., 1995). Sequences homologous to the NIT-2 binding motif in Fd-VI have been found in maize. This is interesting because, in contrast to Fd III which is a constitutive form, Fd-VI is an nitrate-inducible isoform (Matsumura et al., 1997). Moreover, in vivo footprinting of the spinach *NII* promoter revealed nitrate inducible binding of proteins to GATA elements (NIT-2 binding sequences) in transformed tobacco plants in which the spinach *NII* promoter was fused to a reporter gene (Rastogi et al., 1997). The putative NIT2 binding site in the spinach *NII* gene promoter was localized in the critical region for nitrate inducibility between –230 bp and –180 bp (Rastogi et al., 1997). A cDNA encoding a NIT-2 like protein in tobacco, was 60% homologous to the NIT-2 sequence in the zinc finger domain, but there was no evidence that this protein was involved in the regulation of N metabolism (Daniel-Vedele and Caboche, 1993). Unfortunately, no GATA boxes were found in the sequences necessary for nitrate inducibility in the tobacco *NII*1 promoter. This would suggest that while GATA sequences are involved, they are not essential for nitrate induction. This result emphasizes the complex nature of the perception of the nitrate signal in plants.

Screening *A. thaliana* roots for nitrate inducible genes revealed a new component, ANR1. This component of the nitrate signal transduction pathway is involved in nitrate dependent stimulation of lateral root proliferation (Zhang and Forde, 1998; see above and Fig. 1). ANR1 is a putative transcription factor with homology to the MADS box family of transcription factors. In contrast to most other MADS box genes, that are usually expressed in flowers, ANR1 is nitrate inducible and root specific. Production of transgenic *A. thaliana* lines with antisense constructs of ANR1 has allowed the analysis of ANR1 function. The stimulatory effect of nitrate was decreased in the transgenic lines. Conversely, the inhibitory effect of high nitrate on root growth was increased in the antisense lines with low *ANR1* expression. Moreover, the suppression of lateral root development that occurs at concentrations above 10 mM nitrate in wild-type *A. thaliana* was observed at lower nitrate concentrations (0.1 and 1 mM) in the transgenic lines. An overlap between the auxin and nitrate response pathways has also been suggested by studies on an auxin-resistant *A. thaliana* mutant. In contrast to other auxin resistant mutants, this mutant is insensitive to the stimulatory effect of low nitrate (Zhang et al., 1999). As illustrated in Fig. 1, a recent model proposes that the control of root architecture by N status involves two regulatory pathways. These are the localized ANR1 pathway mediating the response to local nitrate availability, and a systemic pathway involving a signal reflecting the general C/N status of the plant. This signal must be translocated from the shoot to the root (Zhang et al., 1999). We foresee that the discovery of ANR1 will initiate other molecular studies of the root nitrate signaling pathway.

Functional complementation of a yeast mutant (defective in the Gln-dependent repression of the expression of genes encoding enzymes involved in N assimilation) with an *A. thaliana* cDNA expression library led to the identification of two cDNAs (RGA1 and RGA2) (Truong et al., 1997). While these gene products, which are not members of the GATA-binding family, seem to be involved in the regulation of root formation (Forde and Zhang, 1998), their function in N metabolism remains unknown.

B. Post-Transcriptional Mechanisms

1. Ser-Protein Kinases/Phosphatases

Signaling cascades involving protein phosphorylation/dephosphorylation events have been described in plant responses to hormones, light and sugar (Jang and Sheen, 1999). In most cases very few of the components in any of the plant signal transduction pathways have been described. Moreover, current understanding of the full sequence of events leading to the regulation of gene expression or to a post-transcriptional modulation of any enzyme remains superficial. This dearth of knowledge is even more profound in the case of N-regulated metabolism. Phosphorylation/dephosphorylation processes in plants post-transcriptionally regulate NR, PEPc, and sucrose phosphate synthase (SPS). This regulation occurs in response to changes in light/dark conditions (or in photosynthetic activity). While a role for the N status of the plant has been described in the post-transcriptional modulation of NR, PEPc and SPS, the nature of the mechanisms involved in this control are far from clear. Kinases that specifically modulate NR, SPS or PEPc have been purified and partially characterized. Recently, a Ca^{2+} independent PEPc protein kinase was described. This is a novel member of the Ca^{2+}/calmodulin-regulated group of protein kinases (Hartwell et al., 1999). This PEPc kinase is regulated by transcription controls that are modulated by light (Hartwell et al., 1999).

Nitrate has been suggested to be a positive effector of PEPc gene transcription and PEPc activity in wheat (Van Quy et al., 1991) and in tobacco (Lancien et al., 1999; Murchie et al., 2000). Nitrate is also involved in the activation of PEPc kinase (Duff and Chollet, 1995). A role for Gln in the modulation of PEPc gene transcription has been described in maize (Sugiharto et al., 1992; Suzuki et al., 1994). In addition, Gln has been suggested to modulate PEPc activation state in wheat (Manh et al., 1993) and tobacco (Li et al., 1996; Murchie et al., 2000). Post-transcriptional regulation of NR also responds to plant organic N status (Gln) (Scheible et al., 1997b) but not to nitrate (Ferrario et al., 1996).

The dark inactivation (involving the high phosphorylation state) of NR is lowest when leaves contain low Gln levels. Partial deactivation of NR was observed when Gln was fed to detached tobacco leaves in the light (Scheible et al., 1997c). Two Ca^{2+} independent protein kinases and two other Ca^{2+} independent protein kinases, which modulate both NR and SPS have been described (McMichael et al., 1995; Sugden et al., 1999). However the regulation of these kinases by metabolites has not been characterized. It is therefore impossible to determine the role of these protein kinases in the modulation of the activation state of NR by metabolites such as Gln.

Pharmacological studies have been widely used to identify the activities of signal-transducing factors such as G-proteins, Ca^{2+} channels, calmodulin, protein phosphatases and protein kinases in plants. Studies with excised barley leaves, incubated with various kinds of inhibitor before the addition of nitrate, have provided evidence that Ca^{2+} channels, protein phosphatases, and Tyr-protein kinases are involved in the nitrate signaling pathways required for the appropriate expression of *NIA* and *NII* (Sueyoshi et al., 1999). Similarly, studies with excised maize leaves pre-treated with EGTA or La^{3+}, have suggested that Ca^{2+} ions are involved in the nitrate signaling cascade leading to the induction of the GS2 and GOGAT, NR and NiR genes (Sakakibara et al., 1997). Furthermore, inhibitors of protein kinases and protein phosphatases prevent the nitrate-dependent accumulation of *NIA, NII* and *GS2* mRNA (Sakakibara et al., 1997). This indicates that phosphorylation of at least some of the proteins involved in the signaling cascade is essential for activation. The induction of Fd-GOGAT by nitrate is insensitive to protein kinase or phosphatase inhibitors (Sakakibara et al., 1997). In contrast, protein kinase inhibitors prevent the induction of NADH-GOGAT by ammonium. The specific inhibitor of protein Ser/Thr phosphatases, okadaic acid, mimics this effect, when applied to rice cell cultures. (Hirose and Yamaya, 1999). However, such inhibitors can affect many diverse intracellular signaling systems. The results of such studies must therefore, be interpreted with caution and this requires additional independent verification by genetic and/or biochemical experiments.

2. 14-3-3 and PII-Like Proteins

The post-transcriptional regulation of the NR protein involves the phosphorylation of the enzyme, followed by the binding of an inhibitor protein. This inhibitor has been identified as a 14-3-3 protein. Recently, plant 14-3-3 binding proteins have been purified and sequenced from cauliflower extracts, and were found to include protein that bound to SPS and GS (Moorhead et al., 1999). While the involvement of 14-3-3 proteins in NR inactivation is well characterized, their role in SPS regulation remains controversial (Toroser et al., 1998; Moorhead et al., 1999) and moreover, the function of 14-3-3 binding to GS is unknown. It is worth noting that there is a 30% sequence identity between the *A. thaliana* 14-3-3 proteins and the cyanobacterial PII-like protein that is implicated in GS regulation in bacteria and cyanobacteria (Moorhead et al., 1999). PII-like proteins from *A. thaliana* and *Ricinus communis* have recently been characterized. These proteins are transcriptionally up-regulated by light and sucrose and down-regulated by some amino acids (Hsieh et al., 1998). However, the function of PII-like proteins in higher plants remains enigmatic since they are localized in the chloroplast (Hsieh et al., 1998). Transgenic *A. thaliana* plants overexpressing the PII-like protein and analyzed in varying C/N conditions, displayed higher sucrose-dependent anthocyanin accumulation in response to added Gln than controls (Hsieh et al., 1998). The authors considered that the PII overexpressors were deregulated in sensing C/N status in relation to anthocyanin accumulation. The PII-like protein has been shown to be a signal transduction protein and a key element in the coordination of C and N metabolism in bacteria (Jiang et al., 1998) and in cyanobacteria (Lee et al., 1999). The signal transduction cascade described in *E. coli* for the control of GS activity at both transcriptional and post-transcriptional levels, involves Gln and 2-OG sensing by a PII protein followed by interaction with a PII uridyltransferase/uridyl removing enzyme as shown in Fig. 3 (Jiang et al., 1998). In cyanobacteria, PII is regulated by phosphorylation instead of uridylation. Moreover, the signal transduction cascade involves PII protein sensing of the Gln/2-OG ratio and controls the rate of nitrate and nitrite uptake (Lee et al., 1998, 1999). The PII protein may be considered to be a direct sensor of 2-OG in cyanobacteria. Recently, 2-OG has been suggested to be involved in the induction of NR gene expression in tobacco and to counteract the inhibitory effect of Gln (Ferrario-Méry et al., 2001, Müller et al., 2001). These observations have encouraged the authors to search for PII mutants by screening a T-DNA tagged *A. thaliana* population available at INRA in Versailles (S. Ferrario-Méry, unpublished). These studies may contribute significantly to the understanding of N signaling in plants.

3. Two-Component Regulatory Systems

Another putative N signal transduction pathway has been recently reported in plants (Sakakibara et al., 2000). The 'two-component regulatory system' or 'multistep His-Asp phosphorelay' is well known in bacteria. It consists of phosphotransfer from a sensor (His-protein kinase) domain to an regulatory phosphorylation (Asp) domain. His-protein kinases, response regulators and His-containing phosphotransfer (HPt) proteins have been cloned from plants and seem to be related to ethylene or cytokinin signaling (Sakakibara et al., 2000). Recently, two cDNAs from maize and five cDNAs from *A. thaliana* have been isolated. These encode response regulator

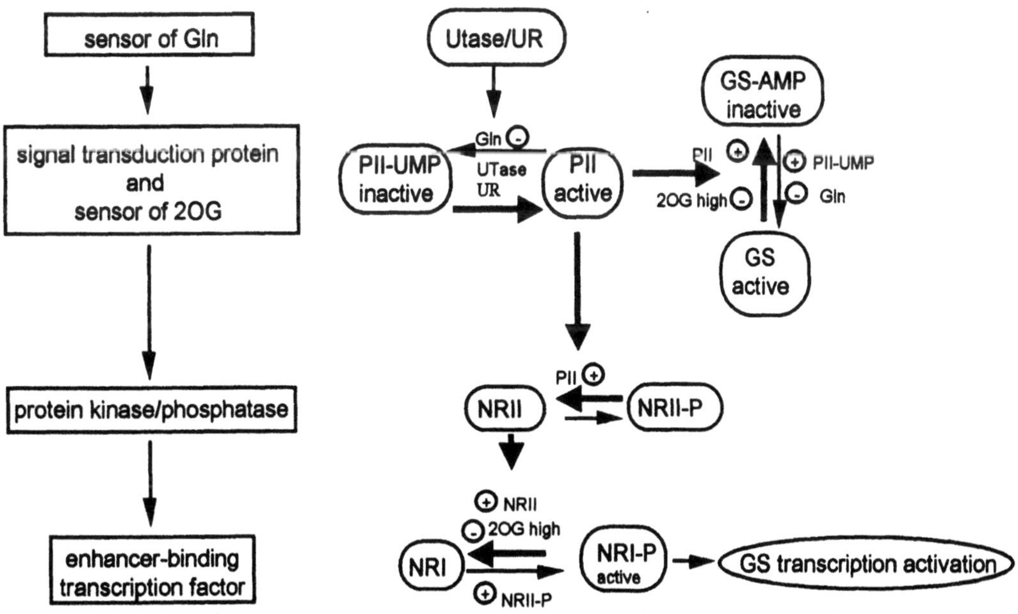

Fig. 3. Signal transduction system controlling glutamine synthetase (GS) adenylation state and the transcription of the GS gene transcription in *E. Coli* (from Jiang et al., 1998). Large arrow = Gln/2OG high; small arrow = Gln/2OG low. Gln, glutamine; GS, glutamine synthetase. 2OG, 2-oxoglutarate; Utase/UR, uridyltransferase/uridyl removing enzyme. The protein PII is involved, via its uridylation status, in the transduction of the Gln signal sensed by Utase/UR sensor, a signal wihich it transmits via the protein kinase/phosphatase, NRII, and the enhancer-binding transcription factor, NRI, leading to changes in GS transcription rates (for further details, see Jiang et al., 1998).

domains and HPt domains (Taniguchi et al., 1998; Sakakibara et al., 1999). One interesting point is that the transcripts of these proteins were induced together with *NIA* transcripts by inorganic N applied to roots after N starvation and also by t-zeatin applied to detached leaves (Taniguchi et al., 1998). This indicates that inorganic N (nitrate or ammonium) is the signal that activates the His-Asp phosphotransfer system in maize and that this may be mediated by cytokinin accumulated in the roots and transferred to shoots.

C. Mechanisms of Nitrogen Sensing

The mechanisms used by plant cells to sense N signals are far from understood. Nitrate itself has been shown to be a signal molecule in several cases, but the nature of the nitrate sensor molecule has not yet been discovered. Glutamine is also a promising candidate for N signaling but to date there is little evidence for the involvement of plant PII homologs in this process, at least in a manner similar to bacteria and cyanobacteria. Putative glutamate-receptors (GluRs) have been discovered (Lam et al., 1998), but possible roles for glutamate in N signaling remain unexplored in plants. Similarities between the plant and animal GluR receptor genes span all the important domains including the ligand-binding domains and the transmembrane segments. GluRs and Glu-activated ions channels in animals are involved in rapid synaptic transmission. In contrast, studies using GluR antagonists in plants have implicated these receptors in light signal transduction (Lam et al., 1998), since they blocked the ability of light to inhibit hypocotyl elongation and to induce chlorophyll synthesis.

IV. Concluding Remarks

Several pathways by which N-metabolites can possibly act as regulators of plant development and metabolism have been discovered recently. However, N-signaling is far from understood. Even when candidate-signaling molecules have been identified the sites of action at a sub-cellular level are largely unknown and the involvement of different pools in the various sub-cellular compartments of the plant cell remains unresolved. For example, there are large (>20-fold) changes in the nitrate content of source leaves during the day (Scheible et al., 1997c). However, microelectrode measurements have shown that the cytosolic nitrate pool is maintained at remarkably constant values, irrespective of nitrate levels in the vacuole or whether nitrate is being accumulated in the cell or released (Miller and Smith, 1996). Furthermore, we need to keep in mind that N signaling is only one part of a large regulatory network that involves multiple interactions with other signaling pathways, such as sugars and hormones.

Acknowledgments

We thank Dr. Brian Forde and Dr. Wolf Scheible for the permission to use Figs. 1 and 2. We are also grateful to Petra Matt, Dr. Hoai-Nam Truong and Dr. Wolf Scheible for suggestions on the manuscript.

References

Amarasinghe BHRR, de Bruxelles GL, Braddon M, Onyeocha I, Forde BG and Udvardi MK (1998) Regulation of *GmNRT2* expression and nitrate transport activity in roots of soybean (*Glycine max*). Planta 206: 44–52

Aoki H, Tanaka K and Ida S (1995) The genomic organization of the gene encoding a nitrate-inducible ferredoxin-$NADP^+$ oxidoreductase from rice roots. Biochim Biophys Acta 1229: 389–392

Aslam M, Travis RL and Huffaker RC (1992) Comparative kinetics and reciprocal inhibition of nitrate and nitrite uptake in roots of uninduced and induced barley (*Hordeum vulgare* L.) seedlings. Plant Physiol 99: 1124–1133

Aslam M, Travis RL and Huffaker RC (1993) Comparative induction of nitrate and nitrite uptake and reduction systems by ambient nitrate and nitrite in intact roots of barley (*Hordeum vulgare* L.) seedlings. Plant Physiol 102: 811–819

Aslam M, Travis RL, Rains DW and Huffaker RC (1996) Effect of ammonium on the regulation of nitrate and nitrite transport systems in roots of intact barley (*Hordeum vulgare* L.) seedlings. Planta 200: 58–63

Bloom AJ, Chapin FS and Mooney HA (1985) Resource limitation in plants—an economic analogy. Annu Rev of Ecol Sys 16: 363–392

Botrel A and Kaiser W (1997) Nitrate reductase activation status in barley roots in relation to the energy and carbohydrate status. Plant J 15: 583–591

Breteler H and Siegerist M (1984) Effect of ammonium on nitrate utilization by roots of dwarf bean. Plant Physiol 75: 1099–1103

Brouwer R (1962) Nutrient influences on the distribution of the dry matter in the plants. Netherl J Agri Sci 10: 399–408

Cao D, Lin Y and Cheng CL. (2000) Genetic interactions between the chlorate-resistant mutant *cr88* and the photomorphogenic mutants *cop1* and *hy5*. Plant Cell 12: 199–210

Causin HF and Barneix AJ (1993) Regulation of NH_4^+ uptake in young wheat plants: Effect of root ammonium concentration and amino acids. Plant Soil 151: 211–218

Chaillou S, Morot-Gaudry JF, Salsac L, Lesaint C and Jolivet E (1986) Compared effects of NO_3^- and NH_4^+ on growth and yield of eggplants (*Solanum melongena* L.). Physiol Vég 24: 679–687

Champigny M-L and Foyer C (1992) Nitrate activation of cytosolic protein kinases diverts photosynthetic carbon from sucrose to amino acid biosynthesis. Plant Physiol 100: 7–12

Cheng C-L, Acedo GN, Dewdney J, Goodman HM and Conkling MA (1991) Differential expression of two *Arabidopsis* nitrate reductase genes. Plant Physiol 96: 275–279

Cheng C-L, Acedo GN, Christinsin M and Conkling MA (1992) Sucrose mimics the light induction of *Arabidopsis* nitrate reductase gene transcription. Proc Natl Acad Sci USA 89: 1861–1864

Chollet R, Vidal J and O'Leary MH (1996) Phosphoenolpyruvate carboxylase: A ubiquitous, highly regulated enzyme in plants. Annu Rev Plant Physiol Plant Mol Biol 47: 273–298

Clarkson DT and Touraine B (1994) Morphological responses of plants to nitrate-deprivation: A role for abscisic acid? In: Roy J, Garnier E (eds) A Whole Plant Perspective on Carbon-Nitrogen Interactions, pp 187–196. SPB Academic Publishing, The Hague

Cogliatti DH and Clarkson DT (1983) Physiological changes in phosphate uptake by potato plants during development of, and recovery from, phosphate deficiency. Physiol Plant 58: 287–294

Cooper HD and Clarkson DT (1989) Cycling of amino-nitrogen and other nutrients between shoots and roots in cereals—a possible mechanism integrating shoot and root in the regulation of nutrient uptake. J Exp Bot 40: 753–762

Crawford NM (1995) Nitrate: Nutrient and signal for plant growth. Plant Cell 7: 859–868

Daniel-Vedele F and Caboche M (1993) A tobacco cDNA clone encoding a GATA-1 zinc finger protein homologous to regulators of nitrogen metabolism in fungi. Mol Gen Genet 240: 365–373

Daniel-Vedele F and Caboche M (1996) Molecular analysis of nitrate assimilation in higher plants. CR Acad Sci Paris Ser III 319: 961–968

Daniel-Vedele F, Filleur S and Caboche M (1998) Nitrate transport: A key step in nitrate assimilation. Curr Opin Plant Biol 1:235–239

Devienne F, Mary B and Lamaze T (1994) Nitrate transport in intact wheat roots. I. Estimation of cellular fluxes and NO_3^- distribution using compartmental analysis from data of $^{15}NO_3^-$ efflux. J Exp Bot 45: 667–676

Diaz A, Lacuestra M and Munoz-Rueda A (1996) Comparative effects of phosphinothricin on nitrate and ammonium assimilation and on anapleurotic carbon dioxide fixation in N-deprived barley plants. J Plant Physiol 149: 9–13

Dorbe MF, Truong H-N, Crété P and Daniel-Vedele F (1998) Deletions of the tobacco *Nii1* promoter in *Arabidopsis thaliana*. Plant Sci 139: 71–82

Drew MC and Saker LR (1976) Nutrient supply and the growth of the seminal root system in barley. II. Localized compensatory changes in lateral root growth and the rate of nitrate uptake when nitrate is restricted to only one part of the root system. J Exp Bot 26: 79–90

Drew MC, Saker LR, Barber SA and Jenkins W (1984) Changes in the kinetics of phosphate and potassium absorption in nutrient-deficient barley roots measured by a solution-depletion technique. Planta 160: 490–499

Duff SDG and Chollet R (1995) In vivo regulation of wheat-leaf phosphoenolpyruvate carboxylase by reversible phosphorylation. Plant Physiol 107: 775–782

Faure JD, Vincentz M, Kronenberger J and Caboche M (1991) Coregulated expression of nitrate and nitrite reductases. Plant J 1:107–113

Feng B and Marzluf GA (1998) Interaction between major nitrogen regulatory protein NIT2 and pathway-specific regulatory factor NIT4 is required for their synergistic activation of gene expression in *Neurospora crassa*. Mol Cell Biol 18: 3983–3990

Ferrario S, Valadier M-H, Morot-Gaudry J-F and Foyer CH (1995) Effects of constitutive expression of nitrate reductase in transgenic *Nicotiana plumbaginifolia* L. in response to varying nitrogen supply. Planta 196: 288–294

Ferrario S, Valadier M-H and Foyer CH (1996) Short-term modulation of nitrate reductase activity by exogenous nitrate in *Nicotiana plumbaginifolia* and *Zea mays* leaves. Planta 199: 366–371

Ferrario-Méry S, Masclaux C, Suzuki A, Valadier M-H, Hirel B and Foyer CH (2001) Glutamine and α-ketoglutarate are metabolite signals involved in nitrate reductase gene transcription in untransformed and transformed tobacco plants deficient in ferredoxine-Gln-α-ketoglutarate aminotransferase. Planta 213: 265–271

Fichtner K and Schulze E-D (1992) The effect of nitrogen nutrition on annuals originating from habitats of different nitrogen availability. Oecologia 92: 236–241

Filleur S and Daniel-Vedele F (1999) Expression analysis of a high-affinity nitrate transporter isolated from *Arabidopsis thaliana* by differential display. Planta 207: 461–469

Forde BG (2000) Nitrate transporters in plants: Structure, function and regulation. Biochim Biophys Acta 11465: 219–235

Forde BG and Zhang H (1998) Nitrate and root branching. Trends Plant Sci 3: 204–205

Foyer CH, Lescure JC, Lefèbvre C, Morot-Gaudry JF and Vaucheret H (1994a) Adaptations of photosynthetic electron transport, carbon assimilation and carbon partitioning in transgenic *Nicotiana plumbaginifolia* plants to changes in nitrate reductase activity. Plant Physiol 104: 171–178

Foyer CH, Noctor G, Lelandais M, Lescure JC, Valadier MH, Boutin JP and Horton B (1994b) Short-term effects of nitrate, nitrite and ammonium assimilation on photosynthesis, carbon partitioning and protein phosphorylation in maize. Plant Physiol 192: 211–220

Fraisier V, Gojon A, Tillard P and Daniel-Vedele F (2000) Constitutive expression of a putative high-affinity nitrate transporter in *Nicotiana plumbaginifolia*: Evidence for post-transcriptional regulation by a reduced-nitrogen source. Plant J 23: 489–496

Gazzarini S, Lejay L, Gojon A, Ninnemann O, Frommer WB and von Wirén N (1999) Three functional transporters for constitutive, diurnally regulated, and starvation-induced uptake of ammonium into *Arabidopsis* roots. Plant Cell 11: 937–947

Glass ADM, Shaff JE and Kochian LV (1992) Studies of the uptake of nitrate in barley. IV. Electrophysiology. Plant Physiol 99: 456–463

Gojon A, Dapoigny L, Lejay L, Tillard P and Rufty TW (1998) Effects of genetic modification of nitrate reductase expression on $^{15}NO_3^-$ uptake and reduction in *Nicotiana* plants. Plant Cell

Environ 21: 43–53

Granato TC and Raper CD Jr (1989) Proliferation of maize roots in response to localized supply of nitrate. J Exp Bot 40: 263–275

Grime JP, Campbell BD, Mackey JML and Crick JC (1991) Root plasticity, nitrogen capture and competitive ability. In: Atkinson D (ed) Plant Root Growth: An Ecological Perspective, pp. 381–397 Blackwell Scientific Publications Oxford

Hartwell J, Smith LH, Wilkins MB, Jenkins GI and Nimmo HG (1996) Higher plant phosphoenolpyruvate carboxylase kinase is regulated at the level of translatable mRNA in response to blue light or circadian rhythm. Plant J 10: 1071–1078

Hartwell J, Gill A, Nimmo HG, Wilkins MB, Jenkins GI and Nimmo HG (1999) Phosphoenolpyruvate carboxylase kinase is a novel protein kinase regulated at the level of expression. Plant J 20: 333–342

Hirose N and Yamaya T (1999) Okadaic acid mimics nitrogen-stimulated transcription of NADH-glutamate synthase gene in rice cell cultures. Plant Physiol 121: 805–812

Hoff T, Truong H-N and Caboche M (1994) The use of mutants and transgenic plants to study nitrate assimilation. Plant Cell Environ 17: 489–506

Hole DJ, Emran AM, Fares Y and Drew M (1990) Induction of nitrate transport in maize roots, and kinetics of influx, measured with nitrogen-13. Plant Physiol 93: 642–647

Hsieh M-H, Lam H-M, Van de Loo FJ and Coruzzi G (1998) A PII-like protein in *Arabidopsis*: Putative role in nitrogen sensing. Proc Natl Acad Sci USA 95: 13965–13970

Huang NC, Chiang CS, Crawford NM and Tsay YF (1996) *CHL1* encodes a component of the low affinity nitrate uptake system in *Arabidopsis* and shows cell type-specific expression in roots. Plant Cell 8: 2183–2191

Huppe HC and Turpin DH (1994) Integration of carbon and nitrogen metabolism in plant and algal cells. Annu Rev Plant Physiol Plant Mol Biol 45: 577–607

Huppe HC, Vanlerberghe GC and Turpin DH (1992) Evidence for activation of the oxidative pentose phosphate pathway during photosynthetic assimilation of nitrate but not ammonium by a green alga. Plant Physiol 100: 2096–2099

Hwang C-F, Lin Y, D'Souza T and Cheng C-L (1997) Sequences necessary for nitrate-dependent transcription of *Arabidopsis* nitrate reductase genes. Plant Physiol 113: 853–862

Imsande J and Touraine B (1994) N demand and the regulation of nitrate uptake. Plant Physiol 105: 3–7

Jang J-C and Sheen J (1999) Sugar sensing in higher plants. Plant Cell 6: 1665–1679

Jensen PE, Hoff T, Stummann BM and Hennigsen KW (1996) Functional analysis of two bean nitrate reductase promoters in transgenic tobacco. Physiol Plant 96: 351–358

Jiang P, Peiska JA and Ninfa AJ (1998) Reconstitution of the signal-transduction bicyclic cascade responsible for the regulation of Ntr gene transcription in *Escherichia coli*. Biochemistry 37: 12795–12801

King BJ, Siddiqi MY, Ruth TJ, Warner RL and Glass ADM (1993) Feedback regulation of nitrate influx in barley roots by nitrate, nitrite, and ammonium. Plant Physiol 102: 1279–1286

Krapp A, Fraisier V, Scheible W-R, Quesada A, Gojon A, Stitt M, Caboche M and Daniel-Vedele F (1998) Expression studies of *Nrt2:1Np*, a putative high affinity nitrate transporter: Evidence for its role in nitrate uptake. Plant J 6: 723–732

Kronenberger J, Lepingle A, Caboche M and Vaucheret H (1993) Cloning and expression of distinct nitrate reductases in tobacco leaves and roots. Mol Gen Genet 236: 203–208

Kronzucker HJ, Glass ADM and Siddiqi MY (1995a) Nitrate induction in spruce: An approach using compartmental analysis. Planta 196: 683–690

Kronzucker HJ, Siddiqi MY and Glass ADM(1995b) Kinetics of NO_3^- influx in spruce. Plant Physiol 109: 319-326

Lainé P, Ourry A and Boucaud J (1995) Shoot control of nitrate uptake rates by roots of *Brassica napus* L.: Effects of localized nitrate supply. Planta 196: 77–83

Lam H-M, Chiu J, Hsieh M-H, Meisel L, Oliveira IC, Shin M and Coruzzi G (1998) Glutamate-receptor genes in plants. Nature 396: 125–126

Lambers H, Cambridge ML, Konings H and Pons TL (1990) Causes and Consequences of Variation in Growth Rate and Productivity of Higher Plants. SPB Academic Publishing, The Hague

Lancien M, Ferrario-Méry S, Roux Y, Bismuth E, Masclaux C, Hirel B, Gadal P and Hodges M (1999) Simultaneous expression of NAD-dependent isocitrate dehydrogenase and other Krebs cycle genes after nitrate resupply to short-term nitrogen-starved tobacco. Plant Physiol 120: 717–725

Lappartient AG and Touraine B (1996) Demand-driven control of root ATP sulfurylase activity and SO_4^{2-} uptake in intact canola. The role of phloem-translocated glutathione. Plant Physiol 111: 147–157

Lauerer M (1996) Wachstum, Kohlenstoff- und Stickstoffhaushalt von *Nicotiana tabacum* mit reduzierter Nitratreduktaseaktivität. Dissertation Universität Bayreuth Bayreuther Forum Ökologie, Band 31, Bayreuther Institut für terrestrische Ökosystemforschung (Hrsg)

Lauter FR, Ninnemann O, Bucher M, Riesmeier JW and Frommer WB (1996) Preferential expression of an ammonium transporter and of two putative nitrate transporters in root hairs of tomato. Proc Natl Acad Sci USA 93: 8139–8144

Lee H-M, Flores E, Herrero A, Houmard J and Tandeau de Marsac N (1998) A role for the signal transduction protein PII in the control of nitrate/nitrite uptake in a cyanobacterium. FEBS Lett 427: 291–295

Lee H-M, Vasquez-Bermudez MF and Tandeau de Marsac N (1999) The global nitrogen regulator NtcA regulates transcription of the signal transducer PII (GlnB) and influences its phosphorylation level in response to nitrogen and carbon supplies in the cyanobacterium *Synechococcus* sp. strain PCC 7942. J Bacteriol 181: 2697–2702

Lee RB (1993) Control of net uptake of nutrients by regulation of influx in barley plants recovering from nutrient deficiency. Ann Bot 72: 223–230

Lee RB and Drew MC (1986) Nitrogen-13 studies of nitrate fluxes in barley roots. II. Effect of plant N-status on the kinetic parameters of nitrate influx. J Exp Bot 185: 1768–1779

Lee RB and Rudge K (1986) Effects of nitrogen deficiency on the absorptions of nitrate and ammonium by barley plants. Ann Bot 57: 471–486

Lejay L, Tillard P, Lepetit M, Olive FD, Filleur S, Daniel-Vedele F and Gojon A (1999) Molecular and functional regulation of two NO_3^- uptake systems by N- and C-status of *Arabidopsis* plants. Plant J 18: 509–519

Lexa M and Cheeseman JM (1997) Growth and nitrogen relations in reciprocal grafts of wild-type and nitrate reductase-deficient mutants of pea. J Exp Bot 48: 1241–1250

Leydecker M-T, Camus I, Daniel-Vedele F and Truong H-N (2000) Screening for *Arabidopsis* mutants affected in the *Nii* gene expression using the Gus reporter gene. Physiol Plant 108: 161–170

Li B, Zhang X-Q and Chollet R (1996) Phosphoenolpyruvate carboxylase kinase in tobacco leaves is activated by light in a similar but not identical way as in maize. Plant Physiol 111: 497–505

Lin Y and Cheng CL (1997) A chlorate-resistant mutant defective in the regulation of nitrate reductase gene expression in *Arabidopsis* defines a new HY locus. Plant Cell 9: 21–35

Lin Y, Hwang C-F, Brown J and Cheng C-L (1994) 5′ proximal regions of *Arabidopsis* nitrate reductase genes direct nitrate-induced transcription in transgenic tobacco. Plant Physiol 106: 477–484

Liu K-H, Huang C-Y and Tsay Y-F (1999) CHL1 is a dual-affinity nitrate transporter of *Arabidopsis* involved in multiple phases of nitrate uptake. Plant Cell 11: 865–874

Manh CT, Bismuth E, Boutin J-P, Provot M and Champigny M-L (1993) Metabolites effectors for short-term nitrogen-dependent enhancement of phosphoenolpyruvate carboxylase activity and decrease of net sucrose synthesis in wheat leaves. Physiol Plant 89: 460–466

Marschner M (1995) Mineral Nutrition of Higher Plants, Second edition. Academic Press, London

Marzluf GA (1997) Genetic regulation of nitrogen metabolism in the fungi. Microbiol Mol Biol Rev 61: 17–32

Matsumara T, Sakakibara H, Nakano R, Kimata Y, Sugiyama T and Hase T (1997) A nitrate-inducible ferredoxin in maize roots. Plant Physiol 114: 653–660

Mattana M, Brambilla J, Bertani A and Reggiani R (1996) Expression of nitrogen assimilating enzymes in germinating rice under anoxia. Plant Physiol Biochem 34: 653–657

McMichael RW, Bachmann M and Huber SC (1995) Spinach leaf sucrose-phosphate synthase and nitrate reductase are phosphorylated by multiple protein kinases in vitro. Plant Physiol 108: 1077–1082

Meyer zu Hörste G (1998) Einfluss von Aminosäuren auf das Wachstum und die Physiologie von *Arabidopsis thaliana.* Diplomarbeit Universität Heidelberg

Migge A, Carroyol E, Kunz C, Hirel B, Fock H and Becker T (1997) The expression of the tobacco genes encoding plastidic glutamine synthetase or ferredoxin-dependent glutamate synthase does not depend on the rate of nitrate reduction, and is unaffected by suppression of photorespiration. J Exp Bot 48: 1175–1184

Miller AJ and Smith SJ (1996) Nitrate transport and compartmentation in cereal root cells. J Exp Bot 47: 843–854

Moorhead G, Douglas P, Cotelle V, Harhill J, Morrice N, Meek S, Deiting U, Stitt M, Scarabel M, Aitken A and MacKintosh C (1999) Phosphorylation-dependent interactions between enzymes of plant metabolism and 14-3-3 proteins. Plant J 18: 1–12

Morcuende R, Krapp A, Hurry V and Stitt M (1998) Sucrose feeding leads to increased rates of nitrate assimilation, increased rates of oxoglutarate synthesis, and increased synthesis of a wide spectrum of amino acids in tobacco leaves. Planta 206: 394–409

Morgan MA and Jackson WA (1988) Inward and outward movement of ammonium in root systems: transient responses during recovery from nitrogen deprivation in presence of ammonium. J Exp Bot 39: 179–191

Muller B and Touraine B (1992) Inhibition of NO_3^- uptake by various phloem-translocated amino acids in soybean seedlings. J Exp Bot 43: 617-623

Muller B, Tillard P and Touraine B (1995) Nitrate fluxes in soybean seedling roots and their response to amino acids: An approach using ^{15}N. Plant Cell Environ 18: 1267–1279

Müller C, Scheible W-R, Stitt M and Krapp A (2000) Influence of malate and 2-oxoglutarate on the *NIA* transcript level and nitrate reductase activity in tobacco. Plant Cell Environ 24: 191–204

Murchie EH, Ferrario-Méry S, Valadier M-H and Foyer CH (2000) Short-term nitrogen-induced modulation of phosphoenolpyruvate carboxylase in tobacco and maize leaves. J Exp Bot 51: 1349–1356

Neininger A, Back E, Bichler A, Schneiderbauer H and Mohr H (1994) Deletion analysis of a nitrite-reductase promoter from spinach in transgenic tobacco. Planta 194: 186–192

Pajuelo P, Pajuelo E, Forde BG and Marquez AJ (1997) Regulation of the expression of ferredoxin glutamate synthase in barley. Planta 203: 517–525

Pouliquin P, Grouzis J-P and Gibrat R (1999) Electrophysiological study with oxonol VI of passive NO_3^- transport by isolated plant root plasma membrane. Biophys J 76: 360–373

Pouliquin P, Boyer J-C, Grouzis J-P and Gibrat R (2000) Passive nitrate transport by root plasma membrane vesicles exhibits an acidic optimal pH like the H^+-ATPase. Plant Physiol 122: 265–273

Pouteau S, Cherel I, Vaucheret H and Caboche M (1989) Nitrate reductase mRNA regulation in *Nicotiana plumbaginifolia* nitrate reductase-deficient mutants. Plant Cell 1: 1111–1120

Quesada A, Galvan A, Schnell RA, Lefebvre PA and Fernandez E (1993) Five nitrate assimilation-related loci are clustered in *Chlamydomonas reinhardtii.* Mol Gen Genet 240: 387–394

Raab TK and Terry N (1994) Nitrogen source regulation of growth and photosynthesis in *Beta vulgaris* L. Plant Physiol 105: 1159–1166

Rastogi R, Bate N, Sivasankar and Rothstein SJ (1997) Footprinting of the spinach nitrite reductase gene promoter reveals the preservation of nitrate regulatory elements between fungi and higher plants. Plant Mol Biol 34: 465–476

Rawat SR, Silim SN, Kronzucker HJ, Siddiqi MY and Glass ADM (1999) *AtAMT1* gene expression and NH_4^+ uptake in roots of *Arabidopsis thaliana*: Evidence for regulation by root glutamine levels. Plant J 19: 143-152

Redinbaugh MG and Campbell WH (1991) Higher plant responses to environmental nitrate. Physiol Plant 82: 640–650

Redinbaugh MG and Campbell WH (1993) Glutamine synthetase and ferredoxin-dependent glutamate synthase expression in the maize (*Zea mays*) root: Primary response to nitrate. Plant Physiol 101: 1249–1255

Ritchie SW, Redinbaugh MG, Shiraishi N, Verba JM and Campbell WH (1994) Identification of a maize root transcript expressed in the primary response to nitrate: Characterization of a cDNA with homology to ferredoxin.$NADP^{(+)}$ oxido-reductase. Plant Mol Biol 26: 679–690

Robinson D (1994) The responses of plants to non-uniform supplies of nutrients. New Phytol 127: 635–674

Sakakibara H, Shimizu H, Hase T, Yamazaki Y, Takao T, Shimonishi Y and Sugiyama T (1996) Molecular identification and characterization of cytosolic isoforms of glutamine

synthetase in maize roots. J Biol Chem. 271: 29561–29568
Sakakibara H, Kobayashi K, Deji A and Sugiyama T (1997) Partial characterization of signaling pathway of nitrate-dependent expression of the genes for nitrogen-assimilatory enzymes using detached maize leaves. Plant Cell Physiol 38: 837–843
Sakakibara H, Hayakawa A, Deji A, Gawronski SW and Sugiyama T (1999) His-Asp phosphotransfer possibly involved in the nitrogen signal transduction mediated by cytokinin in maize: Molecular cloning of cDNAs for two-component regulatory factors and demonstration of phosphotransfer activity in vitro. Plant Mol Biol 41: 563–573
Sakakibara H, Taniguchi M and Sugiyama T (2000) His-Asp phosphorelay signaling: A communication avenue between plants and their environment. Plant Mol Biol 42: 273–278
Sattelmacher B and Marschner H (1978) Nitrogen nutrition and cytokinin activity in *Solanum tuberosum*. Physiol Plant 42: 185–189
Scheible W-R, Lauerer M, Schulze E-D, Caboche M and Stitt M (1997a) Accumulation of nitrate in the shoot acts as a signal to regulate shoot-root allocation in tobacco. Plant J 11: 671–691
Scheible W-R, Gonzales-Fontes A, Lauerer M, Müller-Röber B, Caboche M and Stitt M (1997b) Nitrate acts as a signal to induce organic acid metabolism and repress starch metabolism in tobacco. Plant Cell 9: 783–798
Scheible W-R, Gonzales-Fontes A, Morcuende R, Lauerer M, Geiger M, Glaab J, Gojon A, Schulze E-D and Stitt M (1997c) Tobacco mutants with a decreased number of functional genes compensate by modifying the diurnal regulation of transcription, post-transcriptional modification and turnover of nitrate reductase. Planta 203: 304–319
Scheible W-R, Krapp A and Stitt M (2000) Reciprocal diurnal changes of phosphoenolpyruvate carboxylase expression and cytosolic pyruvate kinase, citrate synthase and NADP-isocitrate dehydrogenase expression regulate organic acid metabolism during nitrate assimilation in tobacco leaves. Plant Cell Environ 22: 1155–1167
Siddiqi MY, Glass ADM, Ruth TJ and Fernando M (1989) Studies of the regulation of nitrate influx by barley seedlings using $^{13}NO_3^-$. Plant Physiol 90: 806–813
Siddiqi MY, Glass ADM, Ruth TJ and Rufty T (1990) Studies of the uptake of nitrate in barley. I. Kinetics of $^{13}NO_3^-$ influx. Plant Physiol 93: 1426–1432
Sivasankar S and Oaks A (1995) Regulation of nitrate reductase during early seedling growth. Plant Physiol 107: 1225-1231
Stitt M and Scheible W-R (1998) Understanding allocation to shoot and root growth will require molecular information about which compounds act as signals, and how meristem activity and cellular growth are regulated: opinion. Plant Soil 201: 259–263
Sueyoshi K, Mitsuyama T, Sugimoto T, Kleinhofs A, Warner RL and Oji Y (1999) Effects of inhibitors for signaling components on the expression of genes for nitrate reductase and nitrite reductase in excised barley leaves. Soil Sci Plant Nutr 45: 1015–1019
Sugden C, Donaghy PG, Halford NG and Hardie DG (1999) Two SNF1-related protein kinases from spinach leaf phosphorylate and inactivate 3-hydroxy-3-methylglutaryl-coenzyme A reductase, nitrate reductase, and sucrose phosphate synthase in vitro. Plant Physiol 120: 257–274
Sugiharto B and Sugiyama T (1992) Effects of nitrate and ammonium on gene expression of phosphoenolpyruvate carboxylase in maize leaf tissue during recovery from nitrogen stress. Plant Physiol 98: 1403–1408
Sugiharto B, Suzuki I, Burnell JN and Sugiyama T (1992) Glutamine induces the N-dependent accumulation of mRNAs encoding phosphoenolpyruvate carboxylase and carbonic anhydrase in detached maize leaf tissue. Plant Physiol 100: 2066–2070
Suzuki I, Crétin C, Omatat T and Sugiyama T (1994) Transcriptional and post-transcriptional regulation of nitrogen-responding expression of phosphoenolpyruvate carboxylase gene in maize. Plant Physiol 105: 1223–1229
Tanaka T, Ida S, Irifune K, Oeda K and Morikawa H (1994) Nucleotide sequence of a gene for nitrite reductase from *Arabidopsis thaliana*. DNA Sequence 5: 57–61
Taniguchi M, Kiba T, Sakakibara H, Ueguchi C, Mizuno T and Sugiyama T (1998) Expression of *Arabidopsis* response regulator homologs is induced by cytokinins and nitrate. FEBS Lett 429: 259–262
Tercé-Laforgue T, Carrayol E, Cren M, Desbrosses G, Hecht V and Hirel B (1999) A strong constitutive positive element is essential for ammonium-regulated expression of a soybean gene encoding cytosolic glutamine synthetase. Plant Mol Biol 39: 551–564
Tillard P, Passama L and Gojon A (1998) Are phloem amino acids involved in the shoot to root control of NO_3^- uptake in *Ricinus communis* plants? J Exp Bot 49: 1371–1379
Tompkins GA, Jackson WA and Volk RJ (1978) Accelerated nitrate uptake in wheat seedlings. Effects of ammonium and nitrite pretreatments and of 6-methylpurine and puromycin. Physiol Plant 43: 166–171
Toroser D, Athwal GS and Huber S (1998) Site-specific regulatory interaction between leaf sucrose-phosphate synthase and 14-3-3 proteins. FEBS Lett 435: 110–114
Touraine B and Glass ADM (1997) NO_3^- and ClO_3^- fluxes in the *chl1-5* mutant of *Arabidopsis thaliana*. Does the *CHL1-5* gene encode a low-affinity NO_3^- transporter? Plant Physiol 114: 137–144
Touraine B, Clarkson DT and Muller B (1994) Regulation of nitrate uptake at the whole plant level. In: Roy J, Garnier E (eds) A Whole Plant Perspective on Carbon-Nitrogen Interactions, pp 11–30. SPB Academic Publishing, The Hague
Trueman LJ, Richardson A and Forde BG (1996) Molecular cloning of higher plant homologues of the high affinity nitrate transporters of *Chlamydomonas reinhardtii* and *Aspergillus nidulans*. Gene 175: 223–231
Truong H-N, Caboche M and Daniel-Vedele F (1997) Sequence and characterization of two *Arabidopsis thaliana* cDNAs isolated by functional complementation of a yeast *gln3 gdh1* mutant. FEBS Lett 410: 213–218
Tsay YF, Schroeder JI, Feldman A and Crawford NM (1993) The herbicide sensitivity gene *CHL1* of *Arabidopsis* encodes a nitrate-inducible nitrate transporter. Cell 72: 705–713
Turpin DH and Bruce D (1990) Regulation of photosynthetic light harvesting by nitrogen assimilation in the green alga *Selenastrum minutum*. FEBS lett 263: 99–103
Turpin DH, Weger HG and Huppe HC (1997) Interactions between photosynthesis, respiration and nitrogen assimilation. In: Plant Metabolism (eds Dennis DT, Turpin DH, Lefebvre DD and Layzell DB) pp. 509–524 Longman, Singapore
Unkles SE, Hawker KL, Grieve C, Campbell EI, Montague P and

Kinghorn JR (1991) *Crna* encodes a nitrate transporter in *Aspergillus nidulans*. Proc Natl Acad Sci USA 88: 204–208

Van der Leij M, Smith SJ and Miller AJ (1998) Remobilization of vacuolar stored nitrate in barley root cells. Planta 205: 64–72

Van der Werf A and Nagel OW (1996) Carbon allocation to shoots and roots in relation to nitrogen supply is mediated by cytokinins and sucrose: Opinion. Plant Soil 185: 21–32

Van Quy L, Foyer CH and Champigny ML (1991) Effect of light and NO_3^- on wheat leaf phosphoenolpyruvate carboxylase activity. Evidence for covalent modification of the C3 enzyme. Plant Physiol 97: 1476–1482

Vaucheret H, Chabaud M, Kronenberger J and Caboche M (1990) Functional complementation of tobacco and *Nicotiana plumbaginifolia* nitrate reductase deficient mutants by transformation with the wild-type alleles of the tobacco structural genes. Mol Gen Genet 220: 468–474

Vidmar JJ, Zhuo D, Siddiqi MY and Glass ADM (2000a) Isolation and characterization of *HvNRT2.3* and *HvNRT2.4*, cDNAs encoding high-affinity nitrate transporters from roots of *Hordeum vulgare*. Plant Physiol 122: 783–792

Vidmar JJ, Zhuo D, Siddiqi MY, Schoerring JK, Touraine B and Glass ADM (2000b) Regulation of high-affinity nitrate transporter genes and high-affinity nitrate influx by nitrogen pools in roots of barley. Plant Physiol 123: 307–318

Vincentz M, Moureaux T, Leydecker M-T, Vaucheret H and Caboche M (1993) Regulation of nitrate and nitrite reductase expression in *Nicotiana plumbaginifolia* leaves by nitrogen and carbon metabolites. Plant J 3: 315–324

von Wirén N, Lauter FR, Ninnemann O, Gillissen B, Walch-Liu P, Engels C, Jost W and Frommer WB (2000) Differential regulation of three functional ammonium transporter genes by nitrogen in root hairs and by light in leaves of tomato. Plant J 21: 167–175

Walch-Liu P, Neumann G, Bangerth F and Engels C (2000) Rapid effects of nitrogen form on leaf morphogenesis in tobacco. J Exp Bot 51: 227–237

Wang MY, Siddiqi MY, Ruth TJ and Glass ADM (1993) Ammonium uptake by rice roots. II. Kinetics of $^{13}NH_4^+$ influx across the plasmalemma. Plant Physiol 103: 1259–1267

Wang R, Liu D and Crawford NM (1998) The *Arabidopsis* CHL1 protein plays a major role in high-affinity nitrate uptake. Proc Natl Acad Sci USA 95: 15134–15139

Wang R, Guegler J, LaBrie ST and Crawford NM (2000) Genomic analysis of a nutrient response in *Arabidopsis* reveals diverse expression patterns and novel metabolic and potential regulatory genes induced by nitrate. Plant Cell 12: 1491–1509

Warner RL and Huffaker RC (1989) Nitrate transport is independent of NADH and NAD(P)H nitrate reductases in barley seedlings. Plant Physiol 91: 947–953

Xiao X, Fu Y-H and Marzluf GA (1995) The negative-acting NMR regulatory protein of *Neurospora crassa* binds to and inhibits the DNA-binding activity of the positive-acting nitrogen regulatory protein NIT2. Biochemistry 34: 8861–8868

Zhang H and Forde BG (1998) An *Arabidopsis* MADS box gene that controls nutrient-induced changes in root architecture. Science 279: 407–409

Zhang H, Hennings A, Barlow PW and Forde BG (1999) Dual pathways for regulation of root branching by nitrate. Proc Natl Acad Sci USA 96: 6529–6534

Zhuo D, Okamoto M, Vidmar JJ and Glass ADM (1999) Regulation of putative high-affinity nitrate transporter (*Nrt2.1At*) in roots of *Arabidopsis thaliana*. Plant J 17: 563-569

Chapter 14

Regulation of Carbon and Nitrogen Assimilation Through Gene Expression

Tatsuo Sugiyama*
RIKEN (The Institute of Physical and Chemical Research) Plant Science Center, 2-1 Hirosawa, Wako 351-0198, Japan

Hitoshi Sakakibara
Laboratory for Communication Mechanism, RIKEN (The Institute of Physical and Chemical Research) Plant Science Center, 2-1 Hirosawa, Wako 351-0198, Japan

Summary 227
I. Introduction 228
II. Physiological and Biochemical Nature of Plant Response to Nitrogen 228
A. Growth and Development 228
B. Assimilation of Carbon and Nitrogen 229
III. Regulation of Nitrogen-Responsive Genes for Carbon Assimilation 230
IV. Regulation of Nitrogen-Responsive Genes for Assimilation and Subsequent Metabolism of Nitrogen 231
A. Genes for Assimilation of Inorganic Nitrogen Sources 232
B. Genes involved in Translocation and Partitioning 233
V. Regulation of Partitioning of Nitrogen into Proteins: A Model for Sensing and Signaling 234
Acknowledgments 235
References 235

Summary

Regulation of the partitioning of nitrogen (N) into proteins is an important mechanism whereby plants can alter metabolism to adjust or acclimate to changes in N availability. Alterations in N assimilation brought about by changes in N availability require regulation of other metabolic processes and re-allocation of nutrients. This requires the mutual coordination and complementation of metabolism and allocation throughout the plant by shuttling substrates, metabolites, and signals. The control of the C/N interaction is particularly important since these elements are abundant in plants and provide the skeletons and moieties for most of the building biomolecules. As a signaling pathway to communicate between plants and their nutritional environment, the 'His-Asp phosphorelay,' concept originally called 'two-component system,' has recently been proposed in higher plants. This chapter focuses on the recent advances that have uncovered genes and mechanisms responsible for the regulation of N partitioning into the machinery of C and N assimilation.

*Author for correspondence, email: sugiyama@postman.riken.go.jp

Christine H. Foyer and Graham Noctor (eds): Photosynthetic Nitrogen Assimilation and Associated Carbon and Respiratory Metabolism, pp. 227–238.

I. Introduction

The mineral nutrient required most abundantly by plants is nitrogen (N). The sources available to plants are usually inorganic forms such as nitrate and ammonia. Their availability changes unexpectedly and rapidly in a natural environment, and limits plant growth and development. The N source functions not only as a substrate for the assimilation, but also as a signal for growth and development by regulating gene expression and thereby metabolism. Metabolic alteration brought about by changes in N availability requires regulation in other metabolic processes and allocation of nutrients, demanding the mutual coordination and complementation of metabolism and allocation through the plant by shuttling substrates, metabolites, and signals. The interaction between C and N is particularly important since these elements are abundant in plants and provide the skeleton and moieties for most of the building biomolecules.

Signals serve to convey information and the message to be transmitted from the outside to the inside of the cell. The general feature of signals at the cellular level may also be extended to a multi-cellular system. From the aspect of the whole plant, inorganic N sources taken up are assimilated in roots and leaves utilizing energy and C skeletons provided by photosynthesis, which takes place in the leaves. The body organization of plants requires regulatory cross-talk in the metabolism of N and C and thereby an elaborate network of signaling pathways at cellular, intercellular, and organ-to-organ levels.

Partitioning of N into proteins is needed to alter metabolism. Changes in the level of key proteins such as enzymes and transporters enable not only the regulation of intermediate pools but also modify the rates of N turnover and transport. Plants must monitor the availability of N, perceptibly sense it, convey the message, and transmit the message to cells at the cellular, intercellular, and organ-to-organ levels, for regulation of gene expression. The correct partitioning of N into proteins requires multi-cellular communication throughout the whole plant.

We will not attempt to review the extensive literature on the regulation and properties of the enzymes involved in the assimilation of C and N and their gene expression. Excellent reviews on these topics are available (Redinbaugh and Campbell, 1991; Hoff et al., 1994; Huppe and Turpin, 1994; Sheen, 1994; Crawford, 1995; Koch, 1996, 1997; Forde and Clarkson, 1999; Stitt, 1999). Instead, this chapter focuses on the recent advances that have uncovered genes and mechanisms responsible for the regulation of N partitioning into the machinery of C and N assimilation.

II. Physiological and Biochemical Nature of Plant Response to Nitrogen

A. Growth and Development

Increasing the supply of N to plants leads to increased growth, accelerated germination of seeds, and morphological changes such as decreased root:shoot ratios, root architecture, delayed flowering, tuber initiation and senescence (Stitt and Krapp, 1999). High nitrate conditions depress root growth, decreasing the root:shoot ratio and the frequency of lateral root formation (Marschner, 1995). Recent research has unequivocally suggested, based on genetic analysis of tobacco mutants deficient in nitrate reductase (NR) genes (*Nia*), that signals derived from nitrate trigger the adaptive changes in root growth and architecture (Scheible et al., 1997). Modification of plant growth and architecture requires integrated changes in metabolism as well as a network of inter- and intra-cellular communication in the allocation of cellular components and signals. General deficiency phenomena observed under lower N, are exacerbated when tissue N contents are seriously decreased due to a reduction in the capacity of N redistribution between tissues (Arp et al., 1998). The mobile nature of N in plants is a key feature of the element to be considered in any analysis of the coordination of N and C metabolism. N availability

Abbreviations: AlaAT – alanine aminotransferase; ARR – *Arabidopsis thaliana* response regulators; AS – asparagine synthetase; Asn – asparagine; Asp – aspartate; AspAT – aspartate aminotransferase; C – Carbon; C_3 – three carbon; C_4 – four carbon; C_4Ppc1 – C_4-type PEP carboxylase; C_4Ppdk – C_4-type PPDK; cAlaAT – cytosolic AlaAT; Fd – ferredoxin; FNR – Fd-NADP oxidoreductase; Gln – glutamine; Glu – glutamate; GOGAT – glutamate synthase; GS – glutamine synthetase; HPt – His-containing phosphotransfer domain; mAlaAT – mitochondrial AlaAT; N – nitrogen; NiR – nitrite reductase; NR – nitrate reductase; PEP – phosphoenolpyruvate; PEPc – PEP carboxylase; PPDK – pyruvate, orthophosphate dikinase; RPP – reductive pentose phosphate; Rubisco – ribulose-1,5-bisphosphate carboxylase/oxygenase; SUMT – *S*-adenosylmethionine-dependent uroporphyrinogen III C-methyltransferase; Z – *t*-zeatin; ZmRR – *Zea mays* response regulators; ZR – *t*-zeatin riboside not only restricts the photosynthetic efficiency of plants but also appears to regulate photosynthate utilization. Typically, N limitation leads to accumulation of products such as non-structural carbohydrates, mainly starch (Stitt and Krapp, 1999), lipids, and carotenoids (Goodwin, 1980). The increase in non-structural carbohydrates can also be seen in plants grown under elevated CO_2 (Stitt and Krapp, 1999). These changes clearly indicate that the assimilation of C and N is closely related to whole plant growth and development. Consequently the supply of either macronutrient results in marked changes in the assimilation of the other. This ultimately reduces the capacity of the whole plant to grow and develop.

B. Assimilation of Carbon and Nitrogen

Of all the essential nutrients, the rate of photosynthesis is most critically limited by the availability of N. A major symptom of N limitation is the loss of chlorophyll. It is frequently accompanied by alterations in photosynthetic capacity (Terashima and Evans, 1988) and energy transduction (Rhiel et al., 1986; Saux et al., 1987; Plumley and Schmidt, 1989). N limitation also decreases the gas exchange capacity of plants that is indicative of decreased carboxylation capacity and decreased ribulose-1, 5-bisphosphate carboxylase/oxygenase (Rubisco) activity and/or protein (Brown, 1978; Stitt and Krapp, 1999). Allocation of N into Rubisco appears to be different in C_3 and C_4 plants (Fig. 1) (Sugiyama et al., 1984). Differences in growth strategies are reflected in the ways that in C_3 and C_4 plants respond to changes in the environment. C_4 plants are considered to utilize light, water and N more efficiently by virtue of the evolution of an ATP-driven CO_2 pump that maximizes CO_2 assimilation. The pump machinery expressed in C_4 leaf mesophyll cells consists primarily of two key enzymes, C_4-type phosphoenolpyruvate (PEP) carboxylase (C4Ppc1) and pyruvate, orthophosphate dikinase (C4Ppdk). These enzymes function in collaboration with the reductive pentose phosphate (RPP) pathway that operates in the chloroplasts of the leaf bundle sheath cells. Leaf levels of the C assimilatory enzymes including Rubisco in leaves are considered to represent an enzymatic limitation on the rate of photosynthesis. They are therefore key to the biomass accumulation and productivity of plants (Sugiyama et al., 1998). N partitioning into C assimilatory enzymes is different in C_3 and C_4 plants in terms of responses to N. The allocation of N into Rubisco is selectively increased by N availability in some C_3 plants, whereas the allocation is decreased in maize (Brown, 1978; Sugiyama et al., 1984; Yamazaki et al., 1986; Sugiharto et al., 1990). Instead, the allocation of N into C4Ppc1 and C4Ppdk is up regulated by N availability, similar to the situation with Rubisco in C_3 plants (Sugiyama et al., 1984). This may reflect the essential requirement of these enzymes for the C_4 pathway of photosynthesis in maize, which is an important appendage to the RPP pathway. The N-responsive accumulation of the two enzymes is regulated at the level of protein synthesis and is accompanied by mRNA accumulation (Sugiharto et al., 1990). A similar mode of regulation has also been demonstrated in the genes encoding carbonic anhydrase (*C4Ca*) in maize (Sugiharto et al., 1992a,b), alanine aminotransferase (*AlaAT*) (Son et al., 1992) and aspartate aminotransferase (*AspAT*) (Taniguchi et al., 1995) in *Panicum miliaceum*, which also function in the C_4 pathway. Further investigations into the nature of N partitioning in C_4 plants will help us to understand how plants regulate N partitioning into the C assimilating machinery.

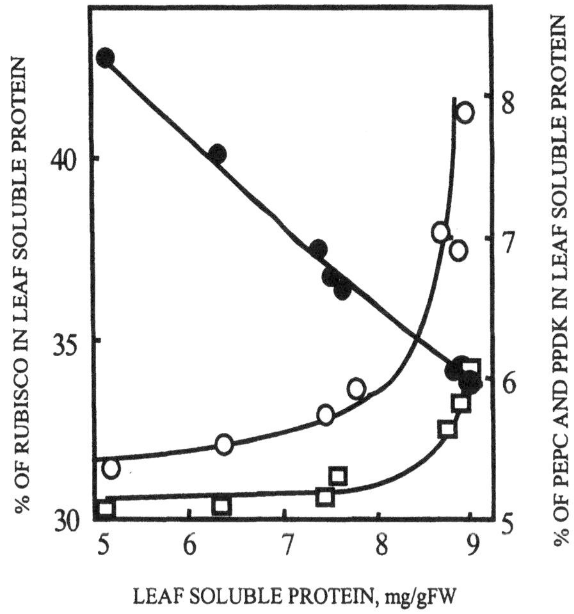

Fig. 1. Regulation by N-availability of N-partitioning into PEPc (open circle), PPDK (open square) and Rubisco (closed circle) in maize leaves (Sugiyama et al., 1984).

Inorganic N sources act as signals for the regulation of expression of genes encoding proteins involved in the assimilation of C and N. Signals from N sources may be derived from the source itself, from intermediates formed during assimilation, or from other cellular constituents derived from cross-talk between metabolic pathways and even between organs and tissues. In addition, signals derived directly or indirectly from N sources may interact with other cellular signals including other nutrients and environmental stimuli. Regardless of the complex nature of signal formation, signaling of N availability can lead to the modification of gene expression. Transcriptional and post-transcriptional controls have been found. Modes of regulation are based on gene expression and involve controls on transcript abundance, as well as translational controls. These form the basis for regulation of N partitioning into protein.

III. Regulation of Nitrogen-Responsive Genes for Carbon Assimilation

The balance of the C/N ratio relies on the interdependent and interconnected regulation of the metabolism. This is achieved through regulatory signals and processes that include metabolites, allosteric effectors, protein phosphorylation, and redox regulation. At the level of gene expression, many genes in photosynthesis are N-responsive (Table 1). These genes include the Rubisco small subunit (*rbcS*) and the light harvesting chlorophyll *a,b*-binding protein (*Cab*) in *Chlamydomonas reinhardtii* (Plumley and Schmidt, 1989), as well as *C4Ppc1*, *C4Ppdk*, and *C4Ca* in maize (Sugiharto and Sugiyama, 1992; Sugiharto et al., 1992b) and *AlaAT* (Son et al., 1992) and *AspAT* (Taniguchi et al., 1995) in *Panicum miliaceum*.

In many C_3 plants, the percentage of leaf protein that is accounted for by Rubisco increases directly in proportion to the leaf N content (Brown, 1978). A similar situation with regard to Rubisco content also occurs in *C. reinhardtii*. In this photosynthetic eukaryote *rbcS* is up regulated by N availability through mechanisms affecting transcription and/or mRNA stability (Plumley and Schmidt, 1989). In maize *rbcS* transcription is also up-regulated by N availability (I. Suzuki and T. Sugiyama, unpublished). Under N starvation Rubisco decreases as a percentage of leaf protein (Sugiyama et al., 1984). In contrast to N, *rbcS* transcription is down-regulated by sugars such as sucrose and glucose in both C_3 (Krapp et al., 1993) and C_4 plants (Sheen, 1990, 1994), supporting the concept of C-mediated feedback or sink-regulated inhibition of photosynthesis. The reciprocal effects of N and C availability as signals in the expression of *rbcS* appears to be primarily important in terms of N partitioning because Rubisco is crucially important from the viewpoints of N economy and enzymatic function as an initial and limiting enzyme of C assimilation.

The C_4 pathway of photosynthesis is considered to be a biochemical appendage to the RPP pathway. This suggests that the expression of some C_4 photosynthesis genes is inducible in nature under certain circumstances, particularly in response to environmental stimuli. Among these, light and N are important as C_4 plants have a higher light use efficiency as well as a higher N-use efficiency than C_3 plants (Brown, 1978). The mechanisms involved in inorganic N-mediated regulation of genes encoding C_4 enzymes have been studied extensively in maize

Table 1. Plant photosynthesis genes regulated by nitrogen availability

Protein (gene)	Nitrogen (effect)	Other factors (effect)
PEPc (*C4Ppc1*)	Nit, Am, Gln (+)	CK, Light (+); Suc (-)
PPDK (*C4Ppdk*)	Nit, Am, Gln (+)	CK, Light (+); Suc (-)
CA (*C4Ca*)	Nit, Am, Gln (+)	CK, Light (+)
AlaAT (*AlaAT-2*)	Nit, Am (+)	
AspAT (*cAspAT*, *mAspAT*)	Nit, Am (+)	
Rubisco (*rbcS*)	Nit, Am, Gln (+)	CK, Light (+); Suc (-)
LHCP (*Cab*)	Nit, Am, Gln (+)	CK, Light (+); Suc (-)

For relevant references, see text. PPDK, pyruvate, orthophosphate dikinase; CA, carbonic anhydrase; LHCP, light-harvesting chlorophyll *a,b*-binding protein; Nit, nitrate; Am, ammonia; CK, cytokinin; Suc, sucrose

(Sugiyama, 1998). There are three isoforms of AlaAT, e.g., AlaAT-1, -2, -3 in *P. miliaceum* (Son et al., 1991). Of these, AlaAT-2, which is light-inducible and expressed most abundantly in the cytosol of the leaf mesophyll and bundle sheath cells, functions in the aspartate/alanine shuttle of the C_4 pathway (Son and Sugiyama, 1992). The AlaAT-2 form selectively accumulates in response to inorganic N availability as a consequence of changes in the level of mRNA (Son et al., 1992). A similar situation is found with cAspAT and mAspAT, which are cytosolic and mitochondrial forms of AspAT, respectively, that participate in the C_4 pathway in *P. miliaceum* (Taniguchi et al., 1992, 1995). cAspAT and mAspAT are developmentally regulated and are expressed during greening in the leaf mesophyll and bundle sheath cells, respectively. These isoforms increase with concomitant accumulation of mRNAs in response to inorganic N sources (Taniguchi et al., 1995).

The expression of *C4Ppc1* is known to be regulated both transcriptionally and post-transcriptionally by N availability (Suzuki et al., 1994). The N-responsive regulation of *C4Ppc1* suggests the existence of a highly organized network for the sensing and transduction of the inorganic N signal. This underlies the preferential allocation of N into the protein. For the identification of processes that are either directly or indirectly regulated by N availability it is of primary importance to define the internal signals that carry information on N availability. In the case of *C4Ppc1* and *C4Ca,* Gln and/or metabolite(s) arising from Gln metabolism, are positive signals for inorganic N-responsive gene expression. While Gln appears primarily to control the stability of mRNAs (Suzuki et al., 1994), it is also a negative signal for *Nia* expression (Deng et al., 1991; Shiraishi et al., 1992; Vincentz et al., 1993). The action of Gln, as a parameter of N nutrition derived from the metabolism of nitrate, is in the opposite direction to nitrate. The inverse relationship between nitrate and Gln with regard to the regulation of N gene expression strengthens the ability of the signal to balance the relative rates of C and N assimilation in plants, although the precise mechanism by which these metabolites modulate gene expression is uncertain.

N availability not only restricts the photosynthetic efficiency of plants but also appears to regulate photosynthate utilization. When plants go through phases of C excess, such as occur for example during N starvation or CO_2 enrichment, starch, lipids and flavonoids accumulate in the plants, as described earlier. Accumulation of such end products of photosynthesis under N starvation may function, at least in part, as a sink for C that helps the adjustment of the C to N balance of the plant. Fine control of the coordination of nitrate assimilation, C assimilation, and sucrose synthesis is evident at the post-translational level. Several key cytosolic enzymes of these pathways, such as NR, sucrose-phosphate synthase and PEPc, are regulated by protein phosphorylation (Huber et al., 1992; Huber and Huber, 1996; MacKintosh and MacKintosh, 1993; Nimmo, 1993; Chollet et al., 1996).

In the regulatory network of protein phosphorylation, nitrate appears to be a key metabolic signal. Nitrate modulates the activities of the protein kinases and protein phosphatases that act on each of the target enzymes and thereby exerts influence on C flow between sucrose and amino acids (Champigny and Foyer, 1992; Foyer et al., 1994).

IV. Regulation of Nitrogen-Responsive Genes for Assimilation and Subsequent Metabolism of Nitrogen

Plant cells, through transporters located in the cytoplasmic membrane, take up nitrate and ammonium ions. Nitrate, the most common inorganic N source, can induce genes for N assimilatory machinery in plants such as nitrate transporters, NR, nitrite reductase (NiR), glutamine synthetase (GS), and glutamate synthase (GOGAT; Table 2). The use of nitrate and ammonium, the substrates of primary N assimilation, as signals for the regulation of gene expression must be advantageous for plants since it favors utilization of the available N sources with a minimum consumption of energy. The expression of genes encoding proteins associated with N assimilation is also regulated by other external and internal signals, such as C and N metabolites, light, hormones and circadian rhythms (Vincentz et al., 1993). Coordination of the expression of genes encoding N assimilatory processes by environmental cues is a basic strategy in plant N acquisition that is necessary for the adjustment of N uptake and use to its availability.

Table 2. Nitrogen-assimilatory genes regulated by nitrogen availability

Protein (gene)	Nitrogen (effect)	Other effectors (effect)
NRT (*Nrt*)	Nit (+); Am, Gln, Arg (-)	
NR (*Nia*)	Nit (+); Gln (-)	CK, Light, Suc, Glc, Frc (+); ABA (-)
NiR (*Nii*)	Nit (+); Gln (-)	Light
GS2 (*GLN2*)	Nit (+)	Light, Suc (+)
Fd-GOGAT (*FdGGT*)	Nit (+)	Light (+)
Fd	Nit (+)	
FNR	Nit (+)	
SUMT (*SUMT1*)	Nit (+)	
GS1	Am (+)	Nod, Scenesence (+)
AS (*ASN1*)	Gln, Glu (+)	Suc (-); Dark (+)

References are shown in the text. Nit, nitrate; Am, ammonia; CK, cytokinin; Suc, sucrose; Glc, glucose; Frc, fructose; ABA, abscisic acid; Nod, nodulation.

A. Genes for Assimilation of Inorganic Nitrogen Sources

The rate of uptake of inorganic N sources by plants can vary by several orders of magnitude in nature depending on availability and composition. The regulation of nutrient uptake is important from the viewpoint of energy budgets. N assimilation is expensive in terms of reducing equivalents and plants need more energy to reduce nitrate than to reduce ammonium ions. Biochemical and physiological studies have revealed that the process of nitrate uptake is multiphasic, involving at least two different transport systems. These are a low-affinity system (Km > 0.5 mM) and a high-affinity system (Km 10 to 300 μM) (Doddema and Telkamp, 1979; Goyal and Huffaker, 1986; Siddiqi et al., 1990; Glass et al., 1992). Many genes encoding nitrate transporters have been identified (Tsay et al., 1993; Lauter et al., 1996; Trueman et al., 1996a,b; Quesada et al., 1997). Some of the genes (*Nrt*) are predominantly expressed in the roots and are induced in response to nitrate.

Regulatory steps are found in the transcription, post-transcriptional events and in post-translational controls governing *Nia* expression in response to changes in environmental conditions (for reviews, see Kleinhofs and Warner, 1990; Solomonson and Barber, 1990; Hoff et al., 1994; Kaiser and Huber, 1994; Crawford, 1995). *Nia* is nitrate-inducible in plants (Cheng et al., 1986; Crawford et al., 1986; Callaci and Smarrelli, 1991; Gowri et al., 1992). It is up-regulated by a variety of signals such as light (Melzer et al., 1989; Lin and Cheng, 1997), plant hormones including cytokinin (Lu et al., 1992) and C metabolites (Deng et al., 1991; Vincentz et al., 1993). It is down-regulated by Gln (Deng et al., 1991).

The expression of the NiR gene (*Nii*) is also enhanced by nitrate in a manner similar to *Nia* (for reviews, see Vincentz et al., 1993; Hoff et al., 1994; Crawford, 1995). In addition to transcriptional regulation of the *Nii* gene, post-transcriptional controls have also been suggested (Crété et al., 1997). Most importantly, the production of siroheme, a prosthetic group of NiR (Siegel and Wilkerson, 1989), appears to be regulated by nitrate. The conversion of the NiR apoenzyme to the holoenzyme depends on the supply of siroheme, which is produced in an N-responsive manner. A gene encoding *S*-adenosyl-methionine-dependent uroporphyrinogen III C-methyltransferase (SUMT) which catalyzes a part of the synthetic pathway of siroheme, has been isolated from maize (Sakakibara et al., 1996b) and *A. thaliana* (Leustek et al., 1997). Like *Nia* and *Nii*, the maize gene *ZmSUMT1* appears to be regulated by nitrate (Sakakibara et al., 1996b). Siroheme is also a prosthetic group of sulfite reductase, an enzyme in sulfur assimilation. This raises the question as to whether the availability of both S and N may regulate expression of SUMT and sulfite reductase. The answer could lead to new insights into the interactions between N and S assimilation.

Some genes encoding proteins participating in the supply of reducing power for the reduction processes of nitrate assimilation are nitrate-inducible. These include genes for FNR (Ritchie et al., 1994; Aoki and Ida, 1994) and Fd in maize (Matsumura et al., 1997). These genes appear to be co-regulated with *Nii* and *SUMT* by nitrate. Such coordination in gene expression implies a concerted genetic program of nitrate reduction processes in plants in response to nitrate availability.

In addition to provision from soil, various metabolic processes in plants can provide ammonia. These include the reduction of nitrate, the fixation of N_2, photorespiration, phenylpropanoid metabolism, and amino acid catabolism. The ammonia provided by such pathways is assimilated into Gln by the GS/GOGAT cycle (Miflin and Lea, 1980; Givan et al., 1988). The ammonia formed by the reduction of nitrate is considered to be assimilated mainly by the plastidic isoform of GS (GS2). GS2 expression is subject to tissue- and cell-specific controls that respond differentially to N availability. It is up regulated by nitrate in pea and maize roots (Emes and Fowler, 1983; Vézina and Langlois, 1989; Sakakibara et al., 1992a; Redinbaugh and Campbell, 1993). In leaves the N sensitivity of GS2 gene expression appears to be species-specific and to depend on N availability. For example, in tobacco and maize, the nitrate-responsive accumulation of GS2 mRNA can be detected only after N starvation (Migge and Becker, 1996; Sakakibara et al., 1997).

GS2 is preferentially accumulated in maize leaf mesophyll cells in response to nitrate, whereas it preferentially accumulated in the bundle sheath cells during the greening period of the etiolated seedlings (Sakakibara et al., 1992a). The differential spatial and environmental responses of GS2 expression supports the hypothesis, based on enzymatic analysis, that nitrate assimilation takes place in the mesophyll whereas the re-assimilation of the ammonia released during photorespiration takes place in the bundle sheath in maize (Ohnishi and Kanai, 1983). Thereby, it is possible to envisage a concept in which the two photosynthetic cell types are functionally distinct in terms of N assimilation as well as C assimilation.

The physiological functions of GS1 are suggested to be the primary assimilation of external ammonium ions (Hirel et al., 1987; Miao et al., 1991; Sakakibara et al., 1996a) and the re-assimilation of ammonium released during N remobilization (Kawakami and Watanabe, 1988; Kamachi et al., 1991). The distribution and localization of GS1 in tissues and organs vary among plant species (McNally et al., 1983). In most plant species, a small multigene family encodes GS1. The expression of each member of the gene family appears to be differentially regulated in different organs (Bennett et al., 1989; Cock et al., 1990). Some GS1 family members also show differential regulation of expression by various N sources (Hirel et al., 1987; Miao et al., 1991; Sakakibara et al., 1992a, 1996a).

The multiplicity of GS1 genes may reflect divergent mechanisms that enable plants to make appropriate adjustments to external and internal environments. Among the five different cytosolic GS forms in maize, for example, GS1c and GS1d are up-regulated by ammonium ions, producing the respective enzyme isoforms in the roots. These have higher specific activities than other isoforms (Sakakibara et al., 1992a,b, 1996a). The superior catalytic properties and N-responsiveness of GS1c and GS1d may be physiologically important as a protective device. By detoxifying excess ammonium ions they minimize the possibility of negative effects of ammonium accumulation that might occur unexpectedly by excessive uptake or generation in root cells. As such, the manipulation of GS isoforms and their distribution might be a useful target for plant improvement because the assimilation of ammonium ion is less expensive, in terms of energy, than nitrate.

Understanding the different N-responses of genes such as *Nia*, *Nii*, *SUMT*, *FNR*, *FdVI*, and *GS*s that are involved in N assimilation will help reveal the mechanisms that underpin the N-responsive accumulation of Gln in plants. Gln has multiple functions as a key product of N assimilation, a transport form of N and presumably as a key parameter of N nutrition, as well as its role as a metabolic signal.

B. Genes involved in Translocation and Partitioning

Like Gln, Asn is also a transport form of N but it has a higher ratio of N to C. The regulation of relative Gln and Asn transport is considered to be rather different particularly in regard to environmental controls (such as light/dark). Since Asn has a higher ratio of N to C, the production and use of the Asn in transport is an important strategy that plants use to achieve efficient N transport under the conditions of limiting supplies of C skeletons. Consistent with this idea, dark treatment enhances the content of Asn in phloem exudates and this is accompanied by an increase in asparagine synthetase (AS) activity (Urquhart and Joy, 1981; Joy et al., 1983; Lam et al., 1995). The expression of *ASN1*, a gene encoding AS has been analyzed in pea (Tsai and Coruzzi, 1990), asparagus (Davis and King, 1993), *A. thaliana* (Lam et al., 1994) and maize (Chevalier et al., 1996). The manner of *ASN1* regulation appears to bear a 'mirror image relationship' to that of the GS2 gene (*GLN2*) in terms of responsiveness to light and sugars. Both light and sugars down-regulate *ASN1* while they up-regulate *GLN2*. Some N-sources such as Gln, Glu, and Asn relieve the sugar-mediated repression of *ASN1* induced during the dark (Lam et al., 1994). This suggests that AS functions to redirect the flow of N from Gln to Asn when the C supply is limiting relative to N.

V. Regulation of Partitioning of Nitrogen into Proteins: A Model for Sensing and Signaling

Transcriptional and post-transcriptional regulation of gene expression forms the basis for N partitioning into proteins but the molecular analysis of the signaling pathway that facilitates these controls in plants is in its infancy. To survive and develop competitively during unexpected changes in N availability, plants must constantly sense changes in their environment and respond appropriately through a variety of signal transduction pathways at the cellular and the whole plant levels. The 'His-Asp phosphorelay,' model provides a mechanism whereby plants can communicate with their nutritional environment. The 'His-Asp phosphorelay,' concept, originally called the 'two-component system,' in bacteria, has recently been proposed to occur in higher plants. We will now outline our ideas on the sensing of N availability and the transduction of that signal that is mediated by cytokinins at the whole plant level. This is the mechanism of regulation of partitioning of N into proteins in plants.

His-Asp phosphorelay is a reversible protein phosphorylation system that was originally elucidated as a mechanism of cellular signal transduction in bacteria. The phosphorelay is typically made up of three functional domains: sensor (His-protein kinase) domain, His-containing phosphotransfer (HPt) domain, and receiver (response regulator) domain. This system was once thought to be restricted to prokaryotes, but has recently been uncovered in diverse eukaryotic species including plants. The basic property of the system has been summarized in recent reviews (Mizuno, 1998; Sakakibara et al., 2000).

At present, 11 genes encoding His-protein kinases, five genes encoding HPt domains and 20 genes encoding response regulators have been identified and characterized in *Arabidopsis* (ARR-series; Imamura et al., 1999; T. Mizuno, personal communication). Three genes encoding His-protein kinases, two genes encoding HPt domains and eight genes encoding response regulators have been isolated in maize (ZmRR-series; Sakakibara et al., 1998, 1999; H. Sakakibara et al., unpublished). The ARRs can be classified into two distinct subtypes, type-A and type-B regulators, based on their structure, biochemical properties and expression profiles (Imamura et al., 1999).

Various phytohormone signals are considered to be transduced by this system. In addition to ethylene, the cytokinin signal has recently been found to use the His-Asp phosphorelay. *CRE1*, a gene encoding a receptor His-protein kinase was isolated as a cytokinin receptor (Inoue et al., 2001). An important step in phosphorelay signaling for cytokinin was identified in maize (Sakakibara et al., 1998, 1999) and *Arabidopsis* (Brandstatter and Kieber, 1998; Taniguchi et al., 1998). This involves the response regulator genes that are involved in early cytokinin recognition. In maize *ZmRR1* and *ZmRR2* are primarily expressed in leaves. These genes are cytokinin-responsive in detached leaves of N-starved plants but they are predominantly N-responsive in the whole plant (Sakakibara et al., 1998, 1999). Similar expression patterns can be seen in the type-A response regulator (Taniguchi et al., 1998). The close relationship between cytokinin and N response implies that cytokinin is an internal signal of N availability. It is possibly a root-to-leaf signal involved in the expression of inorganic N-responsive genes. Also, cytokinins (which are considered to be synthesized in roots) are known to accumulate in roots in response to N availability (Takei et al., 2000). The detailed study of the N-responsive accumulation of cytokinins in roots, xylem sap, and leaves of N-starved maize plants has revealed that isopentenyladenine 5′-monophosphate, an initial product of cytokinin metabolism, accumulates in roots within the first 2 h following treatment. This precedes accumulation of *t*-zeatin riboside-5′-monophosphate, *t*-zeatin riboside (ZR) and *t*-zeatin (Z). In the xylem, both the exudation rate and the concentration of cytokinin increase in response to N, and ZR is the dominant form of cytokinin. In leaf tissue, Z accumulates as the dominant form of cytokinin. It starts to increase 4 h after nitrate is supplied to N-deficient plants and an enhanced level is maintained for at least 24 h. This evidence suggests that cytokinin is transported from the roots to the shoots in response to N re-supply, and that Z and/or its derivatives trigger the induction of *ZmRRs*.

The N-responsiveness of cytokinin accumulation and transport and the induction of response regulators allows us to depict a scheme for the sensing of inorganic N and the transduction of the resultant signal through the His-Asp phosphorelay system to modify the transcription of N-responsive genes (Fig. 2). The scheme can be extended to all plants. For example, N-responsive accumulation of cytokinin has been observed in the roots of *A. thaliana* (T. Takahashi et al., unpublished) and barley (Samuelson and Larsson, 1993). In this model Gln is considered as a metabolic signal of external N-availability that acts in concert with cytokinin. The function of this parameter of N nutrition is considered to be primarily in the stabilization of mRNA. The fact that accumulation of *C4Ppc1* mRNA has an absolute requirement for Gln (Sugiharto et al., 1992b) supports this view although the mechanism by which this is achieved remains to be resolved.

The following issues that remain to be resolved are raised by the sensing model: (1) What is the receptor that perceives cytokinin and phosphorelays the hormone signal to the response regulators/HPt? To date, several sensor kinase genes, whose function is yet to be determined, have been identified in plants including *Arabidopsis* (T. Mizuno, personal communication) and maize (H. Sakakibara, unpublished data). A prime candidate for such a receptor, called CKI1, has been identified in *Arabidopsis*. (2) What component(s) function down stream of the response regulators/HPt? The answer to this question will lead to a much improved understanding of the physiological and biochemical regulation of the His-Asp phosphorelay signaling system in plants. (3) What are the target genes for the N-sensing phosphorelay?

In conclusion, there is an increasing body of information that provides evidence for the existence of a large number of plant genes that are regulated by both cytokinin and N (Sakakibara et al., 2000). These include photosynthesis genes and genes that are involved in abundant N-requiring processes, such as *C4ppc1*, *C4Ca*, *rbcS* and *cab*, *cycD3*, and *polI*. These are important candidate targets for the analysis of the mechanism of cytokinin-mediated His-Asp phosphorelay signaling systems in plants.

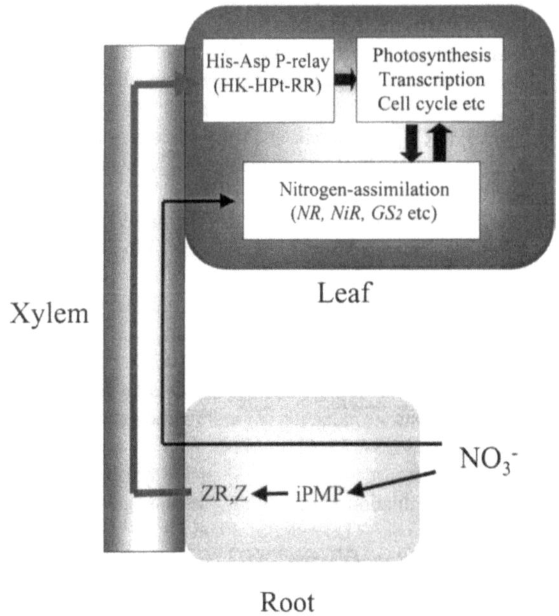

Fig. 2. A current scheme for inorganic nitrogen signal transduction from root to shoot. iPMP, isopentenyladenosine 5′-monophosphate; HK, His-protein kinase; RR, response regulator. Other abbreviations are described in the text.

Acknowledgments

Work in the authors' laboratories was supported by Grant-in-aid for Scientific Research on Priority Areas (09274101 and 09274102 to TS) from the Ministry of Education, Science and Culture, Japan.

References

Aoki H and Ida S (1994) Nucleotide sequence of a rice root ferredoxin-NADP$^+$ reductase cDNA and its induction by nitrate. Biochim Biophys Acta 1183: 552–556

Arp WJ, Mierlo JEM, Berendse F and Snijders E (1998) Interactions between elevated carbon dioxide concentration, nitrogen and water: Effects on growth and water use of six perennial plant species. Plant Cell Environ 21: 1–11

Bennett MJ, Lightfoot DA and Cullimore JV (1989) cDNA sequence and differential expression of the gene encoding the glutamine synthetase γ-polypeptide of *Phaseolus vulgaris* L. Plant Mol Biol 12: 553–565

Brandstatter I and Kieber JJ (1998) Two genes with similarity to bacterial response regulators are rapidly and specifically induced by cytokinin in *Arabidopsis*. Plant Cell 10: 1009–1019

Brown RH (1978) A difference in N use efficiency in C_3 and C_4 plants and its implications in adaptation and evolution. Crop Sci 18: 93–98

Callaci JJ and Smarrelli JJ (1991) Regulation of the inducible nitrate reductase isoform from soybeans. Biochim Biophys Acta 1008: 127–130

Champigny M-L and Foyer C (1992) Nitrate activation of cytosolic protein kinases diverts photosynthetic carbon from sucrose to amino acid biosynthesis. Plant Physiol 100: 7–12

Cheng C-L, Dewdney J, Kleinhofs A and Goodman HM (1986) Cloning and nitrate induction of nitrate reductase mRNA. Proc Natl Acad Sci USA 75: 6110–6114

Chevalier C, Bourgeois E, Just D and Raymond P (1996) Metabolic regulation of asparagine synthetase gene expression in maize (*Zea mays* L.) root tips. Plant J 9: 1–11

Chollet R, Vidal J and O'Leary MH (1996) Phosphoenolpyruvate carboxylase: A ubiquitous, highly regulated enzyme in plants. Annu Rev Plant Physiol Plant Mol Biol 47: 273–298

Cock JM, Mould RM, Bennett MJ and Cullimore JV (1990) Expression of glutamine synthetase genes in roots and nodules of *Phaseolus vulgaris* following changes in the ammonium supply and infection with various *Rhizobium* mutants. Plant Mol Biol 14: 549–560

Crawford NM (1995) Nitrate: Nutrient and signal for plant growth. Plant Cell 7: 859–868

Crawford NM, Campbell WH and Davis RW (1986) Nitrate reductase from squash: cDNA cloning and nitrate regulation. Proc Natl Acad Sci USA 83: 8073–8076

Crété P, Caboche M and Meyer C (1997) Nitrite reductase expression is regulated at the post-translational level by nitrogen source in *Nicotiana plumbaginifolia* and *Arabidopsis thaliana*. Plant J 11: 625–634

Davis KM and King GA (1993) Isolation and characterization of a cDNA clone for a harvest induced asparagine synthetase from *Asparagus officinalis* L. Plant Physiol 102: 1337–1340

Deng M-D, Moureaux T, Cherel I, Boutin J-P and Caboche M (1991) Effects of nitrogen metabolites on the regulation and circadian expression of tobacco nitrate reductase. Plant Physiol Biochem 29: 239–247

Doddema H and Telkamp GP (1979) Uptake of nitrate by mutants of *Arabidopsis thaliana* disturbed in uptake or reduction of nitrate. Physiol Plant 45: 332–338

Emes MJ and Fowler MW (1983) The supply of reducing power for nitrite reduction in plastids of seedling pea roots (*Pisum sativum* L.). Planta 158: 97–102

Forde BG and Clarkson DT (1999) Nitrate and ammonium nutrition of plants: Physiological and molecular perspectives. Advances in Botanical Research, Vol 30, pp 1–90, Academic Press, New York

Foyer CH, Noctor G, Lelandais M, Lescure JC, Vadadier MH, Boutin JP and Horton P (1994) Short-term effects of nitrate, nitrite and ammonium assimilation on photosynthesis, carbon partitioning and protein phosphorylation in maize. Planta 192: 211–220

Givan CV, Joy KW and Kleczkowski LA (1988) A decade of photorespiratory nitrogen cycling. Trends Biochem Sci 13: 433–437

Glass ADM, Shaff JE and Kochian LV (1992) Studies of the uptake of nitrate in barley. Plant Physiol 99: 456–463

Goodwin TW (1980) The Biochemistry of the Carotenoids. Chapman and Hall, New York

Gowri G, Kenis JD, Ingemarsson B, Redinbaugh MG and Campbell WH (1992) Nitrate reductase transcript is expressed in the primary response of maize to environmental nitrate. Plant Mol Biol 18: 55–64

Goyal SS and Huffaker RC (1986) A novel approach and a fully automated microcomputer-based system to study kinetics of NO_3^-, NO_2^-, and NH_4^+ transport simultaneously by intact wheat seedlings. Plant Cell Environ 9: 209–215

Hirel B, Bouet C, King B, Layzell D, Jacobs F and Verma DPS (1987) Glutamine synthetase genes are regulated by ammonia provided externally or by symbiotic nitrogen fixation. EMBO J 5: 1167–1171

Hoff T, Truong H-N and Caboche M (1994) The use of mutants and transgenic plants to study nitrate assimilation. Plant Cell Environ 17: 489–506

Huber JL, Huber SC, Campbell WH and Redinbaugh MG (1992) Reversible light/dark modulation of spinach leaf nitrate reductase activity involves protein phosphorylation. Arch Biochem Biophys 296: 58–65

Huber SC and Huber JL (1996) Role and regulation of sucrose-phosphate synthase in higher plants. Annu Rev Plant Physiol Plant Mol Biol 47: 431–444

Huppe HC and Turpin DH (1994) Integration of carbon and nitrogen metabolism in plant and algal cells. Annu Rev Plant Physiol Plant Mol Biol 45: 577–607

Imamura A, Hanaki N, Nakamura A, Suzuki T, Taniguchi M, Kiba T, Ueguchi C, Sugiyama T and Mizuno T (1999) Compilation and characterization of *Arabidopsis thaliana* response regulators implicated in His-Asp phosphorelay signal transduction. Plant Cell Physiol 40: 733–742

Inoue T, Higuchi M, Hashimoto Y, Seki M, Kobayashi M, Kato T, Tabata S, Shinozak K and Kokimoto T (2001) Identification of CRE1 as a cytokinin receptor from *Arabidopsis*. Nature 409: 1060–1063

Joy KW, Ireland RJ and Lea PJ (1983) Asparagine synthesis in pea leaves, and the occurrence of an asparagine synthetase inhibitor. Plant Physiol 73: 165–168

Kaiser WM and Huber SC (1994) Posttranslational regulation of nitrate reductase in higher plants. Plant Physiol 106: 817–821

Kamachi K, Yamaya T, Mae T and Ojima K (1991) A role for glutamine synthetase in the remobilization of leaf nitrogen during natural senescence in rice leaves. Plant Physiol 96: 411–417

Kawakami N and Watanabe A (1988) Senescence-specific increase in cytosolic glutamine synthetase and its mRNA in radish cotyledons. Plant Physiol 88: 1430–1434

Kleinhofs A and Warner RL (1990) Advances in nitrate assimilation. In: Miflin BJ and Lea PJ (eds) The Biochemistry of Plants, Vol 16, pp 89–120, Academic Press, San Diego

Koch KE (1996) Carbohydrate-modulated gene expression in plants. Annu Rev Plant Physiol Plant Mol Biol 47: 509–540

Koch KE (1997) Molecular crosstalk and the regulation of C- and N-responsive genes. In: Foyer C and Quick P (eds) A Molecular Approach to Primary Metabolism in Plants, pp 105–124, Taylor and Francis Inc. London

Krapp A, Hofmann B, Schäfer C and Stitt M (1993) Regulation of the expression of rbcS and other photosynthetic genes by carbohydrates: A mechanism for the 'sink regulation' of photosynthesis? Plant J 3: 817–828

Lam H-M, Peng SS-Y and Coruzzi GM (1994) Metabolic regulation of the gene encoding glutamine-dependent asparagine synthetase in *Arabidopsis thaliana*. Plant Physiol 106: 1347–1357

Lam H-M, Coschigano K, Schultz C, Melo-Oliveira R, Tjaden G, Oliveira I, Ngai N, Hsieh M-H and Coruzzi G (1995) Use of *Arabidopsis* mutants and genes to study amide amino acid biosynthesis. Plant Cell 7: 887–898

Lauter F-R, Ninnemann O, Bucher M, Riesmeier JW and Frommer WB (1996) Preferential expression of an ammonium transporter and of two putative nitrate transporters in root hairs of tomato. Proc Natl Acad Sci USA 93: 8139–8144

Leustek T, Smith M, Murillo M, Singh DP, Smith AG, Woodcock SC, Awan SJ and Warren MJ (1997) Siroheme biosynthesis in higher plants: Analysis of an S-adenosyl-L-methionine-dependent uroporphyrinogen III methyltransferase from *Arabidopsis thaliana*. J Biol Chem 272: 2744–2752

Lin Y and Cheng C-L (1997) A chlorate-resistant mutant defective in the regulation of nitrate reductase gene expression in *Arabidopsis* defines a new *HY* locus. Plant Cell 9: 21–35

Lu J-L, Ertl JR and Chen C-M (1992) Transcriptional regulation of nitrate reductase mRNA levels by cytokinin-abscisic acid interactions in etiolated barley leaves. Plant Physiol 98: 1255–1260

MacKintosh RW and MacKintosh C (1993) Regulation of plant metabolism by reversible protein (serine/threonine) phosphorylation. In: Battey NH, Dickinson HG and Hetherington AM (ed) Post-Translational Modifications in Plants, pp 197–212. Cambridge University Press

Marschner M (1995) Mineral Nutrition of Higher Plants. Academic Press, London

Matsumura T, Sakakibara H, Nakano R, Kimata Y, Sugiyama T and Hase T (1997) A nitrate-inducible ferredoxin in maize roots: Genomic organization and differential expression of two non-photosynthetic ferredoxin isoproteins. Plant Physiol 114: 653–660

McNally SF, Hirel B, Gadal P, Mann F and Stewart GR (1983) Glutamine synthetases of higher plants. Plant Physiol 72: 22–25

Melzer JM, Kleinhofs A and Warner RL (1989) Nitrate reductase regulation: Effects of nitrate and light on nitrate reductase mRNA accumulation. Mol Gen Genet 217: 341–346

Miao G-H, Hirel B, Marsolier MC, Ridge RW and Verma DPS (1991) Ammonia-regulated expression of a soybean gene encoding cytosolic glutamine synthetase in transgenic *Lotus corniculatus*. Plant Cell 3: 11–22

Miflin BJ and Lea PJ (1980) Ammonia Assimilation. In: Walker JM (ed) The Biochemistry of Plants, Vol 5, pp 169–202, Academic Press, New York

Migge A and Becker TW (1996) In tobacco leaves, the genes encoding the nitrate-reducing or the ammonium-assimilating enzymes are regulated differently by external nitrogen-sources. Plant Physiol Biochem 34: 665–671

Mizuno T (1998) His-Asp phosphotransfer signal transduction. J Biochem (Tokyo) 123: 555–563

Nimmo HG (1993) The regulation of phosphoenolpyruvate carboxylase by reversible phosphorylation. In: Battey NH, Dickinson HG and Hetherington AM (ed) Post-Translational Modifications in Plants, pp 161–170. Cambridge University Press

Ohnishi J and Kanai R (1983) Differentiation of photorespiratory activity between mesophyll and bundle sheath cells of C_4 plants I. Glycine oxidation by mitochondria. Plant Cell Physiol 24: 1411–1420

Plumley FG and Schmidt GW (1989) Nitrogen-dependent regulation of photosynthetic gene expression. Proc Natl Acad Sci USA 86: 2678–2682

Quesada A, Krapp A, Trueman LJ, Daniel-Vedèle F, Fernandez E, Forde BG and Caboche M (1997) PCR-identification of a *Nicotiana plumbaginifolia* cDNA homologous to the high-affinity nitrate transporters of the crnA family. Plant Mol Biol 34: 265–274

Redinbaugh MG and Campbell WH (1991) Higher plant responses to environmental nitrate. Physiol Plant 82: 640–650

Redinbaugh MG and Campbell WH (1993) Glutamine synthetase and ferredoxin-dependent glutamate synthase expression in the maize (*Zea mays*) root primary response to nitrate. Plant Physiol 101: 1249–1255

Rhiel E, Krupinska K and Wehrmeyer W (1986) Effects of nitrogen starvation on the function and organization of the photosynthetic membranes in *Cryptomonas maculata* (Crypophyceae). Planta 169: 361–369

Ritchie SW, Redinbaugh MG, Shiraishi N, Vrba JM and Campbell WH (1994) Identification of a maize root transcript expressed in the primary response to nitrate: Characterization of a cDNA with homology to ferredoxin-$NADP^+$ oxidoreductase. Plant Mol Biol 26: 679–690

Sakakibara H, Kawabata S, Hase T and Sugiyama T (1992a) Differential effects of nitrate and light on the expression of glutamine synthetases and ferredoxin-dependent glutamate synthase in maize. Plant Cell Physiol 33: 1193–1198

Sakakibara H, Kawabata S, Takahashi H, Hase T and Sugiyama T (1992b) Molecular cloning of the family of glutamine synthetase genes from maize: Expression of genes for glutamine synthetase and ferredoxin-dependent glutamate synthase in photosynthetic and non-photosynthetic tissues. Plant Cell Physiol 33: 49–58

Sakakibara H, Shimizu H, Hase T, Yamazaki Y, Takao T, Shimonishi Y and Sugiyama T (1996a) Molecular identification and characterization of cytosolic isoforms of glutamine synthetase in maize roots. J Biol Chem 271: 29561–29568

Sakakibara H, Takei K and Sugiyama T (1996b) Isolation and characterization of a cDNA that encodes maize uroporphyrinogen III methyltransferase involved in the synthesis of siroheme, which is a prosthetic group of nitrite reductase. Plant J 10: 883-892

Sakakibara H, Kobayashi K, Deji A and Sugiyama T (1997) Partial characterization of signalling pathway of nitrate-dependent expression of the genes for nitrogen-assimilatory enzymes using detached maize leaves. Plant Cell Physiol 38: 837–843

Sakakibara H, Suzuki M, Takei K, Deji A, Taniguchi M and Sugiyama T (1998) A response-regulator homolog possibly involved in nitrogen signal transduction mediated by cytokinin in maize. Plant J 14: 337–344

Sakakibara H, Hayakawa A, Deji A, Gawronski S and Sugiyama T (1999) His-Asp phosphotransfer possibly involved in a nitrogen signal transduction mediated by cytokinin in maize: Molecular cloning of cDNAs for two-component regulatory factors and demonstration of phosphotransfer activity in vitro. Plant Mol Biol 41: 563–573

Sakakibara H, Taniguchi M and Sugiyama T (2000) His-Asp phosphorelay signaling: A communication avenue between plants and their environment. Plant Mol Biol 42: 273–278

Samuelson ME and Larsson C-M (1993) Nitrate regulation of zeatin riboside levels in barley roots: Effects of inhibitors of N assimilation and comparison with ammonium. Plant Sci 93: 77–84

Saux, C., Lemoine Y, Marion-Poll A, Valadier MH, Deng M and Morot-Gaudry JF (1987) Consequence of absence of nitrate reductase activity on photosynthesis in *Nicotiana plumbaginifolia* plants. Plant Physiol 84: 67–72

Scheible W-R, Lauerer M, Schulze E-D, Caboche M and Stitt M (1997) Accumulation of nitrate in the shoot act as a signal to regulate shoot-root allocation in tobacco. Plant J 11: 671–691

Sheen J (1990) Metabolic repression of transcription in higher plants. Plant Cell 2: 1027–1038

Sheen J (1994) Feedback-control of gene-expression. Photosynth Res 39: 427–438

Shiraishi N, Sato T, Ogura N and Nakagawa H (1992) Control by glutamine of the synthesis of nitrate reductase in cultured spinach cells. Plant Cell Physiol 33: 727–731

Siddiqi MY, Glass ADM, Ruth TJ and Rufty TW (1990) Studies of the uptake of nitrate in barley. Plant Physiol 93: 1426–1432

Solomonson LP and Barber MJ (1990) Assimilatory nitrate reductase: Functional properties and regulation. Annu Rev Plant Physiol Plant Mol Biol 41: 225–253

Son D and Sugiyama T (1992) Molecular cloning of an alanine aminotransferase from NAD-malic enzyme type-C_4 plant *Panicum miliaceum*. Plant Mol Biol 20: 705–713

Son D, Jo J and Sugiyama T (1991) Purification and characterization of alanine aminotransferase from *Panicum miliaceum* leaves. Arch Biochem Biophys 289: 262–266

Son D, Kobe A and Sugiyama T (1992) Nitrogen-dependent regulation of the gene for alanine aminotransferase which is involved in the C_4 pathway of *Panicum miliaceum*. Plant Cell Physiol 33: 507–509

Stitt M (1999) Nitrate regulation of metabolism and growth. Curr Opin Plant Biol 2: 178–186

Stitt M and Krapp A (1999) The interaction between elevated carbon dioxide and nitrogen nutrition: The physiological and molecular background. Plant Cell Environ 22: 583–621

Sugiharto B and Sugiyama T (1992) Effects of nitrate and ammonium on gene expression of phospho*enol*pyruvate carboxylase and nitrogen metabolism in maize leaf tissue during recovery from nitrogen stress. Plant Physiol 98: 1403–1408

Sugiharto B, Miyata K, Nakamoto H, Sasakawa H and Sugiyama T (1990) Regulation of expression of carbon-assimilating enzymes by nitrogen in maize leaf. Plant Physiol 92: 963–969

Sugiharto B, Burnell JN and Sugiyama T (1992a) Cytokinin is required to induce the nitrogen-dependent accumulation of mRNAs for phospho*enol*pyruvate carboxylase and carbonic anhydrase in detached maize leaves. Plant Physiol 100: 153–156

Sugiharto B, Suzuki I, Burnell JN and Sugiyama T (1992b) Glutamine induces the N-dependent accumulation of mRNAs encoding phospho*enol*pyruvate carboxylase and carbonic anhydrase in detached maize leaf tissue. Plant Physiol 100: 2066–2070

Sugiyama T (1998) Nitrogen-responsive expression of C_4 photosynthesis genes in maize. In: Satoh K and Murata N (ed) Stress Responses of Photosynthetic Organisms, pp 167–180. Elsevier, Amsterdam

Sugiyama T, Mizuno M and Hayashi M (1984) Partitioning of nitrogen among ribulose-1,5-bisphosphate carboxylase, phospho*enol*pyruvate carboxylase and pyruvate, orthophosphate dikinase as related to biomass productivity in maize seedlings. Plant Physiol 75: 665–669

Suzuki I, Cretin C, Omata T and Sugiyama T (1994) Transcriptional and posttranscriptional regulation of nitrogen-responding expression of phospho*enol*pyruvate carboxylase gene in maize. Plant Physiol 105: 1223–1229

Takei K, Sakakibara H, Taniguchi M and Sugiyama T (2001) Nitrogen-dependent accumulation of cytokinins in root and the translocation to leaf: Implication of cytokinin species that induces gene expression of maize response regulator. Plant Cell Physiol 42: 85–93

Taniguchi M, Sawaki H, Sasakawa H, Hase T and Sugiyama T (1992) Cloning and sequence analysis of cDNA encoding aspartate aminotransferase isozymes from *Panicum miliaceum* L. a C_4 plant. Eur J Biochem 204: 611–620

Taniguchi M, Kobe A, Kato M and Sugiyama T (1995) Aspartate aminotransferase isozymes in *Panicum miliaceum* L., an NAD-malic enzyme-type C_4 plant: Comparison of enzymatic properties, primary structures, and expression patterns. Arch Biochem Biophys 318: 295–306

Taniguchi M, Kiba T, Sakakibara H, Ueguchi C, Mizuno T and Sugiyama T (1998) Expression of *Arabidopsis* response regulator homologs is induced by cytokinins and nitrate. FEBS Lett 429: 259–262

Terashima I and Evans JR (1988) Effects of light and nitrogen nutrition on the photosynthetic apparatus in spinach. Plant Cell Physiol 29:143–155

Trueman LJ, Onyeocha I and Forde BG (1996a) Recent advances in the molecular biology of a family of eukaryotic high affinity nitrate transporters. Plant Physiol Biochem 34: 621–627

Trueman LJ, Richardson A and Forde BG (1996b) Molecular cloning of higher plant homologues of the high-affinity nitrate transporters of *Chlamydomonas reinhardtii* and *Aspergillus nidulans*. Gene 175: 223–231

Tsai F-Y and Coruzzi GM (1990) Dark-induced and organ-specific expression of two asparagine synthetase genes in *Pisum sativum*. EMBO J 9: 323–332

Tsay YF, Schroeder JI, Feldmann KA and Crawford NM (1993) The herbicide sensitivity gene *CHL1* of *Arabidopsis* encodes a nitrate-inducible nitrate transporter. Cell 72: 705–713

Urquhart AA and Joy KW (1981) Use of phloem exudate technique in the study of amino acid transport in pea plants. Plant Physiol 68: 750–754

Vézina LPV and Langlois JR (1989) Tissue and cellular distribution of glutamine synthetase in roots of pea (*Pisum sativum*) seedlings. Plant Physiol 90: 1129–1133

Vincentz M, Moureaux T, Leydecker M-T, Vaucheret H and Caboche M (1993) Regulation of nitrate and nitrite reductase expression in *Nicotiana plumbaginifolia* leaves by nitrogen and carbon metabolites. Plant J 3: 315–324

Yamazaki M, Watanabe A and Sugiyama T (1986) Nitrogen-regulated accumulation of mRNA and protein for photosynthetic carbon-assimilating enzymes in maize. Plant Cell Physiol 27: 443-452

Chapter 15

Intracellular And Intercellular Transport Of Nitrogen And Carbon

Gertrud Lohaus*
Albrecht-von-Haller Institut für Pflanzenwissenschaften, Biochemie der Pflanze, Untere Karspüle 2, 37073 Göttingen, Germany

Karsten Fischer
Botanisches Institut der Universität zu Köln, Lehrstuhl II, Gyrhofstr. 15, D-50931 Köln, Germany

Summary 239
I. Introduction 240
II. Transport Processes of Plastids 240
A. Export of Fixed Carbon from Chloroplasts 240
B. Import of Carbon into Starch Storing Plastids 242
C. Transport Processes Involved in Amino Acid Biosynthesis 245
III. Transport Processes Involved in Phloem Loading 246
A. Transport from the Mesophyll to the Vicinity of the Phloem 247
B. Composition and Concentrations of the Exported Carbon and Nitrogen Compounds 247
C. Models of Phloem Loading 251
1. Apoplastic Phloem Loading 251
2. Symplastic Phloem Loading 256
3. Loading of Sugar Alcohols 257
IV. Concluding Remarks 257
Acknowledgments 257
References 258

Summary

Partitioning of carbon (C) and nitrogen (N) assimilates and export of photoassimilates play an essential role in efficient growth and reproductive success of the plant as well as in crop yield. Sink (net importing) organs need to be supplied with energy and fixed C from the source (net exporting) organs of the plant, e.g. green leaves. During the day, the triose phosphate/phosphate translocator located in the inner membrane of chloroplast envelopes catalyzes the export of triose phosphates, the main product of photosynthesis, to the cytosol of the plant cell where they are used in sucrose synthesis. Some sucrose is stored in source tissues, but the bulk is exported. Sucrose is the major form of exported C from leaves. When the rates of sucrose synthesis and export fall behind that of CO_2 fixation, fixed C is retained in the chloroplasts and directed into the synthesis of

*Author for correspondence, Email: glohaus@gwdg.de.

Christine H. Foyer and Graham Noctor (eds): Photosynthetic Nitrogen Assimilation and Associated Carbon and Respiratory Metabolism, pp. 239–263.

transitory starch. At night, starch is degraded to glucose that is exported from chloroplasts via a glucose transporter. Triose phosphates also provide skeletons for amino acid synthesis. From the source organs organic C and N metabolites are transported via the phloem to sink organs. The most abundant sugar in the phloem sap of several plant species is sucrose, with concentrations being about 1 M. Total amino acid concentrations are between 50 and 500 mM. Two principal routes for the delivery of metabolites into the sieve-element-companion cell complex (SE-CCC) have been proposed. These are (i) transporter-mediated export from mesophyll cells, diffusion through the apoplast, and subsequent transporter-mediated uptake into the SE-CCC, and (ii) direct symplastic cell-to-cell diffusion via plasmodesmata. Several sucrose and amino acid transporters have been cloned which mediate the uptake of the photoassimilates from the apoplast into the symplast. This chapter gives an overview of the current state of knowledge on the functions of intracellular and intercellular metabolite transport in leaves.

I. Introduction

In plants the only pathway of net CO_2 fixation is the reductive pentose phosphate (RPP) pathway. In the RPP pathway ATP and NADPH from the light reactions of the photosynthetic membranes are used to reduce CO_2 to carbohydrate. This pathway quantitatively represents one of the most important biosynthetic routes since about 120 billion tons of CO_2 are converted into organic substances by higher plants annually. The RPP pathway and a major part of nitrate assimilation are confined to chloroplasts of mesophyll cells of leaves, stems and siliques. Leaves that fix more CO_2 and nitrate than they need for metabolism and export surplus photoassimilates to sink organs are called 'source' leaves. Organs that depend on the photoassimilate import and also, at least in part, on exported nitrogen (N) metabolites from the source organs, are called 'sink' tissue. These include growing leaves, flowers, seeds or roots. The production and transport of carbon (C) and N compounds involves a cooperation of various cell types and requires several transport steps across membranes, e.g. for the export of assimilates from the plastids to the cytosol of the source cells, and for the transfer from the source cells into the phloem. In this review we focus on metabolite transporters located in the inner envelope membrane of plastids and on processes involved in phloem loading.

Abbreviations: 2-PGA – 2-phosphoglycerate; 3-PGA – 3-phosphoglycerate; ADP-Glc – ADP glucose; AGPase – ADP glucose pyrophosphorylase; C_3 – three carbon; C_4 – four carbon; CCCP – carbonyl cyanide *m*-chlorophenyl hydrazone; D-Glc – D-glucose; DIT1 – oxoglutarate/malate translocator; D-Man – D-mannose; E4P – erythrose 4-P; FBPase – fructose 1,6-bisphosphatase; Fru2,6bP – fructose 2,6-bisphosphate; Glc1P – glucose 1-phosphate; Glc6P – glucose 6-phosphate; GPT – Glc6P/phosphate translocator; OAA – oxaloacetate; OPPP – oxidative pentose phosphate pathway; PCMBS – p-chloromercuribenzenesulfonic acid; PEP – phosphoenolpyruvate; PGI – phosphoglucoisomerase; pGlcT – plastidic glucose translocator; PGM – phosphoglucomutase; P_i – Inorganic phosphate; PPT – PEP/phosphate translocator; R5P – ribose 5-P; RPP – reductive pentose phosphate (RPP pathway = Calvin cycle); SE-CCC – sieve-element-companion cell complex; SEL – size exclusion limit; TP – triose phosphate; TPT – triosephosphate/phosphate translocator

II. Transport Processes of Plastids

A. Export of Fixed Carbon from Chloroplasts

The first product of CO_2 fixation in C_3 plants is 3-phosphoglycerate (3-PGA) that is reduced to triose phosphates (TPs) in the stroma. TPs serve as substrates for starch and fatty acid synthesis, both exclusively located in the chloroplasts. They are also the substrates for cytosolic sucrose biosynthesis and provide the C skeletons for amino acid biosynthesis. Part of the TPs has to be transported from the stroma to the cytosol. Initially, Baldry et al. (1966) and later Cockburn et al. (1967) and Bassham et al. (1968) demonstrated with isolated spinach chloroplasts that TPs and 3-PGA are released from the organelles and that photosynthesis depends on exogenous P_i. Following these observations, the transport processes between isolated chloroplasts and the surrounding medium were measured directly, showing that the above mentioned substrates are transported by the same protein (Heldt and Rapley, 1970; Fliege et al., 1978). This protein, known as the triose phosphate/phosphate translocator (TPT; Flügge et al., 1989, 1996), transports P_i, TPs and 3-PGA in a strict counterexchange.

cDNAs encoding the TPT from different C_3- and C_4-plants have been isolated (Flügge et al., 1989; Willey et al., 1991; Fischer et al., 1994) and the protein from spinach has been expressed in

Schizosaccharomyces pombe (Loddenkötter et al., 1993). Analysis of the transport properties revealed identical substrate specificities of the recombinant TPT compared with the authentic plant protein.

The main function of the TPT is to facilitate export of fixed C in the form of TPs from the chloroplast (Fig. 1). The phosphate released during synthesis of sucrose and amino acids is shuttled back from the cytosol via the TPT into chloroplasts, thus keeping the total pool of P_i and phosphorylated compounds constant (Flügge and Heldt, 1984).

The physiological function of the TPT was further confirmed by analysis of transgenic potato and tobacco plants with a reduced activity of this transporter (Riesmeier et al., 1993a; Häusler et al., 1998). The reduced transport activity results in increased accumulation of transitory starch during the day but also causes enhanced degradation of starch at night (Heineke et al., 1994). In tobacco, this alteration in carbohydrate metabolism is accompanied by a rise in the transport capacity for glucose (Häusler et al., 1998). Similarly, transgenic plants that overexpress the TPT show higher sucrose synthesis in the light and lower starch turnover (Häusler et al., 2000). Obviously, plants can cope with the deficiency in TPT activity by exporting the assimilated C via a glucose carrier at night (see below).

In light, a significant amount of the fixed C is retained in chloroplasts for the synthesis of transitory starch, i.e. starch that is synthesized during photosynthesis and degraded in the following dark period (Stitt et al., 1978; Trethewey and Smith, 2000). Transitory starch serves as an overflow for assimilated C when assimilation exceeds the demand for sucrose (Stitt and Quick, 1989). This starch also provides a source of C and energy during the night. Starch degradation proceeds via two different pathways, a phosphorolytic pathway leading to glucose 1-phosphate (Glc1P) and, due to the activity of phosphoglucomutase (PGM), to glucose 6-phosphate (Glc6P) and, secondly, a hydrolytic pathway producing maltose and glucose (Trethewey and Smith, 2000). There is good evidence for glucose as the main degradation product which is exported via a glucose transporter: (i) In C_3 plants, starch degradation proceeds mainly via the amylolytic (hydrolytic)

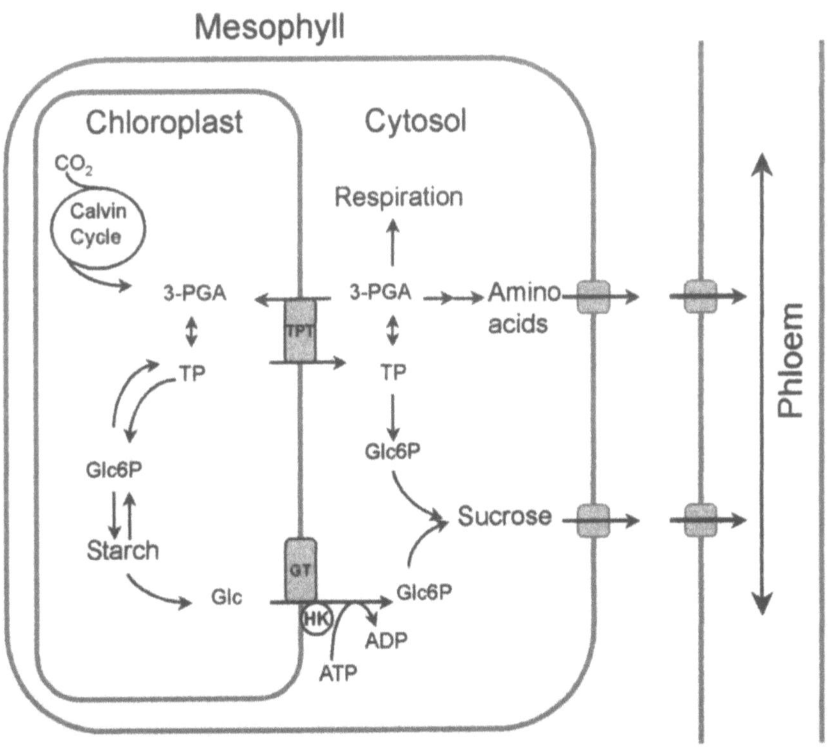

Fig. 1. Export of assimilated carbon from chloroplasts. Glc, glucose; Glc6P, glucose 6-phosphate; GT, glucose transporter; HK, hexokinase; 3-PGA, 3-phosphoglycerate; TP, triose phosphate; TPT, triose phosphate/phosphate translocator. Calvin cycle = reductive pentose phosphate pathway.

pathway (Schleucher et al., 1998; Zeeman et al., 1998). (ii) In leaves of several plants a high concentration of fructose 2,6-bisphosphate (Fru2,6bP) has been found at night, i.e. the synthesis of sucrose from TPs is prevented due to inhibition of the cytosolic fructose 1, 6-bisphosphatase (FBPase; Stitt, 1990). (iii) A glucose transporter has been characterized in spinach chloroplasts that facilitates transport of glucose and other sugars such as xylose and mannose (Schäfer et al., 1977). An *Arabidopsis* mutant, *sex1* (Caspar et al., 1991), characterized by a severely reduced rate of starch degradation, was shown to be deficient in glucose uptake (Trethewey and ap Rees, 1994).

Recently, cDNAs encoding a plastidic glucose translocator (pGlcT) have been obtained from several plants (Weber et al., 2000). A comparison of the amino acid sequences with entries in the databases revealed significant homology with hexose transporters from mammals and bacteria. The plant glucose transporter protein contains an N-terminal presequence directing the protein to the inner envelope of spinach chloroplasts and, in the mature part of the protein, twelve membrane spanning regions, which are typical for most translocator proteins. The pGlcT catalyzes the transfer of D-glucose and D-mannose across the envelope membrane whereas other hexoses (i.e. fructose) or pentoses are not accepted (Weber et al., 2000). Surprisingly, the spinach pGlcT expressed in the *Arabidopsis* mutant *sex1* did not complement the mutant phenotype. Moreover, the pGlcT mapped to a different locus than *sex1* in the *Arabidopsis* genome (Weber et al., 2000). Both findings suggest that *sex1* does not encode the plastidic glucose transporter but a protein with a different function in starch metabolism. Recently, the *sex1* gene has been cloned and analyzed (Yu et al., 2001). The gene encodes the R1 protein that functions as an overall regulator of starch mobilization.

Our present understanding of carbohydrate export from chloroplasts indicates that several pathways exists (Fig. 1): During the day, most of the assimilated C is exported by the TPT in the form of TPs while at night cytosolic sucrose synthesis depends on the export of glucose by the glucose carrier. The glucose is converted to Glc6P by a hexokinase located in the outer envelope (Wiese et al., 1999). This reaction results in a steep concentration gradient for glucose across the chloroplast envelope membrane, which drives efficient export of glucose into the cytosol.

In addition, Glc6P produced through phosphorolytic degradation of starch is converted to TPs either via glycolysis or the oxidative pentose phosphate pathway (OPPP). The resulting TPs are transported into the cytosol by the TPT, thus sustaining dark respiration (Stitt et al., 1985).

A maltose transporter characterized by Rost et al. (1996) could provide another route of carbohydrate export into the cytosol. The authors demonstrated that there is no competition between maltose and glucose transport activities indicating that chloroplasts possess two different transport routes for neutral sugars.

C could potentially also be exported as hexose phosphates but this seems not to be the case. Chloroplasts normally show only very low rates of transport of hexose phosphates (Flügge, 1995). The reason for this is that the Glc6P/phosphate translocator (GPT, Section II.B) that catalyzes the exchange of Glc6P against P_i is not expressed in leaves (Kammerer et al., 1998). In addition, mutants from *Clarkia xantiana* (Jones et al., 1986) and *A. thaliana* (Yu et al., 2000) with moderately (50%) or severely (98%) reduced activities of plastidic phosphogluco isomerase (PGI), which converts Fru6P into Glc6P, exhibit corresponding reductions in their leaf starch content. Evidently, chloroplasts in these plants lack the hexose phosphate transporter that could complement the deficiency of plastidic PGI for starch synthesis. Intriguingly, amyloplasts from roots of these mutants possess a GPT and show normal starch content compared with wild type plants (Yu et al., 2000).

However, a Glc6P transport activity was induced in chloroplasts by feeding leaves with glucose for several days through the petiole (Quick et al., 1995). This induced GPT is involved with import rather than export of carbohydrates for starch synthesis, i.e. this artificial system resembles the metabolism in non-green (sink) tissues but not that of leaves under physiological conditions.

B. Import of Carbon into Starch Storing Plastids

In contrast to chloroplasts, plastids from heterotrophic tissues, e.g. amyloplasts and leukoplasts, rely on the supply of photosynthates synthesized in source tissues. In most plant species, assimilated C is translocated to sink tissues as sucrose (Section III.A), then cleaved by apoplastic invertase and/or by cytosolic sucrose synthase (Sturm and Tang, 1999 and references therein) and converted to hexose phosphates. For many years, the question of what substrate is taken up by heterotrophic plastids, as a precursor for starch synthesis (and as substrate for the OPPP), has been a matter of debate. A first hint came from the enzymic capacities of these organelles. Most non-green plastids contain only low FBPase activities (Journet and Douce, 1985), or have no activity of this enzyme (Entwistle and ap Rees, 1988, 1990; Borchert et al., 1993; Neuhaus et al., 1993b), which is involved in gluconeogenesis. Thus, these organelles are not able to incorporate TPs into starch but, instead, rely on the import of hexose phosphates for starch synthesis. This has been documented in plastids from developing pea embryos (Hill and Smith, 1991), tomato fruit plastids (Büker et al., 1998), plastids from cauliflower buds and maize endosperm (Neuhaus et al., 1993a), wheat endosperm (Keeling et al., 1988; Tyson and ap Rees, 1988; Tetlow et al., 1994) and potato tubers (Hatzfeld and Stitt, 1990; Naeem et al., 1997). One exception to this rule is the case of etioplasts from barley leaves that possess high activities of FBPase. They also use TPs, but not external hexose phosphates, as substrates for starch synthesis (Batz et al., 1992; Neuhaus et al., 1993b).

The first convincing evidence for hexose phosphate uptake into plastids was presented by Borchert et al. (1989). These authors showed that Glc6P was imported by pea root plastids in strict counter exchange for P_i or TPs indicating that a phosphate translocator with substrate specificities different from the TPT was involved in this transport process. Subsequently, a similar transport system was identified in non-green plastids from cauliflower (Flügge, 1995; Möhlmann et al., 1995), potato tubers (Schott et al., 1995), maize endosperm (Flügge, 1995) and sweet pepper (Thom et al., 1998).

The hexose phosphate transporter has been purified from maize endosperm (Flügge, 1995). On the basis of peptide sequences obtained, corresponding cDNA clones have been isolated from maize, pea, potato and cauliflower (Kammerer et al., 1998). Analysis of the deduced protein sequences revealed a low but significant amino acid identity to the TPTs. Thus, these translocators represent a distinct class of the phosphate translocator family. The purified recombinant transporter from pea exhibited a high affinity for Glc6P and P_i and also for TPs and 3-PGA. Interestingly, other hexose phosphates like Glc1P and Fru6P are not accepted by the GPT. Therefore, the biochemical properties of this new class of phosphate translocators are compatible with their proposed physiological functions, that of Glc6P import preceding starch synthesis and as a substrate for the OPPP (Fig. 2). Firstly, P_i released during starch synthesis serves as the substrate for counter exchange (Borchert et al., 1989). Secondly, Glc6P that is fed into the OPPP is converted to TPs, which are subsequently exported via the GPT (Borchert et al., 1989). The main function of the OPPP in plastids is to supply redox equivalents (NADPH) for biosynthetic processes such as nitrite reduction, ammonia assimilation, amino acid and fatty acid synthesis (Bowsher et al., 1989, 1992). As predicted from these physiological functions of GPTs, GPT-specific transcripts are barely detectable in photosynthetic tissues but are abundant in heterotrophic tissues, for example potato tubers, maize kernels and pea roots (Kammerer et al., 1998). Thus, the expression pattern of the GPTs is different to that of the TPTs, which are expressed only in photosynthetic tissues (Schultz et al., 1993). Remarkably, a different situation has been found in plastids of green tomato and pepper fruits. These chloroplasts exhibit photosynthetic activities but also rely to a great extent on the import of carbohydrates for starch synthesis. Intriguingly, two phosphate translocators with overlapping substrate specificities, most probably a TPT and a GPT, have been identified in these tissues (Quick and Neuhaus, 1996; Büker et al., 1998).

Conflicting data have been published with regard to whether Glc1P, exclusively or in addition to Glc6P, is taken up by some non-green plastids. Schünemann and Borchert (1994) provided evidence for Glc1P and Glc6P transport in tomato fruit plastids but later, these authors failed to detect any starch synthesis from Glc1P (Büker et al., 1998). In contrast to data provided by Schott et al. (1995), Naeem et al. (1997) showed that amyloplasts from potato tubers use Glc1P rather than Glc6P to support starch synthesis. In contrast, considerable evidence has been presented that amyloplasts from wheat import Glc1P instead of Glc6P (Tyson and ap Rees, 1988; Tetlow et al., 1994, 1996). However, the molecular nature of this translocator is thus far unresolved. Definitive proof of the identity of the imported compound has been obtained through the analysis of starchless mutant lines from pea (Harrison et al., 1998, 2000), tobacco (Hanson and McHale, 1988) and *Arabidopsis* (Caspar et al., 1986; Kofler et al., 2000). These mutants have

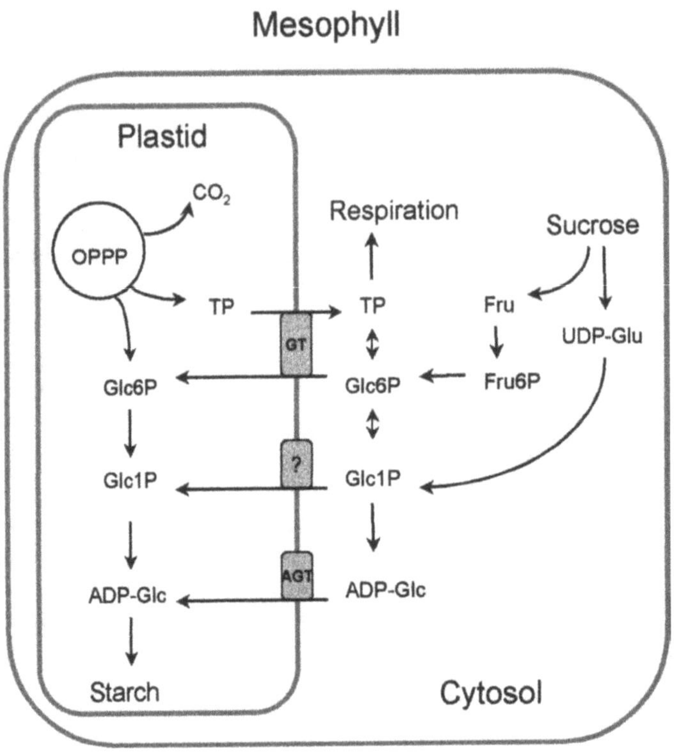

Fig. 2. Import of carbon into starch-storing, heterotrophic plastids. ADP-Glc, ADP-glucose; AGT, ADP-glucose translocator; Fru, fructose; Fru6P, fructose 6-phosphate; Glc1P, glucose 1-phosphate; Glc6P, glucose 6-phosphate; GPT, glucose 6-phosphate/phosphate translocator; GT, glucose transporter; OPPP, oxidative pentose phosphate pathway; TP, triose phosphate; TPT, triose phosphate/phosphate translocator; UDP-Glc, UDP-glucose.

been shown to lack any activity of plastidic PGM, i.e. the enzyme that catalyzes the conversion of both hexose phosphates. In addition, transgenic potato plants with significantly reduced plastidic PGM activity exhibited a dramatic decrease in tuber starch content (Tauberger et al., 2000). These data led to the conclusion that Glc6P is the sole precursor imported for starch synthesis in these plants.

In contrast to pea and *Arabidopsis*, ADP-Glc rather than Glc6P seems to be taken up by amyloplasts as precursor for starch biosynthesis in the monocots maize and barley (Emes and Neuhaus, 1997). ADP-Glc is provided by ADP-glucose pyrophosphorylase (AGPase), a heterotetrameric enzyme composed of two small and large subunits (Preiss, 1991). Multiple isoforms of AGPase have been detected in various plants but their physiological significance remains unclear (Cognata et al., 1995). One reason for the occurrence of these isoforms could be their differential cellular localization. In photosynthetic cells (Okita, 1992), and in heterotrophic tissues from various plants, AGPase is largely (Chen et al., 1998) or exclusively (Kang and Rawsthorne, 1994) confined to the plastids. However, in endosperm from maize (Denyer et al., 1996) and barley (Thorbjørnsen et al., 1996a) 80–90% of total AGPase activity is located in the cytosol while only a residual activity is localized to the stroma. Maize mutants lacking the large (shrunken-2, Sh-2) or the small subunit (brittle-2, bt-2) of AGPase contain only 20% of wild type AGPase activity and show a severe reduction of starch in the kernels (Giroux and Hannah, 1994). Because the proteins encoded by *SH-2* and *BT-2* lack presequences it seems likely that they represent cytosolic isoforms of AGPase (Giroux and Hannah, 1994). In barley, the same gene encodes both the cytosolic and plastidic small subunits. Two different mRNAs are synthesized via an alternative splicing mechanism leading to two different proteins, one bearing a plastidic presequence, the other lacking this signal sequence (Thorbjörnsen et al., 1996b). Thus, in both plants the cytosolic AGPase delivers most of the ADP-Glc that is incorporated into starch. This indicates that maize and barley amyloplasts should be able to import this metabolite. Indeed, such a transport of ADP-Glc has been shown for amyloplasts from sycamore cell cultures (Pozueto-Romero et al., 1991), wheat (Tetlow et al., 1994) and maize (Möhlmann et al., 1997). Most probably, the BT-1 protein, the cDNA of which has been cloned (Sullivan et al., 1991) serves the function of an ADP-Glc translocator. The BT-1 protein shows a significant homology to the mitochondrial translocator family but is localized to the amyloplast inner envelope (Sullivan and Kaneko, 1995). Mutations at the BT-1 locus result in a reduction of starch accumulation and in an increase of the ADP-Glc content of immature endosperm (Shannon et al., 1998). However, direct proof of the identity of BT-1, for example by determining the transport properties of the recombinant protein, is still lacking. Whether in maize both the GPT and BT-1 are delivering precursors for starch synthesis or whether the GPT serves a different function remains to be determined.

C. Transport Processes Involved in Amino Acid Biosynthesis

Plants synthesize and translocate all amino acids (Section III.A) commonly found in proteins and in addition produce hundreds of non-protein amino acids. Chloroplasts are the major site of their synthesis in leaves, particularly of nutritionally essential amino acids (Wallsgrove et al., 1983). These include those of the aspartate family (i.e. threonine, lysine) or the branched chain amino acids like valine. Some steps of methionine and arginine synthesis are located in the cytosol. In addition, the plastid-located shikimate pathway leads to the formation of aromatic amino acids (tyrosine, phenylalanine, tryptophan). Surprisingly, $^{14}CO_2$ incorporation into amino acids by isolated chloroplasts is low (Kirk and Leech, 1972; Bagge and Larson, 1986). This is attributed to the limitation of the chloroplast to synthesize the metabolites used as C skeletons for amino acid synthesis. These are pyruvate, phosphoenolpyruvate (PEP), oxaloacetate (OAA), oxoglutarate, erythrose 4-P (E4P) and ribose 5-P (R5P), i.e. compounds that are part of glycolysis, tricarboxylate cycle and the pentose phosphate pathway. Chloroplasts are able to synthesize E4P and R5P only, whilst the other metabolites have to be taken up from the cytosol due to incomplete plastidic glycolysis (Stitt and ap Rees, 1979; Fig. 3). OAA synthesized from PEP in the cytosol is transported by a high affinity OAA carrier (Hatch et al., 1984) that has not yet been described on the molecular level. Pyruvate enters plastids either by diffusion through the membrane or translocation through a specific pyruvate carrier. Such a pyruvate translocator has been identified in mesophyll chloroplasts from C_4 plants (Huber and Edwards, 1977a; Flügge et al., 1985; Ohnishi and Kanai, 1990). In contrast, chloroplasts from C_3 plants take up pyruvate mainly by diffusion (Proudlove and Thurman, 1981). Also in non-green plastids, evidence for both pyruvate uptake systems has been obtained (Eastmond et al., 1997; Eastmond and Rawsthorne, 2000). Thus, it is likely that only plastids with a demand for high pyruvate transport rates possess a carrier mediated pyruvate uptake system. The molecular nature of this carrier is unknown. Oxoglutarate import in exchange for stromal malate is mediated by an oxoglutarate/malate translocator (DIT1; Weber et al., 1995; Flügge, 2000 and references therein). Most, if not all, plastid types are able to transport PEP across their envelope membranes (Fischer et al., 1997). Chloroplasts from C_4 plants show especially high rates of PEP transport (Huber and Edwards, 1977b; Day and Hatch, 1981). In these plants, the export of PEP is part of C_4 photosynthetic metabolism. Data from Heldt and Rapley (1970) and Fliege et al. (1978) already provided evidence that chloroplasts from C_3 plants also possess a low PEP transport activity. Later it was shown that non-green plastids from different C_3 plants have the capability of a PEP/P_i exchange as well (Borchert et al., 1993; Schünemann and Borchert, 1994; Flügge, 1995; Schott et al., 1995). By means of specific inhibitors of PEP transport, a 30 kD protein was identified as the PEP translocator in maize and *Panicum miliaceum* (Thompson et al., 1987; Ohnishi et al., 1989).

cDNAs from different plants encoding the PEP/phosphate translocator (PPT) have been cloned and sequenced (Fischer et al., 1997). These cDNAs exhibit high homology to each other but only 30% identity to the TPTs and GPTs indicating that the PEP transporter represents a third class of the phosphate translocator family. The recombinant PPT protein mediates transport of PEP and 2-PGA in counter exchange with phosphate, i.e. only C_3 compounds phosphorylated at C-atom 2 are accepted as substrate (Fischer et al., 1997). Thus, the transport characteristics are quite different from the other phosphate translocator classes.

The physiological function of the PPTs in C_3 plants is to supply plastids with PEP, which is fed

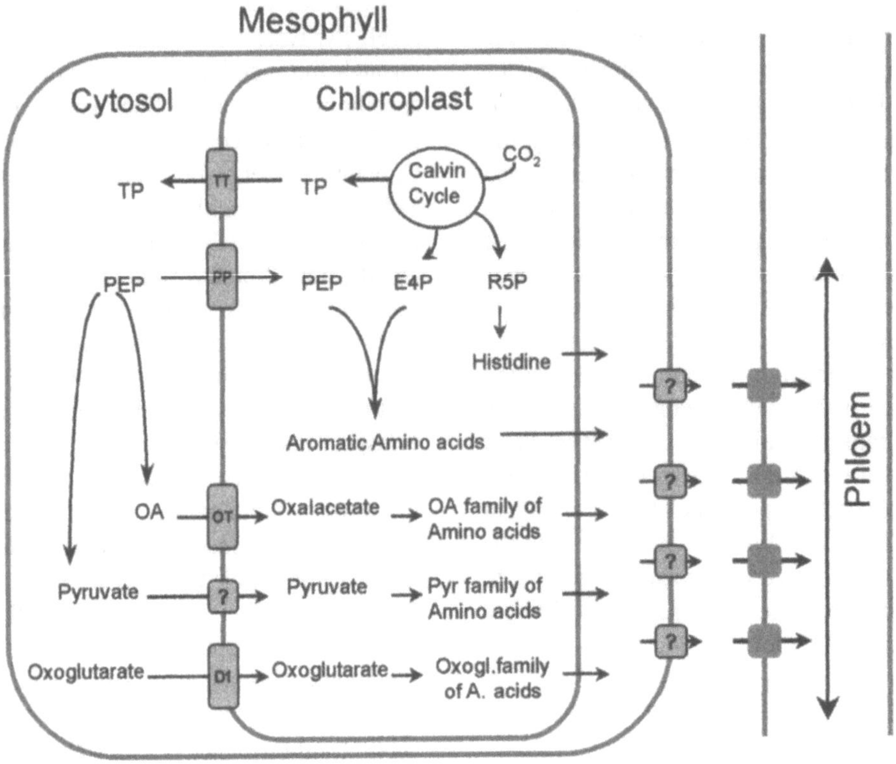

Fig. 3. Transport of carbon skeletons for amino acid biosynthesis. D1, oxoglutarate/malate transporter; E4P, erythrose 4-phosphate; OA, oxaloacetate; OT, oxaloacetate transporter; PEP, phosphoenolpyruvate; PP, phosphoenolpyruvate/phosphate translocator; Pyr, pyruvate; R5P, ribose 5-phosphate; TT, triose phosphate/phosphate translocator (TPT). Calvin cycle = reductive pentose phosphate pathway.

into fatty acid synthesis and, most important, into the shikimate pathway (Bagge and Larsson, 1986). This leads to the synthesis of aromatic amino acids and a large number of secondary metabolites (Schmid and Amrhein, 1995; Herrmann and Weaver, 1999). The proposed physiological function was further validated through analysis of PPT mutants from *Arabidopsis* (Li et al., 1995; Streatfield et al., 1999) showing a reticulate phenotype in which interveinal regions of the leaves are visibly pale, whereas paraveinal regions are green. Intriguingly, secondary metabolites that are synthesized via the shikimate pathway are clearly reduced in these mutants. Furthermore, the reticulate leaf phenotype of these mutants can be rescued by feeding with the three aromatic amino acids together (Streatfield et al., 1999). Thus, the PPT represents an important link between primary and secondary plant metabolism.

The amino acids synthesized in leaf chloroplasts are mainly exported into the cytosol for cytosolic protein synthesis and for allocation to other parts of the plant. Unfortunately, nothing is known about plastidic amino transporters.

III. Transport Processes Involved in Phloem Loading

After synthesis in the source tissues of the plant, the C and N assimilates have to be translocated to the sink tissues. The phloem of higher plants forms an extensive conduit for this long-distance transport of a diverse range of compounds, including metabolites, ions and macromolecules. Several cell types, including sieve elements, companion cells, and phloem parenchyma cells form the phloem. During cell division of their common mother cell, sieve elements and companion cells remain in close contact by numerous pore-plasmodesmata units and behave as a single functional unit, which has led to the term sieve element-companion cell complex (SE-CCC). Recent reviews by Sjölund (1997) and Oparka and Turgeon (1999) should be consulted for additional information about the structure of the phloem.

According to the pressure-flow hypothesis of Münch (1930) long-distance solute movement through the phloem is driven by the pressure gradient between source and sink. This gradient depends on the localized solute accumulation in the source tissues. The term 'phloem loading' describes the active accumulation of solutes against a concentration gradient in the SE-CCC. As a consequence of loading, solute concentration and osmotic pressure are elevated at the source end of the phloem and in this way the solution of the phloem will flow to regions of low pressure.

A. Transport from the Mesophyll to the Vicinity of the Phloem

The transport of sucrose, amino acids, sugar alcohols or other solutes from the mesophyll to the vicinity of the sieve element companion cell complex are expected to be the same in all plant species, independent of the type of phloem loading. Mesophyll cells are highly interconnected with each other and with bundle sheath and vascular parenchyma cells by plasmodesmata, allowing the passage of assimilates along this route. The importance of plasmodesmata for assimilate export is reflected in the formation of secondary plasmodesmata upon the transition of maize leaves from importing sink tissues to exporting source tissues (Evert et al., 1996). These findings were supported by the study of Russin et al. (1996), who described a mutant maize, termed sucrose export deficient 1 (sxd1), in which tissue sucrose was increased and phloem export was decreased. An ultrastructural examination of wild-type and mutant leaf tissues revealed that in the mutant line the plasmodesmata interconnecting the bundle sheath and vascular parenchyma cells had been sealed by wall material.

B. Composition and Concentrations of the Exported Carbon and Nitrogen Compounds

Since phloem transport plays a very important role in the growth of sink organs, including roots, storage tissues, fruits, seeds, developing leaves and meristems, much effort has gone into the analysis of the composition of the phloem sap and the determination of the concentrations of transported solutes. Knowledge of the transported substances may also give some indications of the phloem loading mechanism (Section III.C; Ziegler, 1975). Various methods have been used to collect phloem sap. Most information on phloem sap composition has been derived from the analysis of phloem exudates. These are obtained either by stem incision (Zimmermann, 1957; Hall and Baker, 1972) or leaf petiole exudation. Chelating agents (EDTA) are added to avoid the sealing of the wounds by callose formation (King and Zeevart, 1974). Interpretation of data obtained by either method is complicated by the presence of stem or petiole metabolites that may contaminate phloem sap samples. Also, normal transport patterns are seriously impaired by the incision. In an excellent survey, Zimmermann and Ziegler (1975) have compiled data on the composition of sugars in the phloem exudates from more than five hundred species. However, since these exudates were collected using the incision method artifacts in the composition can not be excluded and concentrations of metabolites had not been determined.

A less invasive although time consuming technique is the aphid (or planthopper)-stylet-technique. The collection of phloem sap from the cut ends of aphid stylets can be accomplished using relatively large aphid species feeding on trees (Kennedy and Mittler, 1953; Weatherley et al., 1959). With such aphids, it is possible to sever the stylets with a razor blade. This is difficult with smaller species and species feeding on soft plant tissue because during the cutting the tiny embedded stylets are dislocated and therefore do not exude. A focused beam from a laser or radio-frequency microcautery have solved this problem (Barlow and McCully, 1972; Fisher and Frame, 1984), allowing the collection of pure phloem sap from a large number of different plant species (Lohaus et al., 1995, 1998; Knop et al., 2001).

The concentrations of the C and N compounds in the phloem sap of several important crop plants collected from aphid (or planthopper) stylet exudation are shown in Table 1. Sucrose is the exclusive sugar present in the phloem of most plant species studied so far, being found at concentrations in the range of 200–1500 mM. Sucrose allows high translocation rates (up to 1 m h^{-1}) because it creates a high osmotic potential per C atom and, in solutions with high concentrations, its viscosity is relatively low. Reducing sugars like glucose or fructose were found in the phloem sap only in very low concentrations or were not detectable (Table 1; Ziegler, 1975). The nature of sugars transported in the phloem can be different from the predominant carbohydrates in source and sink tissues. In *Alonsoa meridionalis* sucrose, glucose and fructose are the predominant sugars in the leaves, whereas stachyose and raffinose are the main transport sugars (Knop et al., 2001). Pertinent questions therefore concern the nature of

Table 1. Concentrations of carbon and nitrogen compounds in the phloem sap of leaves from rice, pea, maize, barley, rape and sugar beet (with the exception of pea, grown on nitrate). The samples were collected with the laser-aphid (plant hopper)-stylet technique.

Plant species	Rice[a] mM	Pea[d] mM	Maize[b] mM	Barley[c] mM	Rape[d] mM	Sugar beet[e] mM
Sucrose	205	350	844	1030	1384	1537
Glucose and fructose		u.d.l.	u.d.l.	u.d.l.	u.d.l.	u.d.l.
Malate	7		4	10		
Nitrate	2	u.d.l.	u.d.l.	u.d.l.	u.d.l.	u.d.l.
Amino acids	103	342	59	186	365	62
Glutamate (%)	14	9	25	48	19	57
Glutamine (%)	20	16	4	4	23	5
Aspartate (%)	20	3	11	14	6	12
Asparagine (%)	6	6	2	1	6	4
Alanine (%)	2	2	21	13	3	2
Serine (%)	13	3	8	6	14	3
Homoserine (%)		42	u.d.l.	u.d.l.	u.d.l.	u.d.l.

Data from [a]Hayashi and Chino (1985); (1990); [b]Lohaus et al. (2000); [c]Lohaus et al. (1995), [d]Lohaus and Moellers (2000); [e]Winzer et al. (1996). u.d.l. = under detection limit (0.5 mM)

Table 2. Concentrations of sugar alcohols, mono-, di- and oligosaccharides in the phloem sap of *Alonsoa meridionalis* (Scrophulariaceae), *Asarina barclaiana* (Scrophulariaceae), *Apium graveolens* (Apiaceae), and *Prunus persica* (Rosaceae). The phloem sap were collected with the laser-aphid stylet-technique.

Plant species Concentration	*Alonsoa meridionalis*[a] mM	*Asarina barclaiana*[a] mM	*Apium graveolens*[b] mM	*Prunus persica*[c] mM
Mannitol	u.d.l.	56	938	
Sorbitol	u.d.l.	u.d.l.	u.d.l.	558
myo-Inositol	10	5	7	
Galactinol	2	u.d.l.	u.d.l.	
Hexoses	u.d.l.	1	u.d.l.	u.d.l.
Sucrose	132	1190	303	142
Raffinose	252	28		
Stachyose	360	30		
Verbascose	u.d.l.	u.d.l.		

Data from [a]Knop et al. (2001); [b]G. Lohaus, unpublished; [c]Moing et al. (1997). u.d.l. = under detection limit

the processes that regulate sugar synthesis in the mesophyll and minor veins relative to those that regulate the synthesis of sugars for storage (or transient storage) and export.

Some plant species translocate other sugars additionally to sucrose (Table 2). These sugars fall into two main groups: the sugar alcohols (mannitol and sorbitol) and oligosaccharides of the raffinose family (raffinose, stachyose and verbascose). The raffinose-oligosaccharides are characterized by the attachment of one or more galactose residues to sucrose and were first demonstrated in the phloem sap of trees by Zimmermann (1957). Other oligosaccharide transporting plant species belong to taxonomically diverse plant families: Cucurbitaceae, Lamiaceae, Oleaceae, Onagraceae, Scrophulariaceae, and at least ten other plant families (Zimmermann, 1957; Ziegler, 1975; Zimmermann and Ziegler, 1975; Flora and Madore, 1993; Knop et al., 2001). In some trees, raffinose-oligosaccharides appear only at a given time, namely in the spring before the leaves appear, and are virtually absent during summer and fall (Hill, 1962; Zimmermann and Ziegler, 1975).

Evidence for sugar alcohol phloem transport comes primarily from labeling studies (Webb and Burley, 1962), although there are some reports on the analysis of aphid stylet exudates (Table 2; Moing et al., 1997; Knop et al., 2001). Mannitol is the most widely distributed sugar alcohol and has been found in more than 100 species of vascular plants, including most

Table 3. Amino acid concentrations in the phloem sap (mM), translocation rate of amino-N in the phloem (μmol N [g fresh weight h]$^{-1}$), and protein contents (% of dry weight) in the seeds of different plant species.

	Phloem sap Sum of amino acids	Phloem translocation rate of amino-N	Seed Protein
	mM	μmol N [g fresh weight h]$^{-1}$	% of dry weight
Zea mays (Illinois low protein)[a]	41	0.5	5
Zea mays (Illinois high protein)[a]	118	1.7	23
Brassica napus (DSV15)[b]	204	1.6	19
Pisum sativum ('Kleine Rheinländerin')[b]	342		21
Brassica napus ('Duplo')[b]	365	2.1	25
Brassica carinata (BRA 1151/90)[b]	497	3.5	39

Data from [a] Lohaus et al. (1998); [b] Lohaus and Möllers (2000)

species of the Oleaceae (olive, privet), Apiaceae (celery, carrot), Rubiaceae (coffee), Cucurbitaceae (pumpkin, squash) and Scrophulariaceae (snapdragon) (Barker, 1955; Zimmermann and Ziegler, 1975). Its synthesis occurs simultaneously with either sucrose synthesis, as in celery (Rumpho et al., 1983), or with raffinose-oligosaccharide synthesis, as in olive (Flora and Madore, 1993). Members of the sorbitol translocating group are species of the Rosaceae subfamilies Spiroideae, Pomoideae and Prunoideae (Webb and Burley, 1962), including all members of the economically important genera *Malus* (apple), *Pyrus* (pear) and *Prunus* (stone fruits such as peach, cherry, plum and apricot) (Zimmermann and Ziegler, 1975; Moing et al., 1997).

All protein amino acids are present in leaves as well as in phloem sap. Many of them are present at high concentrations, but depending on the plant species and the N supply, amino acid contents show large variations. The concentration of the sum of amino acids in the phloem sap differs between 60 mM in maize or sugar beet and about 400 mM in rape seed (Table 1). In most of the plant species listed in Table 1, glutamate, glutamine, and aspartate dominate. Other abundant amino acids are alanine in maize and asparagine and homoserine in the legume pea. Some woody plant species have been shown to contain special nitrogenous substances in their phloem exudate. These can be putrescine (formed by decarboxylation of ornithine) as found in *Yucca flaccida*, canavanine (a derivative of guanidine) in *Robinia preudoacacia*, allantoin and allantoic acid in species of the genera *Acer*, *Platanus* or *Aesculus* and citrulline in species of the genera *Betula* or *Alnus* (Ziegler, 1975). Nitrate is normally absent from phloem sap (Table 1; Ziegler, 1975). Low concentrations of nitrate have been detected only in rice phloem sap (Hayashi and Chino, 1985).

Mobilization of N from leaves and export of amino acids via the phloem usually contribute most of the N requirements of seeds. Earlier studies with different plant species suggested that the pattern of substances translocated from the shoot to the developing seeds may affect the relative content of protein and C compounds in the seeds (Lohaus et al., 1998). In different crops the amino acid concentrations and the amino-N translocation rate in the phloem varied considerably and corresponded well to the seed protein contents (Table 3).

Although the enucleate sieve elements of the phloem probably are incapable of protein synthesis, phloem sap samples were found to contain more than 100 soluble polypeptides (Fisher et al., 1992). Protein concentrations in the phloem exudate from non-cucurbits are in the order of 0.2 to 2 g l^{-1} (Kenneke et al., 1971) and are probably higher in *Cucurbita* species (Eschrich et al., 1971). A large number of phloem sap proteins were shown to move from wheat leaves to the apex, as well as into sink tissues, such as the grains (Fisher et al., 1992). The role of these proteins, their involvement in long-distance signaling and their movement into and out of the sieve elements poses important questions for phloem physiology and for cell-to-cell protein movement via plasmodesmata (Fisher et al., 1992).

In some cases, severed aphid stylets exude phloem sap at a relatively high rate for relatively long periods. This allows continuous measurements on intact plants. A series of samples were collected over a period of 30 h from a single stylet embedded in a barley leaf (Fig. 4). During the illumination period the exudation rate of the phloem sap was about twice that found during the dark (Fig. 4; Winter et al., 1992). Whereas the sucrose concentration in the

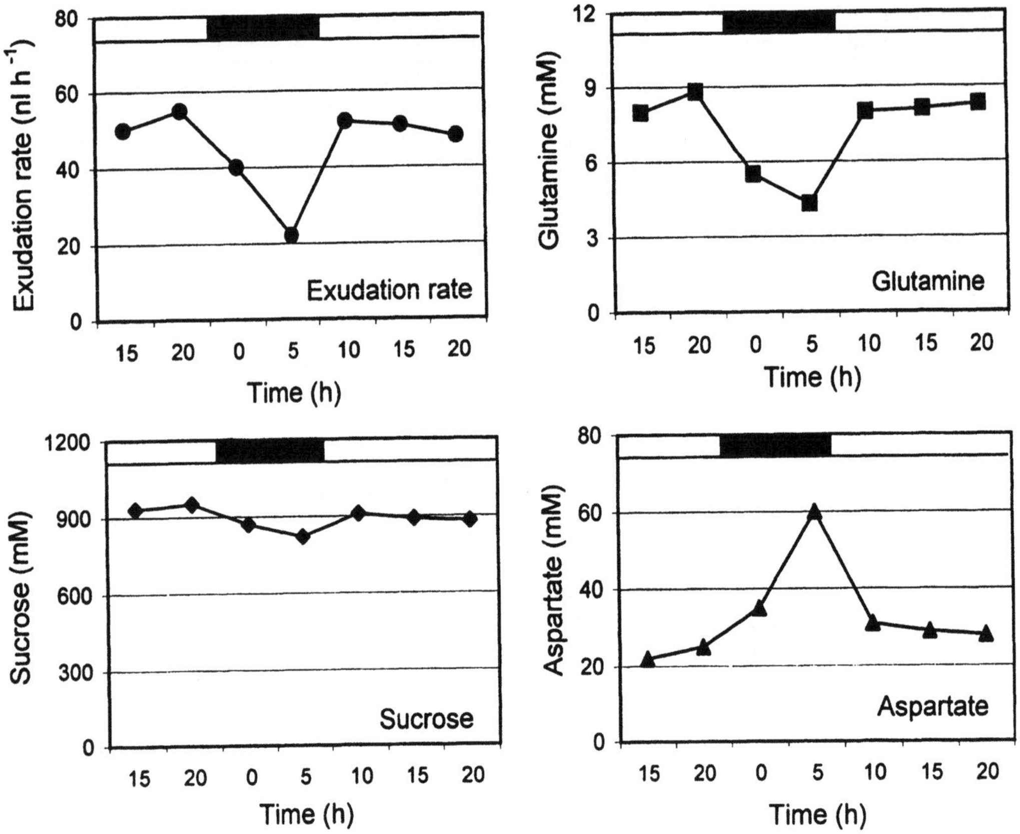

Fig. 4. Diurnal changes of the exudation rates and the concentrations of sucrose, glutamate and aspartate in the phloem sap of barley leaves. Samples were taken from one served aphid stylet. Data from Winter et al. (1992)

phloem sap decreased only by a maximum of 20% during darkness, marked alterations in the diurnal concentrations of amino acids were observed (Winter et al., 1992). The phloem sap concentration of glutamine decreased during the dark period whereas the concentration of aspartate increased (Fig. 4). The diurnal changes of the concentrations of many amino acids in the phloem sap reflect changes in their concentrations in the leaves (Winter et al., 1992).

In order to understand the function of phloem transport in supplying metabolites to sink tissues, one needs to know the relative rates of metabolite export from the source leaves during the day and night periods. Starch and sucrose have an essential role as C assimilate storage pools in the leaf (Section II.A). Between 10 and 45% of the C-assimilation products are often stored in leaves during the day (Table 4) in the form of starch and sucrose. In some species malate, amino acids, fructans or other sugars also accumulate (Riens et al., 1994; Heineke et al., 1994; Lohaus et al., 1998; Lohaus and Möllers, 2000). In fully expanded source leaves the remaining portion of the C-assimilation products, between 55 and 90%, are exported via the phloem during the day (Table 4). The accumulated assimilates are exported from the leaves during the following dark period. In spinach leaves the translocation rate of assimilates during the night was found to be 42% of the translocation rate observed during the day. The corresponding values were 39% in barley (Riens et al., 1994), about 35% in cotton (Hendrix and Huber, 1986), 75% in potato (Heineke et al., 1994), and 58% in rape (Lohaus and Möllers, 2000). In maize leaves, however, the export rate during darkness was only one-seventh of the export rate during the day (Kalt-Torres et al., 1987; Lohaus et al., 1998). The above mentioned results for barley concur with data from the measurement of phloem sap exudation (Fig. 4), where the rate of exudation during the night was found to be about half of that observed during daytime. Considering that a strict correlation between reduced translocation via the phloem and a reduced exudation via a served aphid stylet is not to be expected, these data may be taken as independent evidence that considerable phloem transport occurs at night in barley leaves, although at a reduced rate.

Table 4. Rates of assimilation and export of carbon in several plant species. Spinach and barley were grown in a climatized chamber with a photon flux density of about 300 μmol m^{-2} s^{-1} and rape, potato, and maize were grown in a green house with a photon flux density of 300 – 900 μmol m^{-2} s^{-1}.

Plant species	Spinach [a]	Barley [a]	Potato [b]	Rape [c]	Maize [d]
1. Amount of carbon assimilate during the light period (15 h)					
a) assimilated during the day μmol CO_2 (g fresh weight)$^{-1}$	702	1050	2061	1460	1770
b) accumulated during the day μmol CO_2 (g fresh weight)$^{-1}$	315 (45%)	294 (28%)	885 (43%)	402 (28%)	176 (10%)
2. Rate of carbon assimilate export					
a) during light period μmol CO_2 (g fresh weight h)$^{-1}$	43	54	98	71	106
b) during dark period μmol CO_2 (g fresh weight h)$^{-1}$	18 (42%)	21 (39%)	74 (75%)	41 (58%)	16 (15%)

Data from [a] Riens et al. (1994); [b] Heineke et al. (1994); [c] Lohaus and Möllers (2000); [d] Lohaus et al. (1998).

C. Models of Phloem Loading

Phloem loading proceeds by at least two different mechanisms: (1) the apoplastic way, in which sucrose and amino acids are first exported into the apoplast and then taken up into the SE-CCC by energy-dependent transport systems and (2) the symplastic way in which the assimilates are transferred from the source cells into the SE-CCC via plasmodesmata. Several features have been used to categorize plant species as apoplastic or symplastic phloem loaders. (i) According to Gamalei (1989) the mechanism of phloem loading in various plants depends on the minor vein configuration, describing the type of the companion cells (Turgeon et al., 1993) and the symplastic connections between the mesophyll cells and the SE-CCC. (ii) The mode of phloem loading may also depend on the type of carbohydrate being loaded (Zimmermann and Ziegler, 1975; Turgeon, 1996). (iii) As a physiological criterion for apoplastic or symplastic phloem loading the sensitivity or insensitivity toward thiol group-modifying agents such as p-chloromercuribenzenesulfonic acid (PCMBS) has been used. Sensitivity to PCMBS has been taken to indicate carrier-mediated transport involved in phloem loading. (iv) In several plant species sucrose transporters have been identified in the phloem of source leaves (Riesmeier et al., 1992; Sauer and Stolz, 1994). These are supposed to be involved in apoplastic phloem loading. (v) Sucrose concentration gradients between the cytosol of mesophyll cells and the phloem may be used as a criterion to discriminate between the mode of phloem loading (Geiger et al., 1973; Lohaus et al., 1995).

1. Apoplastic Phloem Loading

In several plant species, there are relatively few plasmodesmata connecting the SE-CCC to surrounding cells. Gamalei (1989) classified these plant groups as 'type 2 (closed).' Some of them show a modification in either companion cells or parenchyma cells relative to transfer cells. Transfer cells are characterized by numerous cell wall invaginations, resulting in an increase in the plasma membrane surface area (Pate and Gunning, 1972). In plant species with such morphology apoplastic phloem loading is expected to be predominant. In apoplastic assimilate export at least two crossings of the membrane are required for solutes to reach the SE-CCC: from the cytosol of bundle sheath cells or minor vein parenchyma cells to the apoplastic space and subsequently from the apoplast to the SE-CCC (Fig. 5). It is still unknown how sucrose and amino acids are transported from the cytosol of source cells into the apoplast. For both sucrose and amino acids, the apoplastic concentrations are much lower than those in the cytosol of mesophyll cells and in the phloem sap (Lohaus et al., 1995). Since these concentration gradients continue to exist under conditions when phloem export is inhibited by cold-girdling (Lohaus et al., 1995), the efflux of sucrose and amino acids into the apoplast appears to be restricted. These data may suggest that regulated proton symport carriers catalyze the export of sucrose and amino acids.

Following the efflux of sucrose and amino acids

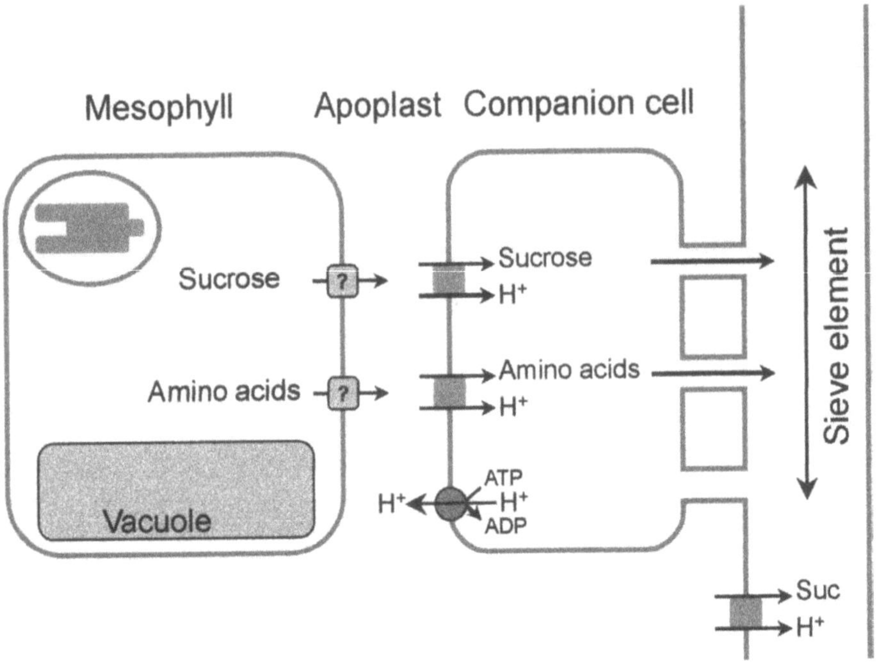

Fig. 5. Model of the apoplastic phloem loading: Transporter-mediated export from mesophyll cells, apoplastic diffusion through the cell-wall continuum, and subsequent transporter-mediated uptake of sucrose and amino acids into the sieve element-companion cell complex (SE-CCC). Suc, sucrose.

Table 5. Comparison of sucrose and amino acid concentrations in the phloem sap compartment with those in the cytosolic compartment of mesophyll cells and in the apoplastic space of leaves.

	Cytosol (mesophyll cell) mM	Apoplast mM	Phloem sap mM	Phloem/Cytosol
	Sucrose concentrations			
Spinach [a]	53	1.3	830	16
Barley (cv. Apex) [a]	232	1.5	1030	4
Maize (Illinois High Protein) [b]	100		945	10
Rape (Duplo) [c]	114		1384	12
Sugar beet [d]	30		1549	52
	Amino acid concentrations			
Spinach [a]	86	3.2	192	2.2
Barley (cv. Apex) [a]	275	3.1	186	0.7
Maize (Illinois High Protein) [b]	78		118	1.5
Rape (Duplo) [c]	140		365	2.6
Sugar beet [d]	60		101	1.7

Data obtained from [a]Lohaus et al. (1995); [b]Lohaus et al. (1998); [c]Lohaus and Möllers (2000); [d]Winzer et al. (1996)

into the apoplast, sucrose and amino acids are loaded into the SE-CCC against a steep concentration gradient (Table 5; Lohaus et al., 1995, 1998; Lohaus and Möllers, 2000; Knop et al., 2001). From transport studies with isolated cells and plasma membrane vesicles it was concluded that sucrose is taken up by active transport from the apoplastic space into the SE-CCC (Geiger et al., 1973; Giaquinta, 1977). This

Table 6. Sucrose transporter genes, isolated from source leaves, and transported sugars in the phloem sap of different plant species. Phloem sap samples were collected from source leaves with the laser aphid stylet technique.

Plant species	Gene name	Phloem transport sugar	References
Apiaceae			
Daucus carota	*DcSUT1,2*	Sucrose (?)	Shakya und Sturm (1998)
Apium graveolens	*AgSUT1,2*	Sucrose, Mannitol	Noiraud et al. (2000) G. Lohaus, unpublished
Brassicaceae			
Arabidopsis thaliana	*AtSUC2*	Sucrose	Sauer und Stolz (1994) C. Knop, personal communication
Chenopodiaceae			
Beta vulgaris	*BvSUT1*	Sucrose	Acc. Nr: X83850 Winzer et al. (1996)
Fabaceae			
Vicia faba	*VfSUT1*	Sucrose (?)	Weber et al. (1997) Harrington et al. (1997)
Pisum sativum	*PsSUT1*	Sucrose	Tegeder et al. (1999) G. Lohaus, unpublished
Euphorbiaceae			
Ricinus communis	*RcSUT1*	Sucrose	Weig und Komor (1996) G. Lohaus, unpublished
Graminaeae			
Oryza sativa	*OsSUT1*	Sucrose	Hirose et al. (1997) Fukumorita und Chino (1982)
Zea mays	*ZmSUT1*	Sucrose	Aoki et al. (1999) Lohaus et al. (1998)
Hordeum vulgare	*HvSUT2*	Sucrose	Weschke et al. (2000) Winter et al. (1992)
Plantaginaceae			
Plantago major	*PmSUC2*	Sucrose (?)	Gahrtz et al. (1994)
Solanaceae			
Lycopersicum esculentum	*LeSUT1*	Sucrose	Barker et al. (2000) E. Valle, personal communication
Nicotiana tabacum	*NtSUT1*	Sucrose (?)	Bürkle et al. (1998)
Solanum tuberosum	*StSUT1*	Sucrose	Riesmeier et al. (1993) Heineke et al. (1992)
Spinacia oleracea	*SoSUT1*	Sucrose Riens et al. (1991)	Riesmeier et al. (1992)
Scrophulariaceae			
Asarina barclaiana	*AbSUT1*	Sucrose, Oligosaccharides	Knop et al. (2001)
Alonsoa meridionalis	*AmSUT1*	Sucrose, Oligosaccharides	Knop et al. (2001)
Vitaceae			
Vitis vinifera	*VvSUT*	Sucrose (?)	Davies et al. (1999) Ageorges et al. (2000)

model was supported by the characterization of translocators involved in such transport processes. Functional complementation of modified yeast strains has enabled the isolation of cDNAs of sucrose transporters from spinach and potato (Riesmeier et al., 1992, 1993b) and heterologous screening of cDNA or genomic libraries was successfully employed to identify homologous genes from several other species (Table 6).

All plant sucrose transporters analyzed thus far are energy dependent and sensitive to protonophores, such as carbonyl cyanide-*m*-chlorophenylhydrazone (CCCP), indicating that they function as proton co-transporters (Boorer et al., 1996b). They have a K_m for sucrose in the range of 1 mM (Riesmeier et al., 1993b) and are able to accumulate sucrose against a steep concentration gradient. The driving force is supplied by a proton ATPase in the plasma membrane of the companion cells (transfer cells of *Vicia faba*; Bouché-Pillon et al., 1994). Hydrophobicity analyses indicate that sucrose transporters belong to a class of transporter proteins that consist of two sets of six membrane-spanning regions separated by a central cytoplasmic loop.

Different pieces of evidence have demonstrated the involvement of these sucrose transporters in phloem loading. The physiological function of SUT1 was confirmed by the analysis of transgenic potato plants in which the activity of this transporter was decreased. Reduced transporter activity resulted in crinkled leaves and increased leaf soluble sugar and starch contents (Riesmeier et al., 1994). RNA *in situ* hybridization studies showed that SUT1 transcripts were phloem-associated (Riesmeier et al., 1993b). Moreover, the promoter of the *Arabidopsis SUC2* gene directed the expression of reporter genes to the phloem of leaves, stems, and roots (Truernit and Sauer, 1995). In *Plantago major* and *Arabidopsis*, immunolocalization studies showed the occurrence of SUC2 in companion cells (Stadler et al., 1995; Stadler and Sauer, 1996). However in tobacco, potato, and tomato SUT1 was located in the plasma membranes of enucleate sieve elements (Kühn et al., 1997). It might be concluded from these different observations that sucrose loading occurs in companion cells as well as in sieve elements, with different transporters operating in each cell type. All transporter proteins are probably synthesized in the companion cells. In the future it will be important to determine the signals that regulate the transport of the transporter proteins to their final destinations.

Sucrose transporters are expressed in all the source leaves of plants that translocate sucrose in the phloem sap (Table 6). Such species include *Apium graveolens* (Noiraud et al., 2000), *Arabidopsis thaliana* (Sauer and Stolz, 1994), *Daucus carota* (Shakya and Sturm, 1998), *Hordeum vulgare* (Weschke et al., 2000), *Lycopersicon esculentum* (Barker et al., 2000), *Nicotiana tabacum* (Bürkle et al., 1998), *Oryza sativa* (Hirose et al., 1997), *Pisum sativum* (Tegeder et al., 1999), *Plantago major* (Gahrtz et al., 1994), *Solanum tuberosum* (Riesmeier et al., 1993b), *Spinacia oleracea* (Riesmeier et al., 1992), *Ricinus communis* (Weig et al., 1996), *Vicia faba* (Weber et al., 1997; Harrington et al., 1997) and *Zea mays* (Aoki et al., 1999). Recently, sucrose transporter cDNAs have also been isolated from leaves of oligosaccharide translocating species (Table 6; Knop et al., 2001). In these species sucrose transporter transcripts were detected in different organs and also in the phloem sap (Knop et al., 2001). One may conclude from these findings that sucrose transporters are involved in phloem loading and/or in retrieval of sucrose from the phloem in oligosaccharide translocating species.

Sucrose transporters are encoded by gene families in higher plants. So far at least seven different sucrose transporter genes have been sequenced in *Arabidopsis* (Sauer and Stolz, 1994). In addition to the presence of sucrose transporters in source leaves, some members of the family are expressed in import zones of sink organs. Thus *DcSUT2* is expressed in storage parenchyma tissues of carrot tap-roots where it seems to be involved in the import sucrose for storage (Shakya and Sturm, 1998). Sucrose transporter transcripts can be detected in the transfer cells of cotyledons from *Vicia faba* and *Pisum sativum*. These sucrose transporters are responsible for sucrose loading into the symplastically isolated seeds (Harrington et al., 1997; Weber et al., 1997; Tegeder et al., 1999) and in the cells of the maternal-filial boundary in developing barley caryopses. In the latter tissues the HvSUT1 transporter controls sucrose unloading from maternal tissues and/or loading into the endosperm (Weschke et al., 2000). Interestingly, these carriers are also expressed in source leaves, indicating a dual function in phloem loading in leaves and in seed sucrose import. Only relatively few sucrose transporters have been identified that are expressed specifically in sink tissues, e.g. *AtSUC1* in anthers and gynoecia (Stadler et al., 1999). Further analysis of the roles of individual members of the gene family is required.

Some recent data indicate that phosphorylation could regulate the sucrose transporter activity (Roblin et al., 1998), but transcriptional regulation seems also important. This was demonstrated by the rapid turnover of the SUT mRNA and protein measured in potato leaves (Kühn et al., 1997). There is also evidence that biotic and abiotic factors, i.e. light, water, salt stress and sugar levels have effects on the expression and activity of certain sucrose transporters (Kühn et al., 1997; Aoki et al., 1999; Noiraud et al., 2000).

In most plants amino acids represent the major transport form of organic N (Table 1). The amino acid concentration in the cytosol of mesophyll cells is similar to the concentration in the phloem (Table 5), whereas a large drop in concentration was observed in the apoplast (Lohaus et al., 1995). The large concentration gradient between the apoplast and the phloem (Table 5) indicates that the phloem loading of amino acids also involves active transport, probably by proton co-transport.

Several amino acid transporter genes have been isolated from *Arabidopsis* by complementation of yeast transport mutants defective in the uptake of certain amino acids. Based on sequence homology, plant amino acid transporters are classified into two superfamilies: the ATF (amino acid transporter) and the APC (amino acid-polyamine-cholin) superfamily (recently reviewed by Fischer et al., 1998). From the analysis of substrate specificity and sequence comparison, the ATF superfamily has been divided into several families.

The first family of related genes was named amino acid permeases (AAP) (Frommer et al., 1993; Kwart et al., 1993; Fischer et al., 1995, 1998). The corresponding proteins are highly hydrophobic and contain 9–12 putative membrane spanning regions. Amino acid transport mediated by the AAPs is pH dependent and occurs against a concentration gradient, suggesting active transport via a H^+-symport mechanism (Boorer et al., 1996a). All AAPs are capable of transporting a large spectrum of structurally diverse amino acids. Based on their differential affinity toward basic amino acids, the AAPs could be divided into two subfamilies: (i) transporters with broad specificity that recognize acidic and neutral amino acids and ureides (AAP1, 2, 4, and 6) and (ii) general amino acid transporters that recognize acidic, neutral, and basic amino acids (AAP3 and 5; Fischer et al., 1998).

The second family of amino acid transporters contains proteins that are highly specific for proline transport and are induced by water or salt stress (Rentsch et al., 1996). This family includes LeProT1 from tomato which, as well as proline, translocates glycine betaine and γ-amino butyric acid and was found to be specifically expressed both in mature and germinating pollen (Schwacke et al., 1999). The third family consists of lysine-histidine transporters (Chen and Bush, 1997). A related family of proteins contains putative auxin transporters (Fischer et al., 1998).

All transporters identified so far show a specific expression pattern in various tissues of the plant. The expression of AtAAP1 and AtAAP2 was found to be associated with the vascular system in cotyledons and developing siliques, indicating its role in supplying developing seeds with amino acids and remobilization of storage N in developing seedlings (Kwart et al., 1993). AtAAP4 and AtAAP5 are expressed in mature leaves and at lower levels in young leaves not capable of export (Fischer et al, 1995). AtAAP3 transcripts were found in roots, where the transporter might function in the uptake of amino acids from the soil. However, apart from cotyledons, where the expression of amino acid transporters was associated with the vascular system (Kwart et al., 1993) none of these transporters has been localized in the SE-CCC of leaves until now.

Table 7. Percentage of the concentration of each amino acid from total amino acid concentrations in the cytosol of mesophyll cells, apoplast and phloem sap of illuminated barley leaves.

	Cytosol %	Apoplast %	Phloem %
Aspartate	12	9	14
Asparagine	1	1	1
Glutamate	32	25	48
Glutamine	9	8	4
Serine	18	20	13
Glycine	9	3	3
Threonine	6	6	4
Alanine	8	18	6
Valine	2	3	2
Isoleucine	1	2	1
Leucine	1	2	1
Lysine	1	2	1
Tyrosine	1	1	1

Data from Lohaus et al. (1995).

General amino acid transporters that transport many different amino acids might be adequate for the situation in source leaves, where all amino acids are synthesized and exported into the phloem. The assumption that transporters with low substrate specificity are involved in phloem loading of amino acids is supported by the finding that the percentage of each amino acid of the total amino acid concentration is rather similar in the cytosol of mesophyll cells, in the apoplast as well as in the phloem (Table 7, Lohaus et al., 1995). Moreover, in spinach and barley the amino acid concentrations in the cytosol of mesophyll cells and in the phloem sap have been found to be nearly identical (Lohaus et al., 1995). During the life cycle of a plant, organic N, synthesized in the form of amino acids or stored in the form of proteins, has to be mobilized, i.e. from storage proteins in leaves or during leaf senescence, and translocated to the sink organs. Therefore general amino acid transporters with low substrate specificity

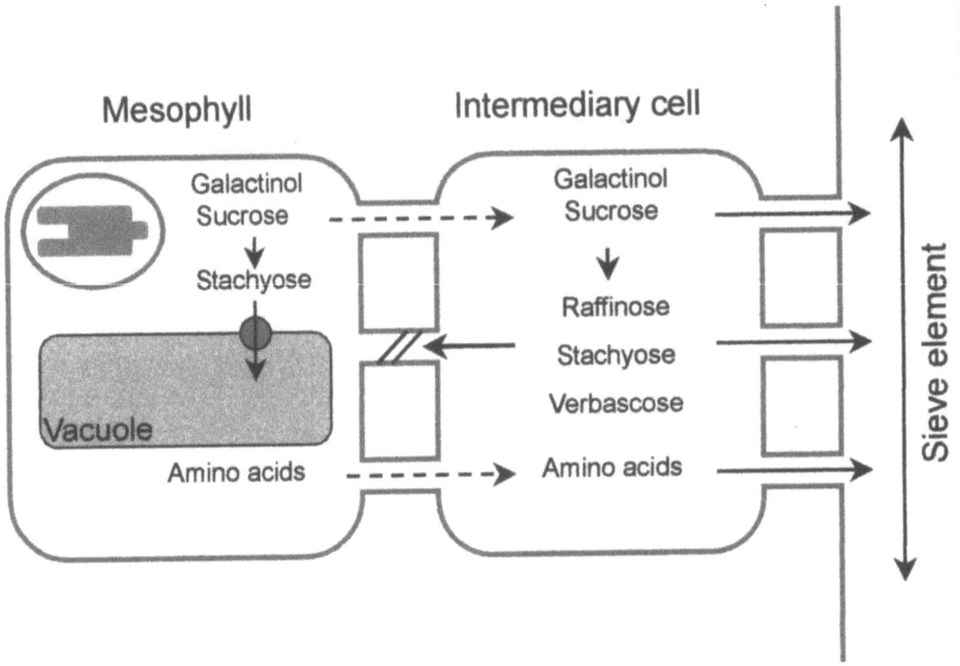

Fig. 6. Model of symplastic phloem loading: According to the 'polymer trap hypothesis' (Turgeon, 1991), sucrose diffuses into the intermediary cells from the mesophyll and the bundle sheath cells and is used in the synthesis of raffinose, stachyose, and verbascose. These sugars are too large to diffuse back through the plasmodesmata.

would be the most efficient system to export all of these amino acids. However, there have been also amino acid transporters found in the plant that recognize only a few amino acids. This might reflect the special requirements of certain cell types under changing environmental conditions.

2. Symplastic Phloem Loading

In plant species regarded as symplastic phloem loaders numerous plasmodesmata between the SE-CCC and the surrounding cells can be found. According to the classification by Gamalei (1989) these species are termed 'type 1 (open).' However, because with plasmodesmata the conductivity (open or closed state) and the size exclusion limit (SEL) varies, data based only on plasmodesmata frequency must be viewed with reservation.

In symplastic phloem loaders the minor vein-companion cells are often specialized as 'intermediary cells.' Intermediary cells have a distinct appearance and are connected to the bundle sheath by dense fields of branched plasmodesmata. Species with intermediary cells include members of the Cucurbitaceae, Lamiaceae, Oleaceae, and Scrophulariaceae (Turgeon et al., 1975; Flora and Madore, 1993; Turgeon et al., 1993). This companion cell type is correlated with the translocation of considerable amounts of raffinose and stachyose in addition to sucrose (Table 2).

The 'polymer trapping' model of phloem loading has been proposed to explain the coincidence of intermediary cell structure and stachyose transport (Turgeon, 1991). It is based on a size discrimination function of the plasmodesmata connecting the intermediary cells with the bundle sheath. Sucrose, which is synthesized in the mesophyll, diffuses through the plasmodesmata between mesophyll cells to the bundle sheath cells and thereafter into the intermediary cells (Fig. 6). Galactinol is synthesized from *myo*-inositol and UDP-galactose in the cytosol of the intermediary cells. It can also be produced in the mesophyll cells of certain plants (Sprenger and Keller, 2000). Inside the intermediary cells sucrose and galactinol are consumed during the synthesis of raffinose-family oligosaccharides. Enzymes involved in raffinose or stachyose synthesis have been localized in these cells (Holthaus and Schmitz, 1991). Since raffinose and stachyose are larger than sucrose, the SEL of the plasmodesmata is thought to inhibit the diffusion back into the bundle sheath cells but not into the sieve tubes. This may explain how oligosaccharides accumulate to high concentrations in the intermediary cells, and after transfer, also in the sieve elements (see *Alonsoa meridionalis*; Table 2). To date, however, there has been no direct experimental demonstration of the hypothesized differences in the SEL of plasmodesmata connecting intermediary cells to the bundle sheath.

The finding of a complete symplastic route from mesophyll cells to sieve elements does not exclude the possibility of apoplastic transport of sugars or amino acids across the plasma membranes between these cell types. Sucrose transporters are found in the leaves of oligosaccharide translocating and putative symplastic phloem loaders (Table 6; Knop et al., 2001). Whether these transporters are involved in phloem loading, similar to the transporters in apoplastic phloem loaders, has still to be examined. Raffinose-induced membrane depolarization indicated the presence of carrier-mediated uptake of the oligosaccharide in *Catharanthus* and *Ocimum* (Van Bel et al., 1996). Further research on symplastic loading is required to elucidate this complex process.

3. Loading of Sugar Alcohols

Current knowledge of the mechanism of phloem loading for sugar alcohols is limited. Export of both sorbitol and mannitol is often related to the synthetic capacity of the source and the resulting concentrations in the mesophyll cells (Moing et al., 1994). Based on studies of proton gradient dependent uptake of mannitol in plasma membrane vesicles isolated from celery phloem tissues, Salmon et al. (1995) concluded that a mannitol carrier exists. Recently, a putative mannitol transporter gene, *AgMaT1*, has been cloned in celery (Noiraud et al., 2001). These findings suggest that apoplastic transport might be involved in this type of phloem loading. The finding that PCMBS inhibited sorbitol and sucrose phloem transport in peach, a sorbitol transporting plant (Moing et al., 1997) supports this view. On the other hand, in *Prunus* species the minor vein configuration could allow symplastic phloem loading according to Gamalei (1989). Up to now, no sorbitol transporter has been identified in the plasma membrane. It might be that mannitol and sorbitol are loaded into the phloem by different routes.

IV. Concluding Remarks

A major goal of modern plant science research is to understand carbohydrate and amino acid partitioning within a plant cell and between different plant tissues (source-sink regulation). This includes the identification of proteins involved in intracellular transport processes as well as in phloem loading. Much progress has been achieved over the last few years in the elucidation of structure-function relationships, regulation of transport, cellular and temporal transporter expression patterns and their physiological functions, by studying the biochemistry and molecular biology of the transport protein. In addition, metabolite concentrations in different subcellular compartments and different cell types have been determined. However, our knowledge of metabolite transport in plants is still rudimentary. In particular, clarification is required on the nature and mechanisms of sugar and amino acid efflux from cells and of phloem loading and unloading. In addition, it is important to understand the relationship between storage and translocation. How do plants assign priority to the multitude of sinks that will utilize these photoassimilates? As mentioned above, many intracellular transport processes, e.g. across the plastidic envelopes, have not yet been characterized. With the recent completion of the *Arabidopsis* and rice genomic sequencing programs the whole set of genes from these plants is available. This should lead to the identification of other proteins involved in transport processes. Central to the elucidation of transporters and their physiological function is the biochemical and molecular analysis of 'knock out' *Arabidopsis* mutants. These can easily be isolated by a PCR-based approach. This strategy has already led to the characterization of numerous plant proteins, involved in diverse cellular functions.

Acknowledgments

This work was supported by grants from the Deutsche Forschungsgemeinschaft to G.L. We thank Katharina Pawlowski, Hans-Walter Heldt, Richard Jensen and Hans Bohnert for helpful discussions and critical reading of the manuscript.

References

Ageorges A, Issaly N, Picaud S, Delrot S and Romieu C (2000) Identification and functional expression in yeast of a grape berry sucrose carrier. Plant Physiol Biochem 38: 177–185

Aoki N, Hirose T, Takahashi S, Ono K, Ishimaru K and Ohsugi R (1999) Molecular cloning and expression analysis of a gene for a sucrose transporter in maize (*Zea mays* L.). Plant Cell Physiol 40: 1072–1078

Bagge P and Larsson C (1986) Biosynthesis of aromatic amino acids by highly purified spinach chloroplasts—Compartmentation and regulation of the reactions. Physiol Plant 68: 641–647

Baldry CW, Bucke C and Walker DA (1966) Incorporation of inorganic phosphate into sugar phosphates during carbon dioxide fixation by illuminated chloroplasts. Nature 210: 793–796

Barker L, Kühn C, Weise A, Schulz A, Gebhardt C, Hirner B, Hellmann H, Schulze W, Ward JM and Frommer WB (2000) SUT2, a putative sucrose sensor in sieve elements. Plant Cell 12: 1153–1164

Barker SA (1955) Acyclic sugar alcohols. In: Peach K and Tracey MV (eds) Modern Methods of Plant Analysis, Vol. 2, pp 158–192. Springer Verlag, Berlin

Barlow CA and McCully ME (1972) The ruby laser as an instrument for cutting the stylets of feeding aphids. Can J Zool 50: 1497–1499

Bassham JA, Kirk M and Jensen RG (1968) Photosynthesis by isolated chloroplasts. I. Diffusion of labeled photosynthetic intermediates between isolated chloroplasts and suspending medium. Biochim Biophys Acta 153: 211–218

Batz O, Scheibe R and Neuhaus HE (1992) Transport processes and corresponding changes in metabolite levels in relation to starch synthesis in barley etioplasts. Plant Physiol 100: 184–190

Boorer KJ, Frommer WB, Bush DB, Kreman M, Loo DDF and Wright EM (1996a) Kinetics and specificity of a H^+/amino acid transporter from *Arabidopsis thaliana*. J Biol Chem 271: 2213–2220

Boorer KJ, Loo DDF, Frommer WB and Wright EM (1996b) Transport mechanism of the cloned H^+/sucrose transporter StSUT1. J Biol Chem 271: 25139–25144

Borchert S, Grosse H and Heldt HW (1989) Specific transport of inorganic phosphate, glucose 6-P, dihydroxyacetone phosphate and 3-phosphoglycerate into amyloplasts from pea roots. FEBS Lett 253: 183–186

Borchert S, Harborth J, Schünemann D, Hoferichter P and Heldt HW (1993) Studies of the enzymic capacities and properties of pea root plastids. Plant Physiol 101: 303–312

Bouché-Pillon S, Fleurat-Lessard P, Fromont JC, Serrano R and Bonnemain JL (1994) Immunolocalization of the plasma membrane H^+-ATPase in minor veins of *Vicia faba* in relation to phloem loading. Plant Physiol 105: 691–697

Bowsher CG, Hucklesby DB and Emes MJ (1989) Nitrite reduction and carbohydrate metabolism in plastids purified from roots of *Pisum sativum*. Planta 177: 359–366

Bowsher CG, Boulton EL, Rose J, Nayagam S and Emes MJ (1992) Reductant for glutamate synthase is generated by the oxidative pentose phosphate pathway in non-photosynthetic root plastids. Plant J 2: 893–898

Büker M, Schünemann D and Borchert S (1998) Enzymic properties and capacities of developing tomato fruit plastids. J Exp Bot 49: 681–691

Bürkle L, Hibberd JM, Quick WP, Kühn C, Hirner B and Frommer WB (1998) The H^+-sucrose co-transporter NtSUT1 is essential for sugar export from tobacco leaves. Plant Physiol 118: 59–68

Caspar T, Huber SC and Somerville CR (1986) Alterations in growth, photosynthesis and respiration in a starchless mutant of *Arabidopsis thaliana* deficient in chloroplast phosphoglucomutase. Plant Physiol 79: 11–17

Caspar T, Lin TP, Monroe J, Benbow L, Preiss J and Somerville CR (1991) Mutants of *Arabidopsis* with altered regulation of starch degradation. Plant Physiol 95: 1181–1188

Chen BY, Wang Y and Janes HW (1998) ADP-Glucose pyrophosphorylase is localized to both the cytoplasm and plastids in developing pericarp of tomato fruit. Plant Physiol 116: 101–106

Chen L and Bush DR (1997) LHT1, a lysine and histidine specific amino acid transporter in *Arabidopsis*. Plant Physiol 115: 1127–1134

Cockburn W, Baldry CW and Walker DA (1967) Some effects of inorganic phosphate on O_2 evolution by isolated chloroplasts. Biochim Biophys Acta 143: 614–624

Cognata UL, Willmitzer L and Müller-Röber B (1995) Molecular cloning and characterization of novel isoforms of potato ADP-glucose pyrophosphorylase. Mol Gen Genet 246: 538–548

Davies C, Wolf T and Robinson SP (1999) Three putative sucrose transporters are differentially expressed in grapevine tissue. Plant Sci 147: 93–100

Davis JM and Loescher WH (1990) [^{14}C]-Assimilate translocation in the light and dark in celery (*Apium graveolens*) leaves of different ages. Physiol Plant 79: 656–662

Day DA and Hatch MD (1981) Transport of 3-phosphoglyceric acid, phosphoenolpyruvate, and inorganic phosphate in maize mesophyll chloroplasts and the effect of 3-phosphoglyceric acid on malate and phosphoenolpyruvate production. Arch Biochem Biophys 211: 743–749

Denyer K, Dunlap F, Thorbjörnsen T, Keeling P and Smith AM (1996) The major form of ADP-glucose pyrophosphorylase in maize endosperm is extra-plastidial. Plant Physiol 112: 779–785

Eastmond PJ and Rawsthorne S (2000) Coordinate changes in carbon partitioning and plastidial metabolism during the development of oilseed rape. Plant Physiol 122: 767–774

Eastmond PJ, Dennis DT and Rawsthorne S (1997) Evidence that a malate/phosphate exchange translocator imports carbon across the leucoplast envelope for fatty acid synthesis in developing castor seed endosperm. Plant Physiol 114: 851–856

Emes MJ and Neuhaus HE (1997) Metabolism and transport in non-photosynthetic plastids. J Exp Bot 48: 1995–2005

Entwistle G and ap Rees T (1988) Enzymatic capacities of amyloplasts from wheat endosperm. Biochem J 255: 391–396

Entwistle G and ap Rees T (1990) Lack of fructose-1,6-bisphosphatase in a range of higher plants that store starch. Biochem J 271: 467–472

Eschrich W, Evert RF and Heyser W (1971) Proteins of the sieve-tube exudate of *Cucurbita maxima*. Planta 100: 208–221

Evert RF, Russin WA and Bosabalidis AM (1996) Anatomical and ultrastructural changes associated with sink-to-source transition in developing maize leaves. Int J Plant Sci 157: 247–261

Fischer K, Arbinger B, Kammerer B, Busch C, Brink S, Wallmeier H, Sauer N, Eckerskorn C and Flügge UI (1994) Cloning and in vivo expression of functional triose phosphate/phosphate translocators from C_3- and C_4-plants: Evidence for the putative participation of specific amino acid residues in the recognition of phosphoenolpyruvate. Plant J 5: 215–226

Fischer K, Kammerer B, Gutensohn M, Arbinger B, Weber A, Häusler R and Flügge UI (1997) A new class of plastidic phosphate translocators: A putative link between primary and secondary metabolism by phosphoenolpyruvate/phosphate antiporter. Plant Cell 9: 453–462

Fischer W-N, Kwart M, Hummel S and Frommer WB (1995) Substrate specificity and expression profile of amino acid transporters (AAPs) in *Arabidopsis*. J Biol Chem 270: 16315–16320

Fischer W-N, André B, Rentsch D, Krolkiewicz S, Tegeder M, Breitkreuz K and Frommer WB (1998) Amino acid transport in plants. Trends Plant Sci 3: 188–195

Fisher DB and Frame JM (1984) A guide to the use of the exuding-stylet technique in phloem physiology. Planta 161: 385–393

Fisher DB, Wu Y and Ku MSB (1992) Turnover of soluble proteins in the wheat sieve tube. Plant Physiol 100: 1433–1441

Fliege R, Flügge UI, Werdan K and Heldt HW (1978) Specific transport of inorganic phosphate, 3-phosphoglycerate and triosephosphates across the inner membrane of the envelope in spinach chloroplasts. Biochim Biophys Acta 502: 232–247

Flora LL and Madore MA (1993) Stachyose and mannitol transport in olive (*Olea europaea* L.). Planta 189: 484–490

Flügge UI (1995) Phosphate translocation in the regulation of photosynthesis. J Exp Bot 46: 1317–1323

Flügge UI (2000) Metabolite transport across the chloroplast envelope of C_3-plants. In: Leegood RC, Sharkey TD and von Caemmerer S (eds) Photosynthesis: Physiology and Metabolism, pp 137–152. Kluwer Academic Publishers, Dordrecht

Flügge UI and Heldt HW (1984) The phosphate-triose phosphate-phosphoglycerate translocator of the chloroplasts. Trends Biochem Sci 9: 530–533

Flügge UI, Stitt M and Heldt HW (1985) Light-driven uptake of pyruvate into mesophyll chloroplasts from maize. FEBS Lett 183: 335–339

Flügge UI, Fischer K, Gross A, Sebald W, Lottspeich F and Eckerskorn C (1989) The triose phosphate-3-phosphoglycerate-phosphate translocator from spinach chloroplasts: Nucleotide sequence of a full-length cDNA clone and import of the in vitro synthesized precursor protein into chloroplasts. EMBO J 8: 39–46

Flügge UI, Weber A, Fischer K, Häusler R and Kammerer B (1996) Molecular characterization of plastid transporters. C R Acad Sci Paris 319: 849–852

Frommer WB, Hummel S and Riesmeier JW (1993) Expression cloning in yeast of a cDNA encoding a broad specificity amino acid permease from *Arabidopsis thaliana*. Proc Natl Acad Sci USA 90: 6160–6164

Gahrtz M, Stolz J and Sauer N (1994) A phloem-specific sucrose H^+ symporter from *Plantago major* L. supports the model of apoplastic phloem loading. Plant J 6: 697–706

Gamalei Y (1989) Structure and function of leaf minor veins in trees and herbs. A taxonomic review. Trees 3: 96–110

Geiger DR, Giaquinta RT, Sovonick SA and Fellows RJ (1973) Solute distribution in sugar beet leaves in relation to phloem loading and translocation. Plant Physiol 52: 585–589

Giaquinta RT (1977) Possible role of pH gradient and membrane ATPase in the loading of sucrose in the sieve tubes. Nature 267: 369–370

Giroux MJ and Hannah LC (1994) ADP-glucose pyrophosphorylase in shrunken-2 and brittle-2 mutants of maize. Mol Gen Genet 243: 400–408

Hall SM and Baker DA (1972) The chemical composition of *Ricinus* phloem exudate. Planta 106: 131–140.

Hanson KR and McHale NA (1988) A starchless mutant of *Nicotiana sylvestris* containing a modified plastid phosphoglucomutase. Plant Physiol 88: 838–844

Harrington GN, Franceschi VR, Offler CE, Patrick JW, Harper JF, Frommer WB, Tegeder M and Hitz WD (1997) Cell specific expression of three genes involved in plasma membrane sucrose transport in developing *Vicia faba* seed. Protoplasma 197: 35–50

Harrison CJ, Hedley CL and Wang TL (1998) Evidence that the rug3 locus of pea encodes plastidial phosphoglucomutase confirms that the imported substrate for starch synthesis in pea amyloplasts is glucose-6-phosphate. Plant J 13: 753–762

Harrison CJ, Mould RM, Leech MJ, Johnson SA, Turner L, Schreck SL, Baird KM, Jack PL, Rawsthorne S, Hedley CL and Wang TL (2000) The rug3 locus of pea encodes plastidial phosphoglucomutase. Plant Physiol 122: 1187–1192

Hatch MD, Dröscher L, Flügge UI and Heldt HW (1984) A specific translocator for oxaloacetate transport in chloroplasts. FEBS Lett 178: 15–19

Hatzfeld WD and Stitt M (1990) A study of the rate of recycling of triose phosphates in heterotrophic *Chenopodium rubrum* cells, potato tubers and maize endosperm. Planta 180: 198–204

Häusler RE, Schlieben NH, Schulz B and Flügge UI (1998) Compensation of decreased triose phosphate/phosphate translocator activity by accelerated starch turnover and glucose transport in transgenic tobacco. Planta 204: 366–376

Häusler RE, Schlieben NH, Nicolay P, Fischer K, Fischer KL and Flügge UI (2000) Control of carbon partitioning and photosynthesis by the triose phosphate/phosphate translocator in transgenic tobacco plants. I. Comparative physiological analysis of tobacco plants with antisense repression and overexpression of the triose phosphate/phosphate translocator. Planta 210: 371–382

Hayashi H and Chino M (1985) Nitrate and other anions in the rice phloem sap. Plant Cell Physiol 26: 325–330

Hayashi H and Chino M (1990) Chemical composition of phloem sap from the uppermost internode of the rice plant. Plant Cell Physiol 31: 247–251

Heineke D, Kruse A, Flügge UI, Frommer WB, Riesmeier JW, Willmitzer L and Heldt HW (1994) Effect of antisense repression of the chloroplast triose phosphate translocator on photosynthetic metabolism in transgenic plants. Planta 193: 174–180

Heldt HW and Rapley L (1970) Specific transport of inorganic phosphate, 3-phospho-glycerate and dihydroxyacetone phosphate and of dicarboxylates across the inner membrane of spinach chloroplasts. FEBS Lett 10: 143–148

Hendrix DL and Huber SC (1986) Diurnal fluctuations in cotton leaf carbon export, carbohydrate content, and sucrose synthesizing enzymes. Plant Physiol 81: 584–586

Herrmann KM and Weaver LM (1999) The shikimate pathway. Annu Rev Plant Physiol Plant Mol Biol 50: 473–503

Hill GP (1962) Exudation from aphid stylets during the period from dormancy to bud break in *Tilia americana* (L.). J Exp Bot 13: 144–151

Hill LM and Smith AM (1991) Evidence that glucose 6-P is imported as the substrate for starch biosynthesis by the plastids of developing pea embryos. Planta 185: 91–96

Hirose T, Imaizumi N, Scofield GN, Furbank RT and Ohsugi R (1997) cDNA cloning and tissue specific expression of a gene for sucrose transporter from rice (*Oryza sativa* L.). Plant Cell Physiol 38: 1389–1396

Holthaus U and Schmitz K (1991) Distribution and immunolocalization of stachyose synthase in *Cucumis melo* L. Planta 185: 479–486

Huber SC and Edwards GE (1977a) Transport in C_4-mesophyll chloroplasts. Characterization of a pyruvate carrier. Biochim Biophys Acta 462: 583–602

Huber SC and Edwards GE (1977b) Transport in C_4 mesophyll chloroplasts. Evidence for an exchange of inorganic phosphate and phosphoenolpyruvate. Biochim Biophys Acta 462: 603–612

Jones TWA, Gottlieb LD and Pichersky E (1986) Reduced enzyme activity and starch level in an induced mutant of chloroplastic phosphoglucose isomerase. Plant Physiol 81: 367–371

Journet EP and Douce R (1985) Enzymic capacities of purified cauliflower bud plastids for lipid synthesis and carbohydrate metabolism. Plant Physiol 79: 458–467

Kalt-Torres W, Kerr PS, Usuda H and Huber SC (1987) Diurnal changes in maize leaf photosynthesis. Carbon exchange rate, assimilate export rate, and enzyme activities. Plant Physiol 83: 283–288

Kammerer B, Fischer K, Hilpert B, Schubert S, Gutensohn M, Weber A and Flügge UI (1998) Molecular characterization of a carbon transporter in plastids from heterotrophic tissues: The glucose-6-phosphate/phosphate antiporter. Plant Cell 10: 105–117

Kang F and Rawsthorne S (1994) Starch and fatty acid synthesis in plastids from developing embryos of oilseed rape. Plant J 6: 795–805

Keeling PL, Wood JR, Tyson RH and Bridges IG (1988) Starch biosynthesis in developing wheat grain. Plant Physiol 87: 311–319

Kennecke K, Ziegler H and De Fekete MAR (1971) Enzymaktivitäten im Siebröhrensaft von *Robinia pseudoaccacia* L. und anderen Baumarten. Planta 98:330–356

Kennedy JS and Mittler TE (1953) A method for obtaining phloem sap via the mouth-parts of aphids. Nature 171: 528

King RW and Zeevart JAD (1974) Enhancement of phloem exudation from cut petioles by chelating agents. Plant Physiol 53: 96–103

Kirk PR and Leech RM (1972) Amino acid biosynthesis by isolated chloroplasts during photosynthesis. Plant Physiol 50: 228–234

Knop C, Voitsekhovskaja O and Lohaus G (2001) Sucrose transporters in two Scrophulariaceae with different types of transport sugar. Planta 213: 80–91

Kofler H, Häusler RE, Schulz B, Gröner F, Flügge UI and Weber A (2000) Molecular characterisation of a new mutant allele of the plastidic phosphoglucomutase and complementation of the mutant with the wild-type cDNA. Mol Gen Genet 263: 978–986

Kühn C, Franceschi VR, Schulz A, Lemoine R and Frommer WB (1997) Localization and turnover of sucrose transporters in enucleate sieve element indicate macromolecular trafficking. Science 275: 1298–1300

Kwart M, Hirner B, Hummel S and Frommer WB (1993) Differential expression of two related amino acid transporters with differing substrate specificity in *Arabidopsis thaliana*. Plant J 4: 993–1002

Li H, Culligan K, Dixon RA and Chory J (1995) CUE1: A mesophyll cell-specific positive regulator of light-controlled gene expression in *Arabidopsis*. Plant Cell 7: 1599–1610

Loddenkötter B, Kammerer B, Fischer K and Flügge UI (1993) Expression of the functional mature chloroplast triose phosphate translocator in yeast internal membranes and purification of the histidine tagged protein by a single metal affinity chromatography step. Proc Natl Acad Sci USA 90: 2155–2159

Lohaus G and Möllers C (2000) Phloem transport of amino acids in two *Brassica napus* genotypes and one *Brassica carinata* in relation to their seed protein content. Planta 211: 833–840

Lohaus G, Winter H, Riens B and Heldt HW (1995) Further studies of the phloem loading process in leaves of barley and spinach: Comparison of metabolite concentrations in the apoplastic compartment with those in the cytosolic compartment and in the sieve tubes. Bot Acta 3: 270–275

Lohaus G, Büker M, Hußmann M and Heldt HW (1998) Transport of amino acids with special emphasis on the synthesis and transport of asparagine in Illinois Low Protein and Illinois High Protein strains of maize. Planta 205: 181–188

Lohaus G, Hussmann M, Pennewiss K, Schneider H, Zhu JJ and Sattelmacher B (2000) Solute balance of a maize (*Zea mays* L.) source leaf as affected by salt treatment with special emphasis on phloem re-translocation and ion leaching. J Exp Bot 51: 1721–1732

Möhlmann T and Neuhaus HE (1997) Analysis of the precursor and effector dependency of lipid synthesis in amyloplasts isolated from developing wheat- or maize-endosperm tissue. J Cereal Sci 26: 161–167

Möhlmann T, Batz O, Maaß U and Neuhaus HE (1995) Analysis of carbohydrate transport across the envelope of isolated cauliflower-bud amyloplasts. Biochem J 307: 521–526

Möhlmann T, Tjaden J, Henrichs G, Quick WP, Häusler RE and Neuhaus HE (1997) ADP glucose drives starch synthesis in isolated maize endosperm amyloplasts. Characterization of starch synthesis and transport properties across the amyloplastidic envelope. Biochem J 324: 503–509

Moing A, Escobar-Gutiérrez AJ and Gaudillère JP (1994) Modeling carbon export out of mature peach leaves. Plant Physiol 106: 591–600

Moing A, Carbonne F, Zipperlin B, Svanella L and Gaudillère JP (1997) Phloem loading in peach: symplastic or apoplastic? Physiol Plant 101: 489–496

Münch E (1930) Die Stoffbewegung in der Pflanze. Gustav Fischer, Jena, Germany

Naeem M, Tetlow IJ and Emes MJ (1997) Starch synthesis in amyloplasts purified from developing potato tubers. Plant J 11: 1095–1103

Neuhaus HE, Henrichs G and Scheibe R (1993a) Characterization of glucose-6-P incorporation into starch by isolated intact cauliflower-bud plastids. Plant Physiol 101: 573–578

Neuhaus HE, Thom E, Batz O and Scheibe R (1993b) Purification of highly intact plastids from various heterotrophic plant tissues. Analysis of enzymic equipment and precursor dependency for starch biosynthesis. Biochem J 296: 395–401

Noiraud N, Delrot S and Lemoine R (2000) The sucrose transporter of celery. Identification and expression during salt stress. Plant Physiol 122: 1447–1455

Noiraud N, Maurousset L and Lemoine R (2001) Identification of a mannitol transporter, AgMaT1, in celery phloem. Plant Cell 13: 695–705

Ohnishi J and Kanai R (1990) Pyruvate uptake induced by a pH jump in mesophyll chloroplasts of maize and sorghum, NADP-malic enzyme type C4 species. FEBS Lett 269: 122–124

Ohnishi J, Flügge UI and Heldt HW (1989) Phosphate translocator of mesophyll and bundle sheath chloroplasts of a C_4 plant, *Panicum miliaceum*. Identification and kinetic characterization. Plant Physiol 91: 1507–1511

Ohshima T, Hayashi H and Chino M (1990) Collection and chemical composition of pure phloem sap from *Zea mays* L. Plant Cell Physiol 31: 735–737

Okita T (1992) Is there an alternative pathway for starch synthesis? Plant Physiol 100: 560–64

Oparka KJ and Turgeon R (1999) Sieve elements and companion cells—Traffic control centers of the phloem. Plant Cell 11: 739–750

Pate J and Gunning BES (1972) Transfer cells. Annu Rev Plant Physiol 23: 173–196

Pozueto-Romero J, Frehner M, Viale AM and Akazawa T (1991) Direct transport of ADP glucose by an adenylate translocator is linked to starch biosynthesis in amyloplasts. Proc Natl Acad Sci USA 88: 5769–5773

Preiss J (1991) Biology and molecular biology of starch synthesis and its regulation. Oxf Surv Plant Mol Cell Biol 7: 59–114

Proudlove MO and Thurman DA (1981) The uptake of 2-oxoglutarate and pyruvate by isolated pea chloroplasts. New Phytol 88: 255–264

Quick WP and Neuhaus HE (1996) Evidence for two types of phosphate translocators in sweet pepper (*Capsicum annuum*) fruit chromoplasts. Biochem J 320: 7–10

Quick WP, Scheibe R and Neuhaus E (1995) Induction of a hexose-phosphate translocator activity in spinach chloroplasts. Plant Physiol 109: 113–121

Rentsch D, Hirner B, Schmelzer E and Frommer WB (1996) Salt stress-induced proline transporters and salt stress-repressed broad specificity amino acid permeases identified by suppression of a yeast amino acid permease-targeting mutant. Plant Cell 8: 1437–1446

Riens B, Lohaus G, Winter H and Heldt HW (1994) Production and diurnal utilization of assimilates in leaves of spinach (*Spinacia oleracea* L.) and barley (*Hordeum vulgare* L.). Planta 192: 497–501

Riesmeier JW, Willmitzer L and Frommer WB (1992) Isolation and characterization of a sucrose carrier cDNA from spinach by functional expression in yeast. EMBO J 11: 4705–4713

Riesmeier JW, Flügge UI, Schulz B, Heineke D, Heldt HW, Willmitzer L and Frommer WB (1993a) Antisense repression of the chloroplast triose phosphate translocator affects carbon partitioning in transgenic potato plants. Proc Natl Acad Sci USA 90: 6160–6164

Riesmeier JW, Hirner B and Frommer WB (1993b) Potato sucrose transporter expression in minor veins indicates a role in phloem loading. Plant Cell 5: 1591–1598

Riesmeier JW, Willmitzer L and Frommer WB (1994) Evidence for an essential role of the sucrose transporter in phloem loading and assimilate partitioning. EMBO J 13: 1–7

Roblin G, Sakr S, Bonmort J and Delrot S (1998) Regulation of a plant plasma membrane sucrose transporter by phosphorylation. FEBS Lett 424: 165–168

Rost S, Frank C and Beck E (1996) The chloroplast envelope is permeable for maltose but not for maltodextrins. Biochim Biophys Acta 1291: 221–227

Rumpho ME, Edwards GE and Loescher WH (1983) A pathway for photosynthetic carbon flow to mannitol in celery leaves. Activity and localization of key enzymes. Plant Physiol 73: 869–873

Russian WA, Evert RF, Vanderveer PJ, Sharkey TD and Briggs SP (1996) Modification of a specific class of plasmodesmata and loss of sucrose export ability in the sucrose export defective maize mutant. Plant Cell 8: 645–658

Salmon S, Lemoine R, Jamai A, Bouché-Pillon S and Fromont JC (1995) Study of sucrose and mannitol transport in plasma-membrane vesicles from phloem and non-phloem tissues of celery (*Apium graveolens* L.) petioles. Planta 197: 76–83

Sauer N and Stolz J (1994) SUC1 and SUC2: Two sucrose transporters from *Arabidopsis thaliana*: Expression and characterization in baker's yeast and identification of the histidine tagged protein. Plant J 6: 67–77

Schäfer G, Heber U and Heldt HW (1977) Glucose transport into spinach chloroplasts. Plant Physiol 60: 286–289

Schleucher J, Vanderveer PJ and Sharkey TD (1998) Export of carbon from chloroplasts at night. Plant Physiol 118: 1439–1445

Schmid J and Amrhein N (1995) Molecular organization of the shikimate pathway in higher plants. Phytochemistry 39: 737–749

Schott K, Borchert S, Müller-Röber B and Heldt HW (1995) Transport of inorganic phosphate and C_3- and C_6-sugar phosphates across the envelope membranes of potato tuber amyloplasts. Planta 196: 647–652

Schulz B, Frommer WB, Flügge UI, Hummel S, Fischer K and Willmitzer L (1993) Expression of the triose phosphate translocator gene from potato is light dependent and restricted to green tissues. Mol Gen Genet 238: 357–361

Schünemann D and Borchert S (1994) Specific transport of inorganic phosphate and C_3- and C_6-sugar phosphates across the envelope membranes of tomato leaf chloroplasts, tomato fruit chloroplasts and fruit chromoplasts. Bot Acta 107: 461–467

Schwacke R, Grallath S, Breitkreuz KE, Stransky E, Stransky H, Frommer WB and Rentsch D (1999) LeProT1, a transporter for proline, glycine betaine, and γ-amino butyric acid in tomato pollen. Plant Cell 11: 377–391

Shakya R and Sturm A (1998) Characterization of source- and sink-specific sucrose/H+ symporters from carrot. Plant Physiol 118: 1473–1480

Shannon JC, Pien FM, Cao H and Liu KC (1998) Brittle-1, an adenylate translocator, facilitates transfer of extraplastidial synthesized ADP-glucose into amyloplasts of maize endosperm. Plant Physiol 117: 1235–1252

Sjölund RD (1997) The phloem sieve element: A river runs through it. Plant Cell 9: 1137–1146

Sprenger N and Keller F (2000) Allocation of raffinose family oligosaccharides to transport and storage pools in *Ajuga reptans*: the roles of two distinct galactinol synthases. Plant J 21: 249–258

Stadler R and Sauer N (1996) The *Arabidopsis thaliana AtSUC2* gene is specifically expressed in companion cells. Bot Acta 109:299–308

Stadler R, Brandner J, Schulz A, Gahrtz M and Sauer N (1995) Phloem loading by the PmSUC2 sucrose carrier from *Plantago major* occurs into companion cells. Plant Cell 7: 1545–1554

Stadler R, Truernit E, Gahrtz M and Sauer N (1999) The AtSUC1 sucrose carrier may represent the osmotic driving force for anther dehiscence and pollen tube growth in *Arabidopsis*. Plant J 19:269–278

Stitt M (1990) Fructose 2,6 bisphosphate in plants. Annu Rev Plant Physiol Plant Mol Biol 41: 153–185

Stitt M and ap Rees T (1979) Capacities of pea chloroplasts to catalyse the oxidative pentose phosphate pathway and glycolysis. Phytochemistry 18: 1905–1911

Stitt M and Quick WP (1989) Photosynthetic carbon partitioning: Its regulation and possibilities for manipulation. Physiol Plant 77: 633–641

Stitt M, Bulpin PV and ap Rees T (1978) Pathway of starch breakdown in photosynthetic tissues of *Pisum sativum*. Biochim Biophys Acta 544: 200–214

Stitt M, Wirtz W, Gerhardt R, Heldt HW, Spencer C, Walker DA and Foyer C (1985) A comparative study of metabolite levels in plant leaf material in the dark. Planta 166: 354–364

Streatfield SJ, Weber A, Kinsman EA, Häusler RE, Li J, Post-Beittenmiller D, Kaiser WM, Pyke KA, Flügge UI and Chory J (1999) The phosphoenolpyruvate/phosphate translocator is required for phenolic metabolism, palisade cell development and plastid-dependent nuclear gene expression. Plant Cell 11: 1609–1621

Sturm A and Tang GQ (1999) The sucrose-cleaving enzymes of plants are crucial for development, growth and carbon partitioning. Trends Plant Sci 4: 401–407

Sullivan TD and Kaneko Y (1995) The maize brittle1 gene encodes amyloplast membrane polypeptides. Planta 196: 477–484

Sullivan TD, Strelow LI, Illingworth CA, Phillips RL and Nelson OE (1991) Analysis of maize Brittle-1 alleles and a defective suppressor-mutator-induced mutable allele. Plant Cell 3: 1337–1348

Tauberger E, Fernie AR, Emmermann M, Renz A, Kossmann J, Willmitzer L and Trethewey RN (2000) Antisense inhibition of plastidial phosphoglucomutase provides compelling evidence that potato tuber amyloplasts import carbon from the cytosol in the form of glucose-6-phosphate. Plant J 23: 43–53

Tegeder M, Wang X-D, Frommer W, Offler CE and Patrick JW (1999) Sucrose transport into developing seeds of *Pisum sativum* L. Plant J 18: 151–161

Tetlow IJ, Blisset KJ and Emes MJ (1994) Starch synthesis and carbohydrate oxidation in amyloplasts from developing wheat endosperm. Planta 194: 454–460

Tetlow IJ, Bowsher CG and Emes MJ (1996) Reconstitution of the hexose phosphate translocator from the envelope membranes of wheat endosperm amyloplasts. Biochem J 319: 717–723

Thom E, Möhlmann T, Quick WP, Camara B and Neuhaus HE (1998) Sweet pepper plastids: Enzymic equipment, characterisation of the plastidic oxidative pentose-phosphate pathway, and transport of phosphorylated intermediates across the envelope membrane. Planta 204: 226–233

Thompson AG, Brailsford MA and Beechey RB (1987) Identification of the phosphate translocator from maize mesophyll chloroplasts. Biochem Biophys Res Commun 143: 164–169

Thorbjørnsen T, Villand P, Denyer K, Olsen OA and Smith AM (1996a) Distinct isoforms of ADP glucose pyrophosphorylase occur inside and outside the amyloplasts in barley endosperm. Plant J 10: 243–250

Thorbjørnsen T, Villand P, Kleczkowski LA and Olsen OA (1996b) A single gene encodes two different transcripts for the ADP-glucose pyrophosphorylase small subunit from barley. Biochem J 313: 149–154

Trethewey RN and ap Rees T (1994) A mutant of *Arabidopsis* lacking the ability to transport glucose across the chloroplast envelope. Biochem J 301: 449–454

Trethewey RN and Smith AM (2000) Starch metabolism in leaves. In: Leegood RC, Sharkey TD and von Caemmerer S (eds) Photosynthesis: Physiology and Metabolism, pp 205–231. Kluwer Academic Publishers, Dordrecht

Truernit E and Sauer N (1995) The promoter of the *Arabidopsis thaliana* SUC2 sucrose-H^+ symporter gene directs expression of β-glucuronidase to the phloem: Evidence for phloem loading and unloading by SUC2. Planta 196: 564–570

Turgeon R (1991) Symplastic phloem loading and the sink-source transition in leaves: A model. In: Bonnemain J-L, Delrot S, Lucas WJ, Dainty J (eds) Recent advances in Phloem Transport and Assimilate Compartimentation, pp 18–22. Quest Editions, Paris

Turgeon R (1996) Phloem loading and plasmodesmata. Trends Plant Sci 1: 403–411

Turgeon R, Webb JA and Evert RF (1975) Ultrastructure of minor veins of *Cucurbita pepo* leaves. Protoplasma 83: 217–232

Turgeon R, Beebe DU and Gowan E (1993) The intermediary cell: Minor-vein anatomy and raffinose oligosaccharide synthesis in the Scrophulariaceae. Planta 191: 446–456

Tyson RH and ap Rees T (1988) Starch synthesis by isolated amyloplasts from wheat endosperm. Planta 175: 33–38

Van Bel AJE, Hendriks JHM, Boon EJMC, Gamalei YV and van de Merwe AP (1996) Different ratios of sucrose/raffinose-induced membrane depolarizations in the mesophyll of species with symplasmic (*Catharanthus roseus*, *Ocimum basilicum*) or apoplasmic (*Impatiens walleriana*, *Vicia faba*) minor-vein configurations. Planta 199: 185–192

Wallsgrove RM, Keys AJ, Lea PJ and Miflin BJ (1983) Photosynthesis, photorespiration and nitrogen metabolism. Plant Cell Environ 6: 301–309

Weatherley PE, Peel AJ and Hill GP (1959) The physiology of the sieve tube. Preliminary experiments using aphid mouth parts. J Exp Bot 10: 1–16

Webb KL and Burley JWA (1962) Sorbitol translocation in apple. Science 137: 766

Weber A, Menzlaff E, Arbinger B, Gutensohn M, Eckerskorn C and Flügge UI (1995) The 2-oxoglutarate/malate translocator of chloroplast envelope membranes: Molecular cloning of a transporter protein containing a 12-helix motif and expression of the functional protein in yeast cells. Biochemistry 34: 2621–2627

Weber A, Servaites JC, Geiger DR, Kofler H, Hille D, Gröner F, Hebbeker U and Flügge UI (2000) Identification, purification and molecular cloning of a plastidic glucose translocator. Plant Cell 12: 787–801

Weber H, Borisjuk L, Heim U, Sauer N and Wobus U (1997) A role for sugar transporters during seed development: Molecular characterization of a hexose and a sucrose carrier in fava bean seeds. Plant Cell 9: 895–908

Weig A and Komor E (1996) An active sucrose carrier (Scr1) that is predominantly expressed in the seedling of *Ricinus communis* L. J Plant Physiol 147: 685–690

Wiese A, Gröner F, Sonnewald U, Deppner H, Lerchl J, Hebbeker U, Flügge UI and Weber A (1999) Spinach hexokinase I is located in the outer envelope membrane of plastids. FEBS Lett 461: 13–18

Weschke W, Panitz R, Sauer N, Wang Q, Neubohn B, Weber H and Wobus U (2000) Sucrose transport into barley seeds: Molecular characterization of two transporters and implications for seed development and starch accumulation. Plant J 21: 455–467

Willey DL, Fischer K, Wachter E, Link TA and Flügge UI (1991) Molecular cloning and structural analysis of the phosphate translocator from pea chloroplasts and its comparison to the spinach phosphate translocator. Planta 183: 451–461

Winter H, Lohaus G and Heldt HW (1992) Phloem transport of amino acids in relation to their cytosolic levels in barley leaves. Plant Physiol 99: 996–1004

Winzer T, Lohaus G and Heldt HW (1996) Influence of phloem transport, N-fertilization and ion accumulation on the sucrose storage in the taproots of fodder beet and sugar beet. J Exp Bot 47: 863–870

Yu TS, Lue, WL, Wang SM and Chen J (2000) Mutation of *Arabidopsis* plastid phosphoglucose isomerase affects leaf starch synthesis and floral initiation. Plant Physiol 123: 319–325

Yu TS, Kofler H, Häusler RE, Hille D, Flügge UI, Zeeman SC, Smith AM, Kossmann J, Lloyd J, Ritte G, Steup M, Lue WL, Chen J and Weber A (2001) The *Arabidopsis sex1* mutant is defective in the R1 protein, a general regulator of starch degradation in plants, and not in the chloroplast hexose transporter. Plant Cell 13: 1907–1918

Zeeman SC, Northorp F, Smith AM and ap Rees T (1998) A starch-accumulating mutant of *Arabidopsis thaliana* deficient in a chloroplastic starch-hydrolysing enzyme. Plant J 15: 357–365

Ziegler H (1975) Nature of transported substances. In: Zimmermann MH and Milburn JA (eds) Encyclopedia of Plant Physiology, Vol 1, pp 59–100. Springer Verlag, Berlin

Zimmermann MH (1957) Translocation of organic substances in trees. I. The nature of the sugars in the sieve-tube exudate of trees. Plant Physiol 32: 288–291

Zimmermann MH and Ziegler H (1975) List of sugars and sugar alcohols in sieve-tube exudates. In: Zimmermann MH, Milburn JA (eds) Encyclopedia of Plant Physiology, Vol 1, pp 480–503. Springer Verlag, Berlin

Chapter 16

Optimizing Carbon-Nitrogen Budgets: Perspectives for Crop Improvement

John A. Raven*
Department of Environmental and Applied Biology, School of Life Sciences, University of Dundee, Dundee DD1 4HN, U.K.

Linda L. Handley
Scottish Crop Research Institute Invergowrie, Dundee DD2 5DA, U.K.

Mitchell Andrews
Ecology Centre, University of Sunderland, Sunderland SR1 3SD, U.K.

Summary 265
I. Introduction 266
II. The Nature of Crops 268
III. What Are We Seeking to Optimize in Carbon-Nitrogen Budgets? 269
IV. How Can We Change Carbon-Nitrogen Budgets? 269
V. What are the Outcomes of Changing Carbon-Nitrogen Budgets? 271
VI. Prospects and Conclusions 272
Acknowledgments 272
References 272

Summary

Crops are photosynthetic organisms cultivated, or otherwise deliberately encouraged to grow, by man. The harvested products of the crops, which are used by man, include food, ranging from the photosynthetic structures themselves, directly as green vegetables and indirectly as animals which eat these structures, to organic stores and vegetative organs, seeds and fruits. Non-food uses include wood, fuel, carbon (C) sequestration, amenity and ornamentation. These uses have very different optimal outputs in terms of their C and nitrogen (N) contents, and also have variable inputs in terms of other resources (e.g. water) and criteria for sustainability (e.g. minimizing habitat degradation). In general, an optimal C and energy budget is one which involves minimal total inputs of C and N per unit of C and/or N in the harvested product; the reason that C is included among the inputs is that C fixation involves transpiratory water loss. To the extent that N in the photosynthetic apparatus enables the organisms to harvest more energy and C (and hence N), it has a catalytic role. The quantities of different N-containing components of the photosynthetic apparatus vary with genotype (via natural or artificial selection) and with acclimation of a genotype to varying environments within its

*Author for correspondence, email: j.a.raven@dundee.ac.uk

Christine H. Foyer and Graham Noctor (eds): Photosynthetic Nitrogen Assimilation and Associated Carbon and Respiratory Metabolism, pp. 265–274.

lifetime, and can also be modified by genetic manipulation. The N form used by the plant, and the site of N assimilation, have a significant impact on the energetics of N assimilation, and these characteristics are amenable to agronomic and genetic manipulation. It is emphasized that negative effects on plant performance of changes in components of the N costs by the photosynthetic apparatus, which aim to maximize harvesting productivity, are sometimes not seen under optimal growth conditions. However, such negative effects can occur under suboptimal and/or varying growth conditions.

I. Introduction

The 300,000 or so described species of O_2-evolving photosynthetic organisms (Falkowski and Raven, 1997) encompass a great phylogenetic and ecological diversity. This is reflected in the diversity of molecular species of micromolecular and macromolecular end-products of metabolism which contain C and N, or C but no N, and of the genes and proteins which determine the production of these molecules. The photosynthetic reactions use a much smaller diversity of molecular species as pigment, redox agent and protein catalysts and as metabolic intermediates. The core of photosynthesis (photoreactions I and II; the cytochrome b_6f complex; plastoquinone; the ATP synthetase complex; Rubisco and the rest of the photosynthetic C reduction cycle) is identical among all O_2-evolvers so far examined (Raven, 1984a). Other catalysts in the photosynthetic process show more variability within the range of O_2-evolvers, with alternative light-harvesting complexes (phycobilins or proteins associated with chlorophylls *a* and *b*, or chlorophylls *a* and *c*, with or without carotenoids), catalysts coupling the cytochrome b_6f complex to the oxidizing end of photoreaction I (plastocyanin or cytochrome c_6), the reducing end of photoreaction I to the $NADP^+$ reductase (ferredoxin or flavodoxin), and the mechanisms by which glycolate is metabolized (Raven, 1984a,b; Falkowski and Raven, 1997; Raven et al., 1999, 2000).

As well as these differences among some of the catalysts of photosynthesis, there are a number of 'add-ons' to the core of photosynthesis e.g. C_4 and Crassulacean acid metabolism (CAM). These 'add-ons' provide a means of biochemically concentrating CO_2 by carboxylation/decarboxylation cycles prior to fixation by Rubisco and a variety of inorganic C accumulation mechanisms which do not depend on carboxylation/decarboxylation cycles (Falkowski and Raven, 1997). These differences among catalysts, and the occurrence of the various CO_2 concentrating mechanisms, have impacts on the quantity of N required in the photosynthetic mechanisms when normalized to a given functional attribute (e.g. photon absorption in a given radiation environment; Raven, 1984b). Furthermore, in different higher taxa there are large differences in the relative quantities of the various protein complexes, e.g. the high ratio of photoreaction I to photoreaction II in organisms (cyanobacteria sensu stricto; Rhodophyta) with phycobilisomes relative to the ratio (generally below one) in organisms lacking phycobilisomes (Raven, 1984a; Raven et al., 1999).

These sources of variation in the qualitative and quantitative occurrence of N-containing components of the photosynthetic apparatus are essentially genetic. In the broad sense, such variations are 'adaptive', i.e. comprise genetically determined differences among organisms which may have, or had at some time in the relevant taxon's evolutionary past, significance in natural selection, provided that the organism under consideration has not been subjected to artificial selection as is the case for many crop plants. An increase in inclusive fitness can occur during natural selection; artificial selection does not necessarily have such an aim or outcome (see below). In addition to these variations there are differences in the quantity of photosynthetic catalysts in a given genotype as a function of the environment in which the organism is growing, i.e. 'acclimatory' responses, shown within a generation time. Clearly, the occurrence and extent of any such acclimatory differences are a function of the genetic constitution of the organism. In general, these acclimatory responses involve analogous (if not homologous) phenotypic differences to the genetically determined ('adaptive') responses, e.g. to variations in the incident photon flux density of photosynthetically active radiation. For both adaptation and acclimation, the response to lower photon flux densities involves more pigment per unit biomass, and often a high ratio of pigment to photoreaction I and photoreaction

Abbreviations: C – carbon; C_3 – three-carbon; C_4 – four-carbon; CAM – Crassulacean acid metabolism; Gln – glutamine; N – nitrogen; Rubisco – ribulose-1,5-bisphosphate carboxylase/oxygenase

II (Raven, 1984a,b; Falkowski and Raven, 1997). Another example is acclimatory changes in the CO_2 concentrating mechanisms of CAM and of many aquatic plants with membrane-based inorganic C pumps. Here CO_2 limitation (or, for terrestrial plants, the surrogate of lack of the water which could otherwise be traded, via transpiration, for atmospheric CO_2) leads to expression of CAM in plants which can express C_3 or CAM, and of inorganic C pumps in those organisms which can vary the expression of these pumps or not express them at all at high external CO_2 levels and rely entirely on diffusive CO_2 entry. These acclimatory responses to low inorganic C supplies are predicted to decrease the quantity of catalytic N required to sustain a given rate of CO_2 fixation at low inorganic C levels, although such decreases are not always found (Falkowski and Raven, 1997). Similarly, the predicted decrease in N cost of catalyzing a given rate of C fixation for C_3 plants, as atmospheric CO_2 levels are increased to simulate the predicted anthropogenic CO_2 increases by late in the present century, is not always observed (Zerihun and BassiriRad, 2000; Marriott et al., 2001).

As well as these variations in the kind and/or quantity of N in photosynthetic catalysts there are variations in the fraction of total plant N which is found in the photosynthetic apparatus relative to the rest of the organism. This fraction is generally highest in algal unicells. In larger acellular or multicellular organisms in polarized environments (soil or sediment with air or water above) there is differentiation, with photosynthesis and nutrient uptake largely confined to (different) mature regions of the organism and growth (cell division and cell expansion to produce more assimilatory structural and reproductive structures) confined to other areas which are linked to the resource acquisition or storage parts of the organism by transport pathways. While such differentiation presumably has selective advantages (e.g. in keeping the meristems of vegetative and reproductive structures in the shoot at lower O_2 concentrations than would occur if the meristems were more (or at all) photosynthetically active; Raven et al., 1994) it might nevertheless demand a greater N commitment to provide a given specific rate of biomass increase (Andrews et al., 1995a, 1999). However, as a measure of 'N use efficiency,' the retention time of N in the organisms is also important (Berends and Aerts, 1987).

A number of authors have attempted to explain the fraction of biomass, and N, allocated to roots as an acclimatory response to the availability of N around the roots and of light to the shoots. Less light and/or more available soil N means relatively more biomass and N in shoots, and vice versa (Ågren and Bosatta, 1996; Sultan, 2000). Such relatively simple ideas can go a significant way toward quantitative modeling of the effect of light and N supply (concentration, molecular species) on the fraction of biomass and N allocated to roots (but see Andrews et al., 1995a,b, 1999). Perhaps more importantly, global and localized changes in N availability alter root architecture including the extent and location of branching, and the ratio of fine roots to more robust roots (Fitter, 1987, 1996; Robinson, 1996; Robinson et al., 1998; Zhang and Forde, 1998; Raven and Edwards, 2001). Other architectural results of variations in N, and light, supply involve the shoot (Meziane and Shipley, 1999; Sultan, 2000).

A further source of variation in the costs in terms of energy and C (and hence water lost in transpiration) in producing the N-containing components of the photosynthetic apparatus is the external source of N and, for nitrate, where it is reduced. Raven (1985) has modeled these costs from biochemical principles, with a prediction of greater water (and energy) costs for growth of a plant of comparable composition with nitrate than with ammonium as N source, a conclusion which is qualitatively independent of the site of nitrate reduction and of any biochemical means of disposing of excess OH^- generated in nitrate assimilation. However, the plants do not always show the predicted differences in the transpiratory costs of growth as a function of N source (Raven et al., 1992b; Yin and Raven, 1998). Andrews et al. (1995b) showed that *Phaseolus vulgaris* under otherwise optimal growth conditions shows greater dry matter per unit N with nitrate than with NH_4^+ or glutamine (Gln) as N source. There is also a greater leaf area per unit N with nitrate than with NH_4^+ or Gln as N source, possibly because there is greater nitrate transport to, and assimilation in, the shoot leading to greater osmoticum supply, and hence leaf expansion and specific leaf area. It is likely that this effect could be mimicked in plants which usually reduce most of their nitrate in the root, by genetic modification to express nitrate and nitrite reductases primarily in the shoots.

The diversity among plants in the N costs of producing the photosynthetic apparatus, and in the effectiveness of the N so used in catalyzing C (and N) metabolism, shows that there is a significant degree of variation among the photosynthetic organisms in the way in which N and C interact in photosynthesis and growth. The genetic diversity in interactions of C and N in plants forms the basis for analysis in the rest of this chapter. We first consider the range of crops, and the extent to which crops are defined by selection and breeding rather than the crop environment. We then consider what we are seeking to optimize in C-N budgets and how these budgets can be achieved. Finally, we consider what outcomes result from such changes followed by a brief consideration of prospects for future work.

II. The Nature of Crops

An inclusive definition of crops is that they are photosynthetic organisms cultivated, or otherwise deliberately encouraged to grow, by man. Most crops are thought of in terms of a product, which is harvested by man directly or, indirectly, via domesticated grazing animals. This definition would include not only the 'obvious' crops such as cereals and 'root crops', but also managed pastures, trees used for wood or fuel, seaweeds cultivated for their wall polysaccharides, microalgae cultivated as food for maricultured invertebrates or as sources of human dietary supplements, and cut flowers. It may be stretching the definition of crops to include amenity plantings and sports turf (unless dislodging divots can be termed harvesting!), or any plantations for C sequestration.

The broad definition of crops includes plants in which the harvested material contains very little N (e.g. wood for construction or fuel) and where what N is present is of little or no significance for the uses to which man puts the material. The lack of N in wood is a result of effective N recycling and retranslocation, since, for example, every aromatic nucleus in lignin was produced from the action of phenylalanine ammonia-lyase on phenylalanine (Raven et al., 1992a). However, such crops are harvested by coppicing or, more usually by sacrifice of all of the above-ground parts of a tree. This means that significant N (in leaves and small stems) is discarded, although N has to a substantial extent been withdrawn from time-expired leaves before they are abscised, and retranslocated to new, growing leaves. At the other extreme are crops for which the harvested part is the main photosynthetic organs, e.g. *Lactuca*, *Spinacia* and pasture grasses. In an intermediate position are perennial crops in which the harvested structures are fruits, leaving the tree or shrub to produce a crop in a subsequent season. Most fruits of perennial plants fix by net photosynthesis much less than half of their C and rely on phloem (to a lesser extent xylem) for most of their N and the remainder of their organic C. Thus, the provisioning of the fruits is to some extent in competition for organic N and C with the assimilatory, and especially the photosynthetic, apparatus. Also in an intermediate position are the annual grain crops with fruits or seeds as the harvested structure. As with the fruits of perennial plants less than half of the organic C in the fruits or seeds of these annuals comes from in situ photosynthesis, so that much of the organic C and almost all the organic N comes from dedicated photosynthetic structures. This can, as for perennials, be regarded as competition between photosynthetic structures and grainfilling, although for the annuals no N (or C) in the vegetative plant is harvested, so that there can be a temporal distinction between vegetative growth and reproductive growth (Cohen, 1966). While Cohen (1966) dealt mainly with organic C, this model also applies to N. Thus, the optimal strategy for the annual as a wild plant in a variable habitat is to follow vegetative growth with reproduction. Such a use of N as a catalyst in photosynthesis followed by transfer to seeds and fruits in crops in producing seed and fruit protein may be the optimal strategy for the wild ancestors of annual grain crops (Cohen, 1966). Man has capitalized on these traits in the breeding of crops.

A good account of the regulation of fluxes between organs in a range of life-forms of higher plants can be found in Stitt and Schulze (1994). Extending the concept of crops to macroalgae and microalgae can use the same models of flux control as are used for higher plants (Stitt and Schulze, 1994), although the structures involved are different. Water flow over macroalgal thalli can influence allocation to wall polysaccharides (Kraemer and Chapman, 1991a,b), just as wind can influence allocation to wall materials in terrestrial vascular plants (Niklas, 1992).

A consideration of the nature of crops requires a consideration of the crop environment as well as of the organisms. This is especially the case where it is only the environment, which distinguishes crop plants from their wild relatives, e.g. in many micro- and macro-algal cultures. The crop environment frequently (in theory at least) involves monocultures, with high plant densities. Any selection, breeding or

Color Plates

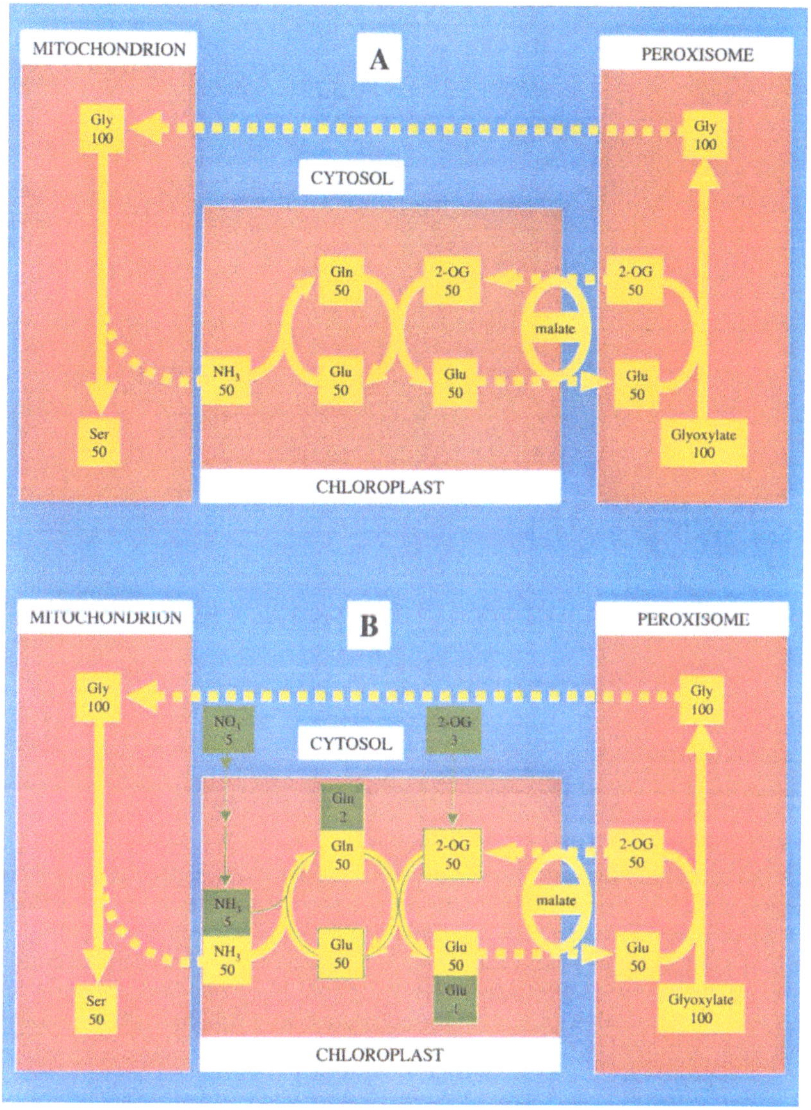

Color Plate 1: Simplified scheme showing possible effects of photorespiratory and non-photorespiratory ammonia generation on the chloroplastic Gln pool during typical rates of photosynthesis and photorespiration at intermediate to high irradiance in air. Numbers in the boxes give typical rates of formation or supply of each compound (μmol.mg^{-1}chl.h^{-1}) during fast photosynthesis. A. All ammonia is produced by photorespiration. Net CO_2 fixation and RuBP carboxylation:RuBP oxygenation are taken to be 200 μmol.mg^{-1}chl.h^{-1} and 2.5 (Sharkey 1988; Keys 1999; Noctor et al. 2002b). The rate of glyoxylate formation is therefore 100 μmol.mg^{-1}chl.h^{-1} and photorespiratory ammonia release is 50 μmol.mg^{-1}chl.h^{-1}. Assuming sufficient energy supply and capacities of enzymes and transporters capacity, Gln only accumulates if the supply of NH_3 from Gly oxidation is faster than the rate of 2-OG regeneration by Glu:glyoxylate aminotransferase (50% of glyoxylate is transaminated by other amino acids, chiefly Ser). B. Influence of a rate of non-photorespiratory NH_3 formation ten-fold lower than photorespiratory ammonia production, if 2-OG supply is not sufficient. For simplicity, metabolism of Glu and Gln in downstream transamination reactions is not considered. Reduction of nitrate to ammonia occurs at 5 μmol.mg^{-1}chl.h^{-1}, equivalent to 2.5% the rate of net CO_2 fixation (shown in green). The hypothetical case shown assumes a rate of anaplerotic 2-OG formation equal to 60% of the rate of nitrate assimilation, i.e., 3 μmol.mg^{-1}chl.h^{-1} (15 μmol.C mg^{-1}chl.h^{-1} or 6 triose phosphate equivalents, i.e., more than 9% of the rate of net CO_2 fixation (net CO_2 fixation will be lower than in (A) if anaplerotic 2-OG generation increases 'day' respiration). Within 1 h, non-photorespiratory ammonia assimilation leads to *net formation* of 1 μmol.Glu and 2 μmol.Gln per mg chl (green boxes). (See Chapter 1, p. 7.)

Christine H. Foyer and Graham Noctor (eds): Photosynthetic Nitrogen Assimilation and Associated Carbon and Respiratory Metabolism, pp. CP-1 – CP-3.

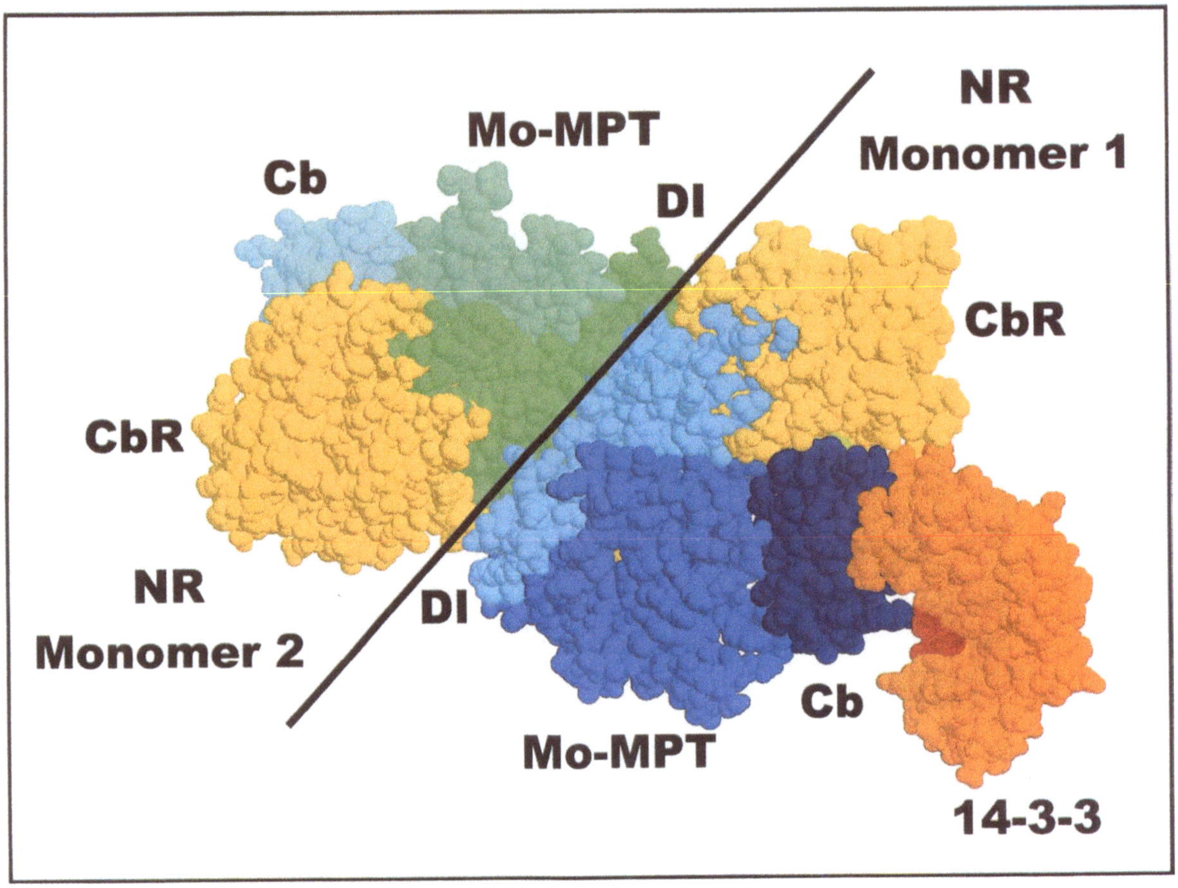

Color Plate 2. Model of 14-3-3 bound to phospho-NR. The homo-dimer model of NR was modified to contain a raf peptide sequence and 14-3-3 was docked (Petosa et al., 1998; Campbell, 1999). Only one of the two binding sites in the 14-3-3 dimer is bound to NR. In each NR monomer, the major domains and fragments are color coded: C-terminal Cyt *b* reductase fragment (CbR) is gold in both monomers; Cyt *b* domain (Cb) is dark blue in monomer 1 and blue-green in monomer 2; the dimer interface domain (DI) is blue-green in monomer 1 and dark green in monomer 2; and the molybdenum-molybdopterin cofactor (Mo-MPT) domain is blue in monomer 1 and light green in monomer 2. The 14-3-3 dimer is shown in orange with the bound peptide in red. The N-terminal region of NR is not shown, while the dimer interface is emphasized by a black line. (See Chapter 3, p. 43.)

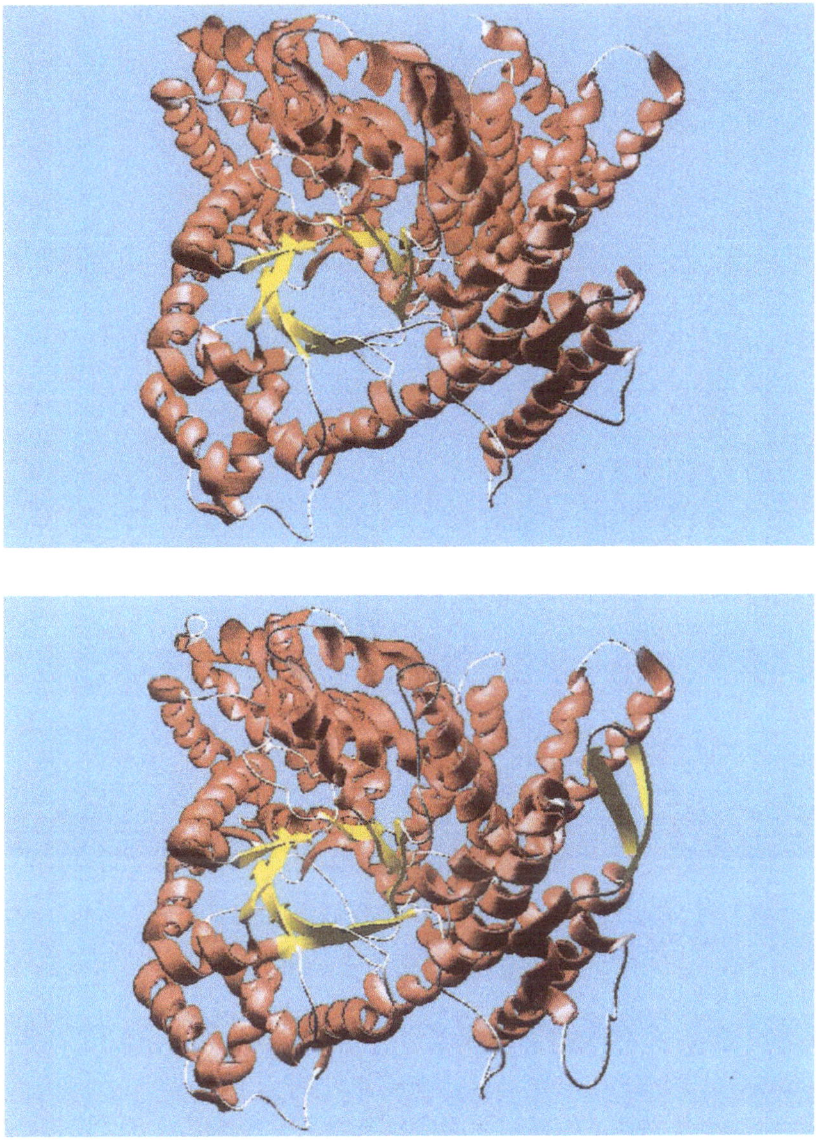

Color Plate 3. Ribbon drawings showing the 3D-structure of the *E. coli* (upper panel, after Kai et al., 1999) and *Sorghum* (lower panel) PEPC monomers. The eight β-strands forming the β-barrel are shown in yellow and the α-helices are shown in red. The *Sorghum* PEPC monomer was modelled using the Swiss Prot modeling service (Guex and Peitsch, 1997). (See Chapter 9, p. 138.)

genetic modification programs related to manipulating C and N budgets must take into account the crop environment. An example is shading by the upper canopy in later growth stages of annual crops which impacts on C acquisition at the individual plant level, perhaps more than at the whole crop level. For N acquisition, the timing of nitrogenous fertilizer application in relation to crop growth can be very relevant to the effectiveness of use of the applied N, with benefits sometimes accruing from split applications and the use of slow-release fertilizers.

III. What Are We Seeking to Optimize in Carbon-Nitrogen Budgets?

The discussion in Section II shows that crops have a wide range of C:N ratios in the harvested portions. On both economic and on environmental grounds the N inputs to, and N losses from, an agroecosystem should be as small as is consistent with economically viable and environmentally sustainable crop production. The very low N content of wood only requires catalytic N in the photosynthetic and nutrient absorption apparatus, and in wood synthesis. Since the lifespan of leaves and fine roots is less than the time taken for a tree to produce useable wood, even in short-rotation coppice, minimizing N requirements and losses would best be achieved by maximizing internal recycling of N (and other nutrients). Moreover, minimization of the quantity of N in catalytic and structural components is consistent with delivering organic C to wood at an economic (to humans) rate.

For crops whose harvested portions are photosynthetic, the requirement for minimal N in the photosynthetic apparatus is less stringent than in woody crops, especially if the consumer organisms obtain a significant fraction of their organic N from the crop. Any manipulations of N in leaf vegetables must be compatible with other nutritional requirements of the human (or other animal) consumer (Grusak and Dellapenna, 1999).

More complex in optimization terms is the allocation of N when the harvested product contains N as a desirable component but the harvested product does not perform much, or any, of the photosynthesis required in provisioning the harvested product with organic C. Here the sorts of models pioneered by Cohen (1966) are useful in indicating optimal N allocation between the photosynthetic apparatus (and other essential components other than the harvested component) and the harvested product as a function of time.

These sorts of considerations, and especially those in which major temporal changes take place in the spatial disposition of N within the plant, are most readily modeled assuming a constant environment. Such assumptions are most reasonable for greenhouse crops, although even here the biotic environment (pests and pathogens) may be rather variable, and the light environment is not always controlled. In less managed crop environments the variability of the habitat is, of course, greater.

The optimization of C-N budgets in a variable environment requires that the organism not only deals with a temporarily restricted supply of a resource but also can deal with a temporary excess. Resource excess is perhaps most obvious with light in the form of photoinhibition (Long et al., 1994; Niyogi, 1999; Marshall et al., 2000). Accordingly, any optimization which focuses on maximizing C fixation per unit plant N in a given constant environment may not achieve the highest crop yields in a variable environment. This is exemplified by Mott and Woodrow (2000) for the large and frequent variations in photon flux density, and by Raven and Glidewell (1981), Cowan (1986), Majeau et al. (1994), Price et al. (1994), Evans and von Caemmerer (1996), Evans (1999) and Evans and Loreto (2000) for CO_2 transport in the liquid phase with varying intercellular space CO_2 concentrations in C_3 plants.

IV. How Can We Change Carbon-Nitrogen Budgets?

Changes in the composition of the harvested components of crops have occurred since the first domestication of particular crops (Evans, 1975). Classic plant breeding has thus been effective in altering the composition of fruits and seeds, increasing the oil and protein content of legumes such as *Glycine* and reducing the content of phytotoxins (many of which contain N) in the seeds of *Lupinus* (Evans, 1975; Grusak and Dellapenna, 1999). By contrast the fraction of protein in cereal caryopses may have decreased during artificial selection by man as the size of the grains increased.

For leaf crops the intensity of artificial selection may have been less, and not immediately directed at the organic N content of *Lactuca* or *Spinacia*. One important aspect of leaf chemistry, which is photosynthetically related to N metabolism, is the accumulation of nitrate (held by some to be a health hazard; Steingröver, 1986). Another is the accumulation of oxalate (a product of acid-base regulation following net OH^- synthesis in nitrate reduction and organic N production; Libert and Creed, 1985; Raven, 1985). The nitrate is a product of an excess of nitrate delivery to the leaves in the xylem over nitrate assimilation and may be minimised by altering nitrate fertilization regimes, or in part by harvesting at the end of the photoperiod, provided that delivery of nitrate to leaves in the xylem stream energized by root pressure or transpiration is less light-dependent than is nitrate reduction and subsequent ammonium assimilation.

Oxalate is the accumulated product of a biochemical pH-stat. A survey of 78 cultivars of *Rheum raponticum* showed a wide range of contributions of oxalate relative to the more costly (in energy and C, and hence water) malate to the total M_r organic anion pool (which showed less variation among cultivars) in the harvested petioles (Libert and Creed, 1985).

In addition to classical plant breeding, relying on selection from naturally occurring genetic variability or from that generated by random mutagenesis, there is now also the possibility of genetic engineering. This technique has been applied to such components of the photosynthetic apparatus as Rubisco (Stitt and Schulze, 1994; Ruuska et al., 2000), the cytochrome b_6f complex (Hurry et al., 1996), Rubisco activase (Mott and Woodrow, 2000), carbonic anhydrase (Majeau et al., 1994; Price et al., 1994) and NAD(P)H dehydrogenase (Raven et al., 1999), while Niyogi (1999) considers the potential for molecular genetic approaches to modifying the content of photoprotective pigments. A reduction in the content of these protein complexes would reduce the N content of leaves; this would especially be the case for components such as Rubisco, which comprises a large fraction of the total leaf N in C_3 plants (Evans and Seemann, 1989). Any such reduction in the content of catalytic proteins as a means of reducing plant N requirement must, of course, be evaluated in the context of overall plant performance.

Plant growth rate under optimal conditions in controlled environments is not reduced by lowered expression (to ~70% or so of wild type) of Rubisco or of cytochrome b_6f complex (Stitt and Schulze, 1994; Hurry et al., 1996; Ruuska et al., 2000), by very substantial (to ~1% of wild type) reduction of expression of carbonic anhydrase (Majeau et al., 1994; Price et al., 1994), or by elimination of expression of plastid NAD(P)H dehydrogenase (Raven et al., 1999). However, such reductions of content of particular catalysts might not result in the anticipated increased rate of C gain per unit N since the expression of other catalysts may be increased. An example is the increased content of Rubisco attendant on reduced expression of Rubisco activase (Mott and Woodrow, 2000). Furthermore, decreased content of catalysts may, as will be discussed later, reduce growth rate under continuously or variously suboptimal growth conditions even if there is no effect on growth under optimal growth conditions.

The same goes for such stratagems as reducing the content of ribosomes in mature photosynthetic structures. By definition such a mature structure has no net protein synthesis, so that the only obvious role for its ribosomes is in the synthesis of proteins which are degraded as part of protein turnover, including any photodamaged D1. Raven (1989,1994) has considered the requirement for ribosomes in the replacement of damaged D1 and concludes that there was apparent overprovision of ribosomes for the maximum rate of D1 synthesis observed in mature leaves, but did not consider turnover of other proteins. We remedy this deficiency with the following calculation.

Raven (1989, 1994) considered mature *Oxalis* leaves with 130 μmol chlorophylls $a + b$ per m^2 of leaf area. Evans and Seemann (1989) cite 3.8 mmol chlorophylls per mol N in C_3 plant leaves, so the *Oxalis* leaves would have 0.48 g N per m^2. Assuming that protein is 5.8 times N in plants (Gnaiger and Bitterlich, 1984; Handley et al., 1989) this gives 2.78 g protein per m^2 of leaf, with 120 g per mol of amino-acyl residue there are 0.0232 mol amino-acyl residues in protein per m^2 of leaf area. Penning de Vries (1975) estimated the specific turnover rate of leaf protein of 0.12 day^{-1}; other reports give similar values for leaf protein turnover (Huffaker and Peterson, 1974; Simpson et al., 1981; Davies, 1982). The protein breakdown and synthesis rate in the mature *Oxalis* leaves is then (0.0232×0.12) or 2.78 mmol amino-acyl residues per m^2 leaf area per day or 32.2 nmol amino-acyl residues per m^2 of leaf area per second. Raven (1989) cites a rate of protein synthesis per g of active RNA at 20 °C of 0.06 mg protein per g active RNA (mainly ribosomal) per second. To achieve the computed rate of protein turnover (32.2 nmol amino-acyl residues or 3.86 μg protein per m^2 leaf area per second) requires 3.86 10^{-6} g protein per m^2 per second divided by 6.10^{-5} g protein per g of active RNA per second, or 0.064 g RNA per m^2 leaf area. Raven (1989) quotes a leaf ribosome content of 1.5 g per m^2 leaf area so that, with most of the 0.064 g RNA per m^2 being in ribosomes, and RNA 60% by mass of ribosomes the ribosomal requirement for protein turnover of some 0.1 g per m^2 leaf area is less than one-tenth of the total ribosome content of leaves. This calculation, even with the ribosome requirement for D1 turnover computed by Raven (1989), suggests that there is significant overprovision of ribosomes in mature leaves.

V. What are the Outcomes of Changing Carbon-Nitrogen Budgets?

Increasing the N (protein) content of grains on a per plant basis can be achieved without an increased net N uptake by each plant if the N required by the non-harvested parts of the plant is decreased, or the fraction of this N transported to the harvested structures is increased, or both. The extent to which increased translocation to grains (or other harvested structure) can occur without impacts on the capacity of photosynthetic structures to fix CO_2 and assimilate nitrate, and on roots to take up nutrients for a sufficient fraction of the growth cycle, remains to be determined. There seem to be upper limits on how much N can be translocated per unit C in the phloem and hence on how rapidly the N component of the harvested product can be supplied, even if the speed at which N can be incorporated into the storage organs is adequate. Evans (1975) discusses the changes in the capacity for phloem transport to wheat caryopses in the case of breeding by classical means.

Manipulating the N content of the major photosynthetic organs of grain or 'root' crops, thus permitting the abovementioned more direct diversion of N to harvested structures, can be achieved by the genetic manipulation of particular protein complexes so that their expression is reduced.

We have seen that modest reductions (~25%) in the content of some of the protein complexes of chloroplasts does not cause a decrease in the photosynthetic rate on a leaf area basis, or on the growth rate. However, full growth analysis of the genotypes with downregulation of a chloroplast protein complex in suboptimal as well as optimal growth conditions is rarely undertaken, and reproductive fitness (not, perhaps, of major concern to crop breeders) has been even more neglected. Much of this work has been carried out on *Nicotiana* spp., for which the leaf is the harvested organ for the major cultivated species *Nicotiana tabacum*, and the leaf protein content is not a major commercial consideration except insofar as a decreased N demand for chloroplast components in leaves may permit more N to contribute to synthesis of the alkaloid nicotine which is synthesized in the roots and is transported to the leaves in the xylem.

As has been mentioned earlier, another deficiency in much of the work with genetically manipulated crops with changes in the content of one (or more) chloroplast polypeptides is that plant performance, and especially crop yield, has not been followed under sub-optimal variability or especially field conditions. There are notable exceptions especially for *Nicotiana* plants with modified contents of Rubisco (Stitt and Schultze, 1994; Ruuska et al., 2000). For this enzyme, responses to low light, low N supply, variable CO_2 and, because Rubisco catalyses reactions in C_3 plants which act in photochemical dissipation of excitation energy, photoinhibitory photon flux densities have been tested (Stitt and Schultze, 1994; Ruuska et al., 2000). Decreased Rubisco content interacted with low light, low N, low water and low CO_2 availability, but did not increase susceptibility to photoinhibition (Stitt and Schultze, 1994; Ruuska et al., 2000). Hurry et al. (1996) had earlier found that increasing excitation energy pressure on Photosystem II by decreased expression of the cytochrome b_6f complex did not increase the sensitivity to photoinhibition of *Nicotiana* leaves. These data show that restriction on electron transport downstream of Photosystem II at either the cytochrome b_6f level (Hurry et al., 1996; Krieger-Liszkay et al., 2000) or the Rubisco level (Ruuska et al., 2000) does not increase the potential for photoinhibition.

Notwithstanding the absence of effect on photoinhibition of the molecular genetic treatments which reduce the potential for photochemical energy dissipation using electron transport through the cytochrome b_6f complex with Rubisco-catalysed reactions as the terminal electron acceptors, it is necessary to reiterate that photodamage to photo-reaction II (D1 protein) can occur in wild-type organisms. The requirement to synthesize replacement D1 protein remains if the photosynthetic capacity has to be maintained. Moreover, there is the need for synthesis of proteins that are unrelated to photodamage but which are degraded in protein turnover. This requires that a minimum level of ribosomes is maintained functional, thus limiting the extent to which N can be economized on by reducing the ribosome content of leaves (Section IV).

It has also been pointed out earlier that variations in photon flux density within a periodicity close to, or less than, the time taken for activation (at light on or when light increases) or deactivation (at light off or when light decreases) might particularly impact on transgenic plants with a decreased Rubisco activase activity, since the rate of activation of Rubisco when light increases would be slower (Mott and Woodrow, 2000).

The conditions, if any, in which a substantial downregulation of carbonic anhydrase expression has a large effect on photosynthetic rate (and growth rate) have yet to be established (Raven and Glidewell, 1981; Cowan, 1986; Majeau et al., 1994; Price et al., 1994; Evans and von Caemmerer, 1996; Williams et al., 1996; Evans, 1999; Gillon and Yakir, 2000). However, there is evidence of an increased impact on photosynthetic rate of targeted inactivation of a component of the NAD(P)H dehydrogenase in plastids (*ndhB*), and thus a decreased rate of dark reduction of plastoquinone (and of cyclic electron transport in the light?) when *Nicotiana* is exposed to low relative humidity and thus moderate stomatal closure (Raven et al., 1999; Horváth et al., 2000).

Variable and suboptimal conditions also relate to the uptake of different N sources and the location of their assimilation within the plants. Andrews (1986) and Andrews et al. (1995a) showed that legume species with root nitrate assimilation are more tolerant of low temperatures than species with shoot nitrate assimilation when both are grown with nitrate as their N source.

VI. Prospects and Conclusions

N is a potentially limiting resource for crop growth. Hence maximization of the N content of harvested parts of the plant where N is a desirable component is an objective of crop plant manipulation, as is decreasing the N content of harvested parts when this is desirable (e.g. in malting barley). A complementary set of objectives concern minimizing N content of non-harvested plant parts which, in annual crops at least, means effective remobilization of N. Another means of reducing N requirements is restricting N allocation to structural and catalytic components of the non-harvested parts which are, under the growth conditions employed, in excess of those required for growth. These strategies taken together can improve the quality of the harvested product while not increasing or even decreasing the overall N requirement of the crop, with possible economic and environmental benefits.

While these benefits can be readily seen in general terms, the details of what would be appropriate manipulations to achieve them need further analysis, e.g. the extent to which lower contents of some N-containing photosynthetic components can be attained without compromising growth in the frequently variable environment of the crop.

Once the desirable changes have been identified, implementation of the changes can be achieved by classical plant breeding techniques using natural variation in the crop or its interfertile wild relatives, or by genetic modification.

Acknowledgments

Work in JAR's laboratory in this area is funded by the Natural Environment Research Council and the Scottish Executive Environment and Rural Affairs Department (SEERAD), and was funded by the Biotechnology and Biological Sciences Research Council. LLH's work is funded by SEERAD.

References

Ågren GI and Bosatta E (1996) Theoretical Ecosystem Ecology: Understanding Elemental Cycles. Cambridge University Press, Cambridge

Andrews M (1986) The partitioning of nitrate assimilation between root and shoot of higher plants. Plant Cell Environ 9: 511–519

Andrews M, Raven JA and Sprent JI (1995a) Site of nitrate assimilation in grain legumes in relation to low temperature sensitivity: An assessment. Proceedings of the 2nd European Conference on Grain Legumes, Copenhagen, 1995, pp 120–121

Andrews M, Zerihun A and Watson C (1995b) N form effects on the partitioning of dry matter between root and shoot of *Phaseolus vulgaris*. Proceedings of the 2nd European Conference on Grain Legumes, Copenhagen, 1995, pp 60–61

Andrews M, Sprent JI, Raven JA and Eady PE (1999) Relationships between shoot to root ratios, growth and leaf soluble protein content of *Pisum sativum*, *Phaseolus vulgaris* and *Triticum aestivum* under different nutrient deficiencies. Plant Cell Environ 22: 949–958

Berends F and Aerts R (1987) N-use-efficiency: A biologically meaningful definition? Funct Ecol 1: 293–298

Cohen D (1966) Optimising reproduction in a randomly varying environment. J Theor Biol 12: 119–129

Cowan IR (1986) Economics of carbon fixation in higher plants. In: Givnish TJ (ed) On the Economy of Plant Form and Function, pp 133–170. Cambridge University Press, Cambridge

Davies DD (1982) Physiological aspects of protein turnover. In: Boulter D and Parthier B (eds) Encyclopaedia of Plant Physiology (New Series) Volume 14A, pp 189–198. Springer Verlag, Berlin

Evans JR (1999) Leaf anatomy enables more equal access to light and CO_2 between chloroplasts. New Phytol 143: 93–104

Evans JR and Loreto F (2000) Acquisition and diffusion of CO_2 in higher plant leaves. In: Leegood R C, Sharkey T D and von Caemmerer S (eds) Photosynthesis Physiology and Metabolism, pp 321–351. Kluwer Academic Publishers, Dordrecht

Evans JR and Seemann JR (1989) The allocation of protein nitrogen in the photosynthetic apparatus: Costs, consequences and control. In: Brigg WR (ed) Photosynthesis, pp 183–205. AR Liss, New York

Evans JR and von Caemmerer S (1996) Carbon dioxide diffusion inside leaves. Plant Physiol 110: 339 –346

Evans LT (ed) (1975) Crop Physiology. Cambridge University Press, Cambridge

Falkowski PG and Raven JA (1997) Aquatic Photosynthesis. Blackwell Science, Malden, MA

Fitter AH (1987) An architectural approach to the comparative ecology of plant root systems. In: Rorison IH, Grime JP, Hunt R, Hendry GAF and Lewis DH (eds) Frontiers of Comparative Plant Ecology, pp 217–233. Academic Press, London

Fitter AH (1996) Characteristics and functions of root systems. In: Waisel Y, Eshel A and Kafkafi U (eds) Plant Roots: The Hidden Half, Second Edition, pp 1–20. Marcel Dekker, New York

Gillon JS and Yakir D (2000) Internal conductance to CO_2 diffusion and $C^{18}OO$ discrimination in C_3 leaves. Plant Physiol 123: 201–213

Gnaiger E and Bitterlich G (1984) Proximate biochemical composition and caloric content calculated from elemental CHN analysis: A stoichiometric concept. Oecologia 62: 289–298

Grusak MA and Dellapenna D (1999) Improving the nutrient composition of plants to enhance human nutrition and health. Annu Rev Plant Physiol Plant Mol Biol 50: 133–161

Handley LL, Mehvan M, Moore CA and Cooper WJ (1989) Nitrogen-to-protein conversion factors for two tropical C_4 grasses, *Brachiaria mutica* (Forsk.) Stopf. and *Pennisetum purpureum* Schumach. Biotropica 21: 88–90

Horváth E M, Peter S, Joët T, Rumeau D, Cournac L, Horváth GV, Kavanagh TA, Schäfer C, Peltier G and Medgyesy P (2000) Targeted inactivation of the plastid *ndhB* gene in tobacco results in an enhanced sensitivity of photosynthesis to moderate stomatal closure. Plant Physiol 123: 1337–1349

Huffaker RC and Peterson LW (1974) Protein turnover in plants and possible means of its regulation. Annu Rev Plant Physiol 25: 363–392

Hurry V, Anderson JM, Badger MR and Price GD (1996) Reduced levels of cytochrome b_6-f in transgenic tobacco increases the excitation pressure on photosystem II without increasing the sensitivity to photoinhibition in vivo. Photosynth Res 50: 159–169

Kraemer GP and Chapman DJ (1991a) Biomechanics and alginic acid composition during hydrodynamic adaptation by *Egregia menziesii* (Phaeophyta) juveniles. J Phycol 27: 47–53

Kraemer GP and Chapman DJ (1991b) Effects of tensile force and nutrient availability on carbon uptake and cell wall synthesis in blades of juvenile *Egregia menziesii* (Turn) Aresch. (Phaeophyta). J Exp Mar Biol Ecol 149: 267–277

Krieger-Liszkay A, Kienzler K and Johnson GN (2000) Inhibition of electron transport at the cytochrome b_6f complex protects photosystem II from photoinhibition. FEBS Lett 486: 191–194

Libert B and Creed C (1985) Oxalate content of seventy-eight rhubarb cultivars and its relation to some other characteristics. J Hort Sci 60: 257–261

Long SP, Humphries S and Falkowski PG (1994) Photoinhibition of photosynthesis in nature. Annu Rev Plant Physiol Plant Mol Biol 45: 633–662

Majeau N, Arnoldo M and Coleman JR (1994) Modification of carbonic anhydrase activity by antisense and overexpression constructs in transgenic tobacco. Plant Mol Biol 25: 377–385

Marriott DJ, Stirling CM and Farrar J (2001) Constraints to growth of annual nettle (*Urtica urens*) in an elevated CO_2 atmosphere: Decreased leaf area ratio and tissue N cannot be explained by ontogenetic drift or mineral N supply. Physiol Plant 111: 23–32

Marshall HL, Geider RJ and Flynn KJ (2000) A mechanistic model of photoinhibition. New Phytol 145: 347–360

Meziane D and Shipley B (1999) Interacting determinants of specific leaf area in 22 herbaceous species: Effects of irradiance and nutrient availability. Plant Cell Environ 22: 447–459

Mott KA and Woodrow IE (2000) Modeling the role of Rubisco activase in limiting non-steady-state photosynthesis. J Exp Bot 51: 399-406

Niklas KS (1992) Plant Biomechanics. An Engineering Approach to Plant Form and Function. The University of Chicago Press, Chicago

Niyogi KK (1999) Photoprotection revisited: Genetic and molecular approaches. Annu Rev Plant Physiol Plant Mol Biol 50: 333–359

Penning de Vries FWT (1975) The cost of maintenance processes in plant cells. Ann Bot 39: 77–92

Price GD, von Caemmerer S, Evans JR, Yu J-W, Oja V, Kell P, Harrison K, Gallagher A and Badger MR (1994) Specific reduction of chloroplast carbonic anhydrase activity by antisense RNA activity in transgenic tobacco plants has minor effects on photosynthetic CO_2 assimilation. Planta 193: 331–340

Raven JA (1984a) Energetics and Transport in Aquatic Plants. AR Liss, New York

Raven JA (1984b) A cost-benefit analysis of photon absorption by photosynthetic unicells. New Phytol 98: 259–276

Raven JA (1985) Regulation of pH and generation of osmolarity in vascular land plants: Costs and benefits in relation to efficiency of use of water, energy and nitrogen. New Phytol 101: 25–77

Raven JA (1989) Fight or flight: The economics of repair and avoidance of photoinhibition of photosynthesis. Funct Ecol 3: 5–19

Raven JA (1994) The cost of photoinhibition to plant communities. In Baker NR and Bowyer JR (eds) Photoinhibition of Photosynthesis, pp 449–464. Bioscientific Publishers, Oxford

Raven JA and Edwards D (2001) Roots: Their evolutionary origins and biogeochemical significance. J Exp Bot 52: 381–401

Raven JA and Glidewell SM (1981) Processes limiting photosynthetic conductance. In: Johnston CB (ed) Physiological Processes Limiting Plant Productivity, pp 109–136. Butterworths, London

Raven JA, Wollenweber B and Handley LL (1992a) Ammonia and ammonium fluxes between photolithotrophs and the environment in relation to the global nitrogen cycle. New Phytol 121: 5–18

Raven JA, Wollenweber B and Handley LL (1992b) A comparison of ammonium and nitrate as nitrogen sources for photolithotrophs. New Phytol 121: 19–32

Raven JA, Johnston A M, Kübler JE and Parsons R (1994) The influence of natural and experimental high O_2 concentrations on O_2-evolving photolithotrophs. Biol Rev 69: 61–94

Raven JA, Evans MCW and Korb RE (1999) The role of trace metals in photosynthetic electron transport in O_2-evolving organisms. Photosynth Res 60: 111–149

Raven JA, Kübler JE and Beardall J (2000) Put out the light, and then put out the light. J Mar Biol Assocn UK 80: 1–25

Robinson D (1996) Resource capture by localized root proliferation: Why do plants bother? Ann Bot 77: 179–185

Robinson D, Hodge A, Griffiths BS and Fitter AH (1998) Plant root proliferation in nitrogen-rich patches confers competitive advantages. Proc Roy Soc Lond B 266: 431–435

Ruuska SA, Badger MR, Andrews JT and von Caemmerer S (2000) Photosynthetic electron sinks in transgenic tobacco with reduced amounts of Rubisco: Little evidence for significant Mehler reaction. J Exp Bot 51: 357–368

Simpson E, Cooke RJ and Davies D (1981) Measurements of protein degradation in leaves of *Zea mays* using [^{3}H] acetic anhydride and tritiated water. Plant Physiol 67: 1214–1219

Steingröver E (1986) Nitrate accumulation in spinach: uptake and reduction of nitrate during a dark or 'low light' night period. Plant Soil 91: 429–432

Stitt M and Schultze E-D (1994) Plant growth, storage and resource allocation: From flux control in a metabolic chain to the whole-plant level. In: Schultze E-D (ed) Flux Control in Biological Systems from Enzymes to Populations and Ecosystems, pp 57–118. Academic Press, San Diego

Sultan SE (2000) Phenotypic plasticity for plant development, function and life history. Trends Plant Sci 5: 537–542

Williams TG, Flanagan LB and Coleman JR (1996) Photosynthetic gas exchange and discrimination against $^{13}CO_2$ and $C^{18}O^{16}O$ in tobacco plants modified by an antisense construct to have low chloroplastic carbonic anhydrase. Plant Physiol 112: 319–326

Yin Z-H and Raven JA (1998) Influences of nitrogen sources on nitrogen- and water-use efficiency, and carbon isotope discrimination, in C_3 *Triticum aestivum* and C_4 *Zea mays*. Planta 205: 574–580

Zerihun A and BassiriRad H (2000) Photosynthesis of *Helianthus annuus* does not acclimate to elevated CO_2 regardless of N supply. Plant Physiol Biochem 38: 897–903

Zhang H and Forde BG (1998) An *Arabidopsis* MADS-box gene that controls nutrient-induced changes in root architecture. Science 279: 407–409

Index

Symbols

14-3-3 proteins 12, 16,35, 42–44, 64–66, 218
  inhibition of NR activity 45
  NR complex 65

A

ABA. *See* abscisic acid
aba1-1 18
aba2-1 18
aba3-1 18
ABI1 17
ABI2 17
ABI3-5 17
ABI4 18
ABI5 18
abiotic stress 185
abscisic acid 2, 17, 213
  synthesis 18
acclimation 266–267
acids *see also:* amino acids
  amino 2, 6, 9, 16–17, 183, 249, 251, 25
  organic 2
aconitase 13, 199
Actinomycetales 97
active oxygen species (AOS) 162–163, 180–181, 184, 187, 197
adenylate control 179, 182, 183
ADP-glucose (ADP-Glc)
  transport 244
  translocator 245
ADP-glucose pyrophosphorylase (AGPase) 244
AGPase. *See* ADP-glucose pyrophosphorylase (AGPase)
agroecosystem 269
agronomic manipulation 266
AICAR. *See* 5-aminoimidazole-4-carboxamide 1-β-D-ribofuranoside
algae
  glycolate metabolism 165
  photorespiration 165
  unicellular 12, 267
Alocasia macrorrhiza 26
alternative oxidase (AOX) 160, 163, 164,.173–188, 200
  abiotic stress 185
  active oxygen species 180–181, 184, 187
  adenylate control 179, 182–183
  amino acid pool 183
  antimycin A 178, 180
  antisense inhibition 182, 186
  biochemical regulation 176–177, 179, 182
  citrate 180, 181
  cysteine residues 177–178
  cytochrome pathway 175– 176, 180, 185–188
  electron transport
    in mitochondrial electron transport chain 175
  fruit development 186
  gene expression 180–182
  gene families 175, 180
  glycolysis 179, 182
  growth 184–186
  inhibitors 175
  low temperature 185
  measurement of activity 175
  mitochondrial electron transport chain 174–176
  monoclonal antibody (AOA) 175
  oxygen isotope discrimination 175, 185–187
  phosphate limitation 181–184
  physiological function 181–188
  pollen development 186
  programmed cell death 187-188
  pyridine nucleotides 176–177, 179, 182
  pyruvate 177–179, 182–183
  pyruvate kinase 179, 182–183
  root development 186
  site-directed mutagenesis 177
  sulfhydryl/disulfide regulatory system 177–178
  TCA cycle 177, 179, 182
  thermogenesis 185–186
  tobacco mosaic virus 186 187
  transgenic plants 175, 177, 181
alternative pathway 161
  for carbon recycling 122
Amaranthus edulis 118
amino acids 2, 6, 9, 16–17, 183, 249, 251, 254
  diurnal concentrations 250
  permeases 255
  signaling 17
  synthesis 1, 245
  transporter 255
    aromatic 246
    minor 11
    total leaf 9
5-aminoimidazole-4-carboxamide 1-β-D-ribofuranoside 65
ammonia 2, 36, 50, 53, 54, 57–58, 125, 207, 267
  accumulation 125,
  assimilation 9, 13, 71, 72–86, 93–109, 270
  compensation point 125
  incorporation 4
  transporter 125
  influx 210
  permease 93
  transport 94
  uptake 213, 215
amt1 95, 105
amt2 95
amt3 95
AmtB permease 95
amyloplasts 242
Anabaena azollae 97
Anabaena sp. PCC 7120 96, 99, 100
Anabaena variabilis 95
Anacystis nidulans 97, 101

Anacystis R2 (*Synechococcus* sp. PCC 7942) 95
anaplerosis 16, 135
carbon flow 6, 12, 145
annual crops 268–269, 272
grain crops 268
anoxia 54, 65, 68
ANR1 18, 208, 217
anthocyanin 219
antimycin A 178, 180
Antithamnium sp. 99
AOA *See* monoclonal antibody (AOA)
AOS *See* active oxygen species (AOS)
AOX *See* alternative oxidase (AOX)
AP2 17
aphid-stylet-technique 247
Apiaceae 249
apoplast 50, 52–53, 251
nitrate reduction 50–54
phloem loading 251
aquatic plants 267
Arabidopsis 14, 39, 54, 56–59, 234, 242
Arabidopsis thaliana 117, 140–141, 208, 210, 242
histidine biosynthesis 17
Arg 11
Arum maculatum 185
AS. *See* asparagine synthetase (AS)
Asn 9, 11, 15
Asp 6
Asp aminotransferase 6
asparagine synthetase (AS) 234
ATP synthetase complex (ATPase) 95, 266
CF_0-CF_1 26, 28–29
H^+-ATPase (plasma membrane) 50
ATP/ADP 161
auxin 17, 18
azaserine 214
Azospirillum brasilense 99

B

bacteria 16
B. subtilis 101
Bacteroidaceae 98
Bacteroides fragilis 97–98
barley 6, 57, 118, 128, 157, 244
bifA 103
biomass 266–267
birch 57
blue light regulation 53
brittle-2 244
BSC. *See* bundle sheath cells (BSC)
bundle sheath cells (BSC) 137
Butyrivibrio fibrisolvens 98

C

C *see* carbon (C)
C_3 plants 2, 24, 30, 135, 137, 154, 229, 267, 269–271
photosynthesis 6
C_4 24, 30, 135–137, 165, 229, 266
aspartate/alanine shuttle 231
characteristics 30
glycine oxidation 165
photorespiration 165
calcium 135
influx 143
calcium-dependent protein kinases 35
Calothrix sp. PCC 7601 97
CAM *See* crassulacean acid metabolism (CAM)
cAMP receptor protein (CRP) 103
Candida utilis 59
CAP family 106
carbohydrates 2, 12, 16, 213, 241
export 13, 242
metabolism 11
synthesis 13
recycling 120–124
carbon
acquisition 269
assimilatory enzymes 229
flow 12, 50, 122
inorganic carbon accumulation mechanisms 266
inorganic carbon pumps 267
metabolism 1, 43, 212, 215
sequestration 265
carbon dioxide *See* CO_2
anthropogenic increases 267
assimilation 5-6, 27
carboxylation/decarboxylation cycles 266
post-illumination burst 164
carbon-nitrogen 1, 93-94, 108, 227, 269
budgets 268–269
mitochondria 152–167
photorespiration 117
carbonic anhydrase 26, 229, 270, 272
carrier (see transporter)
carrot 249
catalase 30, 118
mutant 120
celery 249
cereal caryopses 269
Chlamydomonas reinhardtii 54, 57, 118, 230
chlorate 57
Chlorella 53
Chlorella sorokiniana 57
Chlorophyceae 98
chlorophyll 13, 266
Chl *a:b* ratio 28
chloroplast 50, 55, 271
membranes 54
stroma 119
CO_2 concentration 119
O_2 concentration 119
circadian control 40, 140
citrate 157, 180–181
transporter 158
citrate synthase 13, 212
Clarkia xantiana 242
Clematis vitalba 59

Clostridiaceae 98
CO_2
$^{12}CO_2$ 120
$^{14}CO_2$ 119–122
assimilation 5, 6, 27
elevated 29, 68
compensation point 126
concentrating mechanisms 266
low inorganic C levels 267
limitation 267
post-illumination burst 129, 164
coffee 249
companion cells 246
compensation point
ammonia 125
CO_2 126
complex I (mitochondrial) 160, 161
confocal microscopy 141
constitutive NAD(P)H nitrate reductase (cNR) 50-55, 195
control strength 119
cotton 26
crassulacean acid metabolism (CAM) 135–137, 139, 266–267
CO_2 concentrating mechanisms 267
crops 265–272
environmentally sustainable production 269
improvement 266–272
perennial 268
CRP. *See* cAMP receptor protein (CRP)
crystallography
X-ray 136
cucumber 128
Cucurbitaceae 248, 249, 256
cyanobacteria 93–94, 99–101, 103, 266
cyanobacterial NiR 57
cycD3 235
cyclic electron transport 272
Cymopsis tetragonoloba 28
cytochrome *b/f* complex 29, 266, 270–271
cytochrome *c* 51, 53
cytochrome c_6 266
reductase activity 38
cytochrome *c* oxidase (CytOX) 200
cytochrome *f* 26
cytochrome pathway 160–161, 175–176, 180, 185–186, 188
cytokinin 18, 207, 234
cytosol 50
ATP/ADP 161
NAD(H) pool 65, 68-69, 156–157, 162
NADP(H) 154, 158
homeostasis 146
nitrate 63, 66–68
nitrate reductase 49– 50
pH 66, 146
pyruvate kinase 212
CytOX. *See* cytochrome *c* oxidase (CytOX)

D

D1 protein 270-272
photodamage 270-272
synthesis 270
turnover 271
DBMIB 107
DCMU (dichlorophenyldimethylurea) 107
Dehydrogenase (NAD(P)H) external 162
Deioncoccales 98
Deionococcus radiodurans 98
development 207, 213
diaphorase 51
diatoms 98
dicarboxylate transport 84, 124
dietary supplements 268
Digitaria 141
Digitaria sanguinalis 141
divalent cations 96, 105
DNA-binding protein 103
Drosophila melanogaster 17
drought 6, 129

E

E4P. *See* erythrose 4-P (E4P)
eIF-2 17
electron paramagnetic resonance 199
electron transport 23, 44, 174–176, 271, 212
enterobacterial NiR 57
ER 51
erythrose 4-P (E4P) 245
Escherichia coli 57, 98–99, 136–137
ethanol formation 68
Euglena gracilis 123
external NAD(P)H dehydrogenase 162
extracellular nitrate reduction 50-54

F

FAD 37–38, 51–52, 99
fatty acid biosynthesis 162
FBPase. *See* fructose 1, 6-bisphosphatase (FBPase)
Fd. *See* ferredoxin
Fd-GOGAT *See* ferredoxin-dependent glutamate synthase (Fd-GOGAT)
Fd:NADP$^+$ oxidoreductase 55–56, 233
Fd:NiR 57 *See* nitrite reductase
ferredoxin 2, 35, 39, 55-56, 233, 266
ferredoxin NADP$^+$ reductase 35, 39
ferredoxin-dependent glutamate synthase (Fd-GOGAT) 5, 14–15, 74–75, 80–83, 93, 99–100, 116, 119, 126–129.
ferricyanide reductase 38
flavodoxin 266
flavonoids 231
flavoprotein 51
flow cytometry 142
FMN 99
FNR *See* fumarate and nitrate reduction (FNR)
FOCA 54
formate transporter 54
folate 162
formate 123

formyl-tetrahydofolate synthase activity 122
Fru2,6bP. *See* fructose 2,6-bisphosphate (Fru2,6bP)
fructose 1, 6-bisphosphatase (FBPase) 25, 242
fructose 2,6-bisphosphate (Fru2,6bP) 242
fruit 186, 265, 268
fuel 265
fumarase 213
fumarate 51, 56, 106, 233

G

G3PDH. *See* glyceraldehyde-3-phosphate dehydrogenase
G6P. *See* glucose 6-phosphate (G6P)
galactinol 256
Galderia partita 30
GATA 59, 217
GCN2. *See* General Control Non-reversible 2
GDC glycine decarboxylase (GDC)
GDH *See* glutamate dehydrogenase (GDH)
gdhA 102
genes 4, 58, 128, 175, 180-182
  nitrogen-responsive 234
  nitrogen-responsive 230–231
GGAT *See* glutamate-glyoxylate aminotransferase (GGAT)
gifA 106
gifB 106
gin6 mutant 17
Glc-6P 55
Glc1P. *See* glucose 1-phosphate (Glc1P)
Glc6P 242. *See* glucose 6-phosphate (Glc6P)
Gln 2, 4–6, 9, 12, 14–15, 96, 97, 99, 108, 124
glnA 97
glnA promoters 104
glnB 105, 108
glnN 98, 105
glsF 99
gltB 100
gltD 100
gltS 99
Glu 2, 4–6, 11, 96, 100
Glu:glyoxylate aminotransferase 6
glucose 16, 18, 241-242
glucose 1-phosphate (Glc1P) 241
  transport 243
glucose 6-phosphate (Glc6P) 137, 241
  transport 243
glucose-6-phosphate dehydrogenase 39
glucose carrier 241
glutamate 117
glutamate dehydrogenase (GDH) 15, 71, 78, 85–86; 93–94, 99, 102
glutamate synthase (GOGAT) 4, 6, 35, 39, 71–73, 79–86; 93, 94, 100, 146, 211, 231
glutamate-glyoxylate aminotransferase (GGAT) 116, 126–127
glutamate-receptors 220
glutamine 2, 117
glutamine synthetase (GS) 4, 6, 35, 39, 53, 71–79, 81, 85, 93, 94, 211, 116, 231. *See also:* GSIII
  GS2 5, 116, 118, 119, 123, 125–130
  GS type I 96
  GS type II 97
  GS type III 93, 97
glutamine synthetase/glutamate synthase cycle (GS/GOGAT) 9, 13, 78, 83-84, 93-94, 99–101, 146, 155–156, 233
glutathione 11, 162
Gly 2, 6, 9, 11, 117, 121, 124, 164-165
Gly decarboxylase (GDC) 116
glyceraldehyde-3-phosphate dehydrogenase 14, 25, 154
glycerate kinase 121
glycerate-3-P 117
glycine decarboxylase (GDC) 116, 119–128, 130, 154, 160–161
glycolate 165
glycolate 2-P 117
glycolate dehydrogenase 165, 166
glycolate oxidase 116, 120, 122, 128, 129
glycolysis 4, 12, 153, 155–157, 179, 182
glycosyl-phosphatidylinositol anchor 51
glyoxylate 101, 121,123,126
GOGAT *See* glutamate synthase
Golgi apparatus 51
GPI. *See* glycosyl-phosphatidylinositol
GPT 242–243
GS *See* glutamine synthetase (GS)
GS/GOGAT *See* glutamine synthetase/glutamate synthase (GS/GOGAT)
GS2. *See* glutamine synthetase (GS2)
GSI 93, 96, 105
GSI-IFs complex 106
GSII 97
GSIII 93, 97

H

HATS. *See* high affinity nitrate transport systems (HATS)
hemoglobin 199
hetC 105
heterocyst 100
hexokinase 16, 242
high affinity nitrate transport systems (HATS) 37–38, 44, 51, 208
His-Asp phosphorelay 16, 148, 227
His-containing phosphotransfer (HPt) 234
His-protein kinase 234
histidine 17
HPt. *See* His-containing phosphotransfer (HPt)
2-hydroxy-3-butynoic acid 119
hydroxypyruvate 117
hydroxypyruvate reductase (NADH-HPR) 116
hypersensitive response 55, 197
hypoxia 55

I

ICDH *See* isocitrate dehydrogenase (ICDH)
IF. *See* inactivating factors (IF)
IF-GSI stoichiometry 107
IF17 93, 106
IF7 93, 106
inactivating factors (IF) 105
induction of photosynthesis 9, 152, 163–164
inositol-1,4,5-trisphosphate 135

iron-regulatory protein (IRP) 199
iron-sulfur clusters, 56, 57, 99
[3Fe-4S] 99
[4Fe-4S] 99
[Fe4S4] 56
[Fe4S4] 57
IRP. *See* iron-regulatory protein (IRP)
isocitrate 100
isocitrate dehydrogenase (ICDH) 2, 13–14, 16, 85, 93, 100–105, 158–159

K

Kalanchoe fedtschenkoi 139–140, 143
Kranz anatomy 137

L

lactate 68
Lactuca 268, 270
Lamiaceae 248, 256
LATS. *See* low-affinity transport system (LATS)
Lb *See* leghemoglobins (Lb)
Lb-NO 199
leaf 269, 270, 271, 272
area 267, 271
chlorotic 57
drought-stressed 129
expansion 267
nitrate reduction 64–70
nitrogen content 270
protein 270
vegetables 269
LEDR. *See* light-enhanced dark respiration
leghemoglobins (Lb) 199
legumes 269
Lepechinia calycin 26
leukoplasts 242
LHC. *See* light-harvesting Chl *a/b* protein complexes
light 4
enhanced dark respiration 152
signal transduction 141-143
stress 129
light-harvesting Chl *a/b* protein complexes 24, 230, 266
lignin 268
lipids 231
low-affinity transport system (LATS) 208
LR. *See* lateral roots
luciferase 59
Lupinus 269
lysine 17
Lysmachia vulgaris 28

M

macroalgae 268
macroalgal thalli 268
MADS-type of transcription factor 40
maize 55, 57–58, 137, 139, 144, 229, 234, 244–245
malate 14–15, 137, 157
malate valve 153–154, 159, 163–164
malate-OAA exchange 153
malonate 51
malting barley 272
maltose 241
transporter 242
mannitol 248, 257
maricultured invertebrates 268
MC. *See* mesophyll cells (MC)
Medicago sativa 99
Mehler reaction 163
meristems 267
Mesembryanthemum crystallinum 140–141, 143
mesophyll cells (MC) 137, 247
metabolism
arrest 17
control 137
cross talk 16
intermediates 266
secondary 246
methemoglobin 199
methionine-DL-sulphoximine (MSX) 95
methyl viologen 53, 68
methylammonium 95
microalgae 268
microarray analysis 39, 211, 213
microsomes 53
minor amino acids 11
mitochondria 152–167
carbon-nitrogen reactions 152–167
electron transport 14, 174–176
function 200
NAD(H) 155, 161
NAD(P)H 155, 161–162
NAD(P)H dehydrogenases 160
NO 200
respiration 65, 152
uncouplers 65
thioredoxin 162
Mo-MPT 37–38, 40, 44
molybdenum-pterin 51
cofactor 210
enzyme 55
MSX. *See* L-methionine-DL-sulphoximine (MSX)
Mustard 55
mutants
catalase 120
photorespiratory 116-118
starchless 243
tobacco 4

N

N^{10}-formyl-tetrahydrofolate (THF) synthetase 123
N^5,N^{10}-methenyl-THF cyclohydrolase 123
N^5,N^{10}-methylene THF 116
N^5,N^{10}-methylene-THF dehydrogenase 123
NADH/NAD 9, 38, 50, 155
cytosolic 69, 156-157, 162
mitochondrial 155, 161

NADPH/NADP
cytosolic 154, 158
mitochondrial 155, 161-162
NAD(P)H dehydrogenase 160, 270
NAD-ME. *See* NAD-malic enzyme (NAD-ME)
NADH-HPR. *See* hydroxypyruvate reductase (NADH-HPR)
$NADP^+$ reductase 266
NADP-dependent isocitrate dehydrogenase (NADP-ICDH 101, 212
NADP-dependent malic enzyme (NADP-ME) 137
NADP-glutamate dehydrogenase (NADP-GDH) 102
NADP-malate dehydrogenase (NADP-MDH) 163–164
NADPH-dependent hydroxypyruvate reductase 122
NAD(P)H nitrate reductase. *See* NR
NarI 54
ndhB 272
Neurospora crassa 17
nia 12, 58, 228
Nicotiana plumbaginifolia 58–59, 208
Nicotiana spp 271
Nicotiana sylvestris 125
Nicotiana tabacum 57, 271
nicotine 271,
nifHKD 105
nii 57-59
NiR *See* nitrite reductase; nitrite reductase (NiR)
nitrate 2, 4, 9, 12, 18, 35, 36–46, 49, 50, 58, 94, 207, 267, 270
accumulation 53
acquisition 50
assimilation 50, 267, 270
regulation 49
availability 55
concentration 64, 66-68
co-ordinate sensing 18
cytosolic 63, 67
delivery 270
detoxification 49
efflux, vacuole 67
external 52
feeding 67
fertilization 36, 270
induction 59
leakage 63, 67
pools 64
reduction 2, 36, 50–54, 63–70, 94, 156, 206, 267, 270
responsive elements 59
sensing 53
signaling 16, 39, 40, 50, 45
transporters 35, 36, 54, 232
uptake 36, 53, 108, 208, 213, 232
nitrate reductase (NR) 1, 2, 16, 35, 42, 49-50, 66, 73, 81, 201, 210, 228
activation state 4, 12, 35, 42, 63–66, 69
biosynthesis 41
regulation 41-43
constitutive 49-55, 195
K_M 66
inhibition 44
promoters 41
turnover 64
nitric oxide (NO) 46, 49, 53–54, 55, 60, 188, 193–202
synthesis 53-54, 194–196
nitric oxide synthase (NOS) 53, 194-196
nitrite 49–50, 54, 57, 94
accumulation 65
detoxification 49, 54, 68
metabolism 54–59
nitrite reductase (NiR) 35, 49, 50, 53, 56–58, 59, 73, 81, 210, 231
bacterial 57
cytosolic form 55
fungal 57
gene expression 58–59
post-transcriptional control 59
overexpression 59
reduction 36, 94, 211
nitrite transporter 54
nitrite:NO oxidoreductase 53
nitrogen 23, 50, 108, 227, 265-266
^{15}N 214
assimilation 2, 43, 64, 266, 270
cytokinins (nitrogen-responsive accumulation) 234
photorespiratory nitrogen cycle 74, 116, 124–127
responsive genes 230–231
retention time 267
signals 16, 101, 108, 206–22 ,234
translocation rate 249
use efficiency 23, 267
nitrogenase 100
nitrous acid 54
NO *See* nitric oxide (NO)
NO_2 55
nodule 196
non-photochemical quenching 130
non-photosynthetic plastids 55
Norflurazon 59
NOS. *See* nitric oxide synthase (NOS)
Nothofagus solandri 28
NO_x 54–55
NR *See* nitrate reductase
NtcA 93, 103
NtcA-activator 106
NtcA-repressor 106
regulon 96
transcription factor 94
NUE. *See* nitrogen use efficiency

O

OAA *See* oxaloacetate
2-OG. *See* 2-oxoglutarate (2-OG)
OH-pyruvate. *See* hydroxypyruvate
oil 269
okadaic acid 218
Oleaceae 248–249, 256
oligomycin 161
oligosaccharides 248, 256
olive 249
Onagraceae 248
Oocytes, Xenopus 209
OPPP *See* oxidative pentose phosphate pathway (OPPP)
organic acids 2, 52
synthesis 4, 13

ornamentation 265
oscillator
circadian 140
oxalate 270
Oxalis 270
oxaloacetate 117, 124, 157-163
oxidative detoxification 55
oxidative pentose phosphate pathway 50, 55
oxidative pentose phosphate pathway (OPPP) 56, 212, 242–243
2-oxoglutarate (2-OG) 6, 13–15, 93–94, 101–102, 117
amination 96
anaplerotic 2-OG formation 6
transporter 84
-malate translocator 245
oxygen isotope discrimination 175, 185, 186, 187
oxygen sensor 199
oxygenase activity of Rubisco 116-119

P

P. boryanum 100
Panicum miliaceum 230, 245
pasture grasses 268
pathogens 60, 269
pathways
alternative 161
cytochrome 160, 161, 176, 180
reductive pentose phosphate (RPP) 116, 138–139, 212, 213
shikimate 245, 246
PDC. *See* pyruvate dehydrogenase complex (PDC)
PDH *See* pyruvate dehydrogenase
pea 118, 243
PEP (see phosphoenolpyruvate)
transport 245
PEP carboxylase (PEPc) 12–14, 16, 30, 136–148, 155–156, 163, 212, 215
regulation 136–148
PEP/phosphate translocator (PPT) 245
PEPc *See* PEP carboxylase (PEPc); phosphoenolpyruvate carboxylase
PEPc protein kinase (PEPcK) 139
PEPc-specific protein-serine/threonine kinase (PEPc 135
PEPcK. *See* PEPc protein kinase (PEPcK); PEPc-specific protein-serine/threonine kinase (PEPc
peroxynitrite 46, 55, 195
petH 105
PGA. *See* 3-phosphoglyceric acid (PGA)
3-PGA. *See* 3-phosphoglycerate (3-PGA)
PGI. See phosphogluco isomerase (PGI)
PGM See phosphoglucomutase (PGM) 244
PGP *See* phosphoglycolate phosphatase (PGP)
Phaseolus vulgaris 27, 129, 267
phenylalanine 268
phenylalanine ammonia lyase 4, 268
phloem 207, 246–257, 268, 271
loading 246–257, 251
sap 214, 247, 249
transport 271
Phormidium laminosum 97
phosphate limitation 181-184
phosphoenolpyruvate 245
phosphoenolpyruvate carboxylase 4, 136–148, 229
phosphorylation 136–148
regulation 136–148
phosphogluco isomerase (PGI) 242
6-phosphogluconate dehydrogenase 39
3-phosphoglycerate (3-PGA) 138, 240
phosphoglycolate phosphatase (PGP) 116, 119–120, 128
phosphoinositide-specific phospholipase C 135
phosphoenolpyruvate carboxylase 12, 64, 135–148, 231
photochemical dissipation of excitation energy 271
photoinhibition 59, 120, 129, 164–165, 269, 271
photorespiration 1, 2, 6, 9, 11, 15, 23, 30, 74–75, 78, 82, 116–130, 152, 154, 157, 160–62, 164, 166
carbon and nitrogen metabolism 116-117
feedback on other processes 127–129
recycling of carbon 120–124
recycling of nitrogen 124–127
in algae 165
in C_4 plants 165
stress 129, 129–130
photosynthesis 4, 6, 9, 24, 27, 50, 55, 93–94, 120, 266–267, 272
apparatus 26, 265, 266-269
efficiency 229
induction 152, 163, 164
Photosystem I 24, 55
Photosystem II 24, 271
phycobilins 102, 266
phycobilisomes 266
phytochrome 40, 75, 81
Pi translocator 153, 154
PII 16, 108, 218
PK. *See* pyruvate kinase (PK)
plant hormones 2
plasma membrane 50, 52, 53
NO formation 53
nitrate reductase 50-53, 60
bound nitrite:NO oxidoreductase in 49, 50
H^+-ATPase 50
plasmodesmata 247, 256
plastocyanin 266
plastoquinone 26, 266, 272
Plectonema boryanum 99
PM-NR *See* plasma membrane-bound nitrate reductase
PNUE. *See* photosynthetic nitrogen use efficiency
polI 235
pollen 186
polyamines 42, 43
Porphyra purpurea 99
potato 6, 9, 11, 26
PPT. *See* PEP/phosphate translocator (PPT)
Prevotella melaninogenica 98
privet 249
Prochlorococcus 96
Prochlorococcus marinus 101
programmed cell death 187-188
protein kinase 12, 17, 35, 42, 44–45, 231
CDPK 44
cascade 143
inhibitors 218
plant 44
SNF1-related 44

protein phosphatase 17, 42
Prunus ilicifolia 26
Prunus persica 28
PS I. *See* Photosystem I
PS II. *See* Photosystem II
PS II function 28
Pseudanabaena sp. PCC 6903 98
pumpkin 249
purines 17
pyridine nucleotides 176, 177, 179, 182
pyruvate 177, 178, 179, 182, 183, 245
pyruvate dehydrogenase complex (PDC) 13–14, 157
pyruvate kinase (PK) 13, 155, 179, 182–183, 212
pyruvate, orthophosphate dikinase 229
pyruvate translocator 245

Q

quenching
  non-photochemical 130

R

R5P. *See* ribose 5-P (R5P)
raffinose 247, 248
redox 14, 50, 51, 104, 212, 266
reductive pentose phosphate (RPP) pathway 116, 138–139, 212, 213
resource allocation 207
respiration 2, 12-13 15, 30, 55, 152, 173, 174–188,
RFLP markers 57
Rheum raponticum 270
Rhizobiaceae 97
Rhodophyta 266
ribose 5-P (R5P) 245
ribulose-1,5-bisphosphate (RuBP) 23, 116–117, 120, 126, 128,163
  oxygenation 116-119
  regeneration 23
ribulose-1,5-bisphosphate carboxylase/oxygenase (Rubisco) 16, 23, 28–30, 116, 119–122, 125, 126, 127, 128–130, 154, 229, 266, 270–272
Rubisco activase 26, 270, 272
rice 29, 30, 57
roots 267
  architecture 267
  branching 208
  development 186
  fine 267, 269
  lateral 17, 207, lateral roots
  pressure 270
  robust 267
  -shoot ratio 16, 207
Rosaceae 249
RPP. *See* reductive pentose phosphate (RPP) pathway
Rubiaceae 249
Rubisco. *See* ribulose-1,5-bisphosphate carboxylase/oxygenase
RuBP *See* ribulose 1,5-bisphosphate
RuBP regeneration 29
Ruminococcus flavefaciens 98

S

S-adenosylmethionine-dependent uroporphyrinogen II 232
Saccharomyces cerevisiae 16
salicylhydroxamic acid (SHAM) 161
salicylic acid 186–187, 201
Salmonella typhimurium 97
Sauromatum guttatum 175
SBPase. *See* sedoheptulose-1,7-bisphosphatase
Scenedesmus minutum 212
Schizosaccharomyces pombe 241
Scrophulariaceae 248, 249, 256
SE-CCC. *See* sieve element-companion cell complex (SE-CCC)
seaweed 268
sedoheptulose-1,7-bisphosphatase 25
seeds 265
sensors of carbon-nitrogen status 1
Ser 2, 5–6, 9, 124
Ser hydroxymethyltransferase (SHMT) 116, 121–124, 127–129, 154
Ser-glyoxylate (SGAT) 116, 119, 121–122, 124–128, 130
Ser-protein kinases/phosphatases 217
sex1 242
SGAT *See* Ser-glyoxylate (SGAT)
SHAM. *See* salicylhydroxamic acid (SHAM)
shikimate pathway 245, 246
SHMT *See* Ser hydroxymethyltransferase (SHMT)
shrunken-2 244
sieve element-companion cell complex (SE-CCC) 246
sieve elements 246
signal 4
  nitrate 206
  nitrogen 206–220
signal transduction 1, 4, 18, 45, 49, 135, 146, 196
SiR. *See* sulfite reductase
siroheme 56–57, 59
snapdragon 249
SNF1 *See* sucrose non-fermenting
SnRKs. *See* SNF1-related protein kinases
Solanum dulcamara 28
sorbitol 248–249, 257
Sorghum 137–139, 141, 144–145
soybean 29, 55, 161, 195
spermidine 42
spinach 26, 56–57, 59, 64, 66, 195, 268, 270
split root experiments 207, 214
SPS *See* sucrose phosphate synthase (SPS)
squash 249
stachyose 247, 248
starch 2, 11, 13, 231
  biosynthesis 213
  degradation 213, 241
  transitory 241
starchless mutant 243
stress 4, 18, 129, 185
succinate 50–52
succinate dehydrogenase 51
sucrose 2, 11, 15, 18, 58, 240, 247, 251
sucrose non-fermenting (SNF1) SNF1-related protein kinases 16, 44
sucrose phosphate synthase (SPS) 16, 218
sucrose transporters 253, 254

sucrose-phosphate synthase 231
sugar alcohols 248
sugar beet 51
sugar sensing 16, 17
sugar-beet 52
sugars 12, 16
sulfhydryl/disulfide regulatory system 11, 177–178
sulfite 57
sulfite oxidase 55
sulfite reductase 55
SUMT. *See* S-adenosylmethionine-dependent uroporphyrinogen II
sun6-2 17
sunflower 29, 59
symplastic phloem loaders 251, 256
Synechococcus sp. PCC 7942 95–96, 98
Synechocystis sp. PCC 6803 93, 95, 97–98, 100–101

T

t-zeatin (Z) 235
t-zeatin riboside (ZR) 235
t-zeatin riboside-5′-monophosphate 235
TCA. *See* tricarboxylic acid (TCA)
Thioredoxin 162
tetrahydrofolate (THF) 116
thermogenesis 185, 186
THF. *See* tetrahydrofolate (THF)
tobacco 4, 11, 12, 14–15, 30, 53, 55, 57, 59, 118, 129, 177, 181, 195, 207, 233
tobacco mosaic virus 186–187
TP *See* triose phosphate (TP)
TPT *See* triose phosphate/phosphate translocator (TPT)
transaldolase 213
transcription
  cascade 136
  factors 17, 40, 94, 216
transfer cells 251
transhydrogenase 155, 162
transhydrogenation 155, 162
transketolase 213
translocation rate 250
translocator
  amino acid 255
  ADP-Glc 245
  citrate 158
  chloroplastic ATP/ADP 154
  dicarboxylate 124
  Gln 124
  Gly 124
  membrane 118
  nitrite 54
  OAA 159
  oxaloacetate 124
  oxoglutarate/malate 84, 245
  PEP/phosphate 245
  Pi 153, 154
  pyruvate 245
  tricarboxylate 158
  triose phosphate/phosphate translocator (TPT) 240–242
transpiration 267, 270
transport
  ammonia 125
  dicarboxylate 84
  maltose 242
  nitrate 232
  OAA 157
  sucrose 253, 254
  tricarboxylate 158
trees 268
tricarboxylic acid (TCA) 158
  TCA cycle 4, 12–13, 101, 157–158, 177, 179, 182
triose phosphate (TP) 240, 242
Triticum aestivum 26
tryptophan 17
tungsten 210
type-A response regulator 234
Tyr 11

U

Uncouplers 65
unicellular algae 12
urea 94
uridylylation 108
uroporphyrin III methyltransferase 56

V

Vacuole 67
Val 11
verbascose 248
VF1. *See* NtcA
Vibrio sp. 101

W

wall (cell) polysaccharides 268
water-use efficiency 27
wheat 6, 9, 11, 26, 29
wind 268
wood 265, 268, 269

X

X-ray crystallography 97, 136
xanthine oxidase 55
xanthophyll cycle 28, 130
Xenopus oocytes 209
xylem 11, 268, 270–271

Y

yeast 16, 36, 136

Z

Z. *See* t-zeatin (Z)
zeaxanthin 28
ZmRR1 234
ZmRR2 234
ZR. *See* t-zeatin riboside (ZR)

MIX
Papier aus verantwortungsvollen Quellen
Paper from responsible sources
FSC® C105338

If you have any concerns about our products,
you can contact us on
ProductSafety@springernature.com

In case Publisher is established outside the EU,
the EU authorized representative is:
Springer Nature Customer Service Center GmbH
Europaplatz 3, 69115 Heidelberg, Germany

Printed by Libri Plureos GmbH
in Hamburg, Germany